UNDERSTANDING WRISTWATCHES

GERMAN ENGINEERING MEETS SWISS TECHNOLOGY—THE HANDBOOK FOR COLLECTORS AND EXPERTS

WITH PHOTOGRAPHS BY PETER MILDE

SCHIFFER PUBLISHING
4880 Lower Valley Road • Atglen, PA 19310

J. MICHAEL MEHLTRETTER

Photo credits: The photographs used to illustrate this book are by Peter Milde unless otherwise noted.

Originally published as *Armbanduhren* ©2021 by Motorbuch Verlag, Stüttgart
Translated from the German by David Johnston

Library of Congress Control Number: 2023931324

All currencies have been approximated to US dollars where applicable.

Edited by Ian Robertson
Type set in Helvetica/Times

ISBN: 978-0-7643-6686-4
Printed in China

Published by Schiffer Publishing, Ltd.
4880 Lower Valley Road
Atglen, PA 19310
Phone: (610) 593-1777; Fax: (610) 593-2002
Email: Info@schifferbooks.com
Web: www.schifferbooks.com

ROLEX
OYSTER
COSMOGRAPH
DAYTONA
PATEK PHILIPPE
GENEVE
JAEGER-LECOULTRE

1. Contents

Foreword

Tempus fugit—Time flies, even on our favorite watch, but its value remains.

However, this statement is true only if you have been able to acquire the watch you are looking for in good technical condition and at a fair price.

This book is intended to help ensure that the reader has sufficient knowledge and thus a solid basis for decision-making when buying his or her next wristwatch. To achieve this, a number of prerequisites had to be fulfilled.

In addition to his work as a journalist and writer, the author is also a mechanical engineer. Development, design, and production are his daily bread.

All technical and functional details related to the production of machines and technical systems are part of his everyday life.

He holds numerous patents in the fields of assembly and process engineering.

The wristwatch may also be considered a machine, albeit a rather small one.

As an important prerequisite for this book, the author observed the international wristwatch market for more than twenty-five years. At the very beginning his focus was on models from IWC, Jaeger-LeCoultre, Longines, Omega, Zenith, and Universal. After just a few years his interest expanded to include the brands Patek Philippe and Vacheron & Constantin, primarily models with automatic winding.

The much-higher price level and the greater risks involved demanded even more-precise expertise. In the end, the knowledge that was still lacking could be acquired only through the practical experience gained during the overhauling of numerous watches. In the past years, the author had well over a hundred watches overhauled at Patek Philippe alone. The number of Vacheron & Constantin timepieces submitted for servicing was more than fifty.

Specialist knowledge acquired in this way cannot be replaced by even the most elaborate study of literature. This fact is also one of the main features that distinguishes this book from other publications.

In addition to his special focus on Patek Philippe and Vacheron & Constantin, the author examined the technical characteristics of numerous timepieces from Audemars Piguet, IWC, Jaeger-LeCoultre, Rolex, and Omega, and they were then handed over to the factory, or the responsible German outlet, to be serviced. Not only were the different prices of interest, but also the different performance levels. Using time graphers produced by Swiss manufacturer Witschi, precise rate measurements could be made and documented both before and after each overhaul. The results achieved will surprise many a reader.

All the above measures required an immense effort, but they were the most effective way to gain comprehensive knowledge about the technology and special features of the respective watches.

As a practical training process, this "training program" turned out to be quite costly in some cases, as can be seen from the servicing costs listed in each case.

There is no comparable case to be found in available literature in which so much effort was expended on behalf of the interested reader.

The author's many professional and international commitments also enabled him to attend auctions in London, New York, and Hong Kong. The knowledge gained in this way quite considerably broadened his horizons concerning luxury wristwatches and their price development.

One of his first cover stories in this field appeared in *Klassik Uhren* ("Classic Watches") as early as 1995. This extensive article dealt with the topic of "Buying and Overhauling Old Automatic Watches." It was not a general report, but one about the practical experiences gained when buying and servicing ten high-value wristwatches. Other articles of this kind followed.

The watch book *Die Noblen aus der Schweiz* ("The Nobles from Switzerland") was published in 2006 and is now out of print. The book *Patek Philippe Armbanduhren—Kultobjekt & Wertanlage* (*Patek Philippe Wristwatches—Cult Object & Investment*) followed in 2011. This work is also available in English. Some of the contents that form an important part of this book were taken from these two publications.

The numerous publications by the author in other fields can be found on the internet at www.mehltretter.info.

Another novelty might be the anonymous procurement of new wristwatches. To be able to compare old and current models, the manufacturers were not asked for test watches, as is usually the case; instead the selected timepieces were obtained in the normal way from specialist retailers. This somewhat unusual method excludes the possibility of manipulation and guarantees the reader the objectivity that is not always provided by tests made by trade journals. The new IWC, Jaeger-LeCoultre, A. Lange & Söhne, Patek Philippe, and Rolex watches purchased from concessionaires were subjected to comprehensive tests and produced interesting and sometimes unexpected results.

The former service director of a well-known Swiss watch brand once said literally: “Writing a watch book without having access to the watches is like a court case that has no witnesses. A judgment can be based only on circumstantial evidence.” That is why the author remained true to his principle of writing only about watches that were in his possession. This means that almost all the watches listed in this book were physically in the author’s possession and could therefore be described and evaluated from a direct point of view. This also applies to the very valuable pieces. This fact gives this book another unique characteristic.

The fact that this enlightening path also led to some surprises is documented in the case of the excellent special version of the caliber C 89 from IWC.

To help the reader get started, all the important topics of this book are explained in more detail in the following introduction.

This book was written for the watch enthusiast who, despite the wide range of available literature, wants to know some things even more precisely than before. Those who want to deepen their knowledge as experts should always have it at hand. And even the aesthete, who takes less pleasure in the technology and more in the design and fame of a watch, will always enjoy this uniquely illustrated work.

Pullach near Munich, summer 2021
J. Michael Mehltretter

Preface— from the Philosopher's Point of View

True Luxury or Luxury as a Commodity?

by Prof. Dr. Michael Bordt, SJ

All the watches presented and described are considered luxury items. They go far beyond what is actually expected of a watch: that it tells us what time it is. Even watches that can be bought for a few dollars in any department store or online can do this. Such watches, which have been available since the 1970s, most with quartz movements, fulfill their purpose of reliably showing us the time, sometimes even better than expensive mechanical watches: they are accurate to the second, at least until the battery has to be replaced at some point.

The luxury watches presented in this volume fulfill the purpose of a normal watch, but no one buys these watches exclusively to tell the time. This is especially true in the age of the smartphone, which reliably shows us what time it is; we really don't need wristwatches any more. So why do we buy such expensive watches? Why is it that the market for luxury goods is growing at an above-average rate these days? An obvious answer is that one hopes that luxury articles will be a safe and effective investment, or that they will bring a pleasing return. In a time of cheap money, people are looking for crisis-proof, inflation-proof investment opportunities and therefore buy luxury goods, which without a doubt include the best mechanical watches. Anyone who buys watches as an investment need not be interested in much of what is described in this volume. The astonishing skills of an engineer, the simple or distinctly sophisticated aesthetics of the dial, the effort that goes into a *tourbillon*—all this is irrelevant to someone who appreciates expensive watches primarily as an investment. They do not need to understand any of this. What they need to be able to assess is the security of their investment. They are well served by very expensive watches, often unique pieces that can cost many thousands of dollars.

Those who seek and purchase luxury watches for investment purposes will rarely wear them on their wrist, but many luxury watches are intended to fulfill precisely this purpose: those who

Illustration 1
The Patek Philippe Ref. 5125 J was released to celebrate the 125th anniversary of the Patek retailer Wempe of Hamburg. This magnificent piece has an automatic movement with annual calendar. The case with protected crown also looks very successful. This watch will never lose its value. Rarely has the subject of wristwatches been interpreted so perfectly as in this model.

Illustration 2
Patek Philippe Ref. 5125 J seen from the back. The case has a screw-down case back with fine thread and six deep-drawn grooves of the very highest quality. Through the sapphire crystal case back, one can see the automatic caliber 315/203 movement numbered 3,360,141 and the beautifully decorated gold rotor. A total of 125 examples of this watch were made in yellow gold, 125 in red gold, and 50 in platinum.

Illustration 3
IWC Automatic Ref. R 817 A and AD (automatics without and with date) in stainless-steel cases with screw-down case backs. This extremely rare model, which was unveiled in the late sixties, is one of the most successful wristwatches made by IWC in Schaffhausen. The caliber C. 8541 B automatic movement installed in the watch was one of the best Swiss movements at that time. The watch on the left almost looks a bit more elegant without the date window. Compared to the yellow-gold Wempe model on the previous page, this watch shines as a noble piece made of stainless steel.

own them want to wear them, to be seen with them. Luxury watches are then not primarily financial investments, but status symbols. And status symbols have to be displayed; otherwise they have no effect. Of course, status symbols must also be understood. Not everyone is able to recognize a luxury watch on someone else's wrist. Whoever appreciates watches as status symbols reveals themselves to a group that in turn can recognize the status symbol as such. With an expensive watch, one then shows more or less conspicuously to the person who can read it that one belongs to the circle that understands luxury watches. One is an insider and distinguishes oneself from others. One identifies oneself with those who can also afford luxury watches. It is, of course, all the more annoying when excellently made replicas come onto the market that cost only a fraction of the original watch and can hardly be distinguished from the originals at a first or even a second glance.

The line between luxury and showing off or ostentation can be very narrow here. Basically, you don't have to enjoy the watch itself. You are happy only that you own it, that you can afford it, and, above all, that others see that you can afford it. Perhaps one is also happy about arousing envy in others. One can only pity such contemporaries, because those who need to show off lack healthful self-confidence. That's why they need to define themselves by belonging to a group and by brand products. And of course the marketing departments of the companies work diligently to make this identification possible.

Be that as it may, anyone who appreciates lavishly produced and expensive watches as an investment or status symbol has not really grasped what luxury is all about. For those who own a watch as an investment or status symbol, one thing matters most: the price. The watch must be expensive. But true luxury is actually something else. True luxury does not have to cost a lot. Affording a bar of soap that is handmade and smells wonderful will not tear a hole in the household budget but can bring a lot of joy in everyday life. Some luxuries cost nothing at all. Let's think of people for whom it is a luxury to be able to take a nap on the weekend. When we talk about luxury, we don't mean that it is impossible to afford on a normal income, but rather that we have an experience. True luxury must be experienced as luxury. But what is meant by that?

There are only a few scientists who have devoted themselves to the topic of luxury more seriously and beyond economic considerations. But there is unanimity in research that luxury is always about something that is not necessary for our lives. Even the first philosopher to think about luxury, the Greek philosopher Plato, distinguished between luxury and the things that we necessarily need for our lives. Like many other authors after him, he was convinced that luxury is bad for humans and that one should be content with what is necessary to live. But one does not have to share this judgment.

It is true that luxury is not necessary, but at the same time it is also more—and this "more" can be of great importance for us humans. To experience luxury is to be removed, at least for a moment, from the everyday necessities of life. It is the experience that there is a life beyond the often gray, everyday routine, or a life on the hamster wheel.

Experiencing luxury is a protest against the fact that the lives of us humans simply have to function. More and more areas of life are permeated by considerations of utility—and luxury breaks through all the cost/benefit considerations in which our everyday lives, our jobs, and often also our friendships are locked, because some people also have the impression that they simply have to function as a spouse or wife, as a father or mother. Working time has to be divided economically; results have to be achieved in a goal-oriented way. We have to act purposefully, "sensibly," and avoid everything that is superfluous. Not only the life of each one of us, but the whole society functions this way. Actually, there is no place for luxury in our lives. But with luxury we break out of the necessities of life. When we experience luxury, we experience that there is something else in life beyond functionality and economization: purposelessness and abundance. Luxury articles always serve a purpose—luxury watches tell us what time it is or what phase of the moon we are in, Louis Vuitton bags can also be used as tote bags, and luxury limousines get us from one place to another just like a simple car—but they do so in a way that is somewhat extravagant, sometimes even exaggerated. The more we know about a luxury item, the more astonishing it becomes: luxury watches are masterpieces of the art of engineering in a field where the art of engineering is not really needed for everyday life, because there are much-simpler and at-least-as-reliable watches available as mass-produced goods. When we enjoy the luxury of watches, it is not because we have found a suitable financial investment or feel that we belong to a more or less elite group, but because we are amazed that there are people who invest so much time and effort in something that is beyond all rationality of purpose. Experiencing luxury can also reawaken in us a longing for a life that has more to offer than pure rationality and the need to function. We humans do not disappear into social rules, moral constraints, and economic laws. The many images in this volume bear witness to this and can remind us of our very own dignity.

Illustration 4
Prof. Dr. Michael Bordt SJ is a member of the Jesuit Order and teaches at the Munich School of Philosophy. He is a board member of the Institute for Philosophy and Leadership and offers consulting, guidance, and workshops for executives in top positions of large middle-class companies and corporations such as the BMW Group, Deutsche Börse AG, and ThyssenKrupp AG. Bordt studied philosophy and theology in Munich and Oxford, where he earned his doctorate, and conducted research at Princeton University. From 2005 to 2011 he was president of the Munich School of Philosophy. He is the author of numerous books, including *The Art of Enduring Yourself* (2013), *The Art of Understanding Yourself* (2015), *The Art of Disappointing Parents* (2017), and *The Art of Following Our Longing* (2020).

Illustration 5
Close-up of the caliber L 951.6 movement of the Datograph Flyback chronograph, reference 405.031, by A. Lange & Söhne of Glashütte in Saxony. In the upper part of the artistically decorated balance cock, one can see the housing of the Gyromax balance wheel equipped with six *masselotte* weights. The balance staff is axially protected by an Incabloc shock protection system. The fine adjustment of the gear is done by a regulator arm and micrometer screw (*right*). The six rotatable, eccentrically mounted inertial weights provide the last "ounce" of precision. If the centers of gravity of these weights are shifted inward, the moment of inertia GD^2 is reduced and the watch runs faster. If the centers of gravity are on the outside, the watch runs more slowly. To the lower right of the shock absorber can be seen a slotted cylinder head screw used to lock the adjustment of the drop (i.e., the symmetry of the oscillation system). Why so much constructive effort was spent here for this simple activity is not understandable from a purely technical point of view. At Patek Philippe—for example, on the caliber 27-460 M PM (p. 147)—the adjustable hairspring stud (French: *le piton mobile*) is much simpler, but no less efficient. Finally the question arises why such an elaborately constructed movement was equipped with a simple flat hairspring instead of a Breguet hairspring. If even a simple manually wound C. 89 movement from IWC was given a blued Breguet hairspring, one would have been suitable here too.

Introduction

Wristwatch Adventure

A spirit of adventure and a passion for collecting, coupled with specialist knowledge and sufficient mobility, are among the important prerequisites for embarking on this adventure. Added to this are a good knowledge of the market, personal connections to the watch industry, and an appropriate budget.

At least as important are the happy coincidences described here, which give this book a special character.

Of these self-experienced "adventures," three will be mentioned here in advance, to give the reader an understanding of some of the peculiarities of an internationally oriented collecting activity.

More than twenty-five years ago, when people were still on the lookout for special watches from Patek Philippe, a self-winding Reference 2585 in stainless steel with movement no. 765,932 and case no. 694,414 (illus. 6) was offered at Sotheby's New York. As lot 256 with an estimated price of 5,500 to 6,000 USD, this watch was sold for 10,000 USD on October 28, 1996. At the time, the bidder was present in person, but out of price discipline he bid only up to 9,000 USD. Almost nine years later the same watch turned up again at Christies in Geneva. On May 16, 2005, it was opened at 40,000 CHF (Swiss Francs) and, after a fierce bidding war, was knocked down at 104,400 CHF.

This story was far from over, and a few years later this rare model continued to soar to incredible heights, as can be read in the chapter on Patek Philippe watches.

In another case, a valuable Patek Philippe Ref. 3448, a sought-after model with a perpetual calendar, came up at an auction house but only in a postsale. This reference had been produced from 1962 to 1982, but only 586 pieces were made. The reason for this purchase was not so much the value or the

Illustration 6
Patek Philippe Ref. 2585 in stainless steel (movement no. 765,932, case no. 694,414). *Photograph by Christie's*

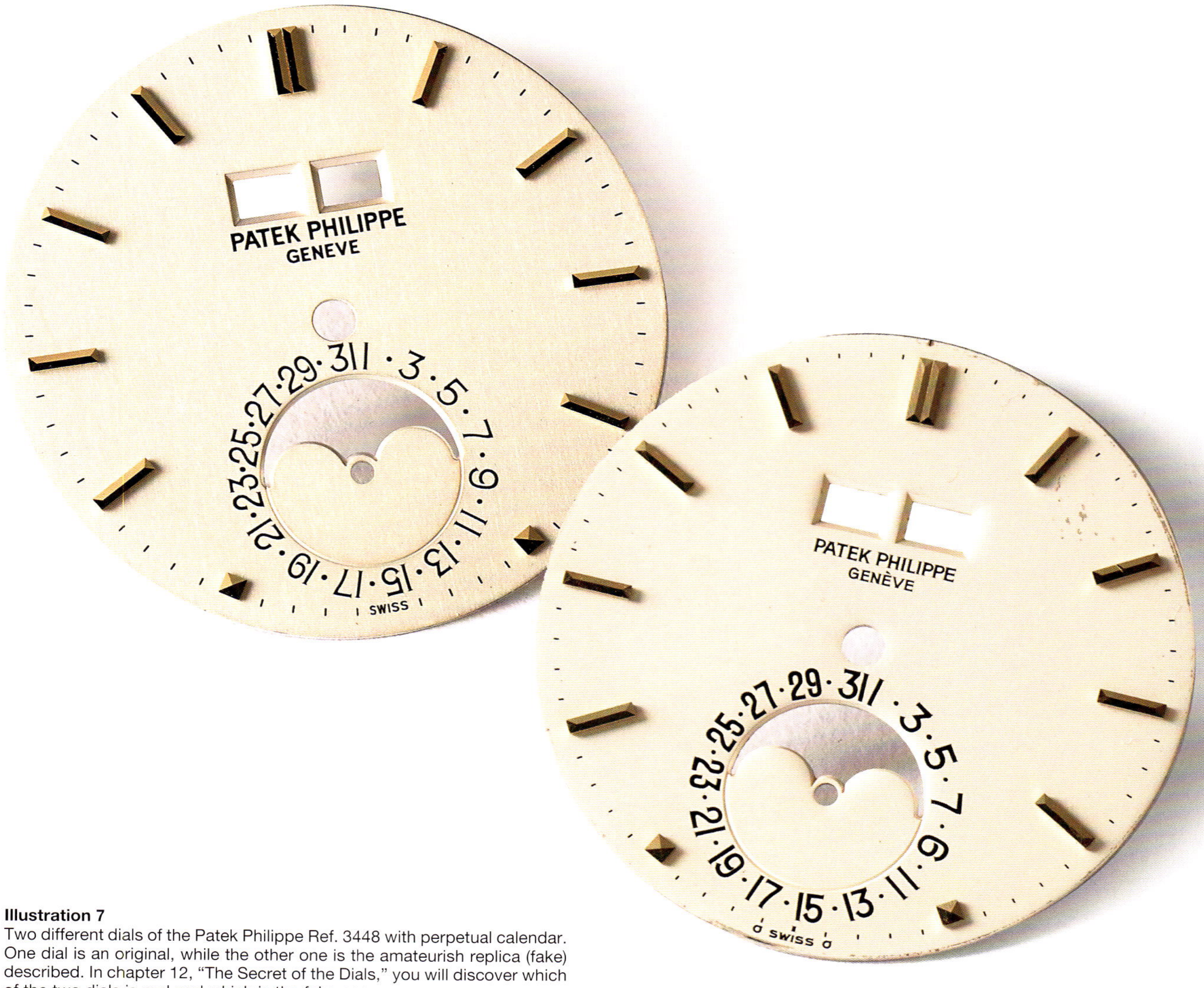

Illustration 7
Two different dials of the Patek Philippe Ref. 3448 with perpetual calendar. One dial is an original, while the other one is the amateurish replica (fake) described. In chapter 12, "The Secret of the Dials," you will discover which of the two dials is real and which is the fake one.

uniqueness of this watch, but the fact that the dial between the varnish was flaking between indices 1 and 2. This led to the suspicion that the dial either had been refurbished or had been made by a third party. This watch was sent to Patek Philippe in Geneva for servicing, including dial replacement. At the same time, the service department was asked to provide both the dial present on the watch when purchased and the new replacement dial, for the purpose of taking photographs. It turned out that the author's previously harbored suspicion was well founded and that the dial mounted on the watch at the time of the sale was a third-party item (i.e., counterfeit). Despite the profound evidence, the service director at Patek Philippe at the time continued to claim to the auction house and the new owner that the dial was an original. The fact that this story almost escalated into a legal dispute because the aforementioned service director refused to respond to the arguments that had been put forward several times and were backed up by evidence, and thus a reasonable settlement of this case no longer seemed possible, was in no way comprehensible even to connoisseurs of the industry. Today watches of this reference are traded for between 108,000 and 220,000 USD.

No one would have imagined that this process could take an unexpected turn. The successful end of this exciting story can be read in chapter 8 on Patek Philippe watches.

The third story should also give this book a special appeal. In the mid-1950s, Vacheron & Constantin in Geneva launched a new chronograph that, unlike its predecessors, was fitted with a screw-down case back and thus received the rating "water resistant." Exactly thirty-six examples of this watch, designated reference 6087, were delivered until the beginning of 1960.

As a confidant at Vacheron & Constantin had already revealed, two prototypes were built during the development

Illustration 8
Vacheron & Constantin Chronograph Ref. 6087 in 750 GG, called the Cornes de Vache (cow horns) (movement no. 470,084, case no. 373,268).

Illustration 9
This is probably one of the two long-lost prototypes in stainless steel built for the chronograph shown on the left.

phase of this new model. In the former, an existing screw-down case back from sister company Jaeger-LeCoultre was used for technical-function testing. In the case of the latter, the intended case shape with the drop-shaped lugs (*cornes de vache*, or cow horns) was already installed to enable a final design assessment of this new watch.

In the following years there were changes in ownership, which finally led to the separation of the former sister companies Vacheron & Constantin and Jaeger-LeCoultre in 1968. As part of the restructuring of Vacheron & Constantin, many employees had to give up their jobs. As a result, much of the know-how of the company, which was founded in 1755, was lost. Along with many other material assets, the same fate probably befell the two aforementioned prototypes.

By chance, in 2011 the author came across an internet offer from Lisbon in which a Portuguese man was offering a Vacheron & Constantin chronograph with a steel case and screw-down case back.

After lengthy discussions, the vendor backed down from his asking price of $10,890 and finally sold this watch for $8,165, of course with a previously agreed-upon right of return. After the arrival and appraisal of this chronograph, it was passed to Vacheron Constantin (now written without the connecting "&") in Geneva, and an overhaul was ordered. Due to the importance of this particular watch, Rudolf Büll, formerly a service manager and later advisor to the president of Vacheron Constantin, took over the internal research and responsibility for the overhaul.

Was this steel chronograph one of the two lost prototypes or not?

Why this exciting discovery came to an almost mysterious but still satisfying end can be read about in chapter 10, covering Vacheron & Constantin.

What Benefit Does This Book Bring to the Reader?

The potential watch buyer who is interested in a certain model of the watch brands listed in this book needs a certain amount of expertise to be able to make an optimal purchase decision. It is not only important whether the watch in question costs $5,500 or $55,000 euros, because the personal inclination and taste of different well-to-do buyers are always in the foreground. The former may have had to save diligently for their watch for several years, while the latter can easily pay the purchase price from their lavishly filled bank account.

Regardless of watch brand and purchase price, this book is intended to serve as a technically in-depth, reliable source of information for all readers and to become a readily used reference work.

Most of the technical contents of this book also apply in principle to all other well-known Swiss watch brands, so this publication can definitely be called a new standard work for wristwatches. With more than four hundred pages and more than seven hundred professionally taken photographs, this claim is credibly substantiated.

If a car tester takes a silky-smooth eight-cylinder from a Bavarian car manufacturer for a test drive and is thrilled by both its performance and its barely perceptible acoustics, he will write a favorable report and thus delight the readers of his magazine.

However, it is far from clear whether this outwardly excellent result was due to a well-designed vehicle or to later improvements; for example, by the subsequent installation of numerous absorbers (the term "absorber" is used to describe weights that reduce the resonance of vibrations). A heavy limousine known to the author had absorbers each weighing 11 pounds (5 kilograms) fitted to both halves of the rear axle, thus compensating for a construction and design error. None of the mostly well-heeled buyers noticed this subsequent improvement—nor did the testers, by the way.

In a similar way this can also happen to employees of watch magazines or authors of watch books. If one does not take a deep look into the technology and is unable to understand it from an engineer's point of view, one is unlikely to understand the real quality of a product and will not be able to evaluate it with any certainty. This is no different in the world of precious Swiss wristwatches.

To fully get to know the watch brands presented here, not only old but also some new models have been presented in the respective chapters. This approach provides an insight into the development history of the brands covered here, and allows us to draw a line from the old watches to the new models. The book *Die Noblen aus der Schweiz* ("The Nobles from Switzerland"), published in 2006, dealt with Swiss wristwatches in detail. Unfortunately this work is now out of print. Due to its timelessness, important contents were taken from it.

Illustration 10
Balance wheel of the Rolex caliber 3135 with Breguet hairspring and Microstella screws

The book *Patek Philippe Armbanduhren—Kultobjekt & Wertanlage* (*Patek Philippe Wristwatches—Cult Object & Investment*), published in 2012, dealt with technical topics, some of which can again be found here in greater depth.

Evaluation of the Design of the Movement and Case

The author has worked for many years not only as a freelance journalist and author, but also as a medium-sized entrepreneur in the field of mechanical engineering and assembly technology.

He can draw on a broad range of knowledge in the fields of new technologies, product development, CAD design, technical mechanics, and manufacturing technology. As a graduate engineer specializing in mechanical engineering, he has worked internationally and supplied well-known companies with his technologies and systems.

This makes it easy for him to understand the conception of a movement and its design in all relevant details. A well-done design is recognizable not only from impressive and colored CAD overview drawings, but also from neatly dimensioned individual part and assembly drawings. Only from these can one see and understand the function and all relevant technical details. High-quality engineering work can easily be distinguished from visually attractive but constructively simple solutions.

The movement of a mechanical wristwatch is in principle no different from a machine, albeit a very small one. With the exception of thermodynamics and electrical engineering, all the technical disciplines listed for engine technology also apply here, with the focus on gear theory, materials science, technical mechanics, and vibration technology. The high-design art for the development of a precisely running movement lies, among other things, in keeping the sum of the resistances of all rotating parts as small as possible, and this value is almost the same both in the horizontal and vertical positions!

Illustration 11
A very rare picture: the automatic bridge with rotor of the Rolex caliber 3135 turned upside down after removal. You can clearly see the red-and-gold anodized gears.

Of the torque applied by the mainspring, only a fraction is needed to operate the three hands of a normal wristwatch; the rest is consumed by the gears and the oscillating system.

Illustration 12
Automatic caliber 2892 A2 from ETA. You can see the balance with eccentric screw for fine adjustment of the rate. This movement is used by numerous watch manufacturers as an *ébauche*.

Illustration 13
Market leader Rolex's excellent automatic caliber 3135. The balance wheel, with its blued Breguet hairspring, is mounted in a bridge, which is adjustable in height. One of the knurled Microstella fine adjustment screws can be seen on the rim of the balance wheel in the foreground.

The installation of additional functions, such as a date display, a perpetual calendar, a chronograph assembly, and other complications, requires further power that the spring movement must also provide.

The heart of a watch is the oscillating system, also known in Switzerland as the regulating system or fine-adjustment mechanism. Whether a simple but proven ETA design with a flat hairspring, simple balance wheel, and rate regulation via an eccentric screw is installed or a high-quality version with a Breguet hairspring and Gyromax balance wheel is fitted is always a question of cost and determined by the respective manufacturer.

Some important technical solutions are presented here, and their characteristics are described.

During the assessment and evaluation of the various designs, surprises came to light that were unknown even to the renowned experts of the concerned manufacturer. In the case of a manually wound caliber C 89 movement in an IWC Yacht Club, the movement had been so refined that—if I may say so—its technical rating was almost superior to a comparable manually wound movement from Patek Philippe (illus. 14)! This statement, which can be made only if you have the watch in question in your hands, can be read in chapter 2, "IWC."

Of particular importance is chapter 13, titled "Physical Principles for the Accuracy of Wristwatches," written by Dr. Gerhard R. Thoma.

This chapter was written to give collectors and experts an understanding of the theoretical basis for the accuracy of a movement. As the head of the acoustics department of a Bavarian car manufacturer for many years, Dr. Thoma ensured that the manufacturer's large limousines in particular always set the right tone and ran without annoying vibration interference. Such an in-depth treatment of this topic on wristwatches is not to be found in the available literature.

One of the greatest challenges in a self-winding watch is material wear. Where metal gears interact there is abrasion, and thus wear.

To be able to analyze the behavior of particularly stressed gears, one must more closely examine the gears of a self-winding module from a watch manufacturer in Geneva. A group of gears that had already been in use for more than twenty years was compared with gears that were identical in design but not yet in use. For this purpose the two groups of gears to be compared were observed with the aid of a scanning electron microscope (SEM) at 200× and 400× magnification and examined for signs of wear.

Illustration 14
Close-up of an IWC manually wound movement caliber C. 89 in a previously unknown special version installed in a Yacht Club model. This movement turned out to be a sensation because of its particularly high technical level. Details can be found in chapter 2, "IWC" (movement no. 1,955,875).

A detailed description of this investigation and the results found—a real surprise—are described in detail with the corresponding SEM images in chapter 14, "Technical Notes—from the Engineer's Point of View." These unexpected results will probably make some watchmakers think again!

In the same chapter, various movement designs are analyzed and particularly successful solutions, or those that could be improved are pointed out. The fact that the rules of technical mechanics and mechanical engineering were not fully observed in all the cases described will probably interest not only some readers, but the watch manufacturers in question as well.

The Anonymous Purchase of Special Wristwatches to Determine Quality and Precision

Here, models from IWC, Jaeger-LeCoultre, A. Lange & Söhne, Patek Philippe, and Rolex were sourced from specialist retailers and comprehensively evaluated.

When watch manufacturers give test watches to journalists, these timepieces are often subjected to extensive checks beforehand and adjusted for the highest possible accuracy. The normal buyer of a watch cannot always expect the same treatment. But they don't need it either, because as a rule they have neither the necessary expertise nor a test laboratory with the necessary testing equipment. They are simply pleased with their newly acquired watch and usually trust the manufacturer's specifications without reservation. Whether this trust is always fulfilled in practice will be examined. Information on quality and precision must be determined and verified by measurement.

To be able to verify the precision promised by market leader Rolex in a statistically secured manner, I went so far as to procure a whole series of watches.

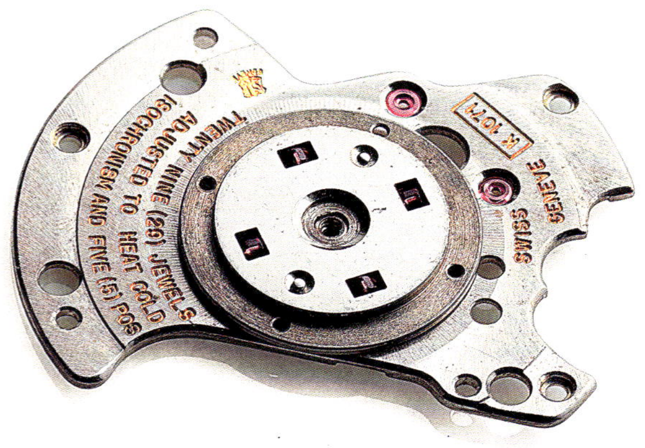

Illustration 16
Automatic bridge of the caliber K 1071 from Vacheron & Constantin. The rotor, which has yet to be mounted, is supported on the four visible ruby rollers. This mint-condition automatic bridge comes from American used-parts dealer francis. antictac in New York and cost $500, plus $30 shipping, for a total of $530 (exactly 396.10 € on August 14, 2014). This part is no longer available from Vacheron Constantin in Geneva.

Such systematic testing has so far been practiced almost only in the automobile industry, but not in the watch industry. Details and results are also summarized in chapter 14, "Technical Notes—from the Engineer's Point of View."

Procurement Channels, Opportunities, and Risks

In building up my early watch collection, knowledge of procurement channels, market prices, and private sales ensured I was usually able to obtain the watches I was seeking at a buyer-friendly price. This often made it possible to acquire some sought-after collector's items at a far-lower price than one would have had to pay at international auctions and collectors' markets.

Illustration 15
Partial view of a gear wheel from the automatic winding mechanism of a Rolex at 200× magnification taken with a scanning electron microscope (SEM). Can signs of wear perhaps be seen here?

Illustration 17
Monthly calendar with moon phase of a watch with a perpetual calendar

Illustration 18
A rarely found Rolex Oyster Day-Date with Ice Blue dial, Roman numerals, and platinum case

Utility Value, Service Costs, and Actual Resale Prices

For all the watches purchased brand new for this book, the focus was on utility value and a realizable resale price. This is of special importance, since there are watches from some Swiss brands whose models have a six-figure price when new but when resold fetch only a fraction of their original price.

As previously mentioned, many of the watches in this book were thoroughly overhauled by the manufacturers.

Overhaul prices of more than $3,200 and delivery times of up to twelve months were not uncommon here. Such valuable information for the collector is unlikely to be found in other publications.

Wristwatches as Investments

Watches as an investment is considered a particularly delicate subject. As we will see, quite considerable increases in value are indeed possible. This assessment applies in particular to watches of the Patek Philippe brand. Here the reader will find numerous concrete examples.

The Secret of the Dials: A Particularly Important Topic

If a reader wants to know more about dials, especially about the markings and hallmarks on the reverse, they will search existing literature in vain.

Thus, in chapter 12, "The Secret of the Dials—the Face of a Watch," we find for the first time an extremely comprehensive dial analysis, in which not only the various manufacturing processes and materials but also the markings and hallmarks applied to the reverse are presented and explained. This ensures that the dial is an original and not one that was fitted later. The markings used by Swiss manufacturers, as well as indications for an authenticity check, are also described. Such analyses can be carried out only if the watch in question is available and a watchmaker has inspected the dial.

Well more than one hundred dials were examined and analyzed for this documentation.

Even if one should not come to the same opinion, a well-known expert has already described this work as half a dissertation.

Illustration 19
A 750 white-gold Patek Philippe Chronograph with perpetual calendar and diamond bezel, Ref. 5270 G. A particularly beautiful model intended primarily for the Arabian market.

Illustration 20
Patek Philippe Chronograph with Perpetual Calendar Ref. 2499 in two case finishes: platinum and yellow gold. One of the two dials was mounted when the watch was made, and is therefore an original. The other is a third-party dial (i.e., an imitation) and is not from the in-house supplier Cadran Stern Frères SA of Geneva. This is a rare, highly sought-after, and valuable watch, but without an original dial its selling price would be considerably lower than expected. Since this topic has not been treated in the common literature so far, nor is it known to the experts of most auction houses, chapter 12, "The Secret of Dials," is of very special importance. After studying this chapter, even the hitherto uninitiated collector will be able to recognize which of the two dials shown is an original and which is a cheap reproduction.

The fact that sinfully expensive watches came up for sale at auctions but no longer had their original dials will come as a surprise not only to the reader.

Decoding the Case Markings

Another innovative feature is the decoding of the hallmarks and the proofs of origin used by the various Swiss watchcase manufacturers. Each supplier had its own personal identification marking that, in conjunction with a number, identified the manufacturer of the case. This official hallmark is valid only for precious metal cases and is on file at the Federal Department of Finance in Bern, Precious Metals Control Division. The allocation of these hallmarks has so far not been disclosed by the Swiss manufacturers' association but has been kept strictly secret. We would like to take this opportunity to thank the Federal Department of Finance for the confidential examination of these documents. Since autumn 2020 these manufacturers' marks have also been available on the internet under Watch-Wiki.

An important change occurred in the marking of precious metals with the Vienna Convention of 1972. For these, too, all the important details are listed.

Finally, there is a small material science section that explains to the reader the relevant properties of the metals used in watchmaking.

Illustration 21
View on the inside of the case back from a Patek Philippe automatic Ref. 3445 with case no. 319,604. The case maker's hallmark is stamped at the very bottom. The key with the no. 4 belonged to the company Antoine Gerlach SA, Genève. This company was registered on September 12, 1934 in the manufacturer list for watchcases and was deleted from the USH-APIC (Association Patronale Suisse des Industries Microtechniques et de L'Habilage Horloger USH-APIC) on August 29, 1977 after the company went out of business.

Photography and Editorial Collaboration

For the first time a Hasselblad 6×6 Digital professional camera was used for the majority of the photographs reproduced here. With the Makro Planar 120/4 and Makro Planar 135/5.6 lenses from Carl Zeiss and the CFV 50 and CFV 50c digital backs from Hasselblad, with 50 million pixels (image points), photos can be taken with a quality of reproduction and depth of field that has scarcely been possible in watch photography before.

Because the author is an engineer and does not claim to be a historian on the subject of watches, Gisbert L. Brunner has taken charge of relating the histories of most of the watch companies listed here, and describes them in his well-known amusing style. No one could have done this better.

Peter Braun, editor in chief of *Armbanduhren* ("Wristwatches"), took over the chapter on the watches produced by A. Lange & Söhne in Glashütte. With his fine sense for a fluent writing style, he greatly supported the author and helped give this book a special touch.

Christian Pfeiffer-Belli, the old master of German watch literature, again advised the author in a friendly manner and gave some valuable tips.

Peter Reiser, a management consultant in Cologne, also helped the author a great deal with his personal advice.

Dr. Gerhard R. Thoma, a physicist and former head of the BMW AG acoustics department, has taken on chapter 16, "Physical Principles for the Accuracy of Wristwatches."

Peter Milde, a renowned watch photographer, was once again able to put all the watches in this publication in the best possible light.

Finally, Daniela Sygulla, a business graduate and authorized signatory in a medium-sized company, took over the optimization of the texts of this work and also supported the author energetically in other ways. She deserves a very special thank-you!

Once again, we would like to express our sincere thanks to all the contributors listed here for their valuable support!

Illustration 22
Daniela Sygulla, Dr. Gerhard R. Thoma, and the author at an in-house exhibition on the subject of assembly technology. As the representative for Bosch Rexroth in southern Bavaria, FMS Montagetechnik GmbH also deals with logistics and manual assembly systems for the production of wristwatches as part of its numerous innovations. *Photo by Stefan Hörnig*

Illus. 24

Illus. 25

Illus. 23

1. Audemars Piguet

1.1. Introduction by Gisbert L. Brunner

Top Watchmaker in Le Brassus since 1875: Audemars Piguet & Co.

There are many luxury watch manufacturers in Switzerland, but only one is older than 125 years and still owned by the founding families: Audemars Piguet. As the name suggests, two ambitious people got together with a view to a successful professional future. This happened in the last quarter of the nineteenth century. Initially, Jules Louis Audemars-Renaud, born in 1851, tried his luck alone. After completing his watchmaking apprenticeship, he hired himself out as a *repasseur*. The aristocrats among the watchmakers had the demanding task of reworking movements (e.g., for chronographs, calendars, and repeater watches) until they functioned perfectly. In 1875 he set up a workshop in Le Brassus to manufacture complicated raw movements. Word of the quality of his work spread to the Geneva manufactories, which were soon sending orders to the "Valley of the Tinkerers." The timely execution of these orders required the employment of additional private watchmakers. One of them was Edward Auguste Piguet. The scion of one of the oldest family lines in this region had first seen the light of day on May 19, 1853, as the son of watchmaker Georges Eugène Piguet. He and Jules Audemars knew each other from school, sports, and singing clubs. When the two met professionally, they were always discussing their craft, the unusual wishes of discerning customers, and ways to optimize fine watches. In the course of fruitful discussions the decision matured to jointly produce and market the complicated timepieces.

The articles of association establishing Audemars, Piguet & Compagnie, comprising fifteen articles, were signed on December 17, 1881, before the notary of the La Vallée district. Edward Piguet provided 10,000 francs in start-up capital for a period of six years without interest. Jules Audemars, on the other hand, contributed eighteen complete or largely complete movements of a highly complicated nature to the new company. A year later the duo registered a trademark with the administration office Technique d'Edition et Publicité in Bern for use on movements and cases.

From the very beginning, Jules Louis Audemars took on the role of technical director, while Edward Auguste Piguet acted as commercial director and financial expert. This division of labor proved so successful that it was maintained even after the deaths of the founding personalities. Members of the Audemars family were primarily responsible for technical matters, while those of the Piguets were more concerned with sales and management.

In 1889, the statistics of the Vaudois watch industry show Audemars, Piguet & Cie. as a company with a total of ten male employees and year-round production. The tenth World's Fair took place that same year. Audemars Piguet presented a whole series of complicated pocket watches in Paris, including the Grande Complication with a split-seconds chronograph, perpetual calendar, and minute repeater. This sensational model in particular earned Audemars Piguet the highest recognition.

After this pleasing trend continued in the early twentieth century, an additional new building with space for seventy employees was built from 1907 onward. Today Audemars Piguet can still be found in this building, which has been enlarged and modernized several times since. The main building serves the factory as a meeting place between the past and the present; in short, as a museum.

The Second Generation Takes the Reins

In May 1917, watchmaker Paul Louis Audemars stepped into the role of chairman of the board of directors and technical director. Paul Edward Piguet succeeded his father in 1919. He subsequently held the position of commercial director until 1962. In a forward-looking move, company management had already given its employees a share in profits or turnover in 1912. In the following years, Audemars Piguet was able to achieve a high international reputation and important customers such as Dent and Frodsham in London, Tiffany in New York and Paris, Cartier in Paris, Bulgari in Rome, Gübelin in Lucerne, and Dürrstein in Glashütte and Dresden, thanks to its consistently high quality and product policy.

The New York stock market crash on October 24, 1929, the world economic crisis it triggered, and the protectionist

Illustrations 23–25
Dürnstein Castle in Dürnstein, Pearl of the Wachau, is today a special five-star hotel managed by its youthful owners Maria-Katharina and brother Christian Thiery. The castle, built in 1630 by Christoph Wilhelm von Zelking, an Austrian nobleman, was acquired by Raimund Thiery in 1937. Later, his son Johann C. Thiery and his wife, Rosemarie, took over the castle and built it up as an outstanding hotel in the 1970s. Today this uniquely situated luxury hotel is run by their children Johannes C., called Christian, and Maria Katharina. Both the front and back of the Audemars Piguet Royal Oak match this noble setting. While the picture at top left (illus. 24) shows the dial indicating not only the time but also the day of the month, day of the week, and phase of the moon, the picture on the right (illus. 25) shows the back with the Royal Oak name and case no. 911. The Porsche 911 Targa, Dürnstein Castle, and the Audemars Piguet Royal Oak with case no. 911—a stylish union of tradition, innovation, and lifestyle.

Illustration 26
Royal Oak Ref. 25594 BA in 18K yellow gold with day of the week, day of the month, and moon phase displays

measures taken by many countries had an unprecedented braking effect. Later, in 1930, all the watchmaking companies in the Joux Valley were forced either to announce reduced hours or close their factories completely for a time. Some never opened their doors again.

When the crisis gradually eased in 1933, watchmaker Jacques-Louis Audemars joined the family business. From 1935 onward, his task was to give new impetus to flagging production. That year, Audemars Piguet produced just 116 watches, most of them wristwatches.

In 1959, the technical management of the manufacture passed to Jacques-Louis Audemars. Three years later, Georges Golay, who had been working for Audemars Piguet as an accountant since 1945, took over the position of commercial director from Paul Edward Piguet. As delegate of the board of directors and general manager, this thoroughbred businessman left his mark on the company's philosophy and image from 1966 until his death in 1987. His name is associated with the long-term success of models such as the Royal Oak (illus. 26) and the Quantième Perpétuel (illus. 27). The world's first series-produced self-winding wristwatch with automatic winding and *tourbillon* can also be attributed to Georges Golay. In addition, the outside boss launched a comprehensive restructuring and reorganization in 1973. In the course of these measures, all workshops underwent a much-needed expansion. At the beginning of the 1980s, Georges Golay also founded a small but highly modern factory for micromechanical and electronic specialties called AP Technologies.

A further step in the diversification of the company's assets was the acquisition in 1978 of a 40% share in the share capital

Illustration 27
AP Quantième Perpétuel Automatic with perpetual calendar in 18K yellow gold, Ref. 5548 BA with caliber 2120/2 QP, movement no. 27,268, and case no. C 1072

of renowned watch manufacturer Jaeger-LeCoultre in Le Sentier, with which Audemars Piguet had enjoyed close and friendly business relations since its foundation. The investment paid off, because in 2000 Richemont International also bought Audemars Piguet's stake in the course of its takeover of IWC, Jaeger-LeCoultre, and Lange Watches. At that time, Jasmine Audemars, great-granddaughter of Jules Louis Audemars, was already acting as chairman of the board of directors, and she is still doing so at present.

Hotbed of Innovation in Le Locle

The opening of a small one-room workshop by Dominique Renaud and Giulio Papi in 1986 would later prove to be a real stroke of luck for Audemars Piguet. Because the two master watchmakers were not having the desired success with the contract manufacture of skeletonized movements, they focused their skills, among other things, on repeater movements. These micromechanisms, one of them with a diameter of just 0.89 inches (22.6 mm), sounded as far away as Le Brassus. And it convinced, since Audemars Piguet acquired it in 1992. Soon afterward, Renaud & Papi SA received an order for a Grande Sonnerie wristwatch. In the course of further joint projects, Audemars Piguet acquired a total of 78.4% of the innovative mechanics forge, which does not shy away from a challenge, by the year 2000. Since then, basic calibers, chronographs, calendar and repeater movements, *tourbillons*, new materials, and the combination of all of these have been among the specialties of the Le Locle–based branch. Without Renaud & Papi there would have been no ultracomplicated cabinet watches in the Tradition d'Excellence collection, which embody their synthesis of traditional additional functions and modern technologies. One example is the Cabinet No. 5, presented in 2006 with the manually wound caliber 2899, whose new precision escapement requires significantly less power than the good old Swiss lever escapement.

Ecology Par Excellence

The sensational Manufacture des Forges in Le Brassus is one of the first Swiss industrial buildings whose design complies with the rigid specifications of the Minergie-Eco quality label for healthful and ecological construction. For its exclusive use of pollutant-free and environmentally friendly materials, the family-owned factory received an award from the Swiss real-estate manager magazine in spring 2010. The building materials include fiber cement for the façades, anodized aluminum for the window frames, FSC-certified wood for the parquet floors, and environmentally friendly water-based paints. A CO_2-neutral woodburning system allows the complete abandonment of fossil fuels such as oil or gas. In the future it will supply heat not only to the Audemars Piguet plant, but also to more than a hundred houses in the village of Le Brassus. Conventional air-conditioning systems were replaced by a "free cooling" system that was optimally adapted to the climatic conditions of the secluded Joux Valley at an elevation of around 3,280 feet (1,000 m) and uses outside air to generate cooling. The modern building offers three hundred employees the greatest possible comfort and a wonderful view of the enchanting Orbe River landscape. The Fondation Audemars Piguet, established in 1992 on the occasion of the twentieth anniversary of the Royal Oak, monitors compliance with the criteria for Minergie-Eco certification. In addition the foundation is committed to saving endangered forests.

Today, visionary watchcases are also created in the aforementioned factory building. After two years of intensive research, the technicians for the first time understood how to produce cases from forged carbon. Until then there had been no precedents in this field. Due to its high mechanical qualities, its extreme chemical stability, its excellent impact resistance, and its special lightness, this material is excellently suited for watchcases, provided that the composition, tools, and process technology are right. In this context the shapes of the octagonal Royal Oak bezel and the no-less-complex case body must have sharp edges and a perfect surface from the outset. After all, workpieces made of forged carbon cannot be reworked. It is precisely this treading of new technical and horological ground that runs through the history of Audemars Piguet like a red thread. The successes associated with this have proven the owners right in every respect.

1.2. The Movements

When we talk about the products of Audemars Piguet, the brand's dominant model, the Royal Oak, inevitably comes to mind. Only insiders know that Audemars Piguet was already producing highly refined wristwatches of special quality in the 1940s, 1950s, and 1960s. In this chapter more than twenty wristwatches are described, some of which are great rarities, such as the early automatics with the calibers P 2498, P 2499, and K 2070. If one takes a look at the production figures, one is astonished by the small numbers of pocket watches and wristwatches produced each year. It should not be overlooked that in the beginning, the workforce consisted of only a small group of watchmakers and assistants and that there was a great depth of production. The comparatively few watches that left the factory in Le Brassus in the 1950s and 1960s are now highly sought after and of great interest to collectors because of their diversity of design. The

Production Year	Number of Watches Made
1892	164
1900	231
1910	310
1930	515
1950	298
1970	5494
1980	14.000
1990	16.000
2000	18.400

following list gives an overview of the production figures. Audemars Piguet did not produce its own calibers for wristwatches, instead sourcing its raw movements (ébauches) from Valjoux and Jaeger-LeCoultre in nearby Le Sentier. From 1960 onward, ébauches from Jaeger-LeCoultre were used almost exclusively, especially since Audemars Piguet had a 40% share in Jaeger-LeCoultre. The table above, compiled by Audemars Piguet for this book, gives an overview of the production figures and periods of the movements presented in this chapter.

The Manually Wound Movements

13''' VZSS and 13''' VZSSC

These two calibers were purchased from the movement supplier Valjoux until 1957. The first version has a movement diameter of 1.18 inches (30.00 mm) and a height of 0.16 inches (4.00 mm). The second has a center second hand (SC stands for *seconde au centre*) and a height of 0.20 inches (5.00 mm). Both versions operate at 18,000 vibrations per hour and are shown on the right.

2001, 2002, and 2003

All three movements were obtained as *ébauches* from Jaeger-LeCoultre. They have a diameter of 0.82 inches (20.80 mm or

9″), eighteen jewels, 18,000 vibrations per hour, and heights of 0.12 inches (2.94 mm) and 0.14 inches (3.44 mm), respectively. The 2001 caliber has a small second hand at six, and the 2002 caliber has a center second hand. Here the seconds wheel is positioned on the gear train bridge. Both of these calibers were used by Audemars Piguet from 1953 onward. The movement numbers from the production year 1953 are listed here:

Caliber 2001: movement nos. 58,000 to 58,699 (700 movements)

Illustration 28
Valjoux caliber 13''' VZSS, which was purchased as a raw movement (*ébauche*) from Valjoux. With an outer diameter of 1.18 inches (30 mm) and a height of 0.16 inches (4.00 mm), there was sufficient volume for a reliable design.

Caliber 2003: movement nos. 59,500 to 59,699 (200 movements)

All three manually wound movements presented are regulated in four positions and temperatures and have chronometer accuracy. The caliber 2003, which is not illustrated in this book, was launched in 1953 and was also an *ébauche* from Jaeger-LeCoultre. With the same diameter of 0.82 inches (20.80 mm), it had an overall height of just 0.06 inches (1.64 mm). Due to its thinness, the seconds display had to be dispensed with.

Illustration 29
Valjoux caliber 13''' VZSSC, where "SC" stands for center seconds. Compared to the variant of the almost identical caliber shown on the left, here one can see the fourth wheel and bridge for the center seconds. Due to these additional components, the movement height increased to 0.196 inches (5.00 mm).

Illustration 30
The caliber 2001 has a diameter of just 0.82 inches (20.8 mm); hence the very wide movement retaining ring.

The Self-Winding Movements

P 2498 and P 2499

This automatic caliber from Audemars Piguet is described in chapter 9, "New Automatic Movements from Switzerland." Once again, it came as a raw movement (*ébauche*) from Jaeger-LeCoultre and was launched in 1953 in two versions. The model with small seconds was given the caliber designation P 2498, the one with center seconds P 2499. The first series of these two

Illustration 31
Caliber P 2498/2 (the "2" stands for ruby rotor ball bearing) with small seconds at six o'clock was supplied as a reserved raw movement (*ébauche*) by Jaeger-LeCoultre from Le Sentier, just 1.8 miles (3 km) away. The rotor has an 18K yellow-gold oscillating weight and winds in both directions. The screw balance wheel has a shock protection system, and the regulator arm is adjusted via a micrometer screw with gooseneck spring.

calibers was listed under the following movement numbers:

P 2498: movement nos. 61,956 to 62,055 (100 movements)
P 2499: movement nos. 61,396 to 61,595 (200 movements)

The basic movements corresponded to their P 1019 and P 1019/1 counterparts from Vacheron & Constantin, except for some details. On the automatic bridge the rotor carrier plate was Geneva striped, and the *guilloché* oscillating weight was made of 18K yellow gold. This difference becomes even more obvious when comparing the automatic caliber P 2498 and P 2499 presented on this page with the Vacheron & Constantin caliber P 1019/1, shown on page 252. The AP calibers P 2498 and P 2499 are of high quality and regulated in four positions and temperatures. Both movement types use a screw balance wheel and a Breguet hairspring. The beat rate is 18,000 beats per hour. The movement diameter is 1.14 inches (29.00 mm), and the movement height is 0.21 inches (5.40 mm). The number of jewels is twenty-one (rubies).

Illustration 32
Caliber P 2499/1; the rotor bearer was always decorated with Geneva stripes on this AP caliber. The P 2498 and P 2499 calibers are regulated in four positions and temperatures and operate at 18,000 vibrations per hour. Both versions have twenty-one jewels.

K 2070 and K 2071

Only six years after the introduction of the first automatic movements, the successor calibers were introduced in 1959. While the basic movement remained almost the same, the automatic assembly was extensively revised. The difference is clearly visible in the rotor bearing, which now consists of four ruby rollers. The oscillating weight is made of 18K yellow gold and has a nice

Illustration 33
Caliber K 2070, with twenty-nine jewels. The rotor is guided by four ruby ball bearings, and the rotor bearer plate is decorated with Geneva stripes. In the foreground, the fine-adjustment mechanism with screw balance wheel, micrometer screw, and gooseneck spring.

guilloché finish. The pawl is on a separate bridge and is screwed to the mainspring barrel bridge from below. The new shock protection system (Parechoc "KIF" Réf. Flector) should also be mentioned. Here are the data on the caliber first launched in 1959:

K 2070: movement nos. 74,761 to 74,770 (100 movements)
K 2071: movement nos. 75,171 to 75,270 (100 movements)

The 2070 and 2071 calibers have the same external dimensions. The movement diameter is 1.14 inches (29 mm), and the height 0.22 inches (5.5 mm). The caliber operates at 18,000 vibrations per hour. Unlike the K 1070 and K 1071 movements from Vacheron & Constantin, these AP calibers are regulated in only four positions and temperatures. A screw balance with Breguet hairspring provides the required precision. However, the number of jewels increased from the twenty-one of calibers P 2498 and P 2499 to twenty-nine rubies.

Illustration 34
Caliber K 2071. Engraved on the automatic bridge is TWENTY NINE (29) JEWELS, ADJUSTED TO HEAT COLD ISOCHRONISM AND FIVE (5) POS.

K 2072

Just two years after the introduction of the calibers K 2070 and K 2071, there followed a version with the date display at three o'clock. The movement diameter remained at 1.14 inches (29.00 mm), but the height increased to 0.25 inches (6.30 mm). The first series had the following data:

K 2072: movement nos. 85,701 to 85,800 (100 movements).

Compared to its sister caliber K 2071, the number of rubies remained at twenty-nine and the number of vibrations per hour at 18,000.

Illustration 35
Caliber K 2072 with screw balance wheel and fine adjustment via micrometer screw and gooseneck spring. With the integration of the date indication, the overall height of the movement increased from 0.22 to 0.25 inches (5.50 to 6.30 mm).

K 2120 and K 2121

As early as the 1960s there was a discernible trend toward increasingly flat watches. This development was given a particular boost by the advent of quartz technology, which made extremely flat cases possible. These fashionable accents of a superflat

Illustration 36
Audemars Piguet was not immune to the trend toward flat watches that emerged at the end of the 1960s, but the designers in Le Brassus were well prepared for this. A new, ultraflat automatic movement (the flattest automatic movement with a central rotor) was already available in 1967, supplied as a reserved raw movement (*ébauche*) by Jaeger-LeCoultre in nearby Le Sentier. Here is the K 2120 version without date indication, with an overall height of just 0.096 inches (2.45 mm).

Illustration 37
Caliber K 2121. The beryllium ring, rotor bearer, and flywheel weight are joined by four screws. The beryllium ring runs on four vertically mounted ruby bearings. The rotor is guided centrally by the inner bore of the drive riveted to the rotor, whose axle consists of a well-dimensioned beryllium hollow flange. In the foreground are the three-spoke balance wheel with adjustable hairspring stud and regulator arm without fine adjustment.

timepiece also demanded the flattest movements, such as the aforementioned manually wound caliber 2003, with an overall height of 0.06 inches (1.64 mm).

In the course of this trend, Jaeger-LeCoultre developed an automatic caliber with a movement diameter of 1.10 inches (28.00 mm) and a total height of just 0.10 inches (2.45 mm). This automatic caliber is discussed several times in chapter 11, "The Noble Sports Watches," and its different variants are explained. From a design point of view, this caliber represents an almost ingenious solution. The basic movement and the changer are housed within a diameter of about 0.83 inches (21 mm). The rotor is guided by the inner bore of the riveted pinion, whose axle consists of a lushly dimensioned beryllium hollow flange. The rotor assembly consists of a beryllium ring, rotor support plate, and gold flywheel weight and is quadruple-bolted. The beryllium ring, held in an angled profile, in turn runs on four ruby rollers mounted vertically and radially in the plate. Further details are the "flying" mainspring barrel, ingeniously simple and responsive rocking-bar changer, self-compensating flat hairspring, and Gyromax balance wheel. Due to lack of space, a second indication had to be omitted. From the production point of view, very tight tolerances are specified for this design, which make high demands on both the single-part production and the assembly. The caliber 2120 was launched in 1967. The K 2121 version with date display followed in 1970. The first series of these two calibers are listed here:

K 2129: 1st production series movement nos. 102,641 to 103,640 (1,000 movements, 1967)

Illustration 38
Caliber 2123, with a case diameter of 1.02 inches (26.00 mm), height of 0.128 inches (3.25 mm), thirty-three jewels, and 21,600 vibrations/h. It is a Jaeger-LeCoultre raw movement based on the caliber 889.

K 2121: 1st production series movement nos. 120,801 to 120,812 (12 movements, 1970)

These two calibers experienced a renaissance at the end of the last century. At that time, Audemars Piguet was able to

Illustration 39
The caliber 3120, a new automatic movement that combines tradition with state-of-the-art technology. Gyromax balance wheel with eight *masselotte* weights, a symmetrical balance bridge, an elegant drop-setting solution, and an aesthetic overall design demonstrate the high innovative power of Audemars Piguet.

purchase the caliber 2120/2121 with the associated tools and drawings and the remaining stock from Jaeger-LeCoultre. Subsequently, the design drawings were converted to CAD (computer-aided design), and the tools as well as the required machining centers for the production of plates, bridges, and other components were newly procured. After the production of the first prototypes, production began in 2005, thus ensuring the continued existence of this impressive automatic caliber. It should not be forgotten that this is a very expensive caliber compared to other automatic movements. According to Audemars Piguet, the widely produced automatic caliber 889.1 by Jaeger-LeCoultre has probably only one-third the production costs of the Audemars Piguet calibers 2120 and 2121, which have been relaunched and are produced only in small series.

1.3. The Reference Numbers

During the production period of the wristwatches presented here, Audemars Piguet did not provide a reference number in its timepieces. It was not until the beginning of the 1960s that the company began to design a classification system and introduced the identification of watches by reference numbers. Watches produced before that time were therefore given this numerical assignment subsequently. Even later (i.e., only in 1973), the identification of the case material was introduced in addition to the reference number. The abbreviation BA 750 means yellow gold; BB 750, rose gold; and BC 750, white gold, and thus the color and fineness of the precious metal used. If an older watch is submitted to Audemars Piguet for service, the respective reference number is listed on the invoice.

Illustration 40
Press-fit case bottom of the Audemars Piguet Ref. 5007 BA, with case number 8363 stamped on the outside

1.4. The Case Numbers

At Audemars Piguet the scheme for case numbers can be easily understood. All cases were numbered consecutively. For example, the steel watch on page 32 has case number 2837 and was produced in 1953. The counterpart with gold case and center second on page 33 has the case number 8363 and comes from the production year 1956. The automatic reference 5115 BA shown on page 35 belongs to the production year 1961 and has case number 12,275. For example, the difference in the case numbers of the watch listed first and that listed last results in 9,445 pieces over a period of eight years.

1.5. The Watches

Manually Wound Watches

References 5006 ST, 5007 BA, 5122 BA, and 5195 BA

Older wristwatches from Audemars Piguet are mostly sought-after collector's items that appear on the vintage market only rarely.

If you open the chapter on wristwatches in the book *Audemars Piguet* (Callwey Verlag Munich), which was published in 1992, you will find only a few watches with self-winding movements, in addition to manually wound timepieces. Consequently, only three of the seven automatic movements are documented photographically.

It is no coincidence that in the book *Automatic Wristwatches from Switzerland*, which was also published in 1992 (also by Callwey Verlag Munich), the automatics, which were already known from the aforementioned book about Audemars Piguet, are also illustrated again. The explanation for this was provided by the editor of the two books, Christian Pfeiffer-Belli, as follows: "There are simply only a very few wristwatches from Audemars Piguet from that period."

In our chapter on this noble brand, you will find not only all the manual-winding and self-winding movements that were in use at that time, but also more than twenty representative wristwatches that document the development of Audemars Piguet timepieces over an era of more than a quarter of a century.

The reference 5006 ST, presented in illustration 41 on page 32, is equipped with the Valjoux caliber 13‴VZSS with small seconds (movement no. 63,324, case no. 2837). This collectible originated in Los Angeles, California, and returned to the Old World in near-scrap condition. This watch spent over half a year in Le Brassus undergoing overhaul. In addition to the professional refurbishment of the movement and case, this watch also received a new dial. Including the crocodile leather strap and pin buckle in 750 white gold (!), the cost of this work was exactly $3,896.20. A considerable investment, considering the purchase price of this simple steel watch was only about one-third of the service costs.

Illustration 41
Wristwatches from the 1940s and also the ’50s carrying the brand name Audemars Piguet may be considered distinct rarities. This is also true of the manually wound Ref. 5006 ST shown, with small seconds and name of the concessionaire, TÜRLER, on the dial. In this watch is the Valjoux caliber 13'' VZSS, which was purchased as a raw movement (*ébauche*) from Valjoux. With an outer diameter of 1.18 inches (30 mm) and a height of 0.16 inches (4.00 mm), there was sufficient volume for a reliable design. This watch in stainless steel was made in 1953. It came from Los Angeles and was extensively overhauled at Audemars Piguet in Le Brassus for $3,931 (movement no. 63,324, case no. 2837).

Illustration 42
Despite the obvious signs of age, this Ref. 5007 AB gold watch still exhibits some of its former beauty. It was made in 1956 and uses the Valjoux caliber 13''' VZSSC, with SC standing for center seconds (movement no. 63,565, case no. 8363).

Illustration 43
On this Ref. 5122 BA manually wound model, one can see the varieties of design on the dial and case that could be found on Audemars Piguet watches in the 1950s and '60s. The use of both arrow and dot indices can be described as interesting, with the latter having the shape of a pyramid.

Illustration 44
Inside this timepiece is a manually wound caliber 2001 movement. Also very elaborately manufactured is the screw-down case back with polygon shape. This design was much more expensive than the one with the press-fit case back, but the case was also water resistant (movement no. 65,209, case no. 7982).

In illustration 42 on page 33, we find the 13''' VZSSC caliber version with center seconds. The SC here stands for *seconde au centre*, thus for center seconds. Both movement versions have a diameter of 1.18 inches (30.00 mm) and an overall height of 0.16 and 0.20 inches (4.00 and 5.00 mm). This height difference of 1 millimeter results from the bridge for the center second hand. Both movement versions are equipped with a screw balance, a flat hairspring, and fine adjustment via micrometer screw and counter spring.

Illustration 45
Flat wristwatches with manual winding have always found their lovers and are still in demand today due to their timeless elegance. This Audemars Piguet Ref. 5195 in 18K yellow gold with a caliber 2001 movement once ticked on the left wrist of American entertainer Frank Sinatra. How it came to be and why this timepiece is shown in this book are described in more detail in the text. Audemars Piguet records show that the watch was completed on June 5, 1961, and delivered to its then New York representative on February 9, 1962. The retail price at the time was 1,245 Swiss Francs (movement no. 77,609, case no. 24,518).

Illustration 46
The caliber 2001 measures 0.82 inches (20.80 mm) in diameter and 0.115 inches (2.94 mm) in height. There was no fine adjustment of the regulator arm.

The reference 5007 BA watch was sold to the Audemars Piguet branch in New York on October 9, 1956, and did not make its return to Europe until 2005 (movement no. 63,565, case no. 8363).

Reference 5122 BA, shown on page 34, has a screw-down case back and is equipped with the caliber 2001 with small seconds (movement no. 65,209, case no. 7892). The screw-down case back is twelve sided and can be opened only with a special key. This case design proves that it is possible to achieve a technically effective screw-down case back despite the low case height. The case middle part is made of one piece (i.e., stamped from a flat gold band). The screw-down case back is beveled on the inside; the inscription and hallmarking were done in the next processing step.

The reference 5195 BA, which we find on page 34, also has the 2001 caliber encased (movement no. 77,609, case no. 24,518). Unexpected sensations are not uncommon in Audemars Piguet watches, either. This noble piece surprisingly rewrote history. As the owner of this timepiece informed us, American entertainer Frank Sinatra was a lover of AP wristwatches. Thus, this timepiece also came into his possession, albeit temporarily. The artist's son from his second marriage, who lived in Beverly Hills, California, was kidnapped in 1967 and freed by the FBI soon after. The FBI agent who personally delivered the kidnapped son to his home received the watch shown here from "Frankie Boy" as a spontaneous gesture of his gratitude. In 1999, this former federal police officer passed away, and the watch passed into the hands of his son. Two years later this watch was offered several times on the internet, without finding a buyer at first. Presumably the asking price of $3,000 was too high, despite the watch's special history. After much negotiation, the watch finally changed hands for $1,250 and could be made available for our book. From Audemars Piguet in Le Brassus, it was learned that this manually wound watch with reference 5195 BA was manufactured on June 5, 1961, and sold to a buyer in New York, USA, on February 9, 1962. The selling price at that time was exactly 1,245 Swiss Francs.

Automatic-Winding Watches

References 4161 BA, 5112 BA, 5115 BA, 5116 BA, 5116 BC, 5122 BA, 5124 BA, 5195 BA, 5205 BA, 5209 BA, 5273 BA, 5279 BA, 5302 BA, 5402 BA, 5548 BA, 25594 BA, and 25594 SA

Automatics with the first calibers P 2498 and P 2499 and water-resistant screw-down case backs are absolute rarities. On this page we find the automatic reference 5115 BA, with small seconds and caliber P 2498/2 (movement no. 62,377, case no. 12,275). According to Audemars Piguet, this watch was produced in 1959. On the next page we see another automatic Ref. 5115 BA with central seconds and caliber P 2499/1 from the production year 1956 (movement no. 61,785, case no. 9129).

Even rarer are the automatics with the successor caliber K 2070. The author searched for these models for many years without success. The fact that two different references are listed here is due to a coincidence. The reference 5112 BA (p. 36) turned out to be an heirloom that had hardly been worn. The caliber K 2070 has small seconds (illus. 52). In illustration 51, we see the watch's inner workings as well as the neatly finished interior of the screw-down case back (movement no. 74,754, case no. 17,756).

The second rarity with the caliber K 2070, a reference 5124 BA, can be seen on page 37, illustration 53. The case of this watch has only a press-fit case back and is therefore not water resistant like the previously mentioned model. Illustration 51 shows us the caliber K 2070 in all its beauty as well as the corresponding press-fit case back. The case number 20,592 can be clearly read. Exactly 140 caliber K 2070 movements were built between 1959 and 1961. Of the calibers K 2071 with center seconds and K 2072 with center seconds and date, 800 and 1,200 pieces were encased, respectively. Production of the two calibers ended in 1965 and 1969, respectively.

Illustration 47
The hour of the self-winding wristwatch struck at Audemars Piguet in 1953 and thus also at the same time as the first automatic from Patek Philippe. This Ref. 5115 BA automatic has a two-piece case with a screw-down case back. Despite its age of more than sixty years, it still looks like new.

Illustration 48
The caliber P 2498/2 ("2" stands for ruby rotor bearings) with small seconds at six o'clock was supplied as a reserved raw movement (*ébauche*) by Jaeger-LeCoultre from Le Sentier, just 1.8 miles (3 km) away. The rotor has an 18K yellow-gold oscillating weight and winds in both directions. The screw balance wheel has a JLC proprietary shock protection system, while the regulator arm is adjusted via micrometer screw and swan neck spring (movement no. 62,377, case no. 12,275).

Illustration 49
The second variant of the AP Automatics from the 1950s had center seconds, and some also had a chronometer cross on the dial. On Audemars Piguet watches the case number is always stamped or engraved on the outside of the case back. Reference numbers, on the other hand, are not found. However, these can be obtained from Audemars Piguet in Le Brassus by providing the case and movement number.

Illustration 50
Rotor bearers were always decorated with Geneva stripes on this AP caliber. The caliber P 2499/1 was adjusted in four positions and temperatures and operates at 18,000 vibrations per hour. Both versions have twenty-one jewels (movement no. 61,785, case no. 9129).

The movement numbers of the two watches shown in illustrations 51 and 54 are just two numbers apart: 74,754 and 74,756!

On page 38 we see another potpourri of different automatics. Illustration 55 shows us a reference 5115 BA in a sales case with the warranty certificate.

Illustration 51
In 1959, the new K 2070 (small seconds at six o'clock) and K 2071 (center seconds) automatic movements were introduced as improved versions of calibers P 2498 and P 2499. The obvious difference is the rotor housing, with four ruby ball bearings integrated into the mainspring barrel bridge. Although the basic design of the movement was retained, the pawl was placed on a separate bridge and screwed to the mainspring barrel bridge from below. The number of jewels was increased to twenty-nine (movement no. 74,754, case no. 17,756).

Illustration 52
This timelessly beautiful Audemars Piguet Automatic Ref. 5112 BA from 1959 looks as if it just left the factory. The original dial is contrasted in two colors, while the hands and indices are 750 yellow gold. APs with the caliber K 2070 and small seconds were produced only in small numbers and have been impossible to find for years (movement no. 74,754, case no. 17,756).

Illustration 57 shows the reference 5205 BA with a case, which is also called a "fried egg design" on other watch brands.

In illustration 56 we see a comparison of the two references: the 5115 BA without date and the 5209 BA with date.

Illustration 53
Another example of this rare caliber K 2070 automatic movement functions in this rare gold watch with TÜRLER signature. Like the dials of Patek Philippe and Vacheron & Constantin watches, the SWISS label below six o'clock is the only proof of origin.

Illustration 54
Here the automatic shown after removal of the press-fit case back. This case is only splash resistant. When one is looking at the aesthetically pleasing caliber, the question arises whether an even more beautiful automatic movement could be made (movement no. 74,756, case no. 20,592).

Illustration 55
AP Automatic Ref. 5115 BA in 750 yellow gold with original case and guarantee certificate. This timepiece with caliber K 2071 was produced in 1961, exported four years later, and sold only on March 18, 1980, by Bobelt & Co. GmbH, Hanover. Why it took nineteen years to find a buyer for such a beautiful watch is hard to understand (movement no. 85,846, case no. 12,282).

Illustration 56
To the right of the automatic Ref. 5115 BA seen above, a version with date display at three o'clock listed under Ref. 5209 BA. This timepiece uses the caliber K 2072 movement with date display. It was unveiled in 1961, thus coinciding with Patek Philippe's caliber 27-460 M. According to Audemars Piguet, a total of 690 watches with this reference number were produced. The specimen shown was made in 1961 and sold in Switzerland two years later (movement no. 85,856, case no. 26,863).

Illustration 57
This automatic is in keeping with the typical design of the late 1960s with the "fried-egg design," which is also known from other watch brands. Even if this watch Ref. 5205 BA is not to everyone's taste, at least its inner values count in the form of the noble automatic caliber K 2072 (movement no. 95,022, case no. 31,500).

Illustration 58
Here is a look at the inner workings of the Ref. 5205 BA automatic shown above. Typical for the caliber K 2072 is the guilloché oscillating weight in 750 yellow gold.

Illustration 58 introduces us to the K 2072 caliber of the reference 5205 BA shown in illustration 57.

A group photo, which can be taken only if all the watches are available to the author, can be found on pages 40 and 41. The references 5209 BA in yellow gold, 5116 BC in white gold, and 5112 BA with small seconds, again in yellow gold, present themselves in harmonious interaction. These three watches demonstrate unity, but that remained only until the three movements were analyzed in detail. The three watches have almost identical case designs with screw-down backs, with only the lugs differing slightly. The automatic movement of the caliber K 2070 on the right, as well as the caliber K 2017 in the center, are regulated in just four positions, but the caliber K 2072 (*left*) is regulated in five positions. However, after the middle watch with the 750 white-gold case was opened, there was a surprise. Engraved on the automatic bridge—the component on which the rotor is mounted and transmits the torque to the winding movement—is Vacheron & Constantin caliber 1072/1. Here, a "freelance" watchmaker had probably replaced the automatic bridge while performing a service but did not have an original Audemars Piguet component at hand. Since the calibers K 2072 from Audemars Piguet and K 1072 from Vacheron & Constantin are identical except for the company engravings, all relevant components are interchangeable. Only the components of the Vacheron & Constantin movements also have the Geneva hallmark stamped on them. This Geneva hallmark, also called the Geneva Seal, is considered proof of origin for products from Geneva and has thus confirmed the highest precision and longevity.

The same case supplier, Ed. Wenger SA of Geneva, was used for all three watches. The case maker's hallmark consists of a symbolic key with the number 1. This company was canceled by the USH-APIC in 1992. This abbreviation stands for "Association patronale suisse des industries microtechniques et de l'habillage horloger," an interest group of the Swiss watch industry that existed only until November 22, 2010.

Had the white-gold watch not been opened, the surprise would not have been revealed at all. For this reason the potential buyer should always ask to be shown the movement of the watch offered for sale, because such surprises can be avoided only if the watches are physically present and the cases are opened to the viewer.

The Ultra-Thin Automatic Caliber 2120 from 1967

Following the trend of the time, the ultraflat automatic caliber 2120 followed in 1967, and the version with date display, under the designation 2121, in 1970. Since both caliber versions had dispensed with a seconds display, they also did not meet the requirements for an official chronometer test. Nevertheless, this caliber is considered an epoch-making best that is still valid today.

The automatic reference 5302 BA (illus. 59, p. 39) has a one-piece case middle. The screw-down case back has been given a ring bead in which the six grooves for the special key are drawn. The automatic caliber 2120 features a flat hairspring, Gyromax balance wheel, and adjustable hairspring stud. According to Audemars Piguet, this watch was made in 1978 and was sold to a buyer in the USA. In November 2002 it returned to Europe and underwent a general overhaul at Audemars Piguet, with a dial swap. Including a new crocodile leather strap and 750 yellow-gold tang buckle, this work cost exactly $1,821.

Illustration 59
This is the dial side of this automatic (Ref. 5302 BA) produced in 1968. The caliber K 2120 did not feature a seconds display. The Roman numerals are emphasized by the black-lacquered but gold baton hands. The case in the "fried-egg design" does not correspond to modern taste, but nevertheless, one cannot deny that this timepiece has a certain elegance.

Illustration 60
Audemars Piguet was not immune to the trend toward flat watches that emerged at the end of the 1960s, but the designers in Le Brassus were well prepared for this. A new ultraflat automatic movement (the flattest automatic movement with a central rotor) was available in 1967, supplied as a reserved raw movement (*ébauche*) by Jaeger-LeCoultre in nearby Le Sentier. Here is the K 2120 version without date display, with an overall height of just 0.096 inches (2.45 mm). The case was also kept very flat, even though it has a screw-down case back (movement no. 107,503, case no. 47,087).

Illustration 61
This trio consists of three different automatics from the 1960s. *On the left*, the Ref. 5209 BA in 750 yellow gold with caliber K 2072 movement; in the middle, the Ref. 5116 BC in 750 white gold with caliber K 2071; and *on the right*, the Ref. 5112 BA in 750 yellow gold with caliber K 2070. All three versions of this automatic caliber are thus represented.

Illustration 62
The watches shown on the left have identical cases with screw-down backs; only the lugs differ slightly. The caliber K 2070 automatic movement on the right and the caliber K 2071 are regulated in four positions only, but the caliber K 2072 (*left*) is regulated in five positions. However, after the middle watch with the 750 white-gold case was opened, there was a surprise. Engraved on the automatic bridge, the component on which the rotor is mounted and transmits torque to the winding movement, was “Vacheron & Constantin caliber 1072/1.” A “freelance” watchmaker had probably replaced the automatic bridge during a service he had performed but did not have an original Audemars Piguet component on hand. Since the calibers K 2072 from Audemars Piguet and K 1072 from Vacheron & Constantin are identical except for the company engravings, all relevant components can be exchanged. However, only the components of the Vacheron & Constantin movements also have the Geneva hallmark engraved on them. This Geneva hallmark, also known as the Geneva Seal, is considered proof of origin for products made in Geneva and has thus confirmed the highest precision and longevity. The same case supplier, Ed. Wenger SA, Genève, provided the cases for all three watches. The case maker’s hallmark consists of a symbolized key with the number 1. This company was canceled by USH-APIC in 1992. If the white-gold watch had not been opened, this fact would not have been discovered at all. For this reason, the potential buyer should always ask to see the movement of any watch they are considering buying, because such surprises can be avoided only if the watches are physically present and also opened.

Illustration 63
This very thin automatic Ref. 5273/5437 made by Audemars Piguet in 1976 is timelessly beautiful. Despite its low height, the case has a screw-down case back and is therefore water resistant.

Illustration 65
This Audemars Piguet Ref. 5279 BA is a beautiful automatic in a tonneau-shaped case finished in the Carré pattern. With a gold bracelet (Ref. 5279/177), in May 1967 the retail price of this watch was exactly 3,810.00 Swiss Francs.

Illustration 64
A look inside the Ref. 5273/5437 reveals that it is driven by the caliber K 2120 movement. The side view shows us that this case has a polygonal screw-down case back (movement no. 103,210, case no. 46,985).

Illustration 66
This is the automatic watch shown above, disassembled into its three important units: *on the right*, the caliber 2120 automatic movement; *on top*, the single-shell case; and *below left*, the case middle, which is designed as a supporting frame. Such case designs not only were typical for the time but also simplified the manufacturing process, because the movement-retaining ring could be omitted. Only the greater gold content of the case back again increased manufacturing costs (movement no. 46,080, case no. 103,060).

The many different variations that Audemars Piguet used for the case design are documented on page 42. While the automatic Ref. 5273/5437 BA shown on the left in illustration 63 has a round case and screw-down case back, the Ref. 5279 BA watch shown in illustration 65 (*on the right*) has a cambered rectangular shape. It has a press-fit case back and is therefore not water resistant. It was learned from Audemars Piguet that this reference 5279 BA was unveiled in 1967. The watch shown here originally had a soldered gold bracelet and was shipped to Japan under reference number 5279/177 the same year. The public price from 1967, which was exactly 3,810 CHF, may serve as a guide. Through a watch collector this timepiece made its way back to Europe in 2001 for a purchase price of $1,300 plus transportation costs, import sales tax, and customs duties. The watch was professionally overhauled, the gold band was desoldered, and a crocodile leather strap with gold buckle was fitted. For this investment a special price of $525 had to be paid.

New Calibers from Jaeger-LeCoultre

As previously stated, Audemars Piguet also had to look for additional flat automatic movements at the end of the 1960s. Only with such calibers was the realization of further flat wristwatches possible. The solution to this problem was again provided by Jaeger-LeCoultre in nearby Le Sentier in the form of the new automatic caliber 888, which was launched in 1982. At Audemars Piguet this new movement was given the caliber designation 2123. With an installation diameter of 1.02 inches (26.00 mm), a height of 0.13 inches (3.25 mm), thirty-three jewels, and 21,600 vibrations per hour, this new design was well able to replace the previously used caliber K 2072. Also, compared to the caliber K 2072, the mounting diameter was reduced by 0.118 inches (3.00 mm) and the overall height by 0.12 inches (3.05 mm). However, the oscillating system of the new caliber 2123 consists only of a flat hairspring, a three-spoke balance wheel, and a Triovis regulator. A Breguet hairspring and Gyromax balance wheel were omitted for reasons of space and cost. That the new caliber's robustness and serviceability were forced to suffer under the dogma of low construction height is beyond question.

On the next page we see a flat Audemars Piguet Ref. 4161 BA/002 (illus. 68) equipped with the caliber 2123 (illus. 69). The reason why a rather complicated press-fit case back with a push-pull crown was used on this beautiful piece instead of a screw-down case back was most likely due to lower manufacturing costs.

Such strange designs can appear only when case makers give priority to the designer's demands and regard functionality and technical mechanics as secondary.

The success story of the Royal Oak model series is described in chapter 11, "The Noble Sports Watches." This wristwatch still plays first fiddle at Audemars Piguet. In the group picture shown (fig.70, p. 45), we see two different generations of this model—two different versions from the younger references. The Royal Oak in stainless-steel Ref. 5402 ST, which is about forty years old, had a public price of CHF 3,200 when it was launched in 1972. Today this sought-after piece cannot be obtained for less than $54,000–$66,000. This eminent price increase also caused an above-average price increase for the previously not-so-sought-after reference 25594 in various versions.

The world's thinnest wristwatch with automatic winding, perpetual calendar, and moon phase display, Ref. 5548 BA was launched in 1978. This model reached the marketplace before its counterpart from Vacheron & Constantin.

The caliber 2120 had received a corresponding calendar module and was then designated caliber 2120 QP. This module increased the original movement height from 0.10 to 0.16 inches (2.45 to 4.05 mm). The caliber 2120/2 QP consists of exactly 292 parts. How beautifully the dial of this automatic with perpetual calendar is made is shown by the close-up of the dial on this page. The collector's item presented here uses the caliber K 2120/2 QP with movement no. 227,268. The case number is C 1072, which is stamped on the outside of the press-fit back. The three auxiliary press switches are used to set the perpetual calendar. A leap year display is not available.

When this book was originally published in 2021, several watches of this type had been offered on the internet. The seller of the watch with case no. 2621 wanted $15,000 in an immediate sale, while another wanted $12,750 for a watch with case no. C 1115. The prices of watches of this reference sold at Christie's and Philipps auction houses in recent years averaged between $10,950 and $13,000. A white-gold version with case number 1434 from 1980 was auctioned at Christie's on June 28, 2016, for $11,250.

Illustration 67
AP Automatic Caliber 2224/2825
This movement is based on the Jaeger-LeCoultre caliber 889/2, with bidirectional winding rotor, and has a power reserve of thirty-eight hours. The escape wheel shaft is mounted in cover plates. Like the case of the caliber K 2121, easy disassembly has been dispensed with.

Illustration 68
This Ref. 4161 BA/002 timepiece is typical of the automatics of the 1970s. Thanks to the flat caliber 2123 automatic movement, again a raw movement (*ébauche*) from Jaeger-LeCoultre, this watch could be kept very flat.

Illustration 69
View of the modern automatic caliber of the automatic shown above. Jaeger-LeCoultre's caliber 888 ushered in a new generation of movements, and its basic design has been preserved until today in the caliber 889/2. The case design still needs some explanation. Although the case has a press-fit case back, it is water resistant due to its design. Before this pressure case back can be opened, the push-pull crown has to be pulled out. A push-pull crown is a crown with a winding stem, which is split and has a centrally placed clutch. Only then can the case back be lifted off. After the case back is reassembled, the outer coupling shaft can be put back in and mechanically connected again (movement no. 215,576, case no. B 49,064).

Illustration 70
Group photo of different versions of the Audemars Piguet Royal Oak. *On the left and in the middle*, the Royal Oak Day-Date Moonphase in steel and gold, Ref. 25594 SA, and in full gold Ref. 25594 BA from the 1980s; *on the right*, the Ref. 5402 ST from the 1970s, known as the Jumbo. The first two watches have the caliber 2224/2825 (basic Jaeger-LeCoultre 889/2 caliber), while inside the watch on the right is the ultraflat caliber K 2120. Four years old at the time of purchase, the Royal Oak Ref. 5402 ST in stainless steel was purchased at a German auction house for a nominal 4,500 DM. With surcharge and delivery the final price was around 5,540 DM. Today, one would have to pay at least $55,000–66,000 for such a sought-after collector's item. It is also interesting to note that the other models of the Royal Oak, which were not so sought after in the past, have also experienced a significant increase in price.

Illustration 71
AP Ref. 5548 BA Automatic with Perpetual Calendar
Watches with a so-called perpetual calendar are immediately noticeable due to their well-filled dials. The day of the week, day of the month, month, and moon phase all are displayed there. On the back of the case, which is made of 18K yellow gold, one can see the three auxiliary pushers for adjusting the calendar, as well as the case no. C 1072. The automatic caliber 2120/2 QP with movement no. 227,268 operates under the press-fit case back.

Illustration 72
Close-up of the dial of the AP Ref.5548 BA

A Royal Oak Day-Date with moon phase Ref. 25594 BA in solid gold and a 25594 SA in steel and gold are shown in illustrations 73 and 74. The latter is an absolute rarity, since it has case no. 911. This popular model of the Royal Oak series was launched in 1983 and was considered to be the little brother of the perpetual-calendar version released the same year.

This watch was designed by Gerald Genta, who also designed the cases for the Nautilus from Patek Philippe and the 222 from Vacheron & Constantin.

The design of the quite elaborate monocoque case is described in chapter 11, "The Noble Sports Watches." The dial is typical Royal Oak, and the case's octagonal shape adds an additional feature to an already eye-catching watch.

The 2224/2825 automatic movement, based on the caliber 2123 derived from the Jaeger-LeCoultre caliber 889/2, is used in the model shown.

Today, Audemars Piguet largely uses its own manufacture calibers, such as the 3120 or the now-reissued and self-produced K 2120 or 2121 with date.

While the Royal Oak in gold bears the number 828, the steel-and-gold version has the already highlighted number 911. With its sporty appearance, its water resistance, and the famous case number, this watch should be the dream of every Porsche driver! Finally, a note on the further price development of these models. We have heard from watch dealers that the imminent price increase for the Jumbo version reference 5402 ST has also caused a price increase for the other versions of the Royal Oak. Even if one cannot afford the coveted Jumbo model in oversize, a normal version of this sought-after model is also pleasing.

Finally, we have a comparison of the two older models of the Royal Oak (illus. 73 and 74, p. 47) and a current model, illustration 75. The new manufacture caliber 3120 has a balance bridge, a Gyromax balance wheel, and an extremely cleverly designed setting of the drop. Here the designer has proven that the innovation and knowledge of technical mechanics can lead to above-average solutions. *Chapeau*!

The fact that a *mariage* (a watch that marries parts from different watches) can also make a good impression if the movement and dial are original Audemars Piguet is shown by this watch (*right*) in illustration 76. As could be determined in Le Brassus, the two main components—the dial and the movement—were from an automatic reference 5423 BA watch with case number B 15.757 from the year 1977. This *mariage* was offered on the internet and changed hands for $1,927. Not a bad price for a utility watch made of steel with a glass bottom and such a special automatic caliber.

Illustration 73
A sought-after model of this watch legend is the Royal Oak with date, day of the week, and moon phase. The different characters of the yellow-gold version (Ref. 25594BA.OO.0789BA.05) and the gold-and-steel version (Ref. 25594SA. OO.0789SA.06) can be seen nicely here. With a diameter of 1.42 inches (36 mm), it is 0.118 inches (3 mm) smaller than the new automatic and thus looks a bit more elegant, to which the moon phase also contributes. Very unusual is the display of the date and day of the week by means of hands on subdials. Ticking away inside is the automatic caliber 2224/2825, which is based on the Jaeger-LeCoultre caliber 889/2.

Illustration 74
The use of geometric shapes has not changed since Gerald Gentas designed it more than fifty years ago: like the screws of the bezel, the crown has a hexagonal shape, whereas the bezel itself is octagonal. The sides of the bezel and the case back are beveled, and the baton-shaped hands and hour markers have rounded ends.

Overview of Caliber Production (according to Audemars Piguet)

Year	1953	1954	1955	1956	1957	1958	1959	1960	1961	1962	1963	1964
P 2498	100	-	100	100	100	90	-	-	-	-	-	-
P 2499	200	-	100	100	200	100	-	-	-	-	-	-
2001	700	500	500	310	590	400	400	700	700	600	600	400
K 2070	-	-	-	-	-	-	100	-	40	-	-	-
K 2071	-	-	-	-	-	-	100	200	300	-	100	-
K 2072	-	-	-	-	-	-	-	-	100	-	200	400
K 2120	-	-	-	-	-	-	-	-	-	-	-	-
K 2121	-	-	-	-	-	-	-	-	-	-	-	-

Year	1965	1966	1967	1968	1969	1970	1971	1972	1973	1974	1975	Total
P 2498	-	-	-	-	-	-	-	-	-	-	-	490
P 2499	-	-	-	-	-	-	-	-	-	-	-	700
2001	400	500	300	300	400	800	1100	600	3100	2700	3000	19.600
K 2070	-	-	-	-	-	-	-	-	-	-	-	140
K 2071	100	-	-	-	-	-	-	-	-	-	-	800
K 2072	100	-	200	-	200	-	-	-	-	-	-	1200
K 2120	-	-	1000	1500	1000	1300	900	1200	600	1900	1100	10.500
K 2121	-	-	-	-	-	1012	1300	1200	2800	2800	1700	10.812

Illustration 75
Audemars Piguet has created the Royal Oak Automatic (Ref. 15300ST.00.1220ST.01), a high point in the line launched in 1972. The new automatic manufacture caliber 3120, successor to the 2120 developed by Jaeger-LeCoultre, is one of the best three-hand movements on the market, both in terms of construction and aesthetics. The bridge for the balance, the perfect proportions of the plates, the balance wheel with variable moment of inertia, the 22K gold rotor with the coats of arms of the Audemars and Piguet families, and the edges, beveled and polished by hand, can be admired through the glass back. At 0.167 inches (4.25 mm), it is not particularly flat, but it is robust and designed for long-term durability. Thus, it is a perfect match for the Royal Oak, because like the classic, it is elegant and sporty robust at the same time. But Audemars Piguet did not stop at the installation of the new movement: it also fine-tuned the dial with its typical tapestry pattern. It appears clearly arranged. But if you take a closer look at the applied hour markers, you'll notice how detailed and ornate everything is, and the high quality of workmanship that prevails here. This also includes the case and the supple steel bracelet with its barely visible clasp and the interplay of satin and polished surfaces. In addition to the steel version with a white, blue, or black dial, a variant in red gold with a leather strap is also available. The leather strap gives it a unique character.

Illustration 76
Mariage with a caliber K 2121 movement and a foreign case made of stainless steel with a see-through case back. Actually, a *mariage* deserves to be in this chapter, though not as part of our display of watches from the noble watch manufacturer Audemars Piguet. Here both the case maker and the watchmaker have come up with something special. As Christoph Guhl, probably Audemars Piguet's foremost watch expert at Le Brassus, discovered in the archives, the dial and caliber K 2121 movement no. 172,270 originally belonged to a Ref. 5423 automatic watch with the case no. B 15757, from 1977. The fine-adjustment mechanism of the caliber 2121/1 consists of a simple balance wheel with flat hairspring and regulator arm. However, the elaborate shock protection is worth mentioning. Today this movement is again equipped with a Gyromax balance wheel with *masselotte* weights.

Illustration 77
Christoph Guhl (*left*) knowledgeably explains the new caliber 3120 automatic movement as Georges-Henri Meylan, then president of Audemars Piguet, and the author listen attentively.

Illus. 79

Illus. 80

2. IWC

2.1. Introduction by Gisbert L. Brunner

IWC: *Probus Scafusia* or Good, Solid Craftsmanship from Schaffhausen

The Good Side of Bankruptcies

You can turn things around however you like: the whole thing started with a bust! And yet, had it not been for the somewhat mysterious Florentine Ariosto Jones, watchmaking would probably never have made its home in the small town of Schaffhausen on the Rhine. But let's take things one step at a time: sometime in 1868, the same F. A. Jones thought about having pocket watches, which were becoming increasingly popular in the New World, produced inexpensively on the other side of the Great Pond. In concrete terms, he had Switzerland in mind. Misjudging the chronometric structures, the budding entrepreneur found himself in Schaffhausen, which until then had been completely unfamiliar with watchmaking. There, in the tranquil little town on the Rhine Falls, he discovered the advantages of the hydroelectric turbines installed by Heinrich Moser to drive modern production machinery. The legal foundation for the new watch company was laid on April 24, 1871, when the International Watch Co. was founded with headquarters in Schaffhausen and New York. Its concept was to deliver at least ten thousand watches a year to the USA. To this end, in 1873 Jones launched a broad-based advertising campaign with the image of an impressive factory, which, however, existed only in the American's fertile imagination. Despite this effort, his targeted goal became ever more distant. Money became increasingly scarce. Then the world economic crisis caused by the Vienna stock exchange and banking crash of 1873 did the rest. To make matters worse, an audit in September 1875 revealed that Jones had provided incorrect production and inventory data and had managed the company's business in a more than dubious manner. In the end, the glittering bubble also burst because in America, the loudly proclaimed motto "Buy American" and a campaign against imported watches were supplemented by 25% import duties. As a result, the first bankruptcy was imminent in 1875. Jones simply escaped the unpleasant consequences by fleeing.

The employees, who were left to suffer, were facing the end. Because the merchant bank was very concerned about the jobs, which were extremely important for Schaffhausen, it acquired all the shares for a mere 143,300 francs. In the course of the purchase the new owners also presented a new director. Ferdinand F. Seeland also came from the USA, unfortunately speaking neither German nor French, but soon after taking office he presented a comprehensive restructuring concept. This included the conquest of attractive markets and the introduction of new movements. However, as was soon to become apparent, this concept was only apparently successful, not least for the reason that the opening of the 1878–79 annual report culminated in the laconic statement that the boss had quietly and secretly left for America without giving any reason. After a conscientious audit of the accounts, the only option was to go back to bankruptcy court.

After this new disaster, the machine manufacturer Johannes Rauschenbach did not hesitate. The delegate of the board of directors personally acquired the ill-fated company for 280,000 Swiss francs. And with that, IWC became largely family owned for the next hundred years or so. Its transformation into a family-owned joint stock company in 1971 merely involved a change in the legal form of the company. It was not until 1978 that the unpleasant consequences of the global quartz revolution demanded a new sole shareholder in the form of Instek AG, a subsidiary of German tachometer manufacturer VDO. The latter came under the umbrella of the German Mannesmann Group in 1991, which, with the help of successful German manager Günter Blümlein, merged its chronometric activities into the holding company Les Manufactures Horlogères (LMH). The fourth and, for the time being, last chapter of the turbulent company history is quickly told: at the end of July 2000, the Richemont-Luxus group, owned by South African billionaire Johann Rupert, bought LMH for 2.8 billion Swiss francs. With consolidated annual sales of just under 350 million Swiss francs, this was a considerable but ultimately justified price. In 1999 IWC itself produced a good 38,000 watches with a workforce of around 360. The Schaffhausen-based company thus achieved sales of 115 million Swiss francs.

Illustration 78–80
Mercedes 300 SL Gullwing of the IWC Racing Team on display at the IWC Flagship Boutique in Zurich (illus. 79). IWC Portugieser Ref. IW503404, with seven-day automatic movement and perpetual calendar in 750 red gold (illus. 80). Rear view of the IWC Portugieser with caliber 52615 movement. For the expert: the rate is adjusted via four weight screws on the outer ring of the balance wheel as well as two *masselotte* weights on the balance bridges. There's probably no better way to do it.

Illustration 81
IWC Ingenieur with perpetual calendar in 18K yellow gold with matching bracelet. This watch, which was unveiled at the 1985 Basel Watch Fair with the reference number 9240, was given the new reference number 9245 in 1987 due to the modifications made to the perpetual calendar and movement. The watch, which was also delivered with an alligator strap from 1987 onward, had the reference number 3540. The automatic movement is based on the ETA caliber 2892 A2; the calendar module was an IWC development under Kurt Klaus (caliber C. 37582, movement no. 2,427,535, case no. 2,432,137).

From the Schaffhausen Manufacture

Despite his sometimes inglorious work, Florentine Ariosto Jones is inextricably linked with IWC history. After all, several important pocket watch calibers bear the name of the company's founder. Their hallmarks: three-quarters plate, very long regulator arm, bimetallic balance wheel, and a Swiss lever escapement. An inventory from 1878 shows that there must have been a total of seven basic "Jones" calibers at IWC. The first prototypes became available in 1870. These were key-wound or crown-wound, manufactured in various quality gradations with and without screwed *chatons*, with Breguet or flat hairsprings, and with different lever shapes and decorations. Jones imposed capital letters engraved in the plates on the different variants. Even after the flight of its namesake, IWC continued to use movements of this type.

Illustration 82
IWC Da Vinci Automatic Chronograph with perpetual calendar. This innovation was also unveiled at the 1985 Basel Watch Fair but attracted much-greater interest than the Ingenieur with perpetual calendar unveiled at the same time (illus. 81). Although the baroque appearance of this model was not to everyone's taste, it did give IWC's image a considerable boost and was a great sales success. At IWC the automatic 7750 movement from ETA was called caliber C. 79261, and the perpetual calendar module was again an IWC development and identical to that of the IWC Ingenieur with perpetual calendar. The version shown looks very elegant, even with a stainless-steel case. It dates from the production year 1998–99 (caliber C. 79261, movement no. 2,584,973, case no. 2,726,246).

On June 17, 1884, Johann Rauschenbach and Salzburg watchmaker Joseph Pallweber signed a contract granting the Schaffhausen company the rights to the so-called Pallweber system. This refers to pocket watches with digital hour and minute indications by means of small dial cutouts. The seconds were still displayed by a hand. IWC sold around 20,000 pocket watches of this type in Savonnette or Lépine designs. From 1888, calibers 52 (Lépine) and 53 (Savonnette) ranked at the top of the sales hierarchy. Together, some 450,000 pieces with a lever escapement and crown winding broke all internal records until they were discontinued in the 1930s. The jury at the Milan World's Fair in 1906 awarded a grand prize for particularly good movement performance. IWC's highly regarded movement classics of the twentieth century also include the 1.49-inch (37.8 mm) × 0.17-inch (4.43 mm) calibers 972 and 982, also in Lépine and Savonnette designs, respectively. The latter also helped the large Portugieser wristwatch achieve chronometric precision. Lovers of historic IWC manually wound wristwatches will enjoy, for example, the ten-line 401 caliber, the equally sized 61 and 86 with center seconds, the tonneau-shaped 87, the twelve-line 64, which inspired the early officers' watches, or the 89, which ticked in the legendary Mark 11 pilot's watch.

Albert Pellaton and the Automatic

In 1944, watchmaker Albert Pellaton from Le Locle joined IWC. In his capacity as technical director, he convinced his Schaffhausen employers of the advantages of an automatic winding mechanism. On June 14, 1946, his first development, a movement with a pendulum oscillating weight, obtained patent protection from the Swiss Federal Office of Intellectual Property, but it lacked the necessary degree of perfection for mass production. Regardless, this design served as the basis for a legendary automatic caliber, for which IWC submitted a patent application on June 7, 1950. The most spectacular difference from his 1946 creation was the use of pawls instead of direct gearing between the rotor and barrel. But the less obvious elements of the self-winding mechanism were also subjected to a rigorous optimization process. The caliber 81 and especially the subsequent 85 series were characterized by the following features:

1. Efficient transmission of rotor movements by means of a cam plate and a pawl system
2. Reduction of friction by means of small bearings at the ends of the pallet fork
3. The design eliminates the need for an intermediate coupling, thus reducing the overall height. It also reduces manufacturing costs and eliminates resulting malfunctions.
4. The rotor rotates freely around a slim, highly polished bearing shaft attached to a spring-loaded steel block. This effectively absorbs vibrations and shocks.
5. The entire caliber impresses with its great ease of maintenance: the rotor can be removed after loosening a single screw and pushing back a retaining latch.

The result was a successful caliber family that was continually developed and continuously manufactured until 1976, with the representatives 852, 8521, 853, 8531, 854, and 8541. In each case the 1 as the final digit indicates the presence of a date window.

Classic wristwatches such as the Aquatimer, Yacht Club, and Ingenieur owe their international reputations to these movements.

After four years of development, the 8541 was replaced by the 1.50-inch (38.2 mm) automatic caliber 5000 with—a purely theoretical—204 hours of power reserve; an adequate successor from the same factory.

Manufacture with Complication

IWC fans had to wait until 2007 for an exclusive chronograph caliber for wristwatches, but the performance of the new 89360 was particularly impressive. The movement, now known as the 89361 or 89365, turned out to be a sophisticated mix of tradition and modernity.
The most complicated wristwatch in IWC's multifaceted collection is currently made only to personal order. This is the Portuguese Sidéral Scafusia, unveiled in 2012, whose manually wound caliber has ninety-six hours of power reserve and indications on both sides of the case. At the front, a large tourbillon dominates alongside the central hands for hours and minutes of civil time. The escapement installed in it acts with constant power (i.e., independent of the current torque of the energy storage). A second, small pair of hands at twelve o'clock indicates the mean sidereal time. Between four and five o'clock there is also a power reserve indicator. On the back you can't miss the artfully designed starry sky; it rotates in time with the sidereal time. Because the visible sky depends on the place of residence, a scientist calculates each image individually. Thanks to two polarizing filters, the sky is gray during the day and blue at night. A yellow zone marks the viewing horizon; a red line shows the ecliptic. Sunrise and sunset are also calculated for the owner's chosen location and are read with the help of red hands on the external twenty-four-hour ring. Finally, the backside again shows the mean solar time as well as daylight saving time. For correct setting there is a perpetual calendar, whose display is limited to the number of days since the beginning of the year and the years of the leap year cycle. Because precious time, even in the twenty-first century, does not respect what was created without it, the development of the complex mechanics, assembled from more than 520 components, took about ten years.

2.2. The Movements

The International Watch Co. Schaffhausen not only is one of the most interesting Swiss watch manufacturers from a technical point of view but can also call itself the only noble watchmaker in the German-speaking world. Since its foundation in 1868, this company has been responsible for numerous technical innovations but has also experienced many ups and downs. On the one hand, this was due to the strategy and market orientation of the respective owners; on the other hand, it was due to the constantly fluctuating world economic situation and changes in the market. In particular, the emergence of quartz technology posed a very serious threat to IWC—also in conjunction with the scarce financial resources available at the time—which could be averted only with a great deal of diligence and luck. Measured against other major watch brands and IWC's own not-always-obvious high standard, one can come to the impression that the Schaffhausen watchmakers have always been somewhat underestimated. The presentation of around fifty different timepieces from Schaffhausen is intended to help provide a brief overview and highlight the genuine value and high standard of IWC's products.

The Manually Wound Movements

Cal. 52 T 19''' H6 S.C. (illus. 83)

This imposing pocket watch movement was created by converting the caliber 52-19‴ H6 from small to central seconds. With a

diameter of 19‴ (1.69 inches, or 42.86 mm), height of 0.236 inches (6 mm), sixteen jewels, and a micrometer screw with counter spring for fine adjustment and stop-seconds, this precise movement fully met the requirements of the German air force (Luftwaffe) in World War II. According to IWC's Jürgen King, a series of 1,200 watches were made in 1940 (movement numbers 1,013,801 to 1,015,000), of which one thousand movements were installed in the German air force's large pilot's watches.

Cal. 61-10''' (Cal. 61–10 lig.) (illus. 84)

With a movement diameter of 0.92 inches (23.35 mm) and a

height of 0.143 inches (3.65 mm), this manually wound movement with center seconds was suitable for all men's wristwatches. It was introduced in 1938 and remained in production for five years. The balance runs at 18,000 vibrations per hour, and there are sixteen rubies in the movement. The dial side of the plate is finely circular grained, and Geneva stripe finish is applied to bridges and cocks. This caliber has no shock protection.

Cal. 83-12''' (Cal. 83–12 lig.) (illus. 85)

This caliber was launched in the version shown here in 1930 and remained in production until 1947. Movement diameter is

1.04 inches (26.5 mm) and height is 0.161 inches (4.10 mm). The small seconds is positioned at six o'clock. The balance wheel runs at 18,000 vibrations per hour, and there are fifteen to sixteen rubies installed in the movement. Beginning in 1943 a second version of this movement was produced, and in the process it received a shock protection system (Incabloc) for the first time. The caliber installed in the Special Watch for Airmen had an antimagnetic oscillating system (antimagnetic balance wheel and Breguet hairspring, part 8374/8374 J in the parts list).

Cal. 87-10''' (Cal. 87–10 lig., Tonneau) (illus. 86)

This type of movement was intended for watches with rectangular cases. The basic dimensions are 0.78 × 0.98 inches (20 × 25 mm), while the height comes to 0.143 inches (3.65 mm). The balance wheel runs at 18,000 A/h, and fifteen to seventeen jewels are used in the movement. The production period was from 1931 to 1947, with 18,600 pieces made. Worth mentioning are the

Illus. 86

three visible gold *chatons* and the fact that the parts list states that the Geneva stripes are made lengthwise, but on the movement shown here they are made crosswise.

C. 88-12''' (Cal. 88–12 lig.) (illus. 87)

This manually wound movement with small seconds was introduced in 1946, and the first movement number was 1,110,401. Constructively, this caliber with small seconds was the predecessor of the C. 89-12''' manually wound movement with center seconds, which was later built in large numbers. The diameter of this caliber is 27.10 mm (1.07 inches or 12'''), the height is 0.151 inches (3.85 mm), and the number of jewels varies between

Illus. 87

fifteen and seventeen, depending on the version. Power reserve is thirty-six hours. The movement, which operates at 18,000 vibrations per hour, has a blued Breguet hairspring, four gold *chatons* are installed, and the lift angle is 41°. Exactly 38,400 pieces of this movement were produced until 1964. Why only 37,000 pieces are listed in IWC's movement directory could not be determined. The last movement bore the number 1,764,400.

The movement shown with the number 1,373,098 from the production year 1955 contains seventeen jewels and an Incabloc shock protection system. Except for the absence of fine adjustment of the regulator arm via micrometer screw and gooseneck spring and the somewhat low power reserve, this high-precision movement was only slightly inferior to the comparable manually wound caliber 12'''-120 from Patek Philippe.

Cal. 89-12''' (Cal. 89–12 lig.) (illus. 88)

The caliber 89 was probably built in greater numbers than any other IWC manually wound movement. A total of 223,800 were produced. This caliber became well known through its use in the British Mk 11 pilot's watch, but this tried-and-tested design could be found in almost all watches with center seconds.

Illus. 88

According to IWC's serial number list, the caliber C. 89 was produced from 1946 to 1974. The case diameter is 1.04 inches (26.5 mm), and the movement height is 0.167 inches (4.25 mm). There are seventeen rubies installed, and the balance runs at 18,000 vibrations per hour. Pilot's watches used movements in which the oscillating system (balance wheel with Breguet hairspring, part 8974 in the parts list) was antimagnetic. In illustration 88 we have a Mark 11 pilot's watch, whose caliber C. 89 can be seen nicely. The balance cock and bridges have the Geneva stripe finish, and the plate on the dial side has a fine *perlage*.

Cal. 89-12''' Special Edition (illus. 89)

The caliber C. 89 in the Yacht Club with manually wound movement (illus. 109, p. 67) turned out to be a sensation. This

Illustration 89

was not only because it was mounted in the case with seven rubber suspension elements for special shock protection, but also because of its other technical refinements. The escapement wheel shaft is mounted with particularly low friction in two cover plates instead of in gold *chatons* as in the otherwise standard design. A special latch ensures that these cover plates are secured safely but can be removed quickly. The screw balance wheel, which uses a Breguet hairspring anyway, has a *masselotte* weight at the end of each of the two bridges and allows very precise, fine adjustment of the escapement. In the beginning not even the manually wound caliber 27 SC from Patek Philippe had this high-quality technology!

The Automatic Movements
Cal. 85-10½''' / 13 lig. autom.
Cal. 852-10½''' / 13 lig. autom.
Cal. 8521-10½''' / 13 lig. autom.—calendrier

IWC entered the age of the self-winding movement in 1950. In addition to the solid basic movement, the Pellaton winding mechanism deserves special recognition. This ingenious technical solution created a bidirectional winding mechanism that is hard to beat in terms of operational reliability and efficiency.

On the caliber 85, which was produced from 1950 to 1951, the case diameter is 1.14 inches (29.00 mm) and the movement height is 0.215 inches (5.45 mm). There are twenty-one rubies installed, and the balance runs at 18,000 vibrations per hour. It is IWC's first automatic movement with center seconds.

As early as 1952 an improved version went into production and was given the caliber designation 852. The beat rate had been increased to 19,800 A/h, and the overall height to 0.220

Illustration 90
IWC automatic caliber C. 852 with twenty-one jewels, production year 1957

inches (5.60 mm). A Breguet hairspring replaced the previous flat hairspring. The version with date indication at three o'clock, which was released at the same time, was designated Cal. 8521, with the height of the movement increased to 0.251 inches (6.4 mm). Here are the movement numbers of the first series:

Cal. 85: Movement numbers 1,216,001 to 1,217.200
(1,200 movements, 1950) (two batches, each of 600)
Cal. 852: Movement numbers 1,247,201 to 1,248,400
(1,200 movements, from 1951)
Cal. 8521: Movement numbers 1,269,401 to 1,270,000
(600 movements, 1952)

Two new versions, the Cal. 853 without and the Cal. 8531 with date indication, followed in 1958 with a number of improvements (new lever, new balance wheel). They remained

Illustration 91
IWC automatic C. 853

Illus. 92
Caliber C. 8531

in production until 1963. The corresponding movement numbers of the first series are as follows:

Cal. 853: Movement numbers 1,459,601 to 1,462,000 (2,400 movements, from 1958)
Cal. 8531: Movement numbers 1,465,601 to 1,466,800 (1,200 movements, 1958)

C. 854 and C. 8541
C. 854 B and C. 8541 B etc.

The caliber notation of the previously listed automatic movements was changed as of 1964 and replaced by the form reproduced here. Due to the increased requirements and relevant advances in movement technology, a fundamental upgrade of this proven caliber series was carried out. The outer diameter was reduced to 28.00 mm and the movement height to 1.10 inches, or 0.232 inches with date mechanism (4.80 mm or 5.90 mm). Crucial for accuracy were the innovations in the oscillating system. Rate and decay could now be adjusted via two eccentric screws integrated on the balance cock. In addition, the three-spoke balance wheel was provided with three adjustable weights with slot and one-sided milling (comparable to the mass balance weights of the Gyromax balance wheel). With this fine adjustment, a second-accurate rate could be achieved without changing the symmetry of the oscillating system. A total of 43,200 of these movements were produced without and 158,400 with date indication. For identification purposes, here are the numbers of the first movements:

C. 854: Movement numbers 1,710,401 to 1,711,000
(600 movements, from 1963)
C. 8541: Movement numbers 1,711,001 to 1,711,600
(600 movements, from 1963)

The caliber C. 854 and C. 8541 (with date display) also received a number of improvements during their production lives. According to IWC the version introduced in 1969 with the letter "B" added to its designation (C. 8541 B) was fitted with a Breguet hairspring with a Greiner hairspring bearing. The suffix "E" indicates that antimagnetic material was used here specifically for the lever, escapement wheel, and bearings in the Ingenieur. Finally, the suffix "S" was the abbreviation for the stop-second mechanism retrofitted to the caliber C. 8541 B. Thus, the movement stops when the crown is pulled into the "pointer setting." Further improvements were also made to the Pellaton winding system. The two ball bearings of the winding mechanism running on the eccentric were made of artificial rubies. Here are the movement numbers of the first movements made for this caliber type:

C. 854 ES: Movement numbers 2,000,201 to 2,000,800
(600 movements, 1970)
C. 8541 ES: Movement numbers 2,005,601 to 2,006,800
(1,200 movements, 1970)

The caliber designation and the movement number are usually engraved on the plate at IWC. IWC watches equipped with calibers C. 854/8541 are among the most-sought-after

Illustration 93
IWC caliber C. 8541 B

timepieces among collectors today. These include models such as the Ingenieur, the Golf Club, the Polo Club, the Yacht Club, and the Aquatimer. Although the caliber C. 8541 B had some minor flaws and the Pellaton winding system needed a lot of space, this automatic movement was one of the best of its time. It is a pity that the options for further improvement were not pursued at that time.

Illustration 94
The IWC automatic caliber C. 8541 B was one of the best movements of its time. An oscillating system with a Breguet hairspring, a balance wheel with *masselotte* weights, special separate adjustment of rate and decay, and the Pellaton winding system helped elevate this automatic caliber to a level that was found only in movements from Rolex and Patek Philippe. Another interesting detail: to the right of the movement number, there is a hole with a pin. The latter is a dial pin, which is fixed with a transverse screw. The second dial pin is used to hold the dial on the other side of the movement.

C. 3254

This automatic movement was the first IWC caliber with a very low profile and 28,800 vibrations per hour. It was based on the Jaeger-LeCoultre caliber 900 automatic movement from

Illustration 95
IWC caliber C. 3254. This movement is based on the Jaeger-LeCoultre caliber 900.

1977, which had a diameter of 1.04 inches (26.5 mm or 11½‴) and a height of 0.171 inches (4.35 mm). The movement had thirty-four jewels installed and ran at 28,800 vibrations per hour. The power reserve was thirty-eight hours. The production number for this movement came to only three thousand pieces. Interesting note: at Jaeger-LeCoultre, the chronometer version of this caliber was 906.

The raw movements, which were also supplied to Vacheron & Constantin, were upgraded there to chronometer movements for the last version of the Chronomètre Royale line and equipped with a gold weight rotor. They were given the caliber designation 1096.

C. 884, C. 884/2 (887, 8872)

When the Jaeger-LeCoultre caliber 900 was discontinued, it was followed by the caliber 889/2, which is still in production today and was introduced by IWC as the caliber 884/2. With a diameter of 1.01 inches (25.70 mm or 11½‴), an overall height of 0.128 inches (3.25 mm), thirty-four (thirty-six) jewels, and 28,800 vibrations per hour, this caliber was universally applicable. Power reserve is thirty-eight hours.

IWC's caliber 884 and 887 movements are almost identical, but the caliber 887 has two more jewels and all visible parts are gold plated, making it a noble version.

Illustration 96
Here, too, the caliber 889/2 from Jaeger-LeCoultre serves as an *ébauche* for the IWC C. 884 and 884/2 movements.

This proven caliber was also used in the IWC Mk XII Pilot's Watch, and its design was clearly superior to the competing 2892 A2 movement from the raw-movement supplier ETA.

2.3. The Reference Numbers

From the beginning, IWC commonly used watch model numbers, references, or model names for its products. However, the assignment of reference numbers is not always clearly comprehensible. It is clear that watches of the Ingenieur series had reference numbers (e.g., R 666 AD). In the beginning these numbers were not stamped on the inner case back. Only on watches produced in the 1960s were reference numbers stamped on the inside of the case back, but not uniformly and not on all

Illustration 97
Screw-down case back of an IWC Automatic Ref. R 820 AD with case number 2,051,332

watches. The reference numbers consisted of three numbers until the new referencing was introduced on October 1, 1971. The previously used R was the abbreviation for reference, while the following letters A and D stood for automatic winding and date display. The IWC Ingenieur and IWC Yacht Club serve as examples. The designation Ingenieur R 666 AD could be understood to mean both the case versions in stainless steel and in 750 gold. It was not until around 1960, when the Ingenieur was also offered with a gold bracelet, that this model was given the reference number 9386, while the version with a 14-karat case middle and steel case back was given the reference R 766 AD. The IWC Yacht Club R 811 AD could also mean a case version in stainless steel or 750 gold. Only the version with a 14-karat case middle and steel case back was given the reference R 911 AD.

In the 1960s, new specific reference or model numbers were often introduced for different countries as well. The Ingenieur R 666 AD was advertised in Austria as the ING. 42, and the Yacht Club was sold as the Ref. 2200. In Britain, one had to order a Ref. 1088 or 1065 to get an Ingenieur R 666 A in steel or gold. Even the short-lived use of five-digit reference numbers (e.g., the Ingenieur R 666 A was sold under Ref. 56662) did not last long.

With the change in reference numbers that took place on October 1, 1971, there was a switch to four-digit numbers. However, the changeover of reference numbers affected only those models that were still in production. The IWC Ingenieur R 866 AD therefore became the 1808 Ingenieur. The IWC Yacht Club R 811 AD and R 911 AD became the 1811 Yacht Club. The IWC Yacht Club R 811 A and R 911 A without date were uniformly changed to 1911 Yacht Club. Watches with gold bracelets were given the initial numerals 99 in conjunction with other numerals. For example, on page 68 we see a gold Yacht Club with a gold bracelet, which is listed differently in the pedigree book with the reference 9205 Yacht Club.

2.4. The Case Numbers

There was also not always a clear system for handling case numbers. As a rule, every IWC watchcase has a number that is recorded in the production lists. If you ask IWC for an extract from your watch register and submit the watch in question, you will generally receive all the relevant data on the watch, including the case number. As you can see on the Ingenieur Ref. R 666 A (p. 71), the case number of this model is between the lugs at six o'clock. On the subsequent Reference R 866 A (and AD), the case numbers were then on the inside back. As an example, let's take a look at the two Ingenieur models on page 72: the blue dial version has the reference number "866 A" engraved on the case back, and the case number "1.183.480" faintly engraved above it. The version with silvered dial on the right shows the identical reference number, "866 A," on the inner case back. The case number 2,012,103 is engraved below the reference number, but here much more clearly and deeper. In addition,

Illustration 98
The inside of the screw-down case back of a gold IWC automatic Ref. 3205 with case no. 2,232,238. Presumably the engraver made a spelling mistake during the first attempt at 223228i. He then engraved the case number over it again and placed it between two zeros as a special identification marking.

the case number is engraved once again on the flat part of the outer case back. But what do you get if you also take the IWC movement number directory and the pedigree book excerpts? The watch shown on page 72 (*left*) has the movement number 1,918,137 from the production year 1969. According to the genealogical extract, it was delivered on January 11, 1971, to Neslein & Co., Christiana, Norway. The Ingenieur shown on the right has the movement number 2,037,883 from the 1971 production year. According to the logbook extract, this watch was delivered to the IWC branch in Frankfurt/Main on November 22, 1974. These two examples show that IWC did not always produce watches according to orders, but often according to market estimates. Incidentally, there are also IWC watches that do not have a case number listed.

2.5. The Watches

First, a few words on my own behalf. Writing about IWC watches is often much more difficult than writing about the timepieces of other manufacturers. On the one hand, this is due to the wide variety of models produced as early as the 1950s and 1960s, and on the other hand to the sometimes incomprehensible system and marking of these watches. Even the publications initiated by IWC itself do not always turn out to be entirely reliable. Thus, in this chapter a reasonably comprehensible report was possible only because, on the one hand, all the watches listed here were available to the author for inspection and, on the other hand, he was also able to coordinate this documentation with reliable, knowledge persons from IWC. Without the support of Jürgen King, former IWC technical director, and Ronald Jäger, head of all production until 2019, this chapter could not have been compiled so accurately.

More than forty watches were selected for the IWC chapter. Among them are not only some of today's highly sought-after military watches, but also other imposing models from the Ingenieur, Yacht Club, and Aquatimer series. Alongside these are timepieces that are not quite as spectacular but that nevertheless belong to the model spectrum and round off the picture.

The Manually Wound Watches
Pilot's and Military Watches
The Large Pilot's Watch for the German Air Force

Our series of photographs begins with the German air force's large pilot's watch. With such a large and impressive dial (diameter a full 1.97 inches [50 mm]!), it was indeed impossible for the pilot to miss the exact time. The three-piece stainless-steel case consists of a bezel, case middle, and press-fit back. Additional protection from dust and moisture is provided by the snugly fitted inner case back. This is also made of stainless steel. Instead of the matte surfaces that the case has, pure luxury reigns here—invisible to the eye. The outside of the inner case back is finely mirror-polished, and the inside is beautifully

Illustration 99
Large Aviator's Watch by IWC, once in service with the German air force. Airmen wore this powerful timepiece on the forearm over the aviator's uniform. The manually wound caliber 52 T 19''' H6 S.C. is regulated in five positions and features a stop-seconds mechanism. It evolved from IWC's own movement, the long-proven Cal. 52 pocket watch movement, which, however, had a small seconds at six o'clock. The aviator's watch shown was sold to Siegfried Heindorf, supplier to the State Aviation Ministry, in Berlin on October 4, 1940 (movement no. 1,013,963, case no. 1,033,363). The Large Aviator's Watch from IWC is now one of the most sought-after pilot's watches.

beaded. Perhaps it was part of IWC's philosophy to ensure not only high precision, but also special aesthetics in the invisible inner workings of a watch. After all, a deep-drawn inner case back made of untreated stainless-steel sheet could have performed the same function and would have been much cheaper!

Now back to our pilot's watch. The IWC caliber 52 T 19''' H6 S.C. has been fitted in this precision instrument. It is regulated in five positions, has a stop-second mechanism, and is based on IWC's own movement—the caliber 52 pocket watch movement, which had been tried and tested over many years. The movement, which originally had small seconds, was converted to central seconds. The watch and movement are illustrated on this page. A fourth wheel bridge was added on top of the mainspring barrel bridge. You can also see the fourth wheel. A screw balance wheel with Breguet hairspring combined with fine adjustment via micrometer screw and counter spring provides the highest precision. According to IWC, exactly one thousand examples of this pilot's watch were delivered to the German air force. If one imagines, for example, that these watches would have gone to the pilots of a dive-bomber squadron equipped with Junkers Ju 88 aircraft, this unit would have received exactly seventy-two watches. If one further imagines that especially during the Battle of Britain, which began in the summer of 1940, many dive-bombers flew attacks on shipping targets in the English Channel and that the German forces suffered considerable losses, one realizes that certainly not all of these thousand watches can still be present today. The large pilot's watch shown in this book was delivered to the Siegfried Heindorf company in Berlin on October 4, 1940, during the Battle of Britain. This company was the address from which the Reich Aviation Ministry ordered IWC watches.

The Special Watch for Aviators

IWC attaches great importance to the correct designation for the following watch. "Special Watch for Aviators" was the official name of IWC's first pilot's watch. Collectors often still use the term Mark IX, or Mk 9. The caliber 83, which is engaged in this watch, is precisely regulated in five positions and temperatures (from −4° to +40° Celsius) to provide the pilot or navigator with an accurate time. In addition, the calibers inside the special watch for aviators are equipped with an antimagnetic oscillating system (balance wheel with Breguet hairspring antimagnetic). This was just the beginning, because at that time the onboard instruments or other internals did not cause major disturbing magnetic fields. The three-piece stainless-steel case of this pilot's watch consists of a bezel, case middle, and press-fit back. The inner case back, also made of stainless steel, further protects the movement against dust and moisture. The bezel has a rotating crown that can be moved in both directions and is equipped with an orientation triangle with luminous material. Thus, even in the dark time, differences or flight segments can be determined accurately. On the subject of the inner case back, here are a few more notes, which we have also included in a similar way in the Jaeger-LeCoultre chapter. Stainless-steel or

Illustration 100
Special Watch for Aviators is IWC's official designation for this first pilot's watch from 1936. Among collectors, it is also known as the Mark IX or Mk IX. Large numerals covered with luminous material and wide hands ensure reliable readability even in the dark. A stainless-steel inner cover provides the movement with additional protection moisture and dust. The case is not water resistant. The special feature of this coveted piece, in addition to small seconds, is the rotating bezel with marker arrow (movement no. 962,693, case no. 1,057,701). According to the IWC pedigree book, this watch was sold to Weinstabel on March 6, 1942.

Illustration 101
Military watch with steel case and press-fit case back known to watch collectors as the Mark × or Mk X. Here, too, the caliber 83-12''' manual wind movement is used. The case is not water resistant (movement no. 1,080,708, case no. 1,133,858). The case back engraving W.W.W. (watch, wrist, waterproof) does not quite apply to this watch because the case design guarantees only splash protection. According to the IWC pedigree book, this watch was sold on February 21, 1945, but the customer at that time could not be identified. According to Jürgen King, the following check is possible: the case number minus the military number must always result in the number 1,118,980 (1,133,858 – 14,878 = 1,118,980).

brass case backs always mean additional protection against dust and moisture. An inner bottom made of soft iron, also called a soft iron dome, serves as a shield against magnetic fields. Since soft iron tends to rust easily, these inner bottoms were usually provided with a protective surface; thus these soft iron domes can be gold- or nickel-plated. If a watch enthusiast without a clear knowledge of materials science wants to know exactly, he needs only to take a small magnet to help him. If the inner base sticks to the magnet, it is soft iron. If there is no reaction, it is normal stainless steel or another nonmagnetizable metal.

The Military Watch

The third military watch has little to do with aviation. Here, too, IWC pays attention to the correct designation. Not Mark X (Mk X) but "Military Watch" is the desired designation. This timepiece was used by the British army. The watch here has a two-piece case, with the case middle, including the lugs, made in one piece. The stainless-steel case is closed with a press-fit case back and is therefore not water resistant. The "Broad Arrow," the engraving "W.W.W.," and the number "M 14878" can be found on the outside of the case back. The three Ws stand for "watch, wrist, waterproof," although in this case it is only splash-proof. According to Jürgen King, about six thousand examples of this model were made (case numbers 1,131,001 to 1,137,000). According to the logbook extract, the military watch shown here was sold on February 21, 1945; however, the customer at that time could not be identified. But some light can be shed on the historical background of this military watch and its delivery date. The sale took place exactly one week after the Allied bombing raids on Dresden, which had been declared an open city. During this air raid, which was carried out over two days, well more than 100,000 civilians, refugees, children, women, and elderly persons needlessly lost their lives.

The Pilot's Watch Mark 11

The Pilot's Watch Mark 11, also called the Mk 11—an uncompromising service watch—was introduced by the British air force (RAF) in November 1949. According to Jürgen King, a first batch of 625 watches were delivered as early as 1948, recognizable by the /48 engraving on the back. It was also the first pilot's watch to have protection against interfering magnetic fields. With the age of jet aircraft now underway, the proportion of electrical equipment in these machines also increased. Onboard power supply systems, auxiliary electromotive drives, radio equipment, and radar sets, to name just a few examples, can generate electromagnetic fields that can permanently interfere with the movement of a watch. For this reason, the Mk 11 pilot watch dial, movement ring, and inner case back were, for the first time, given a fully enclosed cage made of protective soft iron. The activated caliber C. 89 therefore requires no special precautions; the soft iron cage takes over the required protective function. Jaeger-LeCoultre also supplied Mk 11 watches to the

Illustration 102
Pilot's Watch Mark 11 or Mk 11, an uncompromising service watch introduced by the Royal Air Force (RAF) in November 1949. Although Jaeger-LeCoultre pilot's watches were also procured under the same Mark 11 designation (see p. 105, Jaeger-LeCoultre), only two thousand pieces were produced, all of which were delivered in 1949. IWC, on the other hand, delivered some 7,400 Mark 11s to the RAF in several series and improved versions by 1953. The caliber C. 89-12''' manually wound movement, adjusted in five positions, has shock protection and is protected against magnetic fields. This protection is achieved by the dial, the movement ring, and the inner back (*calotte*) made of soft iron. The required water resistance is ensured by the seal of the two-piece case with sturdy screw-down back (movement no. 1,251,890, case no. 6B/343, 633/52).

same RAF pilot's watch specifications. Details of this can be found on page 105. For all the joy of Mark 11 type pilot's watches that have served with various Commonwealth air forces, there are still models that were supplied as civilian versions by IWC under reference no. 2612. In addition there are also unidentifiable replicas that were assembled from spare parts.

The Large IWC Manually Wound Watches with 1.5-Inch (38 mm) Case from Portugal

In addition to the world of the much-sought-after pilot's watches, IWC also has "sleeping beauty" watches that are waiting for a wake-up call. This refers to timepieces with oversized case diameters that were popular in Portugal in the past. Two examples of this clientele are shown on this page. We don't know the delivery date to Portugal, but from the IWC movement number directory by Jürgen King, we can see the year of manufacture

Illustration 103
These two beauties are actually from Portugal, although they are not the Portugieser type of watch and do indeed appear large with their 1.5-inch (38 mm case) diameter. Also not to be overlooked are the partially protected winding crowns, the one on the left watch probably being a replacement crown. The dial on the left watch is in two different shades of copper and features Arabic numerals. The one on the right is silvered and features Roman numerals. Stainless-steel watches with this case size were rather rare at IWC and were exported only to select countries. The watch on the left uses the caliber C. 83-12''', the one on the right the caliber C. 61-10''' with center seconds (*left*: movement no. 1,031,891, case no. 1,072,593; *right*: movement no. 1,004,051, case no. 1,040,089).

of the manually wound movements. The caliber C. 61-10‴ with movement no. 1,004,051 was produced in 1939, the caliber C. 83-12‴ with movement no. 1,031,891 in 1941. The older watch was discovered on the internet in July 2002 and was purchased for $783.40, including shipping costs. The second IWC was offered a year later by a Munich art dealer at the Munich Watch Fair and changed hands for $1,368. IWC watches from early times with case diameters of 1.5 inches (38 mm) are very rare indeed, but very impressive on the left wrist!

The IWC Reference 521 (Shark Fin) Manually Wound Watches

Another IWC model that actually should not be missing in any collection has the rare X-shaped lugs and is also called the

Illustration 104
A "must have" for any IWC collector is this type of IWC watch with the lugs arranged in the shape of an X, also called "shark fin" in watch jargon. This model was produced in numerous variants. In addition to the dash-shaped indices shown, Arabic and Roman numerals were also used in the dial design. The watch on the left was overhauled by IWC and received a new dial in the process. The model on the right was left in its original form. Only on closer inspection can signs of corrosion be seen on the silvered dial. Both watches use the well-known caliber C. 89-12''' movement. Not all IWC watches have the respective reference designation stamped on the case back. But the overhauled watch, which has the reference number 521, does (*left*: movement no. 1,627,763, case no. 1,908,943, Ref. 521; *right*: movement no. 1,722,653, case no. 1,790,603). The pedigree book extract states that the watch on the left was sold to Siber & Hegner, Zurich, on August 21, 1969.

"Shark Fin" in watch jargon. Two of these examples can be found on this page, one of which is in unserviced condition, and the other of which has been overhauled by IWC.

Illustration 105
This rectangular men's watch in 750 rose gold could be called an absolute beauty that needs no further comment. Even the seconds display is rectangular (manually wound caliber 87-10''', movement no. 1,086,130, case no. 1,163,005). According to the IWC pedigree book extract, this watch was sold to the E. Bloch company on October 10, 1950.

The IWC Manually Wound Watch with Form Movement

Among the most-beautiful wristwatches made by the watchmakers from the Upper Rhine are the rare models with a form movement. Here we find one such beauty. Inside a rose-gold case is a caliber C. 87 form movement. From the pedigree book extract we can see that this watch was sold to the E. Bloch company on October 10, 1950. The IWC movement number directory tells us additionally that the movement, with the number 1,086,130, was manufactured in 1944.

Illustration 106
IWC Manually Wound Watch with Small Seconds and caliber C. 88, Ref. 6095
The case is made of stainless steel and has a screw-down case back. The movement has the number 1,373,098, the 1.32-inch (33.5 mm) case the number 1,391,998. The manually wound movement was produced in 1955.

The IWC Manually Wound Watch with Screw-Down Case Back and Small Seconds

The following screwed-back, manually wound watch is a minor sensation. The steel case is waterproof and extremely robust. This model with small seconds was the predecessor of the English pilot's watch Mk 11. There is no mention of this watch in the current literature. One more reason to keep an eye out for it.

The IWC Ref. 810 Manually Wound Watch

This beautifully designed wristwatch with anthracite dial and stainless-steel case has a special significance. Beneath the screw-down case back with six deep-drawn grooves is the proven caliber C. 89 movement with the number 1,932,243, while the case bears the number 1,978,003. This watch was launched in IWC's anniversary year of 1968 and gained special significance as an inexpensive option for the anniversary watches that were common in Switzerland. Here is the original text of the IWC advertisement: "The main reason we created this new watch is the increasing popularity of the long-lasting IWC Anniversary Watch. The new starting price range makes it possible to now enter into conversation with companies that

Illustration 107
Ref. 810 with caliber C.89 movement in stainless steel, an inexpensive anniversary watch with screw-down case back (movement no. 1,932,243, case no. 1,978,003)

previously could not give their employees celebrating anniversaries an IWC due to budget constraints." IWC's 1970 price list for Germany lists the stainless-steel model with

Illustration 108
IWC Yacht Club manually wound watch Ref. R 811 (movement no. 1,955,875, case no. 2,074,175)

Illustration 109
Inside the Yacht Club reference R 811 manually wound watch is a refined version of the caliber C. 89 movement. Instead of a third gold *chaton*, the escapement wheel shaft is housed in two cover plates, the upper of which has a removable catch. This design improves oil retention and reduces friction. The two-spoke screw balance wheel with Breguet hairspring has a mass balance weight at the end of each spoke for precise fine adjustment of the rate. In addition, this caliber is mounted in the case on seven rubber elements that IWC called "Suspense." This provides even better protection against shocks acting on the watch from the outside. According to IWC's Ronald Jäger, this version of the C. 89 caliber is a sensation.

crocodile leather strap shown here at exactly 265 DM. This amount was equivalent to one-fifth of a young engineer's starting salary at the time and was certainly not too high for rewarding a deserving, hardworking employee. The IWC Ref. 810 was also available as an automatic watch with the designation IWC R 810 A or R 810 AD (automatic and date)! However, the case of this model had only a press-fit case back.

IWC Yacht Club with Reference R 811 Manually Wound Movement

The popular IWC Yacht Club was also offered as a so-called value-for-money watch (*see picture on the left*). Instead of the

Positions of the "Suspense" blocks

Illustration 110

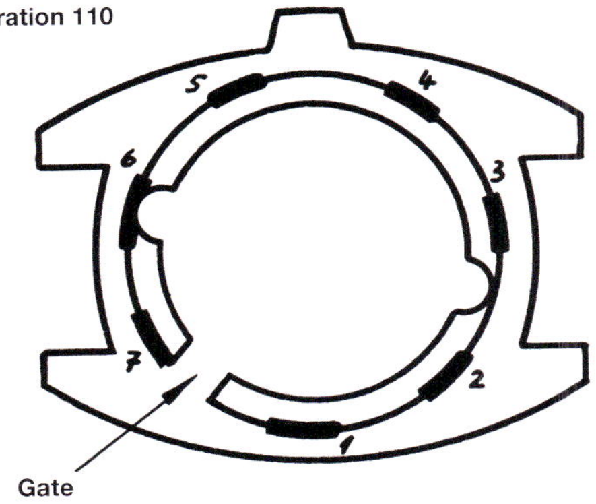

usual C. 854 or C. 8541 (with date) automatic movements, the manually wound C. 89 movement was used here. This model was listed neither in the IWC price list for Switzerland nor in the Deutschmark price list. In the 1970 price list the Yacht Club in stainless steel with automatic movement, with no date display and a crocodile leather strap, was listed at 495 CHF. The price with manual winding might have been around 300 CHF.

2.6. The Self-Winding Watches
A Very Rare Automatic with Fluted Bezel

Illustration 111
The special features of the dial and bezel distinguish this IWC automatic. The dial features two different shades of cream and a circular guilloché pattern across the width of the indices. The bezel is fluted, except directly opposite the indices. The case has a press-fit case back (movement no. 1,229,921, case no. 1,329,344). According to a prominent IWC wristwatch collector, this model would not have appeared in the literature or IWC's catalogs. Movement no. 1,229,921, which is installed in this watch, is listed as a caliber 85 in IWC's movement directory, although 852 is engraved on the plate in the upper right corner.

Here is an automatic with a bezel unusual for IWC watches. The dial of this timepiece is also remarkable. The cream-colored dial has two different shades and a *guilloché* pattern in the circular portion of the indices. Such a beautiful watch actually belongs in the IWC museum, said a distinguished watch enthusiast from Schaffhausen.

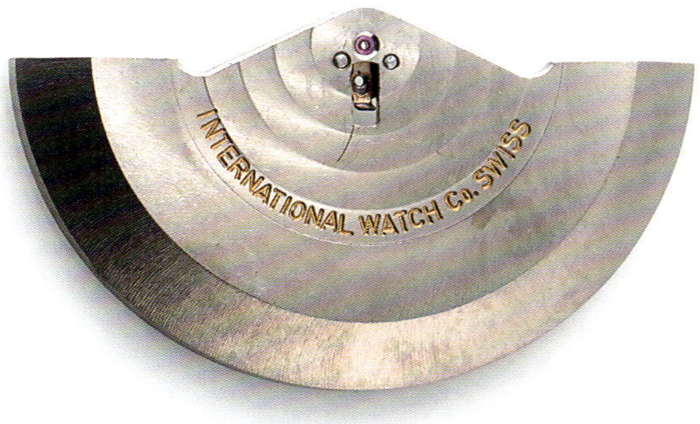

Illustration 112
The rotor of this IWC movement can be removed simply by loosening one screw.

Illustration 113
Caliber C. 852 10½''' movement with twenty-one jewels of the automatic shown on the left. Setting the fine-adjustment mechanism is accomplished by means of a micrometer screw and counter spring (movement no. 1.429.841, case no. 1,382,281).

Illustration 114
The IWC automatics of the 1950s and '60s generally had cases with press-fit case backs that were cheaper to manufacture. Only the Ingenieur, Yacht Club, Golf Club, Polo Club, and Aquatimer models listed had elaborate cases with sturdy screw-down case backs. One of the rare exceptions, also with a screw-down case back, is the automatic shown. Another rarity is the dial. It is made in two colors with elaborate indices (movement no. 1,429,841, case no. 1,382,281).

Another Rarity, Even with a Screw-Down Back

This watch is yet another special feature. On the one hand, this applies to the dial, which is set off in two colors and has beautifully shaped indices. On the other hand, the case also catches the eye because it has a screw-down back, which is an exception to the rule. This waterproof case construction was usually reserved for the line of sporty watches, including the Ingenieur, Yacht Club, Aquatimer, and other models.

Illustration 115
The condition of this watch's case, dial, and hands may be described as mint, even though the IWC automatic is already well more than seventy years old. The 750 white-gold case harmonizes with the silvered dial and the dauphine hands, which are also made of white gold.

The IWC Automatic with 750K White-Gold Case

Wristwatches with cases made of 750K white gold were almost considered exotic in the 1950s and 1960s and were extremely rare. Here is a quote from an IWC employee who has already passed away: "People who buy our watches in white gold must have a pronounced sense of self-confidence. After all, at first glance it could be a steel watch!"

Illustration 116
After the case was opened, the inner workings of this watch also revealed themselves to be in almost-new condition. The rubber ring seal and the special design of the press-fit case back ensure that the case can be considered water resistant (cal. 8531, movement no. 1,590,676, case no. 1,646,776).

The rarity presented here with the caliber C. 8531 movement has no reference number stamped on it, but the movement no. 1,590,676 and case no. 1,646,776 are known to us. The IWC, which was purchased in almost-new condition, probably spent the last sixty years in a safe. Also the movement does not show its great age. According to the IWC movement list, this caliber C. 8531 was produced in 1961. There was no comparable timepiece offered on the internet. This watch's minimum value may be set at around $3,280.

Illustration 117
An IWC automatic with an 18K rose-gold case should be quite rare in the collecting world. After I took possession of this watch, my suspicion that it was a fake was quickly confirmed. Instead of having a legal right of return, the buyer purchased this watch for $2,417.

An IWC Ingenieur Forgery Attempt

Why even some auction houses do not hesitate to call out obvious fakes is shown by this example of a would-be Ingenieur Reference 666. This watch was offered by a Viennese auction house and identified as an IWC Ingenieur in 750K rose gold. Since only the picture of the front shown here was available on the internet,

Illustration 118
After the case was opened, which has only a press-fit instead of a screw-down case back, this model turned out to be a very simple IWC automatic with no reference number. The stylish 750 rose-gold case still has a special charisma, although the tasteless green synthetic alligator strap should be replaced soon (cal. C. 8531, movement no. 1,715,495, case no. 1,625,125). As you can see in the comparison with the picture on the left, the rubber ring seal is missing from this case.

Illustration 119
Group photo of three IWC Ingenieure of the first version in different variants. *Top left*, a Ref. R 766 AD with a 14K yellow-gold case and steel case back (caliber C.8531, movement no. 1,610,885, case no. 1,564,149); *bottom center*, a Ref. R 666 A in stainless steel (caliber C.853, movement no. 1,600,927, case no. 1,591,467); *top right*, a Ref. R 9033 in 18K yellow gold (caliber C.8521, movement no. 1,413,706, case no. 1,395,858; gold case by Wyss & Cie., Biel/Bienne, collective responsibility mark no. 1, hallmark no. 149). Two of these watches still have original dials, but the third has had its dial replaced. Who can tell which one? The production years of these three calibers might be interesting too. The movement of the watch on the right is from 1957, the movement of the middle watch was produced in summer 1961, and the movement of the Ingenieur at the top left was produced in late 1961.

it was not possible to see the back of the case. The experts were struck by the straight lugs as well as the hands. The IWC Ingenieur Reference 666 has stouter, slightly tapered lugs and, with few exceptions, dauphine hands.

After the hammer fell, the buyer and auction house were able to agree on a special price for this watch and avoid a possible legal dispute.

Illustration 120
Reference R 666 A Ingenieur. The steel case has a diameter of 1.44 inches (36.5 mm), and the spring bars are seated in through bores. According to an extract from the company register, this watch was sold to the Bloch company on January 31, 1964 (movement no. 1,981,677, case no. 1,388,522).

Illustration 121
The IWC Ingenieur reference R 666 A is now considered a "cult watch" and is a much-sought-after collector's item.

The IWC Ingenieur Ref. 666

The former Ingenieur series has developed into a cult watch without equal. In the following, we present five different watches

Illustration 122
The Ingenieur reference R 666 A was available with and without a date display at three o'clock. The complete reference designation of the watch on the right is R 666 AD (R for reference, A for automatic, and D for date). The model on the left has indices and hands in rose gold, while those of the one on the right are stainless steel (*left*: caliber C. 852, movement no. 1,381,677, case no. 1,388,522; *right*: caliber C. 8521, movement no. 1,334,253, case no. 1,342,793).

from this popular model series. The Ingenieur Reference R 666 A with steel case and hands (*left*) originates from the USA. Neither the reference number nor the case number is written on the case back. However, the latter can be found between the lugs in the six o'clock position.

Shown next are two more Ingenieure, a Reference R 666 A and an R 666 AD. The first comes from Florida and was purchased by a watch enthusiast in December 2003 for $2,500.

The case of this Ingenieur is made of stainless steel, the hands and the indices of 585 rose gold. The version with date display seen on the right side is from the original owner, who received this watch from his parents in 1958. From the IWC movement number directory it can be seen that the watch's caliber C 8521 movement was manufactured at the end of 1954.

Illustration 123
View of the opened Ingenieur R 866 AD in 750 yellow gold. The inner case back is nicely beaded, and the hammer signet with the number 183 identifies the case supplier as Nouvelle Piquerez SA, Bassecourt. The soft-iron back, which provides protection against external magnetic fields, is stylishly gold plated. The case number 1,835,019 is only faintly engraved and not stamped between the lugs on the outside at six o'clock as it used to be. The large crown is also striking.

Illustration 124
The successor model to the IWC Ingenieur was unveiled in 1964 with the reference R 866 A. Protected in the same way against water and magnetic fields, the case design was slightly modified and the improved automatic caliber C. 8541 movement (C. 854 without date display) was adopted. This was followed from 1969 by the again-modified caliber C. 8541 B (B stands for Breguet hairspring with Greiner hairspring collet). While the Ingenieur at the top still has the original silvered dial, the blue dial shown on the left belongs to those examples of replacement dials that were reissued by IWC in a small series in the late 1990s. While being serviced at IWC, this Ingenieur received one such replacement dial. As with the dials from the first four years of production, the IWC signet is missing below twelve o'clock. From 1969 on, the IWC Ingenieur had its case and reference number stamped on the inside of the screw-down case back (*left*: caliber C.8541, movement no. 1,918,137, case no. 1,813,480; *above*: caliber C. 8541 B, movement no. 2,037,883, case no. 2,012,103). According to the IWC pedigree book extract, the watch on the left was sold to Neslein & Co, Christiana, Norway, on January 14, 1971. This was also confirmed by the owner. The Ingenieur on the right was delivered to the IWC branch in Frankfurt am Main on November 22, 1974.

The IWC Ingenieur Ref. 866

The release of the new automatic caliber C. 8541 in 1964 resulted in the new Ingenieur References R 866 A and R 866 AD. The case was slightly modified for this purpose, especially the screw-down case back. Instead of the six grooves, a fourteen-sided key (IWC tool 8 14 P) was now used to remove the back.

The dials were somewhat more concise, and the date display was larger and easier to read. Also the date window was moved a bit more toward the "3."

The new caliber was protected against magnetic fields by a dial, movement ring, and inner case back made of soft iron. When opening the watch, you can also see that the movement ring and the inner case back were protected against corrosion by a thin layer of gold. Protection against magnetic fields was increased to 1.000 Oersted (80.000 ampère/m). Water resistance at that time was specified as 12 atm (120 m underwater). From 1969 onward this Ingenieur received a stop-seconds, since it was already known in principle from the Pilot's Watch Mk 11. According to the IWC price list from 1970, the reference R 866 AD was listed with the following prices: stainless-steel version with stainless-steel bracelet 618 CHF; with 750 gold case, gold numerals, and crocodile leather strap 1,400 CHF; and the same model as before, but with a 750 gold bracelet, from 3,540 CHF (because the daily price of gold entered here).

Illustration 125
IWC Ingenieur Ref. R 866 AD in 18K yellow gold (from October 1971, Ref. 1808). This particular watch is rarely seen. The version shown with silvered dial, 18K yellow-gold indices, alligator strap, and 18K yellow-gold pin clasp was listed in the 1970 IWC price list at 1,400 CHF or 1,155 DM (caliber C.8541, movement no. 1,831,052, case no. 1,835,019, gold case by Nouvelle Piquerez SA, Bassecourt). At the time this chapter was written (early 2021), there were no examples of this particular watch on offer on the internet or at any known auction house.

Illustration 126
The mighty Ingenieur SL reference 9232 in 750 yellow gold is considered a heavyweight and weighs exactly 10 ounces (285 grams) with the integrated gold bracelet. To protect against magnetic fields, the dial, movement ring, and *calotte* (cap) are also made of soft iron. The movement itself is shock resistant thanks to seven rubber segments (Suspense system) in the case and is designed to be water resistant to a depth of 197 feet (60 m; steel version to 328 feet or 100 m). The crown is pulled out to activate the stop-seconds function (movement no. 2,178,613, case no. 2,259,374).

IWC Ingenieur SL, Overview of Variants Produced (according to Ronald Jäger)

Reference	Case	Band	Movement	Caliber	Dial Variants	Number Sold
1832	steel, steel / gold	steel, steel/gold	automatic	8541 ES	silver-plated or black	543
3003	steel, steel / gold	steel, steel/gold	Quartz	2405	2405 rhodium-plated 335 or black	335
9232	yellow gold	yellow gold	Automatic	8541 ES	gold-plated or black	55 (45)
9503	yellow gold	yellow gold	Quartz	2405	gold-plated or black	45 (55)
3303	steel	steel	Quartz	2250	rhodium-plated or black or white	149
3303	steel / yellow gold	steel / yellow gold	Quartz	2250	gold-plated or white	160
9701	yellow gold	yellow gold	Quartz	2250	gold-plated or black or white	56
9720	yellow gold with 35 diamonds	yellow gold	Quartz	2250	gold-plated or black or white on the bezel	4

The IWC Ingenieur SL, Nickname Jumbo

The next model of the Ingenieur was launched in 1976 as part of the new SL collection. The case and bracelet designs were created by Gerald Genta, whom we already know from Audemars Piguet, Patek Philippe, and Vacheron & Constantin. The new Ingenieur wears mighty and was available only with an integrated metal bracelet. This Ingenieur SL (SL stands for "Steel Line") model was immense in size and was soon nicknamed Jumbo. The steel version got the reference number R 1832 AD, the gold one R 9232 AD. In the price lists and other IWC documentation, the letters were omitted from the reference designations. This model is illustrated on the facing page. The noble piece with gold bracelet brings it to 10 ounces (285 grams) total weight! This Ingenieur also has soft-iron protection against interfering magnetic fields up to 1.000 Oersted. In addition, the automatic caliber C. 8541 E is elastically and shock-resistant-mounted in the case via seven rubber segments (Suspense system). This design was already in use for the Yacht Club model. The reference R 1832 AD was also available as a *mixte* version in steel and 585 yellow gold. From the table on page 74, you can also see that besides the automatic calibers, quartz movements were also used for this model. The quartz version of the golden Jumbo got the reference number 9503. Only one hundred pieces of both versions were available, whereas forty-five with automatic and fifty-five with quartz movements were planned.

According to Jürgen King, the versions with automatic movements were hard to sell at the time. IWC took an unknown number of pieces off the market and retrofitted them with quartz movements. Because of dimensional differences between the two calibers, special adapters and dial underlays had to be made. An additional effort was required to increase the height of the hands. As Walter Baumann, master craftsman at IWC and responsible for the production of these watches at the time, recalls, the watches that had been converted to quartz came back at some point later and were subsequently converted back to mechanical movements again. Ronald Jäger was able to identify nine of these watches and had the corresponding case numbers recorded. They should not be withheld from our readers: case numbers 2,238,075; 2,247,629; 2,247,630; 2,256,769; 2,259,367; 2,259,368; 2,259,369; 2,259,374; and 2,259,380. These watches have Ref. No. 9503 listed on the screw-down case back, although they are equipped with an automatic movement. From the 1979 IWC price list, we can also quote some public prices for the SL collection. For example, the Ingenieur 1832 in stainless steel was priced at 2,800 DM, and the *mixte* (stainless steel and 14K yellow gold) at 5,260 DM. The 750 yellow-gold version was listed at 18,300 DM under reference number 9232.

The IWC Diver's Watches

Other objects of desire are the diver's watches from IWC. This is especially true of the Aquatimer Ref. R 812 AD model, which we present here in two versions. This diver's watch was launched in 1967 and remained in the program until 1982. The task of this watch was the safe management of exact diving time. The crown at the four o'clock position controlled the whole movement, while the crown at two o'clock adjusted the revolving ring dial for decompression times. Accuracy and water resistance were vital criteria if the diver went into the depths. The IWC price list from 1970 tells us how much money the amateur diver had to put on the counter to purchase an Aquatimer reference R 812 AD. The price for the stainless-steel version with stainless-steel bracelet was 620 CHF, and the price for the version with tropical strap (plastic) was 560 CHF.

Illustration 127
IWC diving watches are now also sought-after collector's items. Here are two beautiful examples of the reference R 812 AD (later Ref. 1812). This model was launched in 1967 and produced until 1982. Water resistance was specified to a depth of 656 feet (200 m). Both timepieces have the caliber C. 8541 B movement. The watch on the left has a plastic strap, the one on the right an original IWC steel bracelet. The lower crown is used to operate the winding and hand-setting positions, while the upper one is used to operate the bezel for setting the diving time. Water resistance was a top priority for diving watches and was taken particularly seriously by the designers at IWC. The caliber C. 8541 B movement of the diver's watch on the left was produced in 1972 and, according to the logbook extract, the watch was sold to Zwyro AG on November 17, 1975. The movement of the diver's watch at right is of the identical caliber but was produced one year earlier. The complete reference number R 812 AD is stamped in the screw-down case backs of both watches; only on the right one are the name Super Compressor and the patent numbers Brevet + 31753 and Brevet + 337462 also present (movement no. 2,091,900, case no. 1,937,708; *right*: movement no. 2,036,237, case no. 1,937,603). The outside of the screw-down case back was decorated with a stylized submarine. In addition to this reference, there were other models of the Aquatimer with the reference numbers 1816 and 1822.

The IWC Yacht Club

Among the successful wristwatches created by our specialists from Schaffhausen are the various models of sporty Yacht Club. Although the case of this timepiece is in the "fried egg" design, which is no longer popular elsewhere, this watch suggests sportiness and dynamism to its owner. With more than 45,000 pieces produced, about 10% of which were gold cased, the Yacht Club was IWC's most widely produced model at the time. In addition to the case variants in stainless steel and yellow and white gold, there were also different dial variants. The dials of the steel version were either silver-plated or covered with blue, anthracite-colored enamel, or black enamel varnish. On the white- and yellow-gold versions, the dials were either silver-plated, gold-plated (champagne colored), or enameled. To pay tribute to this successful model, seven different variants are listed in this book (pp. 76, 77, and 78). Among them are two special rarities. One belongs to a prototype series with a special dial in a copper hue, only about thirty-five of which were intended for export to the Far East, according to René Schwarz of IWC. The other represents a luxury version in 750 yellow gold and dial with forty-two diamonds, of which a maximum of twenty-four were produced, according to the recollection of an IWC employee. The former Yacht Club came from antique watch dealer Ralph Silvester, New Haven, Connecticut. Around 2,700 DM had to be paid for the purchase, including import sales tax and transport. To these costs was added the cost of overhaul by IWC at Schaffhausen (485 CHF). We have both the original warranty booklet and the IWC pedigree booklet of this diamond-decorated gem. According to these documents, the watch was sold by the Metropolitan

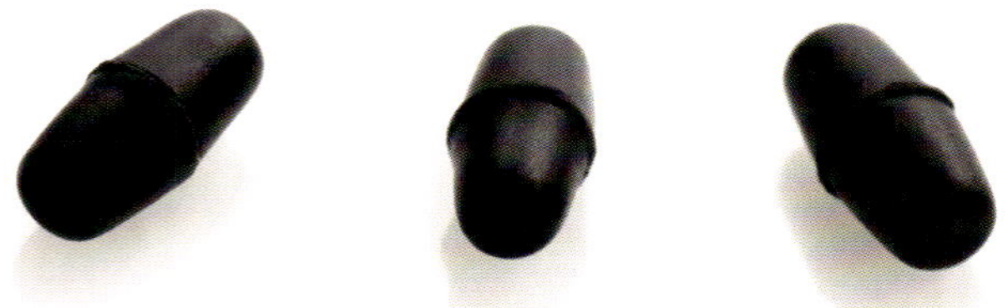

Illustration 129
IWC Suspense System (greatly enlarged)
Seven of these rubber elements, whose exact position is described in the assembly specifications, support the caliber 8541 movement in the case of the Yacht Club. This also provides the movement with enhanced shock protection (see also fig. 110, p. 67).

Illustration 128
Parade of three different versions of the IWC Yacht Club: *on the left*, a stainless-steel version Ref. R 811 AD with blue dial and metal bracelet; *in the middle*, a special version without reference indication with an experimental copper-colored dial; *on the right*, the model Ref. R 811 A with no date display. Only about thirty-five pieces of the center version were made as an experimental series for export. This statement was confirmed by a retired IWC employee who had visited the IWC service workshop by chance and was involved in this project in the late 1960s (*left*: movement no. 1,941,227, case no. 1,928,597; *center*: movement no. 1,833,256, case no. 2,166,151; *right*: movement no. 1,774,258, case no. 1,841,886). We have IWC's pedigree records for two of these sporty timepieces. They reveal that the Yacht Club on the left went to IWC Frankfurt am Main on October 5, 1970, and the one in the middle to Steiner in Lagos on September 30, 1968. Originally the case of this watch bore the number 1,852,779. On the occasion of an overhaul carried out at IWC Schaffhausen, the complete case was replaced and the new case number 2,166,151 was assigned.

Illustration 130
Among IWC's most successful wristwatches are the various models of the sporty Yacht Club. A total of around 45,000 watches were built from this line, around 10% of them in the precious metal gold. The versions in 750 gold and stainless steel bear the reference number 811 (later 9205 and 1811), while those in 585 yellow gold with steel case back bear the number 911 (later 1911). We still have the original warranty booklet for the Yacht Club in 750 yellow gold with blue dial and forty-two diamonds. It shows that this Yacht Club was sold by Metropolitan Store in Alkhobar, Saudi Arabia, on December 5, 1974. The logbook extract issued by IWC further tells us that this watch had gone to Metropolitan in Alkhobar on May 17, 1974. The numbers of the movement and case also matched. The entry in the pedigree book reads "Men's wristwatch Ref. 9205 Yacht Club." After opening this watch, one could find the caliber designation C. 8541 B, as well as movement and case numbers, but not the reference number! The Yacht Club on the right is a reference R 911 AD whose case middle is made of 585 yellow gold, and the screw-down case back of stainless steel. The caliber C. 8541 B movement is in good condition, as you can see. Also the data in the screw-down case back are easy to read. The pedigree book extract shows that this Yacht Club was sold to Andreas Huber in Munich on April 18, 1971. In 2001 it turned up in a Munich pawnshop and changed hands again for $476.

Illustration 131
The C. 8541 automatic movement's shock-absorbing spring.
This shock-absorbing spring, later also called the automatic bridge, forms the ingenious and decisive element of the IWC Pellaton winding system. The axle journal, with two different bearing diameters, accommodates the rotor and allows the suspension to be sprung in three dimensions. This means that shocks acting on the watchcase from outside can be absorbed and cushioned in the X, Y, and vertical Z axes. René Schwarz, who was responsible for all the group's brands for many years, first as service manager in Schaffhausen and later at Richemont in Munich, could not recall a single breakage of the rotor axis in a caliber 8541 movement.

Illustration 133
Axle journal of the current Rolex caliber 3135.
The design of this axle journal is functionally identical to that of the IWC caliber 8541. On the Rolex, the axle journal is mounted in the rotor, and the two synthetic ruby bearings are fixed in the automatic bridge. Thus, the axle journal rotates in the fixed bearings. In the IWC watch, the two synthetic ruby bearings are fixed in the rotor and rotate in the stationary axle journal. Strictly according to the textbook, the larger-diameter bearing is on the inside, the smaller on the outside. Between them is the collar for the rotor lock. Why IWC abandoned this design almost forty-five years ago, but Rolex still practices this technical solution today, is incomprehensible to an engineer. In principle this mounting system corresponds to the axle of an automobile.

Illustration 132
Two different versions of the IWC Yacht Club
This comparison shows two different versions of the IWC Yacht Club with different dials and gold cases. The reference R 811 AD, *on the left*, has a silvered dial and an applied IWC signet (caliber C.8541 B, movement no. 1,976,194, case no. 1,996,659), and the case is made of 18K yellow gold. On the right is a Yacht Club reference R 911 AD, whose case top is made of 14K yellow gold and the screw-down case back of steel. On the gold-plated dial without the IWC signet, one can recognize the earlier year of manufacture (caliber C.8541 B, movement no. 1,919,194, case no. 1,905,506). According to the IWC production list, the movement of the watch on the left was made at the end of 1969, the one in the watch on the right at the beginning of 1969. The two movement numbers are 70,688 examples apart.

Store in Alkhobar, Saudi Arabia, on December 5, 1974. What this watch experienced in the following years we do not know. In fall 2002, this rare piece finally turned up at the antique-watch dealer Lippincott in Narberth, Pennsylvania, and was purchased by a German watch enthusiast for $3,640. Import tax and a proper overhaul at IWC added another $1,314 to the price. The Yacht Club Ref. R 911 AD (p. 77, *right*) was positioned between the stainless-steel and 750 gold versions in terms of price at that time. On this version the bezel and case middle were made of 585 yellow gold, and the screw-down case back was stainless steel. According to IWC's pedigree book, this watch was sold to the Andreas Huber company in Munich on April 18, 1971. Some thirty years later it turned up in a Munich pawn shop and changed hands again for 850 DM. IWC charged 420 CHF for the value-preserving general overhaul. If you take a look at the IWC price list from 1970, this model Ref. R 911 AD with crocodile leather strap is priced at 950 CHF. In addition to the various versions with automatic movements, there was also a price value version of the Yacht Club with the manually wound caliber C. 89. This model is described on page 66.

The following example shows how advertising and reality can sometimes go in different directions. In an advertising brochure dated February 1967, the various advantages of the IWC Yacht Club are highlighted. In addition to the crown with automatic pressure compensation (CAT), the bumper (YC) is also presented. The latter describes the principle of suspension of the movement in the case, which is based on plastic cushions in the form of small cylinders. The original text reads as follows: "The plastic cushions lie, in groups of three, at three points between the case and the movement." In contrast, IWC's assembly instructions for watchmakers on mounting the "Suspense" system in the Yacht Club case read, among other things, as follows: "The Yacht Club case has seven Suspense blocks." A drawing specifies that these seven "Suspense" blocks are to be evenly distributed around the entire circumference of the case. This drawing is shown on page 67 (illus. 110). Thus, advertising and assembly make different statements both as to the number of spring elements and their position. If you open a Yacht Club and the watchmaker has worked correctly, you will see the assembly instructions confirmed. For the interested collector it should be mentioned that few watches of the Yacht Club model were also equipped with a quartz movement.

The IWC Golf Club and Polo Club from the SL Line

The IWC SL line presented so far (SL standing for "Steel Line") included other sporty models in addition to the Ingenieur SL. Among these are the Golf Club SL Ref. R 1930 A and 1830 AD versions and the Polo Club SL Ref. R 1931 A and 1831 AD. They also used the C. 854 B and C. 8541 B caliber series, but in contrast to the Yacht Club and Ingenieur SL, the previously described "Suspense" shock protection system was not used due to lack of space. Instead, scratchproof sapphire crystals were mounted, just like on the Ingenieur R 1832. One special feature of the Golf Club has to be mentioned. It is the only IWC sports watch to have its name engraved radially on the screw-down case back: "IWC GOLF CLUB." The IWC price list from 1976, the year in which these watches were launched, reveals the pricing of these two models at that time. The Golf Club SL Ref. 1830 with stainless-steel bracelet is listed at 1,250 CHF, the Polo Club SL Ref. 1931 with stainless-steel bracelet at 1,190 CHF. In both cases the metal straps were detachable so that a leather strap could also be used. The design of the stainless-steel cases took some getting used to, although they were very elaborate from a production-engineering point of view.

Illustration 134
IWC's sporty SL line (SL stands for "Steel Line") also includes the Golf Club Ref. R 1829 AD (*left*) and Polo Club Ref. R 1931 A (*on right, no date*) models. The latter's design came from Gerald Genta. However, these watches were robustly designed, and with their caliber C. 8541 B and C. 854 B movements, respectively, as well as water-resistant screwed cases, they were definitely predestined for sporty ambitions. In the foreground the Golf Club's automatic movement, and on the left the corresponding screw-down case back. The name Golf Club is engraved on the outside (*left*: movement no. 2,095,508; the watch on the right remained unopened).

IWC Reference R 817 AD, One of the Most Noteworthy Automatics from Schaffhausen

In this author's opinion, the IWC Automatic Reference R 817 AD is one of the most pleasing watches IWC produced at that time. Only upon closer inspection does one appreciate what an elaborate case was built here. The case middle is stamped from a gold band (semifinished) and formed many times. The lugs are rounded at the end and satin-finished on both sides in the longitudinal direction. The 45° break of the lugs, on the other hand, has been diamond-cut. The conical lug on the bottom of the case is diamond-cut. The screw-down case back was double

Illustration 136
IWC Ref. 817 AD Reverse
At the well-known English fashion house Marks & Spencer Ltd., London, it was customary in the 1970s to present deserving executives with a gold wristwatch to mark twenty-five years of service. The fact that on several occasions the choice had fallen on a gold IWC automatic reference R 817 AD came to light only after more than forty years. The Fellows auction house in Birmingham, England, offered this model several times in the past years. On the watch shown, the dedication reads as follows: "PRESENTED TO T. WILLIAMSON in Appreciation of 25 Years' Service 1946–1971 MARKS AND SPENCER LTD." Below the dates, looking rather small, is the case number 1,788,768. Since this anniversary was more than forty-nine years ago, this watch was probably sold by the heirs of the jubilarian (movement no. 1,976,636, case no. 1,788,768; supplier of the gold case was the company Oréacle, Manufacture du boîtes SA, La Chaux-de-Fonds).

Illustration 135
IWC Automatic Ref. R 817 AD in 18K Yellow Gold
This again proves the previously made statement that the reference R 817 AD automatics are among the most-beautiful timepieces from Schaffhausen. The simple but stylish dial, the massive case, elaborately designed to the smallest detail, and the high-precision caliber C.8541 B automatic movement prove that the owner of this watch has thought of everything. He could hardly have made a better choice. In a 1971 IWC catalog for Japan, the Ref. 1817 (817) in stainless steel was listed at 98,000 yen. You can't tell from the neatly turned face of the screw-down case back that an extensive engraving was lasered over here, and the screw-down case back was then professionally refinished.

conical set and has six grooves for waterproof closure. This model was available in 750 yellow gold as well as in stainless steel. Depending on your preference, it was supplied with a leather strap or metal bracelet. In the 1968 IWC price list the stainless-steel version of the reference R 817 AD was priced at 430 CHF, while the 750 yellow-gold version was priced at 1,040 CHF. In the German IWC price list, these watches are listed at 485 and 1,055 DM in the same year. However, this price difference is not so much the result of the currency parity at the time, but rather of the value-added tax (VAT) that had just been introduced in Germany. So why should the IWC reference R 817 AD be considered something special? After all, this model is not listed in the relevant literature. Research at IWC and with former employees of the house led only to the statement that the reference R 817 AD was nothing special, nor a rarity. This watch was presumably envisaged as a "fighting model" that would gain market share as a lower-priced variant in the increasing competitive market. It was first listed in the IWC price lists in 1968 but had disappeared just two years later. Its production life was thus only about two years; exact production figures are not known. IWC estimates that the total number of pieces produced was between one thousand and two thousand. But where was the marketing target for this model at that time? Compared to the Yacht Club and Ingenieur product lines, which were equipped with almost-identical automatic calibers and similarly designed dials, the reference R 817 AD had a far more elaborately designed case. Apart from the different movement bearing of the Yacht Club, where the caliber is also housed elastically by seven rubber segments, there is no convincing reason for the widely differing prices. For example, the 1968 price list has the golden Yacht Club with date (R 811 AD) at 1,495 DM and the golden reference R 817 AD at 1,155 DM. Compared to the simple shape of the Yacht Club case, that of the reference R 817 AD was very elaborately designed and bore a strong resemblance to the case of the Rolex Datejust. With the same calculation basis, the price of the reference R 817 AD should have been at least as high as that of the Yacht Club, or higher. Even IWC's production manager at the time, Ronald Jäger, could not give us an answer to this question. Despite the increasing popularity of antique IWC wristwatches, the reference R 817 AD had not been offered at the renowned auction houses such as Christie's, Sotheby's, and Antiquorum in the last ten years. Only Henry's auction house offered a gold IWC R 817 AD in its March 1996 auction, which went to the highest bidder, a Japanese watch collector, for 3,780 DM, including buyer's premium. At the Dr. Crott auction house, several models of the Ref. R817 AD came up for auction in 2010. They came from a collection liquidation.

IWC Yacht Club II

The follow-up version to the first Yacht Club model, the Yacht Club II, did not enjoy the same success as its predecessor.

The case design, created in collaboration with Gerald Genta, was not to everyone's taste and did not appear to be "all of a piece," as we know it from the Royal Oak from Audemars Piguet, for example. The Yacht Club II was offered in three versions and four sizes. The watch, which came with a leather strap or metal bracelet, was available in stainless steel, *mixte* (stainless steel and 585 yellow gold), and 750 yellow gold. Instead of using their own automatic movement, IWC incorporated the new Jaeger-Le-Coultre caliber 900/901 into the largest version of this watch, Ref. 3312. At just 0.128 inches (3.25 mm) in overall height (C.8541 B at 0.232 inches [5.90 mm]), it met the demands of its time. IWC's internal caliber designation was C. 3254. According to IWC, the Yacht Club IIs with mechanical movements remained the minority, since the bulk of this model series was delivered with quartz movements. The yellow-gold version with yellow-gold bracelet and automatic winding had the reference number 9232 and cost exactly 9,500 CHF in 1976.

The IWC Automatic Ref. R 820 with Stainless-Steel Case

This IWC automatic with a case diameter of 1.42 inches (36 mm) and a screw-down case back was not to be found in the Swiss or German IWC catalogs from the late 1960s to the early 1970s. The screw-down case back was identical to that of the Yacht Club model series and the Ingenieur Ref. R 866 AD. To open the screwed back, the IWC fourteen-sided case key no. 8 14 P with 1.28-inch (32.5 mm) width across flats is required. This rare IWC was purchased via the internet on January 4, 2008, in Hong Kong. Including the import sales tax calculated

Illustration 137
The Yacht Club II reference R 3212 AD was designed by Gerald Genta, who also created the case shapes of the famous Royal Oak from Audemars Piguet and Nautilus from Patek Philippe. In contrast to the Royal Oak, this bulky, mighty watch was found to take a lot of getting used to. It never achieved the success of the Royal Oak, which continues to this day, and remained in production for only a few years. In addition, only a few watches were equipped with a mechanical movement, and most had a quartz movement. The automatic caliber came from Jaeger-LeCoultre (Cal. 900/901) and was given the caliber designation C. 3254 by IWC. With an overall height of just 0.128 inches (3.25 mm), this met the requirements of the time but had nowhere near the robustness of its predecessor, the caliber C. 8541 B. A screw-in crown and quick date adjustment were among the innovations of this caliber. The Yacht Club II was available with an integrated metal bracelet, as well as with a normal leather strap. IWC offered this model in three versions and four different case sizes. At the top right is the stainless-steel version, and below it is the *mixte* (mixed) version in 585 yellow gold and steel. Not shown is the version in 750 yellow gold (*left*: movement no. 2,297,338, case no. 2,255,836; *right*: no details, since this watch remained unopened). According to the IWC pedigree book extract, the watch on the left was sold to IWC in Frankfurt am Main on June 22, 1979.

Illustration 138
IWC Automatic Ref. 820 AD with 36 mm stainless-steel case, an absolute rarity that was probably sold only in the Far East

at that time with 19%, customs duty plus transport, a total of $1,910 had to be paid (movement no. 1,982,052, case no. 2,051,332).

At the beginning of 2021, a copy of this reference was once again offered on eBay Japan for $3,200 USD. This confirms the author's assumption that this model was probably intended only for the markets in the Far East as well as in the USA.

The IWC Automatic Ref. R 1872 AD, Also Called the "TV Watch" (p. 84)

This IWC has the case with the highest gold weight of all automatics offered by IWC at that time, with the caliber C. 8541 B movement. The 18K yellow-gold version shown was purchased on December 26, 2007, via eBay from a dealer in the USA. Including import sales tax, customs duty, and transport, a total of $2,150 had to be paid (movement no. 2,094,650, case no. 2,007,530). The case of this watch, weighing about 1.7 ounces (48 grams), would alone be worth $1,970 at today's gold price (January 2021)!

In a 1971 IWC catalog for Japan, the reference 1872 was listed at 101,000 yen in stainless steel, 330,000 in yellow gold, and 360,000 in white gold.

In early 2021, only two watches of this type were for sale on the internet. A German private individual wanted a fixed price of $4,402 for his watch with a leather strap (movement no. 2,018,406, case no. 2,007,248). The second seller, a dealer from Italy, wanted to name only his asking price (case no. 2,007,573, movement no. unknown). In both cases the gold case with the identification number 109 in the hammerhead comes from the company Oréacle, Manufacture de boîtes SA, La-Chaux-de-Fonds.

The IWC Ingenieur SL with Perpetual Calendar Ref. 9240

This model was unveiled at the 1985 Basel Watch Fair. The perpetual-calendar module was designed by Kurt Klaus at IWC and was also used for the Da Vinci chronograph, with perpetual

Illustration 139
IWC automatic in stainless-steel Ref. R 820 AD with caliber C. 8541 B, movement no. 1,982,052, case no. 2,051,332. *On the upper right*, a 10× magnification of the fine adjustment mechanism with oscillating balance. Despite the motion blur, you can still see a *masselotte* weight on the balance wheel bridge.

Illustration 140
IWC Automatic Ref. R 1872 AD, also called the "TV Watch"
Weighing about 1.7 ounces (48 grams), given the current price of gold, the gold case was already worth $1,900 in 2021!

calendar unveiled at the same time. However, many IWC friends were disturbed by the fact that this new model, which was initially available only with a gold bracelet, featured the automatic caliber 2892 A2 from ETA (see also p. 420). The case of this watch had a diameter of only 1.36 inches (34.5 mm) and could also be delivered with a crocodile leather strap as of 1987. This watch never became a hit and remained in the program only until 1992. In that year's price list, the watch, later designated Ref. 9245 (with the second version of the perpetual-calendar module), was listed at 18,000 CHF (about $19,871). The watch presented here was delivered with a crocodile leather strap and comes from an Italian dealer. It changed hands on November 4, 2007, for $4,415. After a thorough servicing at IWC in Schaffhausen, it received a matching gold bracelet (movement no. 2,427,535, case no. 2,432,137).

At the beginning of 2021, a Munich dealer offered this model (new reference IW 3540 1) with a crocodile leather strap for $6,230. A version with a gold band could not be found anywhere in the world at that time.

The IWC Da Vinci Chronograph with Perpetual Calendar Ref. 3750 in Stainless Steel

The launch of this chronograph with perpetual calendar at the 1985 Basel Watch Fair marked IWC's entry into the *haute horlogèrie*, or high art of watchmaking. Inside this watch ticks the modified chronograph caliber 7750 from ETA, designated by IWC as C. 79261. The built-in perpetual-calendar module is identical to the one in the IWC Ingenieur with perpetual calendar Ref. 9240. With its somewhat baroque appearance, the IWC Da Vinci is not to everyone's taste, especially when compared to wristwatches with perpetual calendars from other manufacturers. This is particularly true of those from Audemars Piguet (p. 25), Patek Philippe (p. 141), and Vacheron & Constantin (p. 287).

Illustration 141
IWC Ingenieur SL with perpetual calendar Ref. 9240 with gold bracelet

Da Vinci models are offered on a larger scale on the internet. Versions with stainless-steel cases are offered for $6,500–$7,600, gold models with 18K gold cases from about $8,700.

By far, the IWC Reference 3750 with chronograph and perpetual calendar is considered the most affordable timepiece in this class. For example, the price of a Blancpain Chronograph Perpetual in 18K white gold starts at $16,400 euros. A Patek Philippe with annual calendar and chronograph cannot be had for less than $43,800. But this IWC Da Vinci stops the time just as precisely and demonstrates the correct day and date all year round. What more could one want (movement no. 2,584,973, case no. 2,726,246)?

Illustration 142
IWC Da Vinci Chronograph with perpetual calendar Ref. 3750 in stainless steel

The IWC Automatic Ref. 3205 in 18K Yellow Gold

Looking at this watch, one has to wonder why a timepiece that was so stylistically and technically accomplished did not get more attention. The caliber 3254 automatic movement is based on the Jaeger-LeCoultre caliber 900 and was at the cutting edge of its time. Supplied by case maker C. R. Spillmann SA, Chêne-Bourg, the screw-down case is extremely flat. The officially required hallmark of this case maker consisted of a hammerhead with the number 136. Comparable cases were also delivered to Vacheron & Constantin, recognizable by the cleverly designed screw-down back with a special key attachment. With a 1.32-

Illustration 143
IWC Automatic Ref. 3205 in 18K yellow gold

inch (33.5 mm) case diameter (movement specification) and an overall height of just 0.287 inches (7.3 mm), including date display and sapphire crystal, this model was fully in line with the trend of the time. For some enthusiasts this model was also used as a ladies' watch.

The gem presented here with movement no. 2,282,701 and case no. 2,232,281 was overhauled at IWC. For this work, despite a small discount, $1,729 still had to be paid—a high price. The reference 3205 with gold case is rarely offered. In spring 2021, only one offer for this model was found on the internet. The German seller wanted a fixed price of $3,833 for his gold watch.

Illustration 144
IWC Automatic Ref. 3205 in 18K yellow gold with screw-down case back removed. The automatic caliber 3254 is based on the Jaeger-LeCoultre 900 movement. The hammerhead with the number 136 reveals that the case was made by C. R. Spiellmann SA, Chêne-Bourg. As previously described in illustration 98 on page 61, the case number was corrected retroactively. The engraver initially applied the number 223228i. When this error was discovered, he made a second engraving above it with the correct number 2232238. To distinguish the two numbers, the engraver placed a zero to the left and right of the correct one.

Illustration 145
IWC Automatic Ref. 3205, a very flat-built wristwatch with a special screw-down case back that requires a special key to open (18 Kt).

The IWC Portofino Automatic Ref. 3205 in Stainless Steel

This IWC from the 1980s shows how elegant a simple steel watch can look. The blue dial is particularly eye-catching, allowing a gentleman to wear this watch with a dark-blue business suit without worry.

IWC's caliber 3254 with date display and small seconds is used in this watch (base caliber 900 from Jaeger-LeCoultre). According to Jürgen King, former IWC technical director, exactly three thousand examples of this automatic movement were made from 1976 onward. This watch remained unopened due to the lack of a suitable key; therefore only the case number 2,219,306 engraved on the outside can be listed here.

The IWC Portofino Automatic Ref. 3205 in Stainless Steel

This IWC from the 1980s shows how elegant a simple steel watch can look. The blue dial is particularly eye-catching, allowing a gentleman to wear this watch with a dark-blue business suit without worry.

IWC's caliber 3254 with date display and small seconds is used in this watch (base caliber 900 from Jaeger-LeCoultre). According to Jürgen King, former IWC technical director, exactly three thousand examples of this automatic movement were made from 1976 onward. This watch remained unopened due to the lack of a suitable key; therefore only the case number 2,219,306 engraved on the outside can be listed here.

Illustration 146
This IWC Automatic Ref. 3205 with stainless-steel case and automatic caliber 3254 movement is shapely and elegant. Because the patented screw-down case back was closed very tightly and no suitable box wrench was available, the case remained unopened.

Illustration 147
At a magnification of more than 10×, the case number 2,219,306 can be read easily.

The IWC Pilot's Watch MARK XII, Reference 3241

A pilot's watch with form, excellent automatic movement, and a robust case could be found only at IWC in the 1990s. Protected against magnetic fields and water resistant to 5 bar, the automatic was launched in 1994 and remained in the range until 1998. The automatic caliber 884/2 was based on the Jaeger-LeCoultre 889/2 movement. With a diameter of 1.01 inches (25.7 mm or 11½'''), a height of 0.128 inches (3.25 mm), thirty-six jewels, 28,800 vibrations per hour, stop-seconds, and thirty-eight-hour power reserve, this movement was "state of the art." This automatic caliber was also used by Audemars Piguet and Vacheron & Constantin. Thanks to the luminous numerals, this watch can also be used at night.

The masculine-looking, three-piece, stainless-steel case brings it to an outer diameter of 1.42 inches (36 mm) and an overall height of 0.41 inches (10.40 mm). The crown is also screwed.

When this watch was launched in 1994, the stainless-steel version was available for 3,850 DM and the 18K yellow-gold version for 6,800 DM. By the time this book went to press, the stainless-steel version was being offered at $4,300–$5,500. The version with gold case and leather strap was offered on the internet with a fixed price of $7,856 from Japan. Another gold

Illustration 148
The IWC Pilot's Watch Mark XII, a very successful model

Illustration 149
The Movement of the IWC Pilot's Watch Mark XII
The main functional components reveal the origin of this IWC C.884/2 caliber from the basic Jaeger-LeCoultre caliber 889/2. However, IWC had omitted the special finishing of the surfaces, the blued screws, and the 21K gold segment on the rotor. At the end of the balance cock, one can still see the Triovis regulator for adjusting the rate. This fine adjustment was later replaced at IWC by four adjustable weight screws on the balance ring (oscillating moment SM = GD^2).

Illustration 150
IWC Romain Automatic Ref. 3208 in 950 platinum, a very distinguished model

model with an original gold bracelet was offered by a Swiss dealer for a fixed price of $13,445.

The timepiece shown comes from a German vintage watch dealer in Düsseldorf and was sold in fall 2019 for this documentation, including box, at a special price of $2,737 euros (caliber 884/2, movement no. 2,795,961, case no. 2,718,164).

Why this outwardly successful Mark XII, which was in demand on the market, remained in IWC's program for only four years probably had its reasons. Both the head of the service department at the time, René Schwarz, and the director responsible for production, Ronald Jäger, agreed that they had not been able to get sufficient control over the production of

Illustration 151
Rear view of the IWC Pilot's Watch Mark XII
The sturdy case back can be opened only with a special tool.

the automatic movement and that numerous customer complaints were no longer acceptable. From a technical point of view, these events cannot be completely understood, because the caliber 889/2 is still in production today and performs satisfactorily in numerous watches from Jaeger-Le-Coultre itself and other well-known brands.

The IWC ROMAIN Automatic Ref. 3208 in 950 Platinum

This automatic radiates a restrained elegance. The dial is somewhat reminiscent of the OMEGA Constellation with pie-pan dial, as was fashionable in the 1960s. The platinum case has a diameter of 1.33 inches (34 mm) and a height of 0.295 inches (7.5 mm). The case back is fixed with six lens head screws. The IWC 884/2 caliber is based on the Jaeger-LeCoultre 889/2 and operates at 28,000 vibrations per hour. The author has encountered this noble watch only once before—namely, the one shown here! In the time frame of early 2021, there were only two examples of this watch on offer. A dealer from Washington, DC, wanted $7,999 for this rare model. A seller from the UK offered this beautiful watch at a fixed price of $6,969.

The watch presented here came from a dealer in California, USA, and cost $5,521.20, including transportation, import VAT, and customs. At that time, this reference was listed in IWC's price list at 17,500 CHF (movement no. not known since watch remained unopened, case no. 2,473,841).

2.7. The IWC-Pellaton Winding System from the Point of View of Technical Mechanics

To be able to understand the design details, the functional sequence, and the technical peculiarities of the Pellaton winding mechanism, it is first necessary to explain some basic rules of technical mechanics and mechanical engineering. Although Albert Pellaton was not a graduate of a technical university, he did everything right, textbook style, in designing his new winding system. This applies first of all to the mounting of the rotor (illus. 152), which in principle is similar to that of a wheel of an automobile (illus. 154). The Pellaton winding system shown as a patent drawing is composed of two dynamically separated functional areas.

The main task of this winding system is to convert the impact energy of all impulses acting on the wristwatch, irrespective of their strength and spatial origin, into winding work in a highly efficient manner.

To be able to fulfill this function reliably, the construction of this ingenious system can be divided into two separate operating areas.

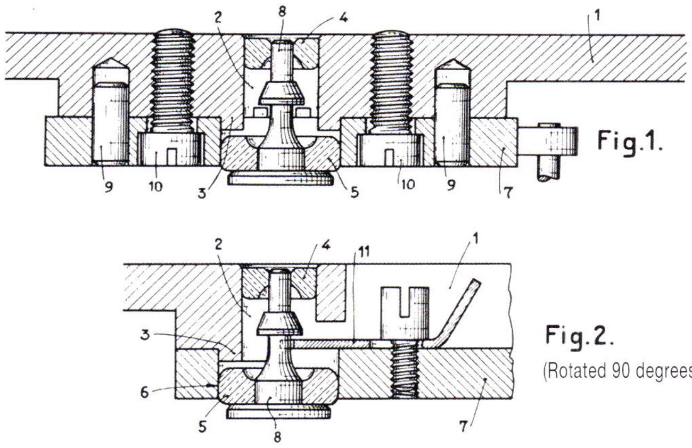

Fig.1.

Fig.2.
(Rotated 90 degrees.)

Illustration 152
Sectional drawing of the IWC automatic caliber 85 winding rotor bearing
After the retaining screw is loosened and the fork latch is pulled out (11), the rotor can be removed upward. Here the two artificial rubies are firmly mounted in the rotor. The axle journal (8) is firmly pressed into the shock absorber spring (IWC drawing).

The first consists of the shock absorber spring, later called the automatic device bridge (20), on whose axle (19) the winding rotor is mounted. The bearing housing (2), which is screwed to the rotor, has two synthetic rubies of different diameters that are precisely matched to the two-part bearing journal.

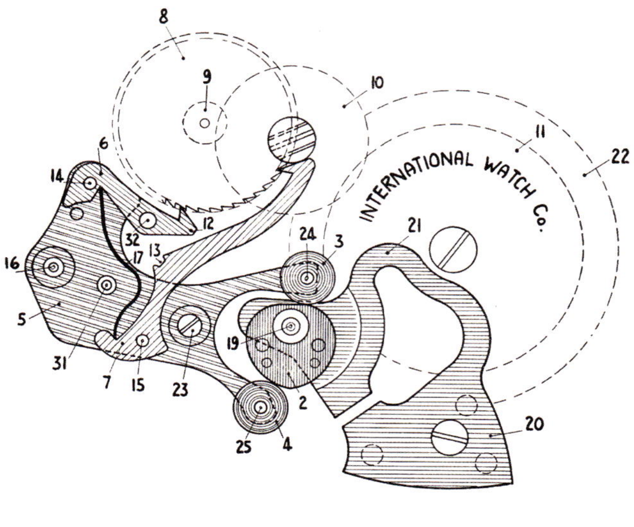

Illustration 153
Schematic representation of IWC's Pellaton drive (IWC drawing)

When asked about the three spring constants in the X, Y, and Z directions of this shock absorber spring, IWC was unable to provide these values. The outer circumference of the bearing housing is designed as an eccentric (2). The rotor and bearing housing form a functional unit that is screwed together and locked on the axis of the shock absorber spring by a fork latch with locking screw. The group, which can move in three dimensions, can thus absorb and intercept pulses or shocks from all directions.

The connection to the clinch carrier (5) with the roller fork, which operates in only one plane, is again formed by the eccentric on the rotor. In simple to-and-fro movements, the roller fork follows the rotations of the eccentric and converts its rotary movements into linear movements. This mechanics thus forms the second dynamic area of this system. If the watchcase receives an impact in the plane of the rotor, the shock-absorbing spring moves with the "excited" rotor first in the direction of the impact and then springs back to its original position.

The roller fork follows both superimposed but linear directions of movement and continues to convert the rotary movements initiated by the rotating rotor into linear movements.

If a perpendicular impact now occurs on the watchcase and the winding rotor, this impulse is absorbed by the shock absorber spring in the axial (i.e., Z) direction. These relative movements in the Z direction can be absorbed by the roller fork, which can move in only one plane, by the two ruby rollers (3, 4) "rolling" spatially on the eccentric. They thus move in curves around the circumference of the eccentric. In this spatially running process, the full function and elevator performance are retained. The two reciprocating linear movements of the roller fork and the pawl carrier are then passed on via the two pawls (6, 7) and the "automatic wheel" (8) and converted back into a rotary movement. From this automatic wheel, this torque is passed on for the winding of the spring movement.

Thus, all impulses acting on the watchcase from different directions can be converted into winding work! An ingenious solution!

Functionally Correct Mounting of a Winding Rotor

A motorist expects the four wheels of his vehicle to operate reliably and with low friction, but to be easily replaceable. One should be able to assume the same of a winding rotor of an automatic watch movement. The fact that comparable functions also lead to similar solutions can also be seen in the Pellaton winding mechanism. In the two images on page 89, one can see the mounting of the winding rotor on the axle pin of the shock absorber spring. The rear bearing (5) has a larger diameter, the front bearing (4) a smaller diameter. Between the two is a collar (2) in the middle, which holds the forked latch for locking the rotor. However, the diameter of this collar is kept smaller than that of the larger bearing seat (8). After retracting the fork bolt, the rotor can be removed in such a way that the two bearing seats do not touch due to the different diameters.

Illustration 154 shows the bearing arrangement of a front wheel of an automobile. The rear axle bearing has a larger diameter, the front one a smaller diameter. After the screws are loosened, the brake drum, which supports the wheel, can be easily pulled off without one bearing seat coming into contact with the other.

In precision mechanics as well as in mechanical engineering, it is considered a "mortal sin" to provide two consecutive bearing seats with the same diameter. You should never place one bearing seat over another of the same diameter! This is also how Pellaton once designed the bearing of the rotor with two rubies of different sizes. The rotor bearing of the excellent Rolex caliber 3135 is designed in the same way today (illus. 157, *top right*).

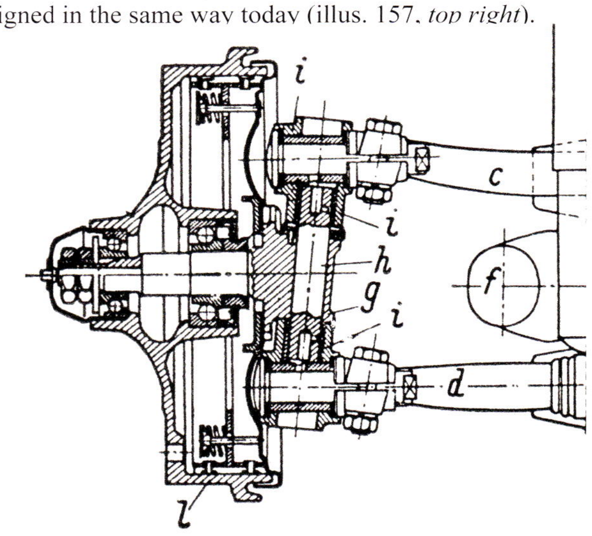

Illustration 154
The front wheel bearing of a VW Beetle. The rear bearing is larger in diameter, the front one smaller. After the two axle nuts are loosened (*far left*), the brake drum can be pulled off. If you turn this drawing clockwise by 90°, in principle the brake drum housing corresponds exactly to that of the rotor of the Pellaton winding system (VW drawing).

Illustration 155
The picture of the IWC caliber C. 8541 B in about triple magnification, with the rotor (*below*) removed. This image corresponds well to the schematic diagram of the Pellaton winding system on the left. The rotor was removed from the movement and positioned below it, rotated by 180°. Now you can see the inside of the rotor with the eccentric, as well as the larger ruby bearing. The eccentric is secured to the rotor by two screws. This eccentric has an outer diameter of 0.118 inches (3 mm) and an eccentric travel of 0.039 inches (1 mm). Purely geometrically, this design consists of two circles, with the center of the smaller one offset by 0.039 inches (1 mm) from the center of the other. This offset is also the distance the roller fork travels back and forth per 360° of rotor rotation. A few years ago, IWC's technical director at the time tried to increase the winding power of the Pellaton system by means of a so-called double pawl drive but failed because of an unnoticed design flaw. If he had increased the outer diameter of the eccentric described here from 0.118 to 0.157 inches (3 to 4 mm), the roller fork would have moved back and forth not by 1 but by 2 mm per rotor revolution. This simple change would have doubled the winding performance.

Illustration 156
The shock absorber spring, later called automatic device bridge, of the caliber 8541 B shown on the left in approximately 9× magnification. On the axle journal, one can see the two different bearing diameters.

Illustration 157
Axle journal on the rotor of the Rolex caliber 3135. Here, too, one can see the two different bearing diameters.

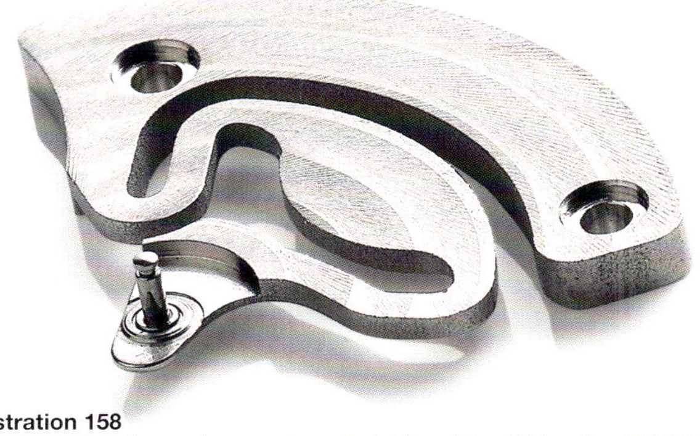

Illustration 158
The shock absorber spring or automatic bridge of the IWC caliber 85110, with an axle journal, which has one continuous bearing diameter

Why IWC and other watchmakers no longer pay attention to this "golden rule" of mechanics but practice the aforementioned "deadly sin" with bravura is hardly comprehensible to a mechanical engineer.

Also, on the marketing side a huge opportunity has been missed in ignorance of the exact kinematics of the Pellaton winding system!

With a simple enlargement of the rotor eccentric from 3 to 4 mm in diameter, the performance of this Pellaton winding system would have been doubled! This banal change could then have been marketed excellently with the promotionally effective statement "By a well-considered constructive improvement, the winding performance of the Pellaton elevator could be doubled!"

But nothing is lost yet. The author and his colleague, Dr. rer. nat. Gerhard R. Thoma, will be pleased to provide IWC with active support in the areas of kinematics and materials science.

Despite this minor criticism of the technology, the watch enthusiast and collector will always take pleasure in the beautiful timepieces from Schaffhausen because they always show him the correct time, even with today's technology.

View of the newer models from IWC

Illustration 159
Two moons circle on the dial of the IWC Portugieser Perpetual Calendar (reference IW502119). The designers and watchmakers didn't look too deeply into the glass and see double, however, but instead thought globally: the moon appears differently in the Southern than in the Northern Hemisphere. Although the moon's phase is the same, it appears in the opposite direction. In addition, the display is extremely accurate: only after 562 years does the moon phase display deviate by one day. But the moon phase is not the only special feature of this perpetual calendar, which also takes leap years into account and adds a February 29 every four years. In addition to the double moon phase, on the traditionally large dial of the Portugieser there are also the displays of the date, day of the week, month, small seconds, hour, minute, power reserve indicator, and four-digit year—a display that is very rarely found. The back is just as exciting as the front: through the glass case back, one can observe the huge IWC caliber 51612 movement with pawl winding by Pellaton and skeletonized rotor. In 2000, IWC returned to the circle of manufactures with the base caliber in a limited-edition Portugieser. In the meantime the rate was increased to 21,600 vibrations per hour, further improving accuracy. Not to be improved upon is the enormous power reserve of seven days, of which the IWC ad declares, "If this watch stops, please call a doctor, because then you haven't moved for seven days." It looks particularly elegant in 18K red gold with a brown alligator leather strap.

The fifty or so noble watches presented in this chapter have made it clear to us the design intent and high-quality technology that lie behind these masterpieces from IWC. The models of the past, which were perhaps intended to be spectacular, such as the Golf Club, the Polo Club, or even the Yacht Club II, often became shopkeepers and do not always command the respect they deserve, even today. In contrast, the military watches and the Ingenieur, Yacht Club, and Aquatimer series have found numerous enthusiasts. The price increases seen in recent years confirm this trend. Therefore, it does not always have to be a Rolex or Patek Philippe that brings joy to the wrist. An impressive Ingenieur, Yacht Club, or even a Ref. R 817 AD from Schaffhausen can do just as well (illus. 135, p. 80). Perhaps not quite 100% identical, but considerably less expensive!

Illustration 160
On land, on water, and in the air: with the right IWC on your arm, you can move in any element. The Ingenieur Automatic with steel bracelet (reference IW322701) is best suited for country outings, and IWC revived its famous family in 2005. Not only because there is an AMG version for lovers of particularly sporty Mercedes cars, but mainly because the newly developed IWC caliber 80110, similar to old IWC calibers, has a spring-loaded rotor that takes the terror out of vibrations and shocks. As with the 5000 calibers, the power reaches the mainspring barrel via a pawl-winding mechanism. Like its historical predecessors, the Ingenieur is protected against magnetic fields: an inner case made of soft iron keeps harmful influences away from the movement. The pilot's watches also feature magnetic field protection. Like all Spitfire versions, the Spitfire Chrono-Automatic (reference IW371702) has a silver dial, which is atypical for pilot's watches. Otherwise it has all the important features, such as the vacuum-proof crystal that stays in place even in the event of a sudden drop in cabin pressure. The caliber used is a heavily revised ETA/Valjoux 7750. Specialized for use in water, the Aquatimer Automatic 2000 (reference IW353804) can also satisfy professional demands. The watch has sufficient reserves, being able to withstand a depth of 6,562 feet (2,000 m), which no diver has ever reached. Its titanium case is lightweight and very resistant to salt water despite its ruggedness. The yellow color of the dial is easy to read even in murky water. The obligatory ring to measure the elapsed dive time is under the crystal and is adjusted via the seconds crown. Inside ticks a modified ETA caliber 2892 A2.

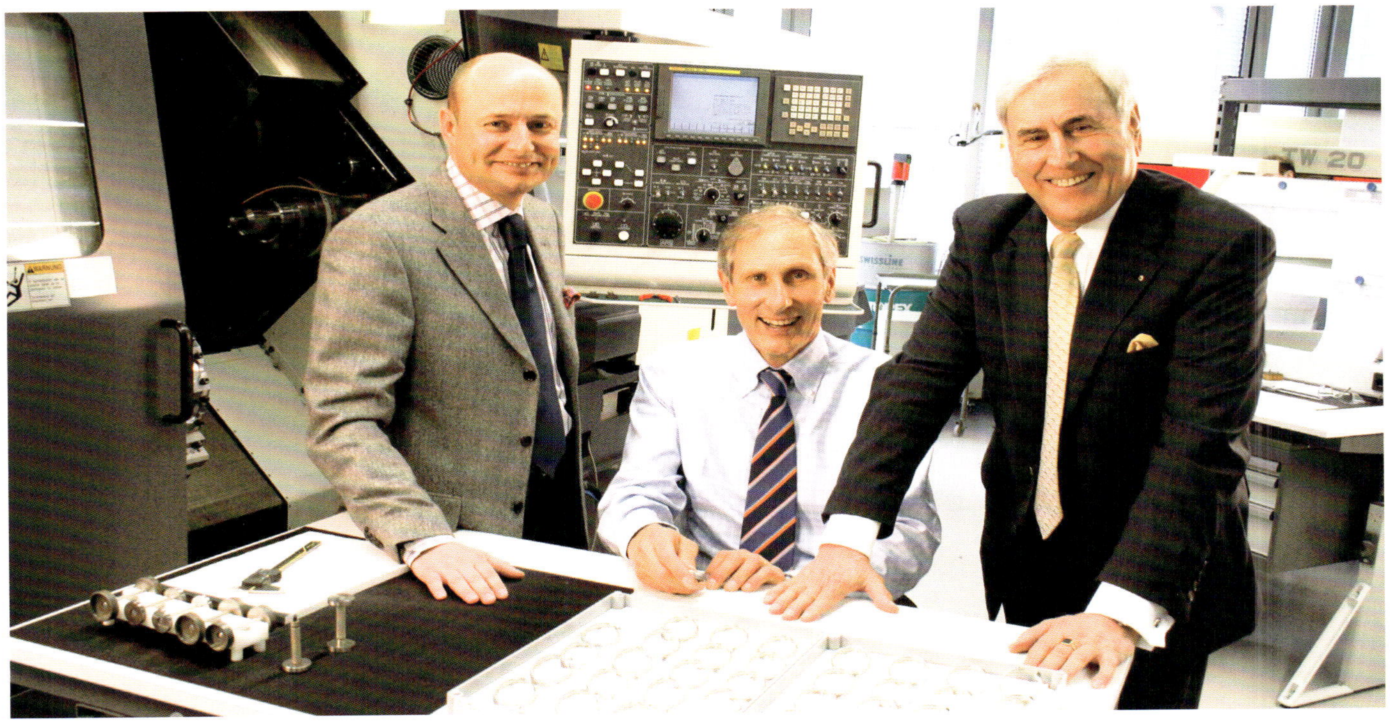

Illustration 161
View of the new case production facility, which is also capable of producing titanium cases, which are difficult to process. In the center is Ronald Jäger, the head of production, who retired April 7, 2020. To his left is Georges Kern, then CEO of IWC, who is now CEO and copartner at Breitling. On the right is the author.

Illustration
163

Illustration
164

3. Jaeger-LeCoultre

3.1. Introduction by Gisbert L. Brunner

Showcase Manufacture in the Vallée de Joux, Time for Watchmaking Pioneers

At the beginning of the nineteenth century, winters were very different in the remote Vallée de Joux. Months of snow and severe cold brought with them a great deal of loneliness, but also the leisure time in which to think a great deal about improvements in the field of watchmaking. The intellectual debate concerned the tools, the raw movements, and even complex additional mechanisms; for example, calendars or striking mechanisms. The resulting creativity ultimately established the outstanding reputation of the Vallée de Joux as the "Valley of the Inventors." And they made an important contribution to the development of the high valley into a veritable "land of milk and honey where there was not a single poor person left." At least, that was how John-César Piguet saw it when in 1881 he looked back on the development of the watchmaking industry in the *vallée*.

A first but significant step toward the mechanized production of watch components was made in 1810 by a man named Abel Golay. Using tools he developed himself, he milled pinions for the first time. Charles-Antoine LeCoultre took this high-level and indispensable craftsmanship to unprecedented perfection from 1826 onward. Together with his brother Ulysse LeCoultre, in 1833 the precision mechanic founded a company in Le Sentier that was to revolutionize watchmaking in the Joux Valley. On the basis of their experiments, which had been going on for seven years, the two of them were initially concerned almost exclusively with the industrial production of the absolutely indispensable gears and pinions. A little later the gifted inventor succeeded in developing new tools and machines with which both could be produced in a largely mechanized way. The year 1844 was marked by the innovative *millionometer*. This precision measuring instrument with an accuracy of 1/1000 of a millimeter led the watch industry into a new age. After all, a thousandth can make a great deal of difference where fractions of a second are at stake. In addition, this instrument led to the introduction of the metric system at LeCoultre & Cie. In 1851, the first World's Fair opened its doors in London. There Antoine LeCoultre exhibited his remarkable products, including gears and pinions, movements, pocket chronometers, and also tools. Impressed by his precision and dedication, the jury awarded Antoine LeCoultre one of the coveted gold medals.

Around 1860 the company, which with almost a hundred employees was already the largest employer in the remote region, slid into severe financial difficulties as a result of impetuous growth, but also due to mismanagement. This was accompanied by a general crisis in the watch industry. The savior for LeCoultre & Cie. was Antoine Borgeaud, who, as a new partner, provided the company with fresh capital and a new watch caliber with rocking-bar winding. The latter provided the company with the orders urgently needed to survive in difficult times. In 1875 the "golden age" in the Vallée de Joux came to a temporary end. The reasons were complex. Increasing mechanization in the production of raw materials had led to overproduction, making watches difficult to sell, and then only at a loss. In addition, the difficult economic situation, especially in Europe, depressed the demand for expensive luxury watches with additional functions. As a result, orders from the Geneva manufacturers failed to materialize. By now it became apparent that the watchmakers in the Joux Valley, among them LeCoultre & Cie., had been wrong to neglect the production and marketing of complete watches.

The Second and Third Generations Show What They Are Made Of

Borgeaud's departure and his demand for a severance payment of 250,000 francs was a slap in the face. But this step was not enough to destroy the optimism of the three sons of the company founder. In 1877, in the midst of the crisis, they were forced to flee to the front. The trio reorganized production, pushing ahead with the design and manufacture of elaborate raw watch movements.

Illustration 162
Heinz Winkler in front of his residence in Aschau. With beautiful watches and classic automobiles, star cuisine can't be far behind. Born in South Tyrol, Heinz Winkler has succeeded in creating a retreat at his residence in Aschau, which may probably be considered singular. The excellent cuisine (2 Michelin stars), the proverbial friendliness of the staff, and the cordiality of Mrs. Evi Winkler make every visit an unforgettable experience (illus. 163, *top left*). The highest art of watchmaking is embodied in the highly complex travel watch from Jaeger-LeCoultre, which bears the name DUOMÈTRE UNIQUE TRAVEL TIME. Not only can two time zones be set on this watch, but there are also world map and day/night displays. The case is made of elegant red gold (illus. 164, *top right*). The glass back allows a view of the complicated movement of this dual-time-zone travel watch. On the top right you can see the two mainspring barrels, and at the seven o'clock position the oscillating system and the fine-adjustment mechanism.

Illustration 165
This chronometer from Jaeger-LeCoultre marked with the name GEOPHYSIC is regarded as a simple and rare watch. Designated with the reference E 168, the timepiece was launched in 1957 and remained in production only until 1959. Of the 1,038 watches with steel cases that were built, one example was called at the Phillips auction in Geneva and was knocked down at 68,750 CHF (movement no. 1,331,979, case no. 748,201). The watch shown has the caliber P 478/BWSbr, movement number 1,391,284. The case number is 748,540. We still have the original sales box and the warranty certificate. The latter was signed by then company president Roger LeCoultre. The date of sale was entered in the warranty certificate as September 7, 1984.

This overdue step ushered in an extremely artful era at LeCoultre & Cie. that was unparalleled in terms of craftsmanship. Complications, the salt in the soup of fine watchmaking, attained paramount importance in the company's biography. Without exaggeration it can be said that there is hardly a raw movement with a complication that LeCoultre has not tackled and mastered. Chronographs, calendars, repeater movements, alarms, and *tourbillons* were part of the broad production spectrum of the venerable *ébauche* factory, which was crowned with international success. According to internal statistics, 60,000 movements with various additional functions were produced between 1860 and 1925. Many of them were truly superlative because they were the smallest or flattest of their kind. In 1903, for example, the conventionally designed caliber 145, just 0.054 inches (1.38

Illustration 166
The Memovox model was launched in 1953 with the advertising slogan "A second memory." While the first versions still had manually wound movements, others followed with automatic winding and date. Intended for the American market, the watches were encased in the USA and marketed under the LeCoultre name. The upper part of the case of this model is gold filled, while the screw-down case back is made of steel.

mm) in height, caused a sensation with the nickname "Razor Blade." A movement with a chronograph was only 0.110 inches (2.8 mm) high, and one with a repeating striking mechanism was just 0.125 inches (3.2 mm) high.

Contacts established in the early twentieth century with Alsatian watchmaker Edmond Jaeger were to have a decisive influence on the further development of the company. Jaeger supplied Cartier and the French navy, among others, with timepieces and instruments. By supplying him with movements, LeCoultre gained immediate access to the luxury timepiece market. In 1914 LeCoultre began manufacturing aircraft tachometers, odometers, and automotive timepieces. But that was not all: the two-story Duoplan caliber 11 BF from 1926 could be exchanged for a fresh one at short notice in case of repair. During the warranty period, the exchange was free of charge. In 1929 the smallest mechanical movement to date was ready for series production. The caliber 101 measured just 0.55 × 0.188 × 0.133 inches (14 × 4.8 × 3.4 mm). Its seventy-four parts, including the dial and hands, weighed just 0.03 ounces (0.9 grams). This time-preserving pipsqueak is still in production today. From 1931 the Reverso made a name for itself; that legendary wristwatch, in which the crystal glass, which was prone to breakage, could be reversed without further ado in delicate situations. And since 1936 the Atmos has had the reputation of being a solitaire among pendulum clocks. Of course this reputation does not come by chance. Air is all that is needed to keep the pendulum moving silently and, above all, continuously. Manual winding is a thing of the past.

From LeCoultre to Jaeger-LeCoultre

In 1937 the watchmaking divisions of Jaeger and LeCoultre merged to form the Jaeger-LeCoultre watch brand, accompanied by the establishment of a joint sales organization based in Geneva. In 1946 Jaeger-LeCoultre presented a model with a simple full calendar and moon phase display using the caliber 486 movement. The following year the company introduced its first automatic with pendulum oscillating weight and power reserve indicator, the caliber 481. When Jacques-David LeCoultre died in 1948, the era of the founding family came to an end. The future was secured by a holding company called SAPIC, under whose umbrella were the raw-movement manufacturer LeCoultre, the watch brand Jaeger-LeCoultre, and parts of the traditional watch manufacturer Vacheron & Constantin. Bernese industrialist and major LeCoultre shareholder George Ketterer acted as delegate on the board of directors. The restructuring had no effect on the company's innovative strength. In the 1950s, Jaeger-LeCoultre was one of the acknowledged alarm pioneers. This was particularly true of the combination of alarm function and self-winding mechanism, which made its debut in 1956 in the Memovox Automatic model.

A Look at the Present

In 1978, German instrument maker VDO acquired 60% of the share capital of Jaeger-LeCoultre SA. Audemars Piguet secured a minority stake with the remainder. This speedometer manufacturer came under the umbrella of the Mannesmann Group in 1991, which later bundled its watchmaking activities in LMH Holding (Les Manufactures Horlogères). As a member of this group of companies, Jaeger-LeCoultre was sold to the globally active Richemont Group in July 2000. This did nothing to diminish the maker's activities. On the contrary: never before in the company's history have there been so many new movements covering the entire spectrum of mechanical timekeeping. From ultrathin calibers to unique combinations of complications—for example, the three-faced Reverso *grand complication à tryptyque* from 2006—Hybris Mechanica stands for the exaltation of the art of mechanical watchmaking. The development processes for the Duomètre watch line took place under the catchy cover name Double Wing because two completely independent gear trains share a common rate regulator.

Finally, there remain those one thousand hours, or the equivalent of forty-one days and sixteen hours, that every Jaeger-LeCoultre timepiece is tested before it leaves the factory without any ifs, ands, or buts. In fact, this long-term test is considered one of the most stringent checks of all. In contrast to the COSC's official chronometer test, the factory sends only fully powered-up examples, naturally equipped with dial and hands, to the tough test track. There, the check is carried out alternately in six positions at different temperatures according to unmistakably defined specifications. To make matters worse, the test candidates are occasionally in motion, then again completely motionlessness. In some cases they are subjected to targeted impacts, alternating magnetic fields that exceed the generally accepted standards, or simply pressure. Thus, the first one thousand hours in the life of a Jaeger-LeCoultre are much harder than later in real life on the owner's wrist.

3.2. The Movements

In the chapters on Audemars Piguet, Patek Philippe, and Vacheron & Constantin, we have often encountered the name Jaeger-LeCoultre as a supplier of movements. In fact, Jaeger-LeCoultre, based in the Swiss Jura village of Le Sentier, has always backed several horses during the period under review in this book. First, it produced sophisticated wristwatches of various designs for the European and overseas markets. Second, business was booming in the USA at that time, but only movements were sent there, in American cases and marketed under the name "LeCoultre." Third, the specialists in Le Sentier served as suppliers of raw movements for various watch manufacturers. Specially reserved raw movements (*ébauche*) of the highest quality were also sent to the noble watch manufacturers Audemars Piguet in Le Brassus and Vacheron & Constantin in Geneva. Even Patek Philippe in Geneva purchased exactly 40,000 examples of the super-flat automatic caliber 28-255 and 28-255 C in the 1970s.

Due to the large number of calibers manufactured during that time, only a representative overview can be given in this book. From the perspective of the time, the 1950s and 1960s were among the most successful in the company's history. The emergence of quartz technology caused Jaeger-LeCoultre's sales to plummet, forcing the company to turn to other areas of business. By the time Jaeger-LeCoultre was part of VDO and incorporated into the French SAPHIR SA holding company, finances had become so tight that in 1983, 40% of the company's shares were sold to Audemars Piguet in neighboring Le Brassus. At that time, Jaeger-LeCoultre had a share capital of twelve million CHF, which was also the basis for the purchase price, at that time a whole 4.8 million CHF! Whether this was the right decision on the part of VDO at that time can perhaps be doubted today.

The Manually Wound Movements

10½'''-480 and 10½'''-800

The caliber 10½'''-480 was regarded as a basic movement with small seconds. The installation diameter is 0.93 inches (23.70 mm), and the overall height is 0.143 inches (3.65 mm). Depending on the version, fifteen to seventeen rubies are installed, and the balance runs at 18,000 vibrations per hour. The caliber 10½'''-800 represents the version with center seconds. This increases the overall height to 0.157 inches (4.00 mm). In total, 39,000 examples of this caliber were produced from 1950 to 1965, of which about 5% were exported to the USA.

12½'''-449 and 12½'''-470

The caliber 12½'''-449 was regarded as a basic movement and was made in different versions. The installation diameter is 1.13 inches (28.80 mm), and the movement height is 0.157 inches (4.00 mm). In the 470 version with center seconds, the height had increased to 0.179 inches (4.55 mm). This manually wound movement was produced from 1943 to 1944 and was used in British military watches. The production total was five thousand movements.

Illustration 167

12½'''-488

This manually wound movement also goes back to the basic 12½'''-449 caliber. Due to the installed stop-seconds mechanism, the movement height is 0.196 inches (5.00 mm). This caliber is used to power the British Pilot's Watch Mark 11 (p. 105). The 12½'''-488 caliber was produced from 1948 to 1953, with a total of 2,700 examples made. Exactly 2,320 of these movements were installed in the watches of the aforementioned Mark 11. The seconds are stopped by pulling the crown via a clearly visible bail and a fine steel spring (*tirette*). This feature can also be found in the caliber P 478/BWSbr of Jaeger-LeCoultre's Géophysic chronometer and as caliber P 1007/BS and P 1008/BS of Vacheron & Constantin's Chronometer Royal.

Illustration 168

12½'''-P 478 / BWSbr

Starting from the basic caliber 12½'''-449, the caliber 12½'''-488 is almost identical in design to the 12½'''-P 478 / BWSbr manually wound movement. The latter was built into the nowadays highly sought-after Géophysic chronometer, reference number E 168. Exactly 83,000 pieces of the 12½'''-478 caliber were made between 1945 and 1965. In the period 1957 to 1959, 2,500 movements were taken from this total and were destined for the E 168 Géophysic and E 2985 Géophysic de luxe chronometers. Of the former reference E 168, 1,038 chronometers with stainless-steel cases, 222 with yellow-gold cases, and thirty with rose-gold cases were produced between 1957 and 1959. The version reference E 2985, which was not water resistant, was produced only 102 times: eighty-two of them with cases in 750 yellow gold and twenty in 750 rose gold. Instead of the reinforced

Illustration 169

crystal, soft-iron dome, and screw-down case back, this model had only a normal crystal and a press-fit case back. The public price of a Géophysic reference E 168 with 750 yellow-gold case was exactly 720 CHF in 1961. All caliber 12½‴-P 478 / BWSbr watches had to have passed the official chronometer test with distinction. Thus, of the 2,500 movements selected, only 1,392 had been granted this honor (a success rate of 56%). Comparing the caliber 12½‴-P 478 / BWSbr (p. 99) with the caliber 12½‴-488 on the same page, the difference in finishing is striking. On the latter the bridges have the Geneva stripe finish.

12½'''-814

The basis for this manually wound movement for the Memovox model was the caliber 12‴-489. Essentially an alarm clock worn on the wrist, the Memovox model was considered a minor sensation and was launched with the following advertising slogans: "Reminds – Warns – Awakens." The movement has an installation diameter of 1.10 inches (28.00 mm) and a height of 0.204 inches (5.20 mm) and runs at 18,000 A/h. It has two mainspring barrels: one for the movement, the other for the alarm function. The crown at two o'clock is used to set the alarm time and to wind the alarm. The crown at four o'clock is used to set the time as well as wind the movement. The movement has seventeen rubies and runs at 18,000 vibrations per hour. Between 1953 and 1964, exactly 115,000 pieces of the 12‴-814 caliber were produced, and 60% of them were exported to the USA.

The Automatic Movements

12'''-476 and 12'''-481 (illus. 170)

The hour of self-winding watches struck at Jaeger-Le Coultre in 1947. For the conversion of kinetic energy, a unilaterally acting oscillating weight with a limited angle of rotation was chosen; hence the movement was popularly known as a "hammer automatic." The case diameter is 1.16 inches (29.56 mm), and the overall height is 0.242 inches (6.15 mm). The movement has seventeen rubies, and the balance runs at 18,000 A/h. Also worth mentioning is the early Jaeger-LeCoultre proprietary shock protection. A variant of the caliber 12‴-476 received a power reserve indicator integrated below twelve, as well as the new caliber designation 12‴-481. In 1948 the well-known Powermatic model was introduced, in which the latter movement was installed. A hint from the watchmakers of Le Sentier might be important to the watch enthusiast. If part 1488 (*roue cliquet complète*) of the caliber 12‴-476 or part 9564 of the modified version 476-12‴ A should be damaged, these parts are no longer available.

Illustration 170

12'''-812

This caliber is a further development of the first automatic movement with a limited rotor and was launched in 1954. This fact may be considered particularly remarkable, however, because Jaeger-LeCoultre had already released two automatic movements with a freely rotating rotor in 1953: the caliber 12‴-498, with small seconds, and 12‴-499, with center seconds. Presumably the watches with "hammer automatics" had proven so successful that they stayed with this technology for a while, despite the alternatives mentioned. Compared to its predecessors, the caliber is a bit more compact. The case diameter is 1.10 inches (28.00 mm) and the height is 0.179 inches (4.55 mm). The number of jewels remains at seventeen, and the balance runs at 19,800 A/h. From 1954 to 1960, exactly 37,000 movements were built, 70% of which were destined for the US market.

880 (p. 101, *top left*) and 881 (p. 101, *center*)

At the end of the 1950s, Jaeger-LeCoultre decided to omit the previously used diameter indication in lines (a French measure of length commonly used by watchmakers, with 1‴ corresponding to 0.087 inches, or 2.223 mm) from the caliber designations. The caliber 880 was launched in 1959 and, with its variant 881 with date display at three o'clock, was incorporated into numerous wristwatches. The basic caliber has an installation diameter of 1.02 inches (26.00 mm) and a movement height of 0.227 inches (5.77 mm). Seventeen rubies are installed. The 881 version with date display has a movement height of 0.241 inches (6.14 mm). This caliber has a bidirectional rotor. The screwless Glucydur balance wheel with Nivarox hairspring operates at 19,800 A/h. Furthermore, it features an adjustable hairspring stud and Parechoc KIF Réf. Flector shock protection.

Illustration 171

A fine adjustment and a regulator arm were omitted. The caliber series 880 to 883 were the most-successful automatic movements of their time, not only at Jaeger-LeCoultre but also in the entire Swiss watch industry.

Illustration 172

881 G (*above right*) and 883 S (*below right*)

These two calibers were specially modified for the Geomatic chronometer models before they were submitted for chronometer testing. The added modifications in the caliber 880 G and 881 G, listed in the parts catalog for the chronometer version, are as follows:

1. increase in the number of rubies from 17 to 23
2. stop-seconds device by pulling the crown
3. fine adjustment with Spirotor Regulator 3

From the parts list for the calibers 882 and 883, only the changes for the 882 S and 883 S versions can be seen. "S" stands here for stop-seconds. Further details are missing. The research requested in Le Sentier led to a typewritten attachment for the

Illustration 173

aforementioned parts list. According to this, there were two versions of the caliber 883 S. The chronometer version received six additional rubies: two of them for the bearings of the minute wheel and four for the mechanical winding mechanism (i.e., for the pawl and transmission wheels). Thus, the caliber 883 S largely corresponded to the caliber 881 G; only the fine adjustment by means of the Spirotor Regulator 3 was omitted. According to Jaeger-LeCoultre, 3,688 examples of the caliber 881 G and exactly 828 movements of the caliber 883 S in chronometer version were built between 1962 and 1971.

Illustration 174

Illustration 175
Screw-down case back of the Jaeger-LeCoultre Master Mariner Ref. E 557 in stainless steel, case no. 1,006,278. Exceptionally, the case number is not stamped on the outside but engraved on the inside. Because the "1" was omitted, the number is 006278.

3.3. The Reference and Case Numbers

As a rule, the men's wristwatches produced in Le Sentier were given a reference number, which was stamped on the inside of the case back. This consisted of the letters E and a three-digit number (see fig. 175 with E 557). Subgroups of a model were given the suffix /1 or .XY. One can see on page 106 that this was not always done uniformly. The gold Futurmatic belongs to the reference E 500 but has this marking listed on the outside, rather than on the inside of the case back. The case numbers were always stamped on the flat surface of the case back. Watches that had the movement installed in the USA had neither reference nor case numbers. The case numbers in the present period under consideration have six or seven digits and are always numbered consecutively.

3.4. The Watches

The Manually Wound, Chronometer, and Pilot's Watches

Of the numerous models launched by Jaeger-LeCoultre during the period under review, here we present some thirty representative collectors' pieces and other watches. With their diversity in case design and sophisticated movements, they provide the collector with an initial overview. On the one hand, the technicians from Le Sentier proved with their Powermatic, Futurmatic, and Memovox models a high level of innovation in their own watches, while at the same time they supplied other watch brands with raw movements of the highest quality that could not be built in their own houses because of this reservation. Customers such as Audemars Piguet from Le Brassus and Vacheron & Constantin from Geneva used movements supplied by Jaeger-LeCoultre in their luxury watches for many decades. And Patek Philippe also once used Jaeger-LeCoultre's raw movements in large quantities until its own new automatic calibers could go into series production.

Illustration 176
Of simple beauty and extremely rare is the Geophysic chronometer, here a model in 750 yellow gold with a ceramic white dial with chronometer cross. The first version of the Chronomètre Geophysic was given the reference E 168, was launched in 1957, and remained in production until 1959. During this period, exactly 1,290 examples of this coveted timepiece left the factory in Le Sentier (1,038 in steel, 222 in yellow gold, and 30 in rose gold).

Illustration 177
Opening the screw-down case back and removing the soft-iron dome permits a view of the officially certified manually wound caliber P 478/BWSbr movement. Above the gear train bridge, one can see a bracket that stops the balance by means of a small steel spring when the crown is pulled. Unlike its counterpart in the Pilot's Watch (see p. 105, Pilot's Watch Mk. 11), the bridges here are not decorated with Geneva stripes (movement no. 1,331,970, case no. 775,561).

Illustration 178
Jaeger-LeCoultre Geophysic Ref. E 168 in stainless steel

Illustration 180
Geophysic chronometer Ref. E 168 from the 1950s, and the new Geophysic 1958 Ref. 800.85.20 from the current collection. While the manually wound caliber P 478/BWSbr movement ticks away in the historical edition of the Geophysic, inside the new model is the 898/1 automatic movement with forty-three hours of power reserve. The case on the left measures 1.38 inches (35 mm) in diameter, the one on the right 1.52 inches (38.50 mm). On the secondhand watch market, also called the secondary market, a representative of the model on the left reached a record price of 68,750 CHF at an auction held by the House of Phillips on November 12, 2016, in Geneva. The new model is not so sought after and on average is offered between $8,800 and $9,880. The steel version of this model was limited to eight hundred copies.

Illustration 179
Chronometer Geophysic, movement view
The chronometer caliber P 478/BWSbr was among the best movements of its time. Under the caliber name P 1008/BS, this movement was also used by sister company Vacheron & Constantin in the famous Chronomètre Royal model. On the gear train bridge you can see an arch-shaped bracket. Pulling the crown causes this bracket to be pulled to the right, stopping the balance via a small bent wire. This mechanism thus causes the seconds stop, which is necessary for the accurate setting of a watch. The inner side of the screw-down case back is beaded, the case number is 748,540, and the movement number is 1,331,284. Above right is the inner case back, which provides additional dust protection for the movement.

Illustration 181
The two screw-down case backs of the old and new Geophysic
While a medallion with name and globe was inserted in the back of the Vintage-Geophysic on the left, the representation of the globe and the company logo on the case back on the right was produced by machining, thus milled and engraved. For the closure of the screw-down case back, the usual Jaeger-LeCoultre twelve-sided shape was retained, albeit with a different key width.

From a marketing perspective, Jaeger-LeCoultre's products were clearly below the level of the aforementioned noble watch manufacturers. Especially in the important US market, Jaeger-LeCoultre placed itself in the midfield and always paid attention to favorable, competitive prices. On the European level, they saw themselves positioned below Rolex, but on the same level as OMEGA and clearly above IWC. In our series of illustrations we begin with the yellow-gold Géophysic, of which just 222 examples were made.

Illustration 182
View of the JLC 898/1 caliber in the new Geophysic
In contrast to the predecessor caliber 889/2, which had an installation diameter of 1.02 inches (26.00 mm) and a total diameter of 1.05 inches (26.60 mm), the new caliber brings it to 1.15 inches (29.2 mm). The overall height of 0.128 inches (3.25 mm) has been retained. Because of the big difference between the case diameter of 1.52 inches (38.5 mm) and the movement diameter of 1.15 inches (29.2 mm; delta 0.37 inches, or 9.30 mm), a perforated intermediate or retaining ring had to be provided.

Illustration 183
Fine-adjustment mechanism of the caliber 898/1 shown at the lower left, magnified about 7×. On the rim of the double-spoke balance wheel there are four weight screws to regulate the rate. The escape wheel shaft is mounted only in synthetic ruby bearings. The usual cover plates were not used. The balance axle's shock protection system is made by KIF.

This now-sought-after chronometer was launched in 1957 and was intended for use by scientists and engineers. Exactly 1,038 examples of the stainless-steel version were made. This chronometer became world famous thanks to the American nuclear submarine *Nautilus* (SSN 571), which under Commander William T. Anderson made the first submerged transit of the North Pole (Operation Sunshine) on August 3, 1958 (23:15 EDT). As a special recognition or award for this historic act, Anderson received a Jaeger-LeCoultre Geophysic in stainless steel from the hand of Admiral Hyman G. Rickover, commander in chief.

The Jaeger-LeCoultre Pilot's Watches

Something that was once essential for survival is now a very popular status symbol. I am referring to pilot's watches, which have always had a special significance at Jaeger-LeCoultre. Most watch collectors are well aware of IWC's Mark 11 (Mk 11) pilot's watches. In parallel, Jaeger-LeCoultre also supplied accurate timepieces to RAF (Royal Air Force) pilots and those of other Commonwealth air forces. Three interesting examples have been selected for this chapter. The first pilot's watch listed is the RAAF (Royal Australian Air Force) Mark 11 (p. 105). It is an heirloom and belongs to an Australian whose father once flew the popular subsonic single-seater fighter Hawker Hunter, and for whom the aforementioned heirloom served as a pilot's watch.

The caliber 488 / Sbr 12½''' seen here should be particularly noteworthy. The abbreviation "Sbr" can be found on the plate

Illustration 184
This RAAF (Royal Australian Air Force) Mark 11 pilot's watch was produced from 1949 onward and was given the reference no. E 161. It consists of a two-piece steel case with screw-down case back and had to meet the special requirements of the RAF (Royal Air Force) (also see Mark 11 by IWC, p. 64). To the right of the watch is the military bracelet made of plastic, with which this timepiece could also be worn over the pilot's uniform.

Illustration 186
The caliber 488 12½''' shown was derived from the base caliber 449 and features a stop-seconds mechanism. When the crown is pulled, the balance is stopped via the clearly visible bail and a thin steel field (see also p. 103, Geophysic). On the beautifully finished bridge, you can also see the Broad Arrow, proving that this watch was owned by her royal majesty. Between 1948 and 1953, exactly 2,700 examples of the caliber 488 and exactly 2,320 of the E 161 case were produced (movement no. 912,529, case nos. G6B/346 and 46/53).

Illustration 185
Two more interesting military or pilot's watches are shown. On the left is a water-resistant model with gold-plated caliber 470 and case inscription A. M. 6B/159 and 4654/56 from 1943. A. M. stands for Air Ministry. Compared to the later Mark 11, this watch has a screw-down case back and a soft-iron dome, but not the high level of protection against magnetic fields. In the upper left corner of the picture you can see this watch from the back, with the previously described markings (movement no. 255,797). The watch on the right comes from a watch collector in Israel whose father served as an airman in the Israeli air force. The case middle is made of nickel silver with a matte nickel-plated finish; the screw-down case back is stainless steel. All relevant components of the caliber 479 movement are gold-plated for enhanced corrosion protection in tropical regions. In addition to this gold plating, there is also the Geneva stripe finish, which gives this movement a particularly noble look. Besides the case number, you can find the Broad Arrow symbol, the marking W.W.W., and the additional number F 17106 on the outside of the screw-down case back. Whether the latter marking originates from the Israeli air force could not be clarified (movement no. 324,933, case no. 287,752).

and stands for "spiral breguet" (i.e., Breguet hairspring). From a purely horological point of view, according to a particularly distinguished watchmaker this manually wound movement should still be somewhat above the level of its counterpart, IWC's caliber C. 89, which was produced in large numbers.

As of yet, we have not found an answer as to why this caliber was not equipped with Jaeger-LeCoultre's own shock protection system. This shock protection had been available since 1947 and was also built into the first automatic movements 12'''-476 and 12'''-481.

The two military watches shown on page 105 (illus. 185) also have some special features worth mentioning. Both have gold-plated movements, as was customary for watches used in tropical regions. The military watch shown on the left with the upper marking A.M. for Air Ministry is also from Australia. Encased in it is the 12½'''-470 caliber with Breguet hairspring (Parts List no. 722). The stainless-steel case has a soft-iron dome and screw-down case back. On this topic, here is a special tip for the watch enthusiast: if you want to know exactly whether the inner case back of a watch serves purely as dust protection or is a soft-iron dome, just find a small magnet. If the inner case back sticks to the magnet, it is a soft-iron dome for shielding against magnetic fields. If, however, there is no reaction, it is a dust protection base made of nonmagnetic stainless steel or brass, or another nonmagnetic metal.

The last watch, seen on the right in illustration 185, again has a nickel silver case with a matte nickel finish, and the screw-down case back is made of stainless steel. The caliber 12½'''-479 movement is derived from the base caliber 12½'''-449 and is equipped with either a flat or Breguet hairspring. It has an installation diameter of 1.13 inches (28.80 mm) and a height of 0.157 inches (4.00 mm). In the current version, seventeen rubies are installed, and the balance runs at 18,000 A/h. As described, the entire movement with all components is gold-plated, and the bridges are Geneva striped. The watch came from Israel and was purchased by the former owner for $535. As the former owner likes to report, this timepiece is a personal heirloom. His father served as an airman in the Israeli air force in the 1950s and received this timepiece as a service watch. Among other types, he flew the French Dassault Super-Mystère fighter aircraft. His long period of service was evident on the watch and made necessary a major overhaul of the movement and case. This work was done by first-class professionals, but including the previously mentioned work as well as the refurbishing of the dial, the price was not more than $547—a good price.

3.5. The Self-Winding Watches

The Jaeger-LeCoultre Futurmatic

The Futurmatic model, which came on the market in 1953, was considered a sensation and was warmly welcomed on the American market in particular. The version intended for the European consumer had either a steel or a gold case with a screw-down case back. The models offered in the US were enclosed mostly in gold-filled cases and were also much cheaper. The Futurmatic illustrated came from a German auction house and was purchased for $2,700. Because both movement and dial needed considerable repair, the new owner could not say that this was a cheap purchase price.

Illustration 187
When Jaeger-LeCoultre unveiled the new Futurmatic automatic model in 1953, it was regarded as a sensation on the watch market. The crown, customarily at three o'clock, was omitted. The watch was wound solely by the movements of the wrist. The version shown belongs to the reference E 500 and has a case in 750 yellow gold. It was made in 1955, and the dial on the right side was replaced during servicing.

Illustration 188
The high-end versions of the Futurmatic had cases with screw-down case backs. The time is set with the flat crown. If you press it in the direction of the center of the back, the second hand stops and the desired time can be set by turning this flat crown. This watch uses the caliber 497 12''', an automatic movement with a limited oscillating weight (movement no. 999,231, case no. 616,322).

Illustration 189
The stainless-steel version of the Futurmatic is also very much sought after (movement no. 1,216,282, case no. 649,501).

The JLC Powermatic

A discerning watch enthusiast wasn't sold on his yellow-gold Powermatic. He had purchased it for about $1,310 on the internet and entrusted the well-preserved watch to his personal watchmaker. By calling in a talented polisher, who brought the case to new condition, the overhaul costs were kept within limits at a total of $380.

Illustration 190
Jaeger-LeCoultre launched its first self-winding watch in 1948, and the model shown here, called the Powermatic, dates from that year. The power reserve can be read below twelve o'clock, which in this picture is exactly ten hours.

Illustration 191
At the top left, the screw-down case in 750 rose gold; in the foreground, the automatic caliber 481. Movement of the rotation-limited oscillating weight (rotor) is regulated by two hairsprings. Also interesting is the screw balance wheel, as well as the early Jaeger-LeCoultre proprietary shock protection system (movement no. 834,183, case no. 588,168).

The Jaeger-LeCoultre Master Mariner and Other Automatics

The Master Mariner series of wristwatches was very popular in the USA and was offered in different quality levels. The models sold in the USA usually had 14-karat gold, gold-filled, or steel cases and also had just a press-fit case back. Therefore these watches were only splash-proof.

Illustration 192
The screw-down case back of this Master Mariner bears the number 780,434. Only the watches produced at Le Sentier are consistently marked with case numbers, which are always stamped on the outside. The automatic caliber 880 (movement no. 1,417,125) can also be seen.

Illustration 193
The automatic models of the Master Mariner line were among the most successful that Jaeger-LeCoultre was able to place on the US market, which was so important to this brand. These timepieces were offered gold filled and in 585 and 750 gold. The watches marketed in the USA had only cases with press-fit backs, but those made in Le Sentier had screw-down case backs. In all cases the LeCoultre name was printed on the dial. The specimen shown is a reference E 385 and was made in Switzerland for the US market in 1959. The dial with Breguet numerals looks like new and may be considered a distinct rarity.

We find a very special model pictured from this successful series. The combination of a silvered dial with Breguet numerals, a screw-down steel case, and a caliber 880 movement gives this model a unique exclusivity. An auctioneer I consulted, who, like the author, saw this model for the first time, immediately estimated its value at no less than $3,800!

The ordinary automatics launched by Jaeger-LeCoultre in the 1960s are also still very popular among watch enthusiasts. On the opposite page we present three such wristwatches. In addition to the contemporary automatic Ref. E 393/1 (illus. 196), on the left there is also a model from the Club series Ref. E 392 (illus. 194), both with screw-down cases in 750 yellow gold. The proven caliber 883 and 880 automatic movements provide the drive. We also see another version of the Master Mariner. This model belongs to the reference E 557 and is driven by the caliber 881 movement. The presence of a steel bracelet from the current Masters series underlines the successful design, which was created at JLC more than forty years ago.

Illustration 194
An automatic line manufactured by Jaeger-LeCoultre was given the name Club and was produced in various versions. The first models with screw-down case backs and manufacture movements are particularly interesting. Later models, especially those with day of the week and date displays, received calibers from ETA and cases with press-fit backs. The Club reference E 392 shown was produced in 1962 (caliber 880, movement no. 1,511, case no. 885,230).

Illustration 195
Jaeger-LeCoultre's program also included models of the Master Mariner line, which were intended not only for the US market. The version seen here, reference E 557, bore the company name Jaeger-LeCoultre on the dial. To show the readers how these timelessly beautiful watches, built about fifty years ago, are still in step with today's design ideas, this timepiece was fitted with a steel bracelet from the current Master line (caliber 881, movement no. 1,781,565, case no. 1,006,278).

Illustration 196
Typical for the automatics of the 1960s, this reference E 393/1 presents itself with date display and screw-down case back in 750 yellow gold. Produced in 1967, this timepiece is equipped with the robust caliber 883 movement and was sold to a buyer in Israel (movement no. 1,877,868, case no. 1,178,936).

Illustration 197
The caliber 880's fine-adjustment mechanism, with the balance cock from the JLC Club model shown above left in about tenfold magnification. The regulator arm is quite plain, and the shock protection system is from KIF.

Illustration 198
Among the most popular Jaeger-LeCoultre watches are those that, for decades, have been sold under the name Memovox. The first model of the Memovox line was launched in 1953 and was still powered by the manually wound caliber 489 12'''. Time indication and alarm mechanism are controlled by two separately operated mechanisms, which also have their own barrels. The crown at four o'clock sets the time and winds the movement; the crown at two o'clock selects the alarm time and cocks the alarm spring. But this "little complication" is useful not only for getting up in the morning. Whether parking time, fast train departure time, or scientific observations, the Memovox was always there. From an automatic model reference 875.42 from the production year 1974, we see a section of the dial. The rotating inner circle, which is used to set the desired time, is reminiscent of a Milky Way design (movement no. 2,149,377, case no. 1,438,935).

12

The JLC Memovox

The Memovox wristwatch alarm clock was launched with the advertising slogan “A Second Memory.” Another advertising slogan was “Reminds–Warns–Awakens.” The first version still had a manually wound caliber 12'''-814 movement and was launched in 1953. Further models followed with automatic movements and were equipped with the calibers 815, 825, and 916. Although the overall design of this series of models would be hard pressed to win a beauty contest, they are very popular among collectors. Three different models are gathered here to demonstrate the different styles of that time.

Illustration 199
The diversity of the Memovox line is illustrated by this group photo. On the left is an early manually wound model with no reference designation with the caliber 814 12''', which was intended only for the US market and was also encased in the USA. The case is only gold plated and closed with a steel back and ring nut (produced 1961, sold 1962, movement no. 1,567,876, no case number). Memovox automatic models were popular and very common all over the world. In the middle is a stainless-steel version Ref. 875.42 with original stainless-steel bracelet that appeared on the two previous pages. The steel model on the right belongs to the reference E 875, is powered by the caliber 916 movement, and dates to the production year 1971. The dial, which is uniformly blue, has an inner disc with luminous indices. Hence the T SWISS MADE T label at six o’clock stands for tritium (movement no. 2,109,706, case no. 1,311,434).

The JLC Geomatic

In addition to the previously described chronometer model the Géophysic, the Chronomètre Géomatic series, launched in 1962, is also one of the most sought-after watches from Jaeger-LeCoultre. They appeared in the new DE LUXE series and bear witness to the highest Swiss watchmaking art. While the two round versions, reference E 398 (gold case) and E 399 (steel case), have been offered here or there at watch fairs, the model E 560, with rectangular case in tonneau shape, has not been offered as far as the author knows. Thus, we are almost dealing with a premiere here. First, we want to introduce the individual models in more detail. There are two references of the model with round case. Cases in 750 yellow and red gold were used for the reference E 398, and only the automatic caliber 881 G movement was installed in it. The two-piece case measures 1.41 inches (36 mm) in diameter and has a screw-down case back with the

Illustration 200
The chronometer model Geomatic timepieces were among the finest watches produced by Jaeger-LeCoultre from 1962 to 1973. They were produced only in small numbers and today are among the most sought-after collector's watches. This Geomatic parade features all three models. From the De Luxe line, in the center we see the Geomatic reference E 399/1 in stainless steel with caliber 883 S movement. Next to it is the gold version reference E 398 with caliber 881 G movement, and third from the left the reference E 560 with caliber 883 S, also in gold.

Illustration 201
Here one can see the backs of the three chronometers with the GEOMATIC logo. The two round cases have screw-down case backs but different case designs. The gold case has a 1.42-inch (36 mm) outer diameter and soldered lugs. For the steel case, the case middle was made in one piece, and the diameter was set at 1.44 inches (36.5 mm). The cambered rectangular case (tonneau shape) of the reference E 560 has a two-piece design and is splash-proof only. The elaborately designed base is fixed via four set screws positioned at a 45° angle.

Illustration 202
Here are the movements of the two round Geomatics. The one in gold has the caliber 881 G movement. The G indicates the increase in the number of jewels from seventeen to twenty-three, stop-seconds, and the Spirotor 3-F regulator for especially fine adjustment. The steel case next to it contains the caliber 883 S movement. S stands for seconds stop; the regulator is of the Spirotor 1 type and has no complex fine adjustment. According to the movement list, the caliber 883 S has only seventeen jewels, although the rotor is marked with "23 Jewels." Research carried out in the company headquarters in Le Sentier revealed that there were two versions of this caliber 883 S. Those self-winding movements that are incorporated in the Geomatic model were specially modified. Of the additional six jewels, two were used for the minute wheel and four for the mechanical winding mechanism (for pawl and transmission wheels). Thus, the caliber 883 S largely corresponded to the caliber 881 G, only the special Spirotor 3 fine adjustment was omitted.

Géomatic logo. The lugs are narrow and soldered. The first series was produced from 1962 to 1968, the second with the reference E 398/1 from 1969 to 1971. In the latter, however, the caliber 883 S movement was used. The stainless-steel version bears the reference number E 399 and has a two-piece case with screw-down case back. On it is the Géomatic logo made of 585 yellow gold. The case middle with lugs is one piece stamped from a piece of flat steel (semifinished). The case diameter is 1.44 inches (36.5 mm). This version was produced from 1962 to 1968, the successor model E 399/1 from 1969 to 1971. The former had the caliber 881 G movement, the latter the caliber 883 S. For the watchmaker, it should be noted that the gold cases can be opened with key number 51, the steel cases with number 52. The chronometers of the reference E 560 with yellow-gold tonneau case are considered to be very rare. In the years 1962 to 1971, production was limited to just 165 pieces. The author got to see this beautiful piece only once. Even the watchmakers in Munich and Le Sentier had never seen this model. Unlike the waterproof cases of the round watches, this tonneau case is only splash-proof. On this model the case back with Géomatic logo is held in place by four set screws positioned at a 45° angle. The case back not only features rectangular rounded shapes but also has multiple arches together with the adapted case. This design places high demands on parts machining and, due to the partially

Chronomètre Géomatic, Numbers and Dates

	Reference	Production Period	Material	Number Made	Calibers
1	E 560 Géomatic	1962–1971	750 yellow gold	165	881 G, 883 S
2	E 398 Géomatic	1962–1968	750 yellow gold, 750 red gold	1,100/773	881 G
3	E 398/1 Géomatic	1969–1971	750 GG	478	883 S
4	E 399 Géomatic	1962–1968	steel	1,700	881 G
5	E 399/1 Géomatic	1969–1971	steel	300	883 S

Illustration 203
Jaeger-LeCoultre Master Mariner Chronometer Ref. 24,000 in stainless steel. Printed on the dial at nine o'clock is the emblem of Saudi Arabia. Exactly 520 pieces of this model were made. This chronometer is from the production year 1975 (movement no. 2,210,461, case no. 1,399,237) and received an extensive overhaul at the German branch of Jaeger-LeCoultre. The retail price of this watch in 1975 was exactly 750 CHF.

Illustration 204
Jaeger-LeCoultre chronometer with leather strap, but without stamped reference and case numbers. In this watch is the caliber 906, movement number 2,208,827. This version has a date and day of the week display. The design of this watch would not necessarily meet today's taste. The counterpart from Vacheron Constantin can be found in illustration 460 on page 284.

spherical surfaces, can be manufactured only by using a three-dimensional jig-boring machine (now a machining center). The author has only very rarely encountered such a complex and cost-intensive design. The caliber 881 G and 883 S movements were used in the reference E 560. The five Géomatic references are summarized with all relevant data in the accompanying table.

Now some information about the origin of these three valuable pieces. The golden E 398 originates from Israel and was purchased at the Munich Watch Fair for $1,800. The steel E 399 was offered on the internet by a watch enthusiast from New Zealand and sold for $1,505. Finally, the golden E 560 comes from Australia and changed hands for $1,302. Even if the latter two prices sound quite reasonable, the overhaul costs still have to be considered. At Jaeger-LeCoultre, the overhaul of such old chronometers should cost between $1,600 and 2,200.

3.6. Jaeger-LeCoultre Watches from the 1970s and '80s

Jaeger-LeCoultre Chronometer Reference 2400 Master Mariner in Stainless Steel

The main dimensions of the then-new caliber 900 (without seconds), 901 (with indirect center seconds), and 902/903 (without calendar) were identical. The case diameter is 1.02 inches (26.00 mm), and the height is 0.127 inches (3.25 mm). From 1973 to 1975, about seven thousand examples of the caliber 906 chronometer were produced, of which 3,060 pieces were installed in the references 24,000, 24,001, and 24,002.

Exactly 520 examples of the Ref. 24,000 shown were made. This model would probably not have received a beauty award. The stainless-steel version has a case diameter of 1.41 inches (36 mm) and a screw-down case back with fine thread and six

Illustration 205
After removing the screw-down case back, we see the ultraflat and frugally constructed chronometer caliber 906. Pulling the winding crown activates the remarkably executed stop-seconds mechanism. The bilious-green plastic injection-molded part, which is pivoted by the cylinder head slotted screw, is turned counterclockwise by a lever mechanism when the crown is pulled, and the end of this injection-molded part presses onto the outer ring of the balance wheel, stopping it. On the balance wheel itself, one can still see the Triovis regulator. The balance staff shock protection system comes from KIF. This movement was also delivered to Vacheron & Constantin, where it was used in the Chronometer Royal model as the caliber 1096. The injection-molded part shown was replaced at Vacheron & Constantin by a rhodium-plated brass part with a steel spring (see p. 284, fig. 461).

grooves. The integrated stainless-steel bracelet consists of rather large, slightly curved links. The watch, which comes with original papers and box, was purchased at a Munich watch fair for $2,300 for this book, especially because of its remarkable stop-second mechanism. Here, a bilious green injection-molded part is used to operate the stop-seconds mechanism on the balance.

At a watch auction held in Hong Kong on November 28, 2017, such a model was sold for the equivalent of $2,098.

Jaeger-LeCoultre Automatic Ref. 5000

As beautiful as this watch looks, what we have here is a "value-for-money" watch. The case material is indeed 18K yellow gold, but only deep-drawn sheet metal and not, as usual, a solid gold block from which the watchcase is created by turning and milling (illus. 260).

When this model was planned and launched, Jaeger-LeCoultre was feeling the adverse effects of the quartz crisis of the time and had been forced to economize and adapt to simple manufacturing processes.

This watch is powered by the ultraslim caliber 900 without seconds. Despite the simple manufacturing process, the design of this model can already be considered appealing, not only because of the tonneau-shaped case, but also because of the dial and the crown successfully integrated into the case.

The case back is also deep-drawn. The stability of the watch is ensured by a steel inner plate, to which the movement and the two case parts are screwed. The watch, which is very rare to find, was purchased at a German auction house for $2,700.

Illustration 206
The reference 5000 certainly makes an impression at first glance, although it is actually a "value for money" watch. The case is made of deep-drawn gold plate, as is the case back. Both parts are screwed together with the clearly visible inner steel plate. In this photo you can also see all relevant components of the caliber 900's oscillating system. Fine adjustment of the rate is accomplished by the Triovis regulator on the balance cock, and the shock protection is from KIF. On the outside, the end of the flat hairspring is screwed to the same component. The escape wheel shaft is mounted in cover plates (*top right of the balance*).

Illustration 207
Detail of the then-new JLC caliber 900/901 movement from the automatic reference 5000 shown on the previous page. The caliber version 901 shown has no center seconds. The fine-adjustment mechanism with balance wheel, flat hairspring, and fine adjustment is clearly visible. The balance staff shock protection system is made by KIF (Elastor BS). Diagonally to the upper right, one sees the small red cover plate of the escape wheel shaft. With an outer diameter of 1.02 inches (26.00 mm, or 11½'''), movement height of 0.127 inches (3.25 mm), and 28,800 A/h, this movement represents a modern design. It has twenty-eight jewels, and the lever angle is 53°.

Jaeger-LeCoultre Automatic Ref. 5003 Vogue

Our automatic presented on this page is from the Vogue series, and with this dial variant is a reference 5003.21. It originates from the USA and was sold by a dealer in Houston, Texas, for $739. Previously they had offered this beautiful piece for weeks at inflated prices on the internet but without success.

Illustration 208
In the fight against the invasion of quartz watches, Jaeger-LeCoultre focused on ultraflat wristwatches. This gold men's automatic Ref. 5003 is from the Vogue line, launched in 1976, and was produced in 1978. Due to the low height, there was no space left for the seconds display. This watch came from Houston, Texas.

Illustration 209
The new generation of automatic movements was heralded at Jaeger-LeCoultre by the caliber 900/901, introduced in 1976. With an outer diameter of 1.02 inches (26.00 mm, or 11½'''), a height of 0.127 inches (3.25 mm), 28,800 vibrations per hour, and date quickset, it met the demands of the time. The construction and design of this movement set standards that can also be found in later Jaeger-LeCoultre automatic movements (movement no. 2,237,910, no case number).

Jaeger-LeCoultre Master Control 1000 h in 18-Karat White Gold

The Master Control 1000 h series, launched in 1992, brought long-awaited success to the factory in Le Sentier. The version presented is a special model for men and women in 750K white gold. Rose-gold indices, numerals, and hands give this watch a very special look. According to Jaeger-LeCoultre, only forty-five examples of this rare model 145.3.89 were made in Le Sentier. With a case diameter of 1.36 inches (34.5 mm) and its special charisma, this watch also fits any lady's wrist (illus. 210).

The JLC Master Control Perpetual in Stainless Steel, Ref. 140.8.80 S.

The crowning glory of the Master Control 1000 h series introduced in 1992 was undoubtedly the various versions with perpetual calendar. The presented watch has a stainless-steel case and the reference number 140.8.80 S engraved on the back. With the steel case and its 1.46-inch (37 mm) case diameter, this watch looks quite masculine. Other variants are available with yellow-, rose-, or white-gold cases.

Illustration 210
The white-gold Master Control 1000 h with its 750K rose-gold numerals and hands is a special watch that one encounters extremely rarely, and of which only forty-five pieces were produced. This outstanding timepiece was professionally overhauled at the main factory in Le Sentier in the Swiss Jura. The main components that were replaced during this overhaul are included with the watch. Here the commissioned watchmaker made a small mistake. When he returned the timepiece, he accidentally enclosed the hands from another watch. Behind the watch you can still see the completely relaxed winder spring.

Illustration 211
Jaeger-LeCoultre Master Control with Perpetual Calendar
Though it has only a stainless-steel case, this successful watch makes a very special impression. The division with date circle on the left, weekdays on the right, month circle at the top, and moon phase on the bottom gives the dial a balanced design. The year can be found in the month circle and is limited to just two digits. The breakouts on the two hands near the hand shafts ensure that the red warning phase can also be seen in the hand positions 11:05 and 12:55, where the calendar may not be adjusted.

Various examples of this model have been offered on the internet. On eBay, an advertiser from the USA was asking $19,378 for a Master Control Perpetual in rose gold. The asking price for a mint steel version from Turkey was $13,888. To these prices would still have to be added import sales tax, customs duty, and shipping fees. As a result, an approximate markup of 24%–27% on the respective purchase price should be expected.

Illustration 212
Thanks to the glass back, one can also see the automatic movement caliber 889-440/2 with fifty jewels. The base model, caliber 889/2, contains just thirty-six jewels. On the fine-adjustment mechanism, one can see the KIF-Ultraflex shock protection system and the Spirofin fine-adjustment device. You can also see the red cover plate for the escape wheel shaft bearing on the upper right side of the balance. This watch remained unopened.

Illustration 213
Despite its somewhat playful and in places baroque appearance, this Geographic can impress. For a CEO, a Rolex GMT-Master might be a better fit for business meetings. On weekends, however, this watch should be worn. Unlike the new Master Geographic model and other Traveller watches, this model still has a twenty-four-hour dial to the left of the second time, which should make travel planning even easier with the day and night distinction. This watch has not been opened.

Jaeger-LeCoultre Geographic Ref. 169.1.92 in 750 GG
Jaeger-LeCoultre Geographic Ref. 169.1.92 in 750 Yellow Gold

Not only the world traveler, but also the deadline-driven manager would enjoy this practical but also tasteful watch. In contrast to the new Master Geographic model, this watch looks a bit more baroque. It belongs to the Odysseus line with Ref. 169.1.92 and has a 750 yellow-gold case. The practical thing about this watch is the time display for the particular place where one is or wants to travel to. Above the 12, one can call up the required cities in a city window by turning the crown and see the time there on the small dial above the 6. Instead of a day/night indication, as the successor model has, this watch still has a twenty-four-hour dial, so that reliable scheduling is always possible. These watches are more commonly found on the secondhand market. In 2021 a yellow-gold model from Germany was being offered on eBay for $8,300, and a rose-gold one from the USA for $8,810.

Illustration 214
The Jaeger-LeCoultre Master Geographic in 750 rose gold illustrates a successful symbiosis between a professional look and a watchmaking work of art. Operation is also quite simple. The right crown is used to wind the movement in the zero position. After first pulling out the crown, one can adjust the hour hand on the subdial. The second position of this crown is used to set the exact time, while the time on the subdial is synchronized. The crown at ten o'clock is used to display the time in a destination city. Nothing can stand in the way of the next long-distance trip!

Jaeger-LeCoultre Master Geographic Roségold Ref. 142.2.92

With the introduction of the new Master Control product line, the previous Geographic model was updated and adapted to the current design standards. The reader can decide for themselves which of the two watch models suits them better. Of course the inner workings of the Master Control have also been redesigned. The watch shown has a caliber 929/3 movement with a speed of 28,800 A/h and thirty-eight jewels. As on the previous model, the destination city can be easily set, and the local time there can be read on the subdial. The professional design of this watch means that the Master Geographic is not as baroque as its predecessor.

Instead of a fixed screw-down case back like that of the stainless-steel version, this model has an elegant hinged case back that reveals the caliber 929/3 movement. A wristwatch with a lot of practical use, but also with style!

3.7. Some New Models from Jaeger-LeCoultre

Illustration 215 *(from left to right)*
It is rather rare to be able to look at the movement of a Jaeger-LeCoultre Reverso. However, the Septantieme (Ref. 300 24 20), limited to five hundred pieces, allows a view of the red-gold-plated, symmetrically designed, manually wound caliber 879 with swan-neck fine adjustment through the sapphire crystal on the back. But the fine guilloché dial with power reserve, day/night indicator, large date, and small seconds also has a lot to offer. The Reverso Grande GMT (Ref. 302 84 20), which is equipped in the same way on the front except for the power reserve, hides a second dial on its back, on which the second time zone with hours and minutes can be read. The pushers on the case allow the time to be adjusted in hourly increments. This makes it easy to correct the time when crossing time zones. A twenty-four-hour display tells whether it is day or night. GMT can be read on another small subdial. The time remaining of the eight-day caliber is indicated by the power reserve indicator. The red-gold Master Calendar (Ref. 151 24 2A) also provides information about power reserve and date. Instead of including a large date in the window, Jaeger-LeCoultre here reverted to the hand date from the center. In addition, the day of the week and month in the window, moon phase, power reserve, and small seconds are displayed in perfect symmetry. The full calendar is powered by the new automatic caliber 924, which can be seen through the sapphire crystal case back. Like all new Master models, the Calendar looks more modern with its larger case, wider bezel, and new delta hands.

Illustration 216
With the Master Compressor Extreme World Chronograph (Ref. 176 84 70), Jaeger-LeCoultre introduced a truly robust revolutionary. Its 1.81-inch (46 mm) two-piece case in stainless steel and titanium (limited model in platinum and titanium) is designed to eliminate shocks and vibrations via an air suspension system. In addition, the crown-locking system ensures ease of use and enhanced water resistance. A quarter turn of the crown key makes the crown and pushers watertight and ensures that they cannot be accidentally operated underwater, allowing water to enter the case. The new strap-changing system also allows the watch to be changed to suit the occasion: so one mounts the fine crocodile leather strap for elegant evening events, or the rubber strap, which is also included, when it comes to sports. On each side of the base, a pusher ensures that the strap parts can simply be released and the new ones snapped in from below. This works very quickly, and yet the band/strap is held very securely. Last but not least, the watch is the first automatic chronograph of modern times. The automatic caliber 752, developed for this purpose, features the elegant ratchet wheel control and allows the stop-seconds hand to start smoothly thanks to a vertical clutch. Under the chronograph counters, instead of small seconds, there is a red-and-white disc as a function indicator, and the world time scale on which the times in all time zones can be read.

Illustration 217
Jérôme Lambert (*center*), then still CEO of Jaeger-LeCoultre in Le Sentier, in conversation with the author. On the left is PR employee Bénédicte Ferlet. Today, Jérôme Lambert is CEO of the Richemont luxury group, which belongs to the South African—and great wine lover—Johann Rupert. In the past financial year (2020) this group generated around fourteen billion in sales.

A. LANGE & SÖHNE
GLASHÜTTE I/SA

4. A. Lange & Söhne

by Peter Braun

4.1. Introduction

The history of the A. Lange & Söhne watch brand is divided into a historical and a modern era, because political circumstances actually forced the Lange family to found their own company twice: on December 7, 1845, and December 7, 1990.

More than 175 years ago, Ferdinand Adolph Lange gave up a secure livelihood as a royal court clockmaker in Dresden and, together with fifteen apprentices, moved to Glashütte, in the economically underdeveloped area on the edge of the Ore Mountains, to establish a watch production facility based on their collaborative effort. Apprentices, who were released from all financial obligations during their apprenticeship period, had to agree to work exclusively for Lange for five years after the end of their apprenticeship, for a relatively modest wage. Thus was created in Glashütte a cottage industry with highly qualified suppliers—the key to the success of the precise timepieces from Germany that at the turn of the last century set a worldwide standard of high quality for pocket watches.

Ferdinand Adolph Lange's sons and grandsons continued to run the company in the spirit of its founder, navigating through the difficult times of World War I and the economic crisis of the 1920s. As a manufacturer of marine chronometers and observation watches for the armaments industry, Lange was dismantled after World War II, and after the founding of the German Democratic Republic, Glashütte disappeared behind the Iron Curtain. The company was expropriated on April 20, 1948, and this ended the independence of A. Lange & Söhne, and with it also the high quality and variety of the watches it produced, which continued to be sold until 1951 under the brand name "Lange VEB." The "modern age" chapter in the history of A. Lange & Söhne began a little more than thirty years ago. Reunification brought about the reestablishment of Lange Uhren GmbH as a sister company of IWC and Jaeger-LeCoultre under the umbrella of the Mannesmann/VDO Group and gave wings to Walter Lange's wildest dreams. The great-grandson of the company's founder was able to assume the family heritage and continue the legacy of the watchmaking dynasty. The German Treuhandanstalt (the agency that privatized East German enterprises) supported the project, Mannesmann/VDO financed the development of a watch collection and the procurement of production equipment, and the sister brands helped out with the production of parts and brought the newly recruited employees from Glashütte and the surrounding area up to date with the latest state of Western watchmaking technology.

Günter Blümlein was responsible for coordinating this unprecedented start-up of the manufacture. He had very clear and detailed ideas about the new Lange wristwatches, their technology, and their quality. He faced a Herculean task: it was about nothing less than the definition—yes, the reinvention—of the German luxury watch. While a workforce was being assembled in Glashütte and young watchmakers were sent to Schaffhausen (IWC) and Le Sentier (Jaeger) for training, Günter Blümlein met with Walter Lange, designers Hartmut Geyer and Annegret Fleischer, and engineer and designer Reinhard Meis to discuss the technical and design cornerstones of the new collection. They could have made things easy for themselves, and, in the general reunification euphoria, they probably would have gotten by with slightly modified Jaeger movements and nostalgic gold cases.

Luckily, Günter Blümlein was of a different opinion: every new model deserved a new movement of its own and an unusual technical feature. He loathed the term "nostalgia," and in his eyes a classically designed detail was valid only if it served a purpose or contained a technical necessity. The design and character of the watches had to stay as far away as possible from the Swiss luxury watch mainstream. To this end he allowed the wristwatches congenially realized by Reinhard Meis to have a little more volume and—above all—a substantial weight. His example of the reassuring massiveness and rich closing sound of a Mercedes door quickly made the rounds and is still often

Illustration 218
Wilhelm Schmid, managing director / CEO of the A. Lange & Söhne Watch Factory, Glashütte, Saxony. He is seen here in his British AC Ace-Bristol two-seat sports car, built in 1957. In the background is the main building of A. Lange & Söhne. Schmid brought industrial management, dynamism, and special innovative strength back to Lange.

cited today when it comes to the difference between Swiss *haute horlogerie* and fine Saxon watchmaking.

And so, on October 24, 1994, the first German luxury watch brand, A. Lange & Söhne, was launched with four solid-gold watches, four newly designed movements, four extraordinary technical details, and four fresh watch faces.

In the less than four years of preparation time between the renewed founding of the company (on December 7, 1990, exactly 145 years to the day after Ferdinand Adolph Lange opened his workshop in Glashütte) and the unveiling of its first watch collection, the successors of the founding father had succeeded in defining a complete catalog of technical and design criteria. Not only is this strictly observed within the Lange watch collections to this day, but it enjoys almost universal recognition as a standard directory of "typical" Glashütte design features.

Yet, before the first new watches from A. Lange & Söhne reached the market, there were neither three-quarter plates, nor screwed gold *chatons*, nor hand-engraved balance cocks in the wristwatch world. Natural nickel silver was completely outdated as a material for the frame parts (today usually made of brass), and gold anchors, polished screwheads, and the sunburst finish on the ratchet and crown wheel were at best still remembered by pocket watch collectors. In the new wristwatch models from A. Lange & Söhne, the decorations and machining forms that had once been highly regarded celebrated a joyous reign and formulated the standard for all future watch collections of all future Glashütte watch brands.

In the beginning, the Saxon ideas were backed up by a great deal of technical production know-how from Switzerland, because the construction of the parts production facility, with its capricious tool machines and the special requirement of the building fabric, dragged on for several years. The group's sister companies IWC and Jaeger-LeCoultre supplied a large portion of the movement parts, which were finished and assembled in Glashütte. The support of Swiss specialty atelier Renaud & Papi was called upon for the Tourbillon Pour le Mérite. But with each generation of models, the young manufacture emancipated itself a little further and secured the know-how for ever-more-complicated designs.

In 1997 the company introduced its first automatic movement, the Langematik, which, true to the guiding principle of product development, featured a "three-quarter rotor" made of embossed gold as an unusual technical detail. The zero-reset system, which automatically resets the second hand when the crown is pulled, became one of the most popular functional details—along with the patented large date, of course, which fueled the general discussion in the mid-1990s about the readability of watch dials and started an irreversible trend toward large date displays.

The next quantum leap in technical development was achieved by the design department in 1999 with the Datograph, which made history as the first completely new mechanical chronograph in the upper class for two generations of watchmakers. Sophisticated isolator mechanisms, precisely jumping totalizers, and drag hand and flyback functions put pressure on the Swiss noble manufactures, who were still operating with ancient Lémania designs in the dawning twenty-first century.

The Saxons scored another coup with the unveiling of a *tourbillon* that could be stopped with a delicate brake lever for precise time setting. This lever does not touch the cage of the bogie because the escapement continues to oscillate for a few beats before it comes to a stop and—most importantly—does not restart immediately when the brake is released. Later the device presented in 2008 in the Cabaret Tourbillon was augmented by a zero-reset function, which resets the small seconds hand in the center of the 1815 Tourbillon as soon as the balance wheel is stopped by pulling the crown.

With the Lange 31, a manually wound watch with a power reserve of a whole month, A. Lange & Söhne made intermediate winding (in French: *remontoir*) salon-worthy again in 2007 and addressed the influence of the power reserve on the isochronism of the mechanical escapement. The Lange Zeitwerk, which was presented in 2009, also has a similar *remontoir*, using the intermediate winding as a power accumulator for the digital jumping minute and hour displays.

The designers at A. Lange & Söhne discovered the moon phase display early on as an interesting field of activity, one that has been repeatedly reinterpreted over the past eighteen years. The improvement of the display accuracy to one day of error in 122 years is as much to their credit as the charming variation of an "upside-down" rotating moon disk to illustrate the moon phases in the Southern Hemisphere. After the transplantation of the moon phase display to the center of the main dial of the Large Lange 1 Moon Phase in December 2013, the presentation of the Richard Lange Perpetual Calendar Terraluna soon followed in January 2014, where the exceptionally descriptive and exact display of the lunar constellation had to be relocated to the back of the movement.

4.2. The Movements

One watch, one movement: this approach sets A. Lange & Söhne apart from all its competitors. Indeed, to this day not a single movement is used in two different models, and the number of manufacture calibers developed since 1990 has now risen to sixty-seven—the equivalent of two new movements per year.

One of the reasons for the success of A. Lange & Söhne watches seems to be that Glashütte had the courage and the patience to build watches with components that had been forgotten elsewhere. No one uses them today because no watch requires them anymore—like gold *chatons* or a swan-neck fine adjustment.

Günter Blümlein, then head of the LMH Group (Les Manufactures Horlogères), which included IWC and Jaeger-LeCoultre in addition to Lange Uhren GmbH, once commented on this in an interview: "Before the war, Lange watches were always handmade, very sophisticated, refined, and individual pieces. We continue this tradition; we make very small series, individual watches. Each movement is assembled by hand and has components with freehand engravings, so it is unique. We have some very anachronistic parts in our movements, such as the three-quarter plate. There used to be a rationale for that. Today, movements are so sophisticated that you don't need such elements. But you love them when you see them." In addition to the unusual manually wound movement of the Lange 1 with double barrel and decentralized displays, the premier collection also featured a more conventionally designed manually wound movement with centrally oriented hands in the Saxonia model, as well as an extremely tightly cut form movement for the Arkade ladies' watch, in which the two-digit large date was particularly prominent. The Tourbillon Pour le Mérite danced a little out of line even then, but it established the tradition of complicated watches from the Saxon manufacture, which were to feature a wide variety of functions and displays over the following decades. Quite a few of them were later taken up and used by other watch manufacturers—often sister brands from the Richemont Group.

Among the technical features and complications with which the watches from A. Lange & Söhne have enriched the watch world are, in addition to the two-digit large date, the astronomical moon phase with a display accuracy of 122 years (Lange 1 Moon Phase, 2002), the display of the moon phase above the Southern Hemisphere (Lange 1 Luna Mundi, 2003), the zero-reset mechanism for automatic zeroing of the second hand when the winding crown is pulled (Langematik, 1997), the precisely jumping chronograph minute counter (Datograph, 1999), the drag hand mechanism for the minute counter (Double Split, 2004) and, for the hour hand (Triple Split, 2018), a manually wound movement with a thirty-one-day power reserve and *remontoir* (Lange 31, 2007), and a digitally indicating mechanical watch with jumping numerals (Zeitwerk, 2009), with quarter and hour strikes (Zeitwerk Striking Time, 2011), or decimal—like the time display—striking minute repeaters (Zeitwerk Decimal Strike, 2015).

On the following pages, six wristwatches from the A. Lange collection are presented as examples, which of course only highlight the complex and extensive oeuvre of the heirs of Ferdinand Adolph Lange.

4.3. Lange 1 W (1994, 2015)

Since its launch in 1994, the Lange 1 has been the flagship of the A. Lange & Söhne manufacture and is regarded as a symbol of the Saxon art of watchmaking.

It is fair to say that a piece of watch history was written with the Lange 1. The special features of this watch are not only its eccentric appearance and, in part, its groundbreaking technology. It was also remarkable that when it was launched in 1994, no one expected the new watches from A. Lange & Söhne to follow on so seamlessly from the quality of A. Lange & Söhne's historic (pocket) watches. Quite as if the enforced pause behind the Iron Curtain had never existed. The pressure of expectation on the watch engineers in Glashütte was enormous, because Walter Lange's ancestors had set the bar at the highest level. This bar had to be cleared again. Anything lower would have been gleefully commented on by the watch industry and enthusiasts.

Unlike a conventional wristwatch, the Lange 1 has an off-center display for hours and minutes that also requires an off-center movement design. Even in the original caliber L901.0 from 1994, the minute wheel does not rotate in the center of the movement as usual but is placed off-center. Far away from it, the small second hand performs its duty at the edge of the dial. Just to make this possible, an additional wheel connection had to be created, since the use of the second wheel in the gear train, which usually bears the small hand, was not possible due to the unusual dial layout.

Just above the subdial for the seconds indication, there is the large hand for the power reserve indication, which covers the distance from UP to DOWN in about three days. Thus, it refers to an important design feature of the Lange 1: its long power reserve. This is achieved by the installation of two barrels whose springs are connected in series and thus more than triple the power autonomy compared to conventional manually wound movements. The large date window, the trademark of the new A. Lange & Söhne, is clearly visible above the power reserve hand and has since found numerous imitators. When looking through the sapphire crystal case back of the Lange 1, the three-quarter plate, a design feature of the early pocket watches from A. Lange & Söhne, catches the eye. With its Glashütte ribbon cut, the flat plate, made of natural German silver, is reminiscent of the carapace of a prehistoric animal protecting the sensitive parts inside. But not only the balance wheel pulsating at 21,600 vibrations and the escapement wheel turning under its own cock, but also the nine gold *chatons* that fix the jewels of the movement wheels suggest what exciting micromechanical processes are

Illustration 219
Lange 1 in 750 yellow gold; an extraordinary watch

probably taking place under the movement plate. Deep-blue screwheads are pleasing to the eye, and the polished cover plate on the upper escapement wheel pivot and the swan-neck fine adjustment on the hand-engraved balance cock also make the movement a feast for the eyes. With a diameter of 1.20 inches (30.4 mm) and a height of 0.232 inches (5.9 mm), the caliber L901.0 is solidly dimensioned.

Illustration 220
Lange 1 in 750 yellow gold with manually wound movement no. 119,533 and case no. 219,026. The case back is fixed with six cylinder head screws. Also worth mentioning is the Gyromax balance wheel with six mass balance weights, as well as an additional fine adjustment of the gear via the regulator arm, micrometer screw, and gooseneck spring. Here they just wanted to know for sure.

Twenty years after the unveiling of the Lange 1, the manually wound movement has been completely revised. The fiftieth newly developed caliber from the Lange watch manufactory bears the designation L121.1 and is a consistent technical evolution of the original caliber in many details. For example, the exact jumping display of the big date now switches on punctually at midnight. The power to advance is built up slowly over a period of several hours. In addition, the movement was given a gear section with—still quite small—an eccentric balance wheel and free-sprung hairspring from its own production. The hairspring still oscillates with a frequency of 21,600 A/h under a hand-engraved balance cock. The proven double barrel with its power reserve of seventy-two hours also remains.

4.4. Lange 1 Daymatic (2010)

The resemblance is truly striking. Only at a second glance does the viewer notice that the characteristic asymmetry of the Lange 1 dial is mirror-inverted and that the small sector hand indicates the day of the week, replacing the power reserve display. The biggest difference remains hidden beneath the dial: the Lange 1 Daymatic has automatic winding.

Since its premiere more than twenty-two years ago, the Lange 1 has been the unmistakable "face" of A. Lange & Söhne. With its off-center dial design and patented large date display, it has become a design icon with high recognition value. At the same time it is one of the most successful timepieces from the Saxon manufacture. Its success spurred the development of a complete Lange 1 family with a total of six models. Until 2010, these always had a manually wound movement with a double barrel and a three-day power reserve—self-winding was reserved for the more conventional, symmetrically designed Langematik and 1815 Automatic models.

The Daymatic is the first and only Lange 1 with automatic winding, and with the design trick described at the beginning with the mirror-inverted redesign of the displays—including the two-digit big date—the designers of the L021.1 caliber prove the much-cited Saxon sense of humor. For the majority of watch enthusiasts, who prefer to wear their timepiece on the left wrist, the arrangement of the hour dial on the right-hand side has the advantage that one's gaze first falls on the most-important information when pushing back the shirt cuff: the time display. In return, the large date has been moved from its usual place at the top right to the left side of the dial. It finds its useful complement in the retrograde weekday display below, which replaces the power reserve display that is dispensable on an automatic watch. The date and day of the week can be conveniently and independently corrected (i.e., advanced) via two pushers.

The mirroring does nothing to change the harmony that the dial of the Lange 1 radiates. On the Lange 1 Daymatic, the center of the large date and the hand axes of the day of the week and seconds displays also lie on a vertical line. This forms the base of an isosceles triangle, the apex of which meets the exact center of the main dial.

Available in yellow gold, red gold, or platinum, the case of the Lange 1 Daymatic is 0.04 inches (1 mm) larger in diameter than the manually wound version. Inside, a completely new development is at work. The automatic caliber with the number L021.1 has only the self-winding mechanism in common with the aforementioned Sax-O-Mat movement of the Langematik.

Illustration 221
Caliber L 0211 with the dial removed. The innovative design of the date display is interesting. The numbers that make up the second digit of the day of the month, from 0 to 9, are on a rotating disc. The numbers that make up the first digit, from 0 to 3, are on a rotating cross, which is slightly raised.

Illustration 222
Lange 1 Dynamic in 750 yellow gold with caliber L 0211, movement no. 88,924, and case no. 196,392, with gold rotor, Gyromax balance wheel, and many other fine details. For the engineer: the escape wheel shaft is guided by a cap jewel whose retaining plate is screwed from the inside (upper-right image). On the caliber 27-460 M from Patek Philippe, the retaining plate is screwed from the top for easy assembly. The reason for this difference is probably known only to the designer. This watch was purchased anonymously from a retailer. It shines with chronometer-capable rates.

Its efficiency is ensured by a heavy central rotor with a platinum oscillating weight embossed with the "A. Lange & Söhne" logo, which covers the movement's entire diameter of 1.24 inches (31.6 mm). When fully wound it has a power reserve of fifty hours.

4.5. Datograph (1999) / Double Split (2010)

Incredible as it may sound, A. Lange & Söhne's 1999 Datograph was the first new chronograph design in decades. The majority of Swiss watchmakers still relied on the Valjoux caliber, developed in the 1970s, which could be completely assembled by ETA, a subsidiary of the Swatch Group, or obtained as a so-called parts kit at reasonable prices. Apart from that, only the automatic caliber called El Primero by Zenith was produced in significant numbers, whereby the caliber variant used by Rolex in the Daytona, modified in some details, must also be counted. At the turn of the millennium, the noble Geneva manufacturers still used Lémania movements as the basis for their complicated chronographs with astronomical displays, the design of which dated back to the period shortly after the Second World War.

The Datograph, unveiled as a world first at the World Watch and Jewelry Show in Basel in 1999, set new standards for chronographs in terms of appearance and technology. As early as 1868, Ferdinand Adolph Lange manufactured the first pocket watch chronographs in what was then known as the column-wheel design. This term is derived from the French word *colonne* (column), which is why people also speak of

Illustration 223
Lange 1 Daymatic with the case back removed

Illustration 224
Automatic caliber L 0211 turned off, here dominated by the rotor made of 21K yellow gold

Illustration 225
After removal of the rotor, one can see the various bridges. The movement is regulated in five positions. Sixty-seven rubies are installed in it.

Illustration 226
The entire inner workings of the caliber L 0211 movement become visible after the gear train and mainspring barrel bridge are removed. The gray-colored mainspring barrel on the right also dominates.

column-wheel chronographs. Today the term "ratchet wheel" has become established. This wheel commands the chronograph clutch to alternately "engage" and "disengage" when the start pusher is pressed, thus directing the engagement between the transmission wheel on the clutch and the chronograph wheel (second-counting wheel).

In most chronographs the minute-counting wheel, which rotates under the thirty-minute totalizer on the dial, is advanced by one graduation by a driver on the second-counter wheel once it has rotated. In the Datograph caliber L951.1, this switching takes place abruptly, so that one can speak of a jumping minute counter. A stepped worm on the shaft of the second-counting wheel continuously raises the switching lever of the minute counter in the course of one revolution (sixty seconds) and then suddenly releases it, whereupon a spring advances the minutes counter by one graduation.

Another chronograph function, which was made presentable again by the Datograph 1999, is the so-called flyback mechanism. The running-time measurement of the L951.1 caliber can be reset to zero "from full run" at any time and automatically restarts after the reset pusher is released. *Retour en vol* is the name of the function in the language of French watchmakers, and their German professional colleagues turned it into *Rückstellung im Flug*. This allows several individual time measurements to be switched in succession without loss, because the cumbersome process of stopping, resetting, and restarting is shortened to a single keystroke.

The functionality of the drag hand chronograph goes a whole step further and is technically considered one of the top achievements in watchmaking, often mentioned in the same breath as the *tourbillon* or the minute repeater. It offers the user the ability to determine split times during a running-time measurement.

For this purpose, the chronograph, which is called "split seconds" in English, has a second second hand, which normally makes its rounds congruently with the main second hand. A third chronograph pusher stops the second second hand while the main hand continues to move. Once the split time has been noted, the button is pressed again and the stopped hand races after its colleague and catches up with it. This is why it is also called a "catch-up hand" or, in French, a *chronographe à rattrapante* (*rattraper* = to catch up).

In the tradition of the Lange company, such a timepiece was once called a "double chronograph." In these watches the chronograph second hands and the associated wheels are always double, with the shaft of the first wheel (chronograph wheel) rotating in the cannulated shaft of the minute wheel and that of the second wheel (*rattrapante* wheel) rotating in the shaft of the first, which is also a tube.

The Double Split, unveiled in 2010, was the first chronograph to offer the useful split-seconds function in two versions—for the second and minute counters. After several years of development work, the designers succeeded in equipping the thirty-minute counter with two hands that can be started, stopped, and

Illustration 227
A. Lange Datograph Flyback chronograph in 750 red gold with large date, seconds, and thirty-minute counters

Illustration 228
Close-up of the Datograph Flyback caliber L 951.1, movement no. 87,294, with forty-six rubies and regulated in five positions. This precision mechanical marvel is a testament to the highest watchmaking skills. At the bottom left is the switching wheel that executes all chronograph functions.

synchronized again independently of each other—but always in sync with the second hands. The double split now allows the usual measurement of thirty-minute intervals, but also the determination of split times up to a duration of twenty-nine minutes and fifty-nine seconds.

When the drag hand function is activated, the heart levers normally drag on the heart piece, leading to a loss of power in the movement and consequently to a reduction of the balance oscillation. Not so with the Double Split: both wheels of the second and minute counters are equipped with a device that separates the heart levers from the heart piece an instant after the drag hand stops. This "isolator circuit" allows the use of the sweep hand function at will without additional loss of power for the gear train, escapement, and balance wheel.

Of course, the caliber L001.1 also has a flyback circuit, due to its design relationship to the Datograph. With the Double Split this function can even be used with stopped drag hands, which retain their position.

The view through the sapphire crystal in the bottom of the Double Split falls on a fascinating spectacle performed by countless levers, springs, and cogs. Almost the entire *rattrapante* mechanism for second and minute counters had to be arranged on the actual chronograph *cadrate*. This resulted in a movement of amazingly airy design and the imposing height of 0.372 inches (9.45 mm).

Viewed from the front, the Double Split's relationship to the Datograph is obvious, although it lacks the distinctive large date display, which had to be dispensed with for technical reasons. The space that has been freed up is taken up by a tiny power reserve display on the two-piece dial, made of solid silver. The dial layout typical of Lange watches, with three displays forming the corners of an equilateral triangle, is thus retained in the Lange Double Split.

4.6. 1815 Tourbillon (2014)

The *tourbillon* (French for vortex) was invented by Abraham-Louis Breguet at the beginning of the nineteenth century. With his invention, Breguet tried to compensate for the negative influence of the earth's gravity on the rate of a watch caused by the unavoidable imbalance of the oscillating system (balance wheel with hairspring). Breguet's idea was ingenious in its simplicity, but also extremely complex in its technical implementation: one must allow this center of gravity to "wander" and thus become ineffective.

The escapement (escape wheel and anchor) and the oscillating system operate in the so-called cage (bogie) of a *tourbillon*, on whose shaft the second drive is located. Since the cage with the second drive rotates around its own axis once a minute, any unbalance that may be present also travels once in a circle within one minute and compensates itself virtually single-handedly.

Illustration 229
A. Lange drag hand chronograph with movement number 867,856. It has forty-six rubies and is regulated in five positions. It allows the determination of split times, which are especially in demand in racing. This caliber is also a masterpiece of watchmaking, which is otherwise known only from the house of Patek Philippe.

Once called *Drehgang* by German watchmakers, the mechanism has played an important role in the young history of A. Lange & Söhne's new wristwatches from the very beginning. Even in the launch collection of 1994, the Tourbillon Pour le Mérite was the showpiece of the collection, and over the years the delicate rotating-gear watch has evolved into a complicated drag-hand chronograph with *tourbillon*, the Tourbograph Pour le Mérite.

Both Pour le Mérite Tourbillons are extremely precious and technically very complex works of art—not only because of their rotating mechanisms, but also because of such extraordinary design finesses as a force regulation mechanism with chain and snail. This mechanism was familiar to watch connoisseurs in connection with high-quality precision pocket watches, but before A. Lange & Söhne, no one had ever reduced the chain and snail to wristwatch dimensions. The snail wheel, which houses a planetary gear consisting of thirty-eight components, has a diameter of just 0.39 inches (10 mm). The 0.59-inch-long (15 cm) steel chain has more than fifty links, whose pins are riveted together with a play that cannot be measured, only felt.

It was only in 2014 that A. Lange & Söhne presented a conventionally designed *tourbillon* built into the no-frills case of the 1815 model—named after the year of birth of the watch's founder, Ferdinand Adolph Lange. But even the supposedly "slimmed down" caliber variant L102.1 can come up with a few extraordinary features.

A movement with *tourbillon* is per se designed for the highest precision, rate stability, and accuracy. Unfortunately, most *tourbillon* movements can't be set precisely because the rotation rate can't be stopped precisely. If you only block the cage, the balance oscillates a few times before it completely swings out and stops. When it is released it takes a few seconds until it has resumed its constant oscillation frequency. In 2008, the designers at A. Lange & Söhne succeeded in braking the balance wheel inside the rotating *tourbillon* cage directly and without delay. Pulling the crown triggers a complex lever mechanism that lowers a movable V-shaped spring onto the balance ring, stopping the balance wheel instantly. Lange watchmakers even went one step further and combined this stop mechanism with the "zero reset" mechanism, which is also patented: when the crown is pulled out, the second hand not only is stopped but is instantly reset to the zero position. In this way, the minute hand is precisely positioned above a minute marker, and the watch can be restarted to the second.

4.7. Lange Zeitwerk (2009)

It is clearly recognizable as a "Lange" at first glance: the case shape, the characteristically typographed scales of the small seconds and power reserve display, and even the A. Lange & Söhne brand lettering leave no room for doubt. Even the large, somewhat squiggly Arabic numerals in the window bezel look familiar—from the large date that caused a sensation eighteen years ago as one of the most important style elements and technical features of the Lange 1. But this watch with the name Zeitwerk has no hands.

Hours and minutes are displayed in two widely spaced display windows framed with a cushion-shaped contour. The mask, which spreads out like two wings and also encloses the small seconds scale, shines in line-cut German silver, since it

Illustration 230
A. Lange Tourbillon 1815 with caliber L 102.1, movement no. 106,760, case no. 207,206, in 950 platinum. The engraved 000/100 tells us that just one hundred examples of this model were made.

is in fact a component of the movement: the right counterpart to the left slotted screwhead is a diamond cap jewel under which the journal of the disk axis rotates. Tens and ones of the minute indication are not separated by a center bridge but show the height offset.

Behind the supposedly simple digital time display is a highly complicated mechanism. A lot of force is needed to move the three display discs, which is fed into the movement by an unusually strong spring. A normal escapement would jam under such a load, because conventional means cannot reduce these brute forces far enough to trigger a filigree pallet.

Illustration 231
Detail enlargement of the balance cock. Here a diamond was used instead of shock protection. The complex mechanism of the *tourbillon* is also clearly visible. Only the designer and the author know the reason for the U-shaped recess on the balance cock.

The designers had already faced a similar problem with the Lange 31 movement, where the spring force had to be large enough to keep the movement running for thirty-one days. Similar to that movement, the digital watch uses a *remontoir*, a kind of intermediate power accumulator. This is arranged in the form of a spiral spring between the two tiers of a double third wheel and drives—with moderate force—the escapement via the lower wheel. The accumulator spring is retightened every minute, whenever the time display switches. This draws its power directly from the barrel at the same moment, and the impulse is so strong that a wind vane wheel—a kind of eddy current brake—must dissipate the kinetic energy remaining after the disc displays have switched. On the other hand, such a strong impulse is needed to switch all three discs simultaneously and precisely when the hour changes. And that happens twenty-four times a day, after all.

Since the two systems are coupled via the double third wheel, it is possible to control the force pulse with absolute temporal precision. Six seconds before the time jump, a large armature preloads the mechanism (recognizable by a tiny forward movement of the one-minute disk) by swinging out in one direction and allowing a small spring-loaded cam to make one revolution. When the armature swings in the other direction, it releases the concentrated force of the spring barrel for a moment until it swings back to its original position, supported by a long spring arm. The whole process takes only a fraction of a second and produces a characteristic, albeit quite discreet, sound that gives the Zeitwerk an added charm.

Illustration 232
The A. Lange Zeitwerk, a visually and technically exceptional wristwatch. The hours are displayed on the left, the minutes on the right. A look at the movement shows the complex design and function of this watch. The hour indication on the left requires a rotating disc with twelve digits, the minute indication on the right two rotating discs with six and ten digits. The caliber L 043.1 movement used here is regulated in five positions and contains sixty-six rubies (movement no. 67,882, case no. 187,454).

In addition to the *remontoir*, the winding mechanism of the caliber L043.1 is also patent pending; more precisely, the design feature that the rollers of the barrel and the spring core were swapped. Not least because of the high spring forces of almost 8.8 pounds (4 kiloponds; usual values are well below 2.2 pounds, or 1 kp), the designers took special care of this highly stressed component and realized that the generally used construction method cannot be the last word. Usually when the watch is wound, the mainspring core rotates, and when the watch is run down, the mainspring barrel rotates. Now, while the mainspring core is mounted in rubies, the mainspring barrel is often mounted only in a steel bushing—problems with smooth running and abrasion are preprogrammed. In the Zeitwerk the situation is reversed: when the mainspring is cocked, the barrel is rotated on its outer teeth, and when the watch runs down, no energy is lost due to the rotation of the cleanly mounted mainspring core.

4.8. Richard Lange (2006)

The collection of simple precision watches named after Ferdinand Adolph's son Richard Lange continues the tradition of Lange's observation watches, for which there are famous models from the period between the world wars. These high-quality pocket watches, mostly handmade, were already in demand in the second half of the nineteenth century for their extraordinary accuracy in science and transportation. Above all, observation watches of this type were also used on ships. There they served as a medium for time transmission between highly precise ship's chronometers in the vessel's interior, according to which they were set, and the officers on the bridge or the quarterdeck (hence the also-common designation "deck clocks"), who were dependent on precise time for navigation. Six precision pocket watches from A. Lange & Söhne were used on the first German expedition to the South Pole between 1901 and 1903.

Later, over a period of twenty years (1917 to 1937), fifteen examples of the "Great Observation Watch" were manufactured in Glashütte, equipped with a movement measuring a proud 2.24 inches (57 mm) in diameter and delivered to such renowned addresses as the Timekeeping Society in Berlin, the Zeppelin shipyard in Friedrichshafen, and the Clausthal-Zellerfeld Mining Academy's Institute of Physics.

The size of the watches was not only to ensure good readability, which had to be guaranteed even under conditions of poor visibility. It also helped ensure that the watch could be adjusted better and more accurately, which is of course facilitated by a large dial area and long hands.

Richard Lange (1845–1932) knew better than almost anyone how to incorporate scientific findings into his design considerations. He was particularly interested in the balance spring as a crucial part of the regulating mechanism. At the end of the 1920s, a scientific study on nickel beryllium alloys was published in which the tempering property of a small addition of beryllium was described. Richard Lange was the first to recognize the usefulness of these research findings for the production of balance springs. He found that adding beryllium to nickel steel alloys reduced the hairspring's sensitivity to temperature fluctuations and magnetic influences, while at the same time increasing its elasticity and hardness. In 1930 he filed a patent application for the result of his research under the name "Metal Alloy for Watch Springs."

Illustration 233
Richard Lange men's wristwatch, a very special precision watch. Consisting of 199 parts, the caliber L 041.2 is regulated in five bearings and has forty-three rubies installed (movement no. 119,533, case no. 219,026).

As a tribute to Richard Lange, the watch named after him was equipped with a special movement in 2006. Consisting of 199 parts, the manually wound caliber L041.2 with base plate, bridges, and cocks made of natural German silver was manufactured according to the highest quality criteria and largely decorated and assembled by hand.

An additional set of wheels that rotate under a bridge on the back three-quarter plate ensures the rotation of the seconds wheel, which is mounted outside the movement (indirect center seconds). The movement can be stopped to accurately set the time.

The solid silver dial is rhodium-plated for the platinum version of the Richard Lange, silver-plated for use in the gold cases, and printed with slender Roman numerals. The minute scale, whose division bears another five graduations, serves the central seconds hand, made of blued steel, to indicate sixths of a second (21,600 vibrations per hour or an oscillation frequency of 3 Hertz correspond to three full or six half amplitudes).

The large balance wheel is regulated by turning eccentrics on its ring, and a hairspring key with regulator was omitted. The outer end of the Lange-developed balance hairspring is fixed to the balance cock by a patent-pending screwed clamping device. The swan-neck fine-adjustment device, which is screwed onto the hand-engraved balance cock, serves solely to precisely adjust the so-called drop, the position of the balance in relation to the escapement, by turning the corresponding fine-thread screw.

Illus. 235
Illus. 236

5. Patek Philippe

5.1. Introduction by Gisbert L. Brunner

Patek Philippe: 175 years and not the least bit old

What would have happened if . . . ?

The fiction of being able to look at the colorful goings-on on our blue planet from various points of view, both pro- and retrospectively, is a very interesting one. Then and only then would we know, for example, what would have happened historically and geographically without the policy toward Poland of Czar Nicholas I, who ruled the Russian Empire from 1825 to 1855. Under these unfortunately completely unrealistic circumstances, there would also be clarity as to whether, even without the political turmoil in so-called Congress Poland, Antoni Norbert Patek de Prawdzic, born in 1812, would have found his way to Geneva, where he launched one of the world's most renowned watch brands. But fortunately—as so often in our unpredictable lives—everything worked out in the best possible way in the end.

In one way or another the twenty-three-year-old, like many other Polish emigrants, found his way to Switzerland in 1835. Patek had gone to the village of Versoix, near Geneva, where he married Marie Adelaide Elizabeth Thomasine Denizart, a merchant's daughter, on July 10, 1839. One of the witnesses to the marriage was named František Czapek. The Polonized Bohemian, who had also fled Poland in 1831, had—because he was a watchmaker by trade—first joined forces in his profession with the uncle of Patek's young wife. Antoni Norbert de Patek, the third member of the group, had already founded Patek, Czapek & Co. in Geneva on May 1, 1839. Czapek contributed his technical expertise to the small company, and the two partners each contributed eight thousand Swiss francs. Patek's contribution was probably his wife's dowry.

Sand in the Gears

Initially, the agreed-on division of labor worked well: Czapek was responsible for watch production, while Patek took care of sales of the approximately two hundred timepieces produced each year. And the uncle kept a very low profile as a silent partner. The clientele, mostly uprooted Poles, felt very much at home in the store at Quai des Berges No. 29 in Geneva. There, in a pleasant atmosphere, they could chat about the good old days in the East and, incidentally, about fine timepieces.

The first discord appeared in 1844, when Czapek, blinded by his remarkable success, cultivated a lavish lifestyle. This and his extensive travels to his former homeland brought the young company close to the brink of ruin. The balance sheets showed a liquid capital of just one franc and eighty-six cents. Other assets: a few finished watches, various accounts receivable, and tools. In this unfortunate situation, requests for urgently needed loans simply fell on deaf ears.

In December 1844, Patek made a forced trip to Paris by carriage. In his luggage were a number of crown-wound watches based on the system introduced by Louis Audemars in 1833. He hoped to sell them during an industrial exhibition on the Champs-Élysées. In the metropolis on the Seine, Patek found out that even good things have their foes. The better presented itself in the form of flat keyless pocket watches by watchmaker Jean-Adrien Philippe. His modern crown-winding system and the awarding of a medal by the jury impressed Patek so much that he offered the Frenchman a joint future on the Rhône.

Strong Partner from France

The call to Geneva was successful. Once there, Jean-Adrien Philippe, as technical director of Patek & Co. from 1845, threw himself with verve into installing machines of his own design. The focus was also on the already patented crown-winding

Illustration 234
Fabrice Kieffer and his charming wife, Katrin, in front of the entrance to their two-star restaurant Les Deux in Munich. Kieffer is Alsatian and comes from a family of winemakers. Not only the excellent cuisine but also patron Fabrice Kieffer's special talent for always giving his guests a real sense of feeling at home makes every visit an unforgettable experience. This high standard is matched by his BMW M550i xDrive, whose V-8 engine embodies the latest in engine technology. The same applies to the drivetrain and chassis, which offer maximum driving stability and the best possible traction.

Illustration 235
The Patek Philippe reference 5270Q G-001 with chronograph, perpetual calendar, and 18K white-gold case bears witness to an equally high technical standard. The small round window between four and five o'clock displays the leap year cycle, the one between seven and eight o'clock the day/night indication.

Illustration 236
The chronograph shown on the left uses Patek Philippe's CH 29-535 PS Q manufacture manually wound movement. This movement measures 1.25 inches (32 mm) in diameter and 0.275 inches (7 mm) in height and consists of 456 individual parts. The regulating system consists of a Gyromax balance, a Breguet hairspring, and an adjustable hairspring stud. There are thirty-three jewels in the movement, and the system runs at 28,800 A/h.

Illustration 237
Among the most-beautiful watches ever made in Switzerland are the reference 2526 three-hand automatics from Patek Philippe. Here are two very rare versions: *on the left*, with a black dial, and, *on the right*, with a cream enamel dial. The model on the right has Breguet numerals, a variant that is not even in the Patek Philippe museum in Geneva.

system, which was regularly optimized over the years. The logical consequence of outstanding commitment was the founding of Patek, Philippe & Co. in 1851.

The break with Czapek was not without risk. With the financial support of Juliusz Gruzewski, a leading figure in the Polish diaspora in Geneva, the deposed man repositioned himself entrepreneurially. In this way Patek lost an important clientele to Czapek. But in the end, the better side won out, because Patek and Philippe, supported by Vincent Gostkowski, another Polish emigrant, were able to convince the hotly contested market with their innovative products. The partnership between Patek and Philippe resulted in great pocket watches that represented the entire spectrum of precision timekeeping at the highest technical and craftsmanship level. The palette ranged from simple flat examples to various grand complications. Like their competitors, Patek, Philippe & Co. did not manufacture their own movements. Raw movements, or *ébauches*, were supplied by Victorin Piguet and LeCoultre from the Vallée de Joux. In Geneva the watches were finished, assembled, regulated, cased, and, of course, sold to all countries, but especially to the New World on the other side of the Great Pond. In 1851, Queen Victoria of England purchased a pocket watch with this signature, and many prestigious customers followed this example. From then on, Patek Philippe pocket watches, with and without any kind of horological complication, were to be found among the aristocracy and the wealthy bourgeoisie.

The Philippes Carry On

After the death of Norbert Antoine de Patek in 1877, neither his twenty-year-old son, Léon Mecyslas Vincent de Patek, nor his sister Marie Edwige, born in 1859, showed any interest in following in their father's very large footsteps. Ergo, the siblings initially contented themselves with a silent partnership. In 1901, the owners transformed Patek Philippe & Co. into a collective

Illustration 238
The first Patek Philippe watch with perpetual calendar and automatic winding was given the reference number 3448. It was unveiled in 1962 and remained in the program until 1984. Initially, this model was not so popular and was almost considered a slow seller. Today it brings hammer prices of up to half a million dollars at auctions.

company called Ancienne Manufacture d'Horlogerie Patek Philippe & Co. SA. Then the two Pateks sold their rights to the name to the company, which had long been dominated by the Philippe family. The latter, in turn, concluded an agreement with the LeCoultre family in the Vallée de Joux in 1904 for the supply of four thousand raw movements per year, including specially designed ones for the famous Chronometro Gondolo. South American jeweler Gondolo & Labouriau in Rio de Janeiro not only was one of the most important customers at the time but also owned shares in Patek Philippe. Many of the fine timepieces also flowed to the USA, where the importer was also a shareholder. After the stock market crash on New York's Wall Street, people in North and South America were very reluctant to buy new watches. And Gondolo & Labouriau did not pay its outstanding bills for several reasons. On the board of directors, which from 1931 included Jacques-David LeCoultre, one crisis meeting followed another. In the wake of growing financial hardship, the board offered the raw-materials supplier and its holding company SAPIC a majority stake in Patek Philippe. However, opinions differed on the valuation of the existing assets, which included not only unfinished watches, but also the magnificent building on Geneva's Rue de Rhône. These facts and other unacceptable demands caused the deal to fall through.

In this unfortunate situation, when the word "bankruptcy" was occasionally mentioned, longtime business partners Charles and Jean Stern, as owners of Fabrique de Cadrans Stern Frères, successively took over all shares. The immediate entrepreneurial decision to entrust Jean Pfister, the CEO with relevant experience, with the development and production of his own raw movements proved to be absolutely right and forward looking in 1933.

Into the Future with the Stern Family

Since then, Patek Philippe and Co. has been run as a family business by members of the Stern family. Until October 2009, Philippe Stern, born in 1938, steered the company's fortunes as the third-generation president. Since then it has been son Thierry's turn to steer the great family business into a glorious future.

Mechanical movements have undergone decisive improvements thanks to Jean Adrien Philippe and his successors. For example, the aforementioned crown-winding system had a decisive influence on the genesis of the wristwatch. The same applies to the so-called "free" mainspring. A loose end prevents the mainspring from breaking off when the watch is fully wound. This system was particularly important in automatic wristwatches. In 1951, Patek Philippe obtained patent protection for the innovative Gyromax variable-torque balance wheel, which allowed the use of a completely free-swinging balance spring. In 1996 a patented annual calendar, which requires manual correction only at the end of February, became the talk of the town. Finally, as of 2005, Patek Philippe was among the pioneers of silicon technology in mechanical watches. This lightweight and nonmagnetic material represents a real revolution in the traditional field of mechanical timekeeping.

Without these and other top achievements, Patek Philippe would never have been able to climb the Olympus of ultimate watchmaking. Many of the most-complicated pocket watches and wristwatches ever made originate from this noble Geneva family business. In addition, it was responsible for important inventions that have had a lasting influence on fine watchmaking. In 1925 the world's first wristwatch with a perpetual calendar and jumping displays was created. In 1937, the close cooperation with ingenious Geneva watchmaker Louis Cottier resulted in the first world time wristwatches. And in 1941, series production of wristwatches with perpetual calendars began. In 1962, Patek Philippe once again presented a world premiere in the perpetual-calendar sector: the first wristwatch of this type with automatic winding. Launched in 1989, the caliber 89 can be considered the most elaborate pocket watch movement ever, with 1,728 components and thirty-three different functions. Since November 2009, all of Patek Philippe's mechanical movements have been manufactured in-house. This also applies to chronographs.

Until the aforementioned year 2009, the Geneva Seal distinguished these ticking timepieces. Since then, the family-owned company has used its own Patek Philippe Poinçon (seal) based on a very comprehensive and much-stricter set of criteria. That, and only that, meets the Stern family's high standards for their watches.

5.2. The Movements

Manually Wound Movements

Until the 1930s this Geneva watchmaker used the raw movements (*ébauches*) of other manufacturers, such as Jaeger-LeCoultre from Le Sentier. It was not until Jean Pfister's seminal decision to produce his own movements in 1933 that this Geneva brand made its major breakthrough. And it is with this decisive era that we begin the following list of manually wound movements.

Illustration 239
Manually wound caliber 12'''-120 with small seconds. This movement has no caliber designation (movement diameter 1.05 inches [26.75 mm], height 0.15 inches [4 mm], 18,000 A/h, eighteen jewels).

12'''-120 (12-120)

Of interest for the collector are those wristwatches that were produced from 1935 and in which ticks the first in-house manually wound caliber 12'''-120 movement. With an outer diameter of 1.05 inches (26.75 mm, or 12'''), height of 0.157 inches (4.00 mm), eighteen jewels, 18,000 vibrations per hour, and about forty hours of power reserve, it was made in four series and remained in production until 1953. Here is a compilation of the four production series:

1st Series: movement nos. 826,892 to 829,999 (3,108 movements, 1935 to 1940)
2nd Series: movement nos. 920,000 to 929,999 (10,000 movements, 1940 to 1950)
4th Series: movement nos. 938,000 to 939,079 (1,080 movements, 1952 to 1953)
3rd Series: movement nos. 960,000 to 969,999 (10,000 movements, 1947 to 1952)

Starting in 1935, the caliber 12'''-120 movement was used in the first men's wristwatch, designated the Calatrava Reference 96, which was launched as early as 1932. When the reference 96 was introduced, it initially used a caliber 12''', a raw movement with fifteen jewels, bimetallic balance wheel, and flat hairspring.

Illustration 240
Manually wound caliber 12''' SC with central seconds. It was a redesign of the caliber 12'''-120 with added center seconds.

12'''SC (12 SC)

Just three years after the introduction of the basic caliber 12'''-120, it was followed by the version with indirectly driven center seconds. This was given the caliber designation 12''' SC and was produced in five series:

1st Series: movement nos. 828,812 to 828,835 (24 movements, 1938 to 1939)
2nd Series: movement nos. 828,920 to 828,943 (24 movements, 1939)
3rd Series: movement nos. 829,238 to 829,261 (24 movements, 1939 to 1940)
4th Series: movement nos. 861,704 to 863,995 (2,292 movements, 1939 to 1948)
5th Series: movement nos. 864,002 to 864,994 (993 movements, 1948 to 1950)

The 12'''SC movement has an elaborate design for the indirect second-hand drive. It originated from the Victorin Piguet company. With the help of the modified gear train bridge and a small attachment bridge, three additional gears bring the seconds beat from the position of the small seconds to the center.

Comparing the respective totals of 24,188 movements for the 12'''-120 caliber and 13,358 movements for the 12''' SC caliber, it may be concluded that watches with small seconds were probably more in demand at that time.

In the course of their production, both calibers underwent numerous changes, including the replacement of the flat hairspring by a Breguet hairspring.

10'''-200 (10-200)

With the manually wound movement launched in 1946, the possibility was created to equip even smaller men's watches with a movement of modern design. With a diameter of 0.89 inches (22.7 mm) and a height of 0.143 inches (3.65 mm), it could be used with or without small seconds. The caliber 10'''-200 had eighteen jewels and a Breguet hairspring and ran at 19,800 vibrations per hour. It was produced in two series and remained in production until 1965:

1st Series: movement nos. 950,000 to 959,999 (10,000 movements, 1946 to 1953)
2nd Series: movement nos. 740,000 to 751,199 (11,200 movements, 1953 to 1965)

For square or rectangular watchcases, the 10'''-200 caliber was given a special movement mount. This consisted of a brass ring placed on a brass plate and soldered in place. This holder subsequently received a protective layer of rhodium.

12'''-400

The manually wound 12'''-400 movement appeared in 1950 as successor to the caliber 12'''-120. With a diameter of 1.05 inches (26.75 mm) and a height of 0.157 inches (4.00 mm), it had the same external dimensions as its predecessor.

Just one series was produced, and it remained in production for eleven years:

Movement nos. 720,000 to 729,999 (10,000 movements, 1950 to 1961).

27 SC

The replacement caliber for the 12''' SC appeared in 1949 and was given the designation 27 SC. While the movement diameter remained unchanged at 1.05 inches (26.75 mm), its height was set at 0.195 inches (4.95 mm). The caliber 27 SC was produced in two series and remained in production until 1970:

1st Series: movement nos. 700,000 to 712,699 (12,700 movements, 1949 to 1970)
2nd Series: movement nos. 888,000 to 888,178 (179 movements, 1959 to 1965)

From a purely visual point of view, the two calibers can be easily distinguished. The 12''' SC movement has a complex design for the indirect second drive that was contributed by the Victorin Piguet company. The caliber 27 SC represents a neatly articulated in-house design, with the fourth wheel elegantly positioned under the gear train bridge. The mainspring barrel bridge was reduced in size, and the crown wheel bridge was executed separately.

27-AM 400

The caliber 12‴-400, the 27-AM 400 variant, was developed for watches that have to withstand the danger of magnetic fields. The Gyromax balance wheel and Breguet hairspring were used in its design, with the balance wheel, pallet, and other components made of amagnetic (Swiss term for antimagnetic) material. The beat rate was raised from 18,000 to 19,800 vibrations per hour. Two series were produced:

1st Series: movement nos. 730,000 to 735,239 (5,240 movements, from 1958)
2nd Series: movement nos. 739,700 to 739,799 (100 movements)

The difference between these two series initially lay in the diameter of the hairspring roller, which was initially 0.021 inches (0.55 mm), then 0.017 inches (0.45 mm). From movement no. 732.700, the thickness of the balance wheel was also reduced from 0.027 to 0.019 inches (0.70 to 0.50 mm). These calibers were mainly used in the Amagnetic models with the references 2570/1 (1958), 3417 (1960; see also p. 186) and 3460 (1962). However, they were also encased in other models, such as references 3420, 3418, and 3419 (see also p. 187).

23-300

Designed for ultraflat wristwatches, the caliber 23-300 was launched in 1956. With a diameter of 0.89 inches (22.70 mm) and a height of 0.118 inches (3.00 mm), it was available with and without small seconds. With eighteen jewels and 19,800 A/h, this caliber was equipped with a Breguet hairspring and Gyromax balance wheel from the beginning, despite its low height. From movement no. 1.148.400 on, an adjustable hairspring stud was added. With this compact design, it was possible to achieve astonishingly good rates. The caliber 23-300 remained in production until 1977 and was produced in the following three series:

1st Series: movement nos. 780,000 to 799,999 (20,000 movements, 1956 to 1966)
2nd Series: movement nos. 1,140,000 to 1,159,999 (20,000 movements, 1966 to 1971)
3rd Series: movement nos. 1,210,000 to 1,228,549 (18,550 movements, 1971 to 1977)

Compared to the five manually wound movements described so far, which together totaled 65,765 pieces, the caliber 23-300, with a total of 58,550 pieces, can be considered the most built manually wound movement produced by Patek Philippe. Due to its small dimensions, it was also used in jewelry watches and in models with square or rectangular cases. In only three examples it was equipped with a perpetual-calendar module and was called caliber 23-300 Q in this design. This movement was also installed in the reference 3449 in 1961.

Illustration 241
Manually wound caliber 9‴-90 with small seconds (dimensions 1.03 by 0.73 inches [26.2 × 18.7 mm], height 0.143 inches [3.65 mm], 19,800 A/h, eighteen jewels)

9‴-90

Finally, the 9‴-90 form movement, which was in production for around thirty-four years, will be described. The installation dimensions for width and length are 0.70 × 1.04 inches (18 × 26.60 mm), and for the overall height, 0.143 inches (3.65 mm). The movement uses eighteen jewels, runs at 19,800 A/h, and on the versions from 1954 onward has a Breguet hairspring with Gyromax balance wheel. The two series are divided as follows:

1st Series: movement nos. 830,000 to 839,999 (10,000 movements, 1934 to 1950)
2nd Series: movement nos. 970,000 to 977,889 (7,890 movements, 1947 to 1967)

This caliber was used in various luxury watches with rectangular cases. The best-known models include the reference 2441 (also called Tour Eiffel or Eiffel Tower) and the reference 2469 (see p. 161, fig. 260).

In this chapter, only manually wound movements from Patek Philippe that are relevant for collectors are described; therefore the calibers 10‴-105 and 10‴-110 have been omitted. Due to their installation diameter of 0.86 inches (21.78 mm), they were used only in models with rather small cases, which are not so much in demand among collectors nowadays.

The Automatic Movements

Compared to for other watch manufacturers, the age of automatic winding began relatively late at Patek Philippe. It was not until April 1953 that the noble brand from Geneva presented its famous "automatic watch" at the twenty-fourth Swiss Watch Fair in Basel, as the sales brochure stated. At that time, the Swiss Watch Fair was still quite manageable and a subfair of the Swiss Samples Fair. In 1953 this fair was held from April 11 to 21. The new automatic caliber named 12‴-600 AT arrived later than those of the other renowned brands, such as IWC, Jaeger-LeCoultre, Rolex, and Vacheron & Constantin, but it came up with so many new and innovative technical features that it immediately beat all the previously mentioned competitors. But what was the reason for the late appearance of this masterpiece? For one thing, the development of the first self-winding movement took several years. Second, at that time, Patek Philippe's manually wound wristwatches were in such high demand that there was no need to produce automatic watches. In the first seven years of production of automatic wristwatches, only 7,100 of this first automatic movement were built and enclosed in watches. Self-winding watches thus only accounted for about 20% of total production.

12'''-600 AT

This masterpiece ushered in the age of automatic watches at Patek Philippe in 1953. The movement included technical attributes such as a Breguet hairspring, Gyromax balance wheel, micrometer fine balance with swan-neck spring, winding with ball-bearing eccentric, guilloché gold rotor, and much more. All this together gives this movement a unique level.

The caliber designation 12‴-600 AT, consisting of numerals and letters—which, by the way, applies to most Patek Philippe movements—already tells the insider the most-important data about the diameter, the construction height, and the time. The most-important data about the diameter, height, and mode of operation: 12‴ means 12 lines, an old French measure of length used in watchmaking, which indicates the diameter of the movement (1‴ = 0.087 inches or 2.223 mm, 12 × 2.223 = 1.050 inches or 26.676 mm). The numeral 600 stands for the (rounded-off) height of 0.236 inches (6 mm). In reality, the movement diameter is exactly 1.06 inches (27.00 mm) and the overall height is 0.212 inches (5.40 mm). Finally, "AT" denotes the automatic-winding mechanism.

An automatic movement can be divided into two subassemblies: namely, the actual movement (base movement)

Illustration 242
Caliber 12'''-600 AT with automatic winding (movement diameter 1.06 inches [27 mm], height 0.212 inches [5.4 mm], 19,800 A/h, thirty jewels)

and the mostly attached automatic assembly. The basic movement has a double-spoke Gyromax balance wheel with eight masselotte weights, a Breguet hairspring, super shock protection, and a ball-bearing-mounted eccentric changer for bidirectional winding, as well as thirty jewels, and runs at 19.800 A/h. The guilloché gold rotor, Geneva stripe finish, beveled edges, and Geneva Seal underline the high horological level of this masterpiece (see also pp. 171–172 and 183).

Although the 12‴-600 AT caliber was designed to meet the highest demands, various improvements were also made to it to improve its accuracy and serviceability. From movement number 760,300, the previously used screw balance wheel made of beryllium was replaced by the Gyromax balance wheel patented by Patek Philippe. Up to movement no. 760,950, there were another ten relevant changes. These included a reinforced rotor axle, a modified rotor fixation, replacement of the eccentric wheel cam by a ball bearing, and an increase in the number of movement fixation screws from two to three. Starting with factory number 760,951, all the previously mentioned improvements were incorporated into production. Movements built up to movement no. 760,950 were upgraded by Patek Philippe to the latest design when they later came to Geneva for overhaul.

The development and production of this masterful caliber took place under the overall responsibility of the then production director, Françoise Cart. The 12‴-600 AT caliber is still one of the most beautiful and best automatic movements ever built. It has the following number range: movement no. 760,000 to 767,099 (7,100 movements, 1953 to 1960). According to a retired Patek Philippe employee, another thirty-one examples of this caliber were kept in stock as spare parts.

The first wristwatch from Patek Philippe equipped with the caliber 12‴-600 AT was also the first example of the coveted reference 2526 with an enamel dial. The movement number is therefore 760,000, and the case number is 674,938. This first piece was produced in March 1953 and, according to the letter from the US concessionaire Henri Stern Watch Agency, Inc., in New York, was delivered to American J. B. Champion in Dallas, Texas, USA, on July 27 of the same year.

The Gyromax balance wheel from Patek Philippe, which has been described several times, needs some explanations here. In contrast to the conventional screw balance wheel, this design needs neither a regulator nor other devices for fine adjustment. The two-spoke balance wheel usually carries eight cylindrical studs on its rim, which have a hole in the center and are slotted on one side. These studs rotate in pins that are positioned

Illustration 243
Caliber 27-460 with automatic winding. Successor to the caliber 12'''-600 AT (movement diameter 1.06 inches [27 mm], height 0.18 inches [4.6 mm], 19,800 A/h, thirty-seven jewels).

vertically in recesses on the balance ring. If one turns the slots of these studs, also called *masselotte* weights, toward the center of the balance, the oscillation moment (GD2) of the balance increases and the watch runs slower. If the slots of the blocks are turned outward, the oscillation moment is reduced and the watch runs faster. This regulation takes place in very fine steps and without changing the balance mass or the functional length of the hairspring. In this way the symmetry of the oscillation system, consisting of hairspring, balance wheel, pallet, and escape wheel, is maintained. This Gyromax balance wheel is a Patek Philippe invention and has been patented several times. The first and basic patent for the Gyromax balance wheel was patent no. 261,431, filed on March 26, 1947, and published by the Swiss Federal Office of Intellectual Property on August 16, 1949. A further patent for the Gyromax balance wheel with the number 280,067 was filed on August 18, 1949, published on April 1, 1952, and concerned various further design improvements to the balance ring, as well as to the mass balance weights.

27-460 and 27-460 M

Despite the resounding success of the 12‴-600 AT caliber, two new trends were evident in wristwatches at the end of the 1950s: on the one hand, there was a tendency toward a flatter overall design, and on the other toward equipping watches with a date display. Patek Philippe followed this trend and launched the successor caliber 27-460 in 1960. One year later it was followed by the version with date and the designation 27-460 M, the "M" standing for Monodate. With the almost unchanged basic movement with a diameter of 1.06 inches (27.00 mm) and 19,800 A/h, the modified automatic assembly made it possible to reduce the overall height to 0.18 inches (4.60 mm) (0.20 inches or 5.30 mm with date indication). Externally the two successor calibers could be distinguished from their predecessor by the fact that they did not have a guilloché gold rotor mounted in slide bearings; instead the solid-gold rotor had a ball bearing and Geneva stripes finish (see also fig. 243, *left side*). The second difference is the design without fine adjustment by micrometer screw and gooseneck spring. The number of jewels was increased from thirty to thirty-seven. The movements of this caliber family also underwent several modifications during their production time. Starting with movement no. 1,116,899, the caliber 27-460 was equipped with a modified Gyromax balance wheel made of beryllium bronze. The eight inertia weights, formerly in the center of the balance ring, were subsequently relocated to the inner side of this ring. The last version of the caliber 27-460 M had an adjustable hairspring stud and the designation 27-460 M PM (PM stands for *piton mobile*) and remained in production until 1975. The following list explains the movement numbers of these calibers:

1. Caliber 27-460, movement nos. 1,110,000 to 1,116,899 (6,900 movements, 1960 to 1970)
2. Caliber 27-460 M, movement nos. 1,120,000 to 1,129,999 (10,000 movements, 1961 to 1970)
3. Caliber 27-460 M, movement nos. 1,116,900 to 1,119,999 (3,100 movements, 1970 to 1986)
4. Caliber 27-460 M PM, movement nos. 1,230,000 to 1,234,999 (5,000 movements, 1970 to 1975)

The caliber 27-460 M was usually produced with, but in some versions without, small seconds. The caliber 27-460 received in the version 27-460 Q a module for a perpetual calendar on top and was installed in the references 3448 and 3450. The latter reference was recognizable by the leap year indicator near 3.

For the third series, the number range of the caliber 27-460 was used in 1970 and continued from movement no. 1,116,900 to 1,119,999.

For the caliber 27-460 M PM, a number range was used that was originally reserved for a pocket watch movement. Just thirty-five examples of this pocket watch movement were built, so the numbers 1,230,000 to 1,230,034 were already assigned. Nevertheless, this duplication of thirty-five movements should not be a problem, because a pocket watch movement can easily be distinguished from a wristwatch movement.

350 and 1-350

With the advent of quartz technology in the 1960s, the design and construction of wristwatches changed decisively. The trend was clearly toward ever-flatter models, and this lowered the price of the expensive automatic movements previously manufactured. Patek Philippe was also not spared this change and began developing a new automatic movement as early as the mid-1960s. The work, under then technical director Eric Jaccard, led to the unveiling of the caliber 350 in 1969. The movement diameter was set at 1.10 inches (28.00 mm), and the

Illustration 244
Caliber 350 with automatic winding, a design with peripheral rotor (movement diameter 1.10 inches [28 mm], height 0.13 inches [3.5 mm], 21,600 A/h, twenty-eight jewels)

Illustration 245
Caliber 1-350 with automatic winding, a revised version of the caliber 350. In addition to the revised caliber designation on the barrel bridge, you can see the difference in the gold segment of the outer rotor. An economy version made of heavy metal was chosen for the 350 caliber.

height at 0.137 inches (3.50 mm). The balance ran at 21,600 A/h and the movement had twenty-eight jewels. To meet the requirements for a flatter watchcase, a completely new design was chosen. Instead of the previous central rotor, an outer rotor with ball bearing and gear ring now took over the winding work. The movement itself was placed inside this housing. The previous crown was omitted, and manual winding, as well as the setting of the hands, was now done by a flat crown mounted in the screw-down case back. The company had basically adopted the Futurmatic from Jaeger-LeCoultre. The omission of the crown at three o'clock allowed the designers to create interesting, appealing case shapes that were initially very popular, especially in the USA. Almost anyone could afford a Jaeger-LeCoultre Futurmatic there, but a Patek Philippe remained the preserve of only a small circle.

After a seemingly good start, the caliber 350 did not achieve hoped-for success. The ball bearing of the rotating ring proved very susceptible to failure, and the changer for bidirectional winding of the rotor switched quite sluggishly and helped the watch only to a very modest power reserve. This market experience, but also the knowledge gained by the Patek Philippe service organization in the meantime, made a further development of this caliber urgently necessary. In 1979 the new version with the designation 1-350 was released. The heavy metal weight was now replaced by an 18-karat gold segment. A new ball bearing, the elimination of the changer, and the improvement of the manual winding and setting mechanism brought advantages not only for the user of the watch, but also for its service conditions. This caliber was never completely satisfactory, and production ended in 1985. In addition, there was the disadvantage that this caliber's design did not permit a date display. The watches were produced in two series:

1st Series, movement nos. 1,180,000 to 1,189,999 (10,000 movements, 1969 to 1979)

2nd Series, movement nos. 1,490,000 to 1,499,999 (10,000 movements, 1979 to 1985)

The models equipped with these calibers had just become famous in the United States under the name Backwinder. René Schwarz, service manager at Richemont in Munich until 2020, had worked at Patek Philippe in New York for twelve years. He knew these two calibers' problems better than anyone else. According to his statements, complaints about the new caliber increased so massively that Patek simply took back new watches and exchanged them for other models.

Caliber 28-255 and 28-255 C

In 1976 Patek Philippe launched its new Nautilus model on the market. No reliable automatic movement was available for this new sporty steel watch designed by Gerald Genta. The caliber 1-350 was also out of the question because it was not sufficiently robust and did not have a date display. A solution was offered by cooperation with movement supplier Jaeger-LeCoultre. They had developed an ultraflat automatic caliber 920 movement, which was not intended for their own watches. The automatic movement, which Patek Philippe used as the caliber 28-255 without and caliber 28-255 C with calendar, had proven itself since 1972 in the sporty watches of the noble brands such as

Illustration 246
Caliber 28-255 C with automatic winding, an ultraflat movement with a peripheral rotor ring (movement diameter 1.10 inches [28 mm], height 0.124 inches [3.15 mm], 19,800 A/h, thirty-six jewels)

the Audemars Piguet Royal Oak, and later also at Vacheron & Constantin in the reference 222. When it appeared in 1967, this caliber was already considered the world's thinnest automatic movement. With a diameter of 1.10 inches (28 mm), an installation height of 0.096 inches (2.45 mm), and a rate of 19,800 A/h, this *ébauche* was the welcome bridge during the period when Patek Philippe still had no reliable automatic movement of its own. The structural design of the caliber 28-255 made it easy to use and service. The rotor was bidirectional and was mounted on a copper-beryllium hoop with four ruby bearings. The 22-karat gold flyweight attached to the outside of the rotor reinforces the winding performance. The excellently designed changer not only allows winding in both rotor directions but also has extremely short switching times and thus a very high efficiency. The mainspring barrel is wound after just a few hours on the winder. The sophisticated design of the movement is proven by the fact that during the ten years of production, only a few modifications were made, such as conversion to an adjustable hairspring stud and the addition of a date display (overall height 0.118 inches or 3.01 mm).

As mentioned, this Jaeger-LeCoultre caliber was considered a reserved raw movement for Vacheron & Constantin and Audemars Piguet. The overall heights of this caliber were 0.096 inches (2.45 mm) without date and 0.120 inches (3.05 mm) with date. Thus, this caliber could call itself the world's flattest automatic movement. In the literature it is sufficiently described that this caliber was delivered as a reserved raw movement (*ébauche*) only to the companies Audemars Piguet and Vacheron & Constantin. This reservation for Audemars Piguet and Vacheron & Constantin initially stood in the way of a business relationship with Patek Philippe. As a retired movement designer revealed to the author, his colleague had the special idea to make the oscillating weight 0.004 inches (0.1 mm) thicker, thus changing the overall heights to 0.10 inches (2.55 mm) without calendar and to 0.124 inches (3.15 mm) with calendar. This elegantly released the movement supplier Jaeger-LeCoultre from its contractual obligation to supply the world's thinnest automatic movement only to the two previously mentioned companies. Although the modified movements were now each a tenth of a millimeter higher than those from Audemars Piguet and Vacheron & Constantin, they ran just as well. The caliber variants bear the following production numbers:

1. Caliber without date display, movement nos. 1,280,000 to 1,299,999 (20,000 movements, 1970 to 1980)
2. Caliber with date display, movement nos. 1,300,000 to 1,319,999 (20,000 movements, 1970 to 1980)

The two calibers listed have proven very successful at Patek Philippe and still represent a worthwhile investment today.

In 1999, Audemars Piguet bought all the rights, drawings, and tools for this caliber from Jaeger-LeCoultre and started its own production in 2005. The first watches with this new movement were unveiled the same year.

5.3. The Case Numbers

A logical scheme for determination of case numbers cannot be identified at Patek Philippe. While the case numbers were still six digits in the 1930s and 1940s, seven-digit numbers were used starting in the mid-1950s. This is related to the ordering procedure in dealing with case suppliers. If Patek Philippe ordered a series of new cases for a specific reference, the case maker had to deliver them according to precise specifications with complete numbering and hallmarking. The reference 2526 (p. 169) can serve as an example: marked with movement no. 763,637 and case no. 696,013, it dates from production year 1956. An identical model produced the following year has movement no. 764,962, but case no. 2,604,085. Watches produced more than ten years later can again have six-digit case numbers, as can be seen on pages 200 and 201 for references 3541 and 3542.

Summary: Case numbers are six or seven digits and are always stamped on the inside case back. Engraved (i.e., machined) case numbers do not exist at Patek Philippe! An example of a manipulated case back can be found on page 156 (illus. 253).

Watches of the references 3439/1 and 3561 have backs with sapphire crystal inserts. How the case number was applied there we will learn on page 202 (reference 3561).

5.4. The Reference Numbers

A reference number may be understood as a type identifier of a watch. Exactly when a systematic identification of wristwatches was introduced at Patek Philippe cannot be easily traced. It is a fact that the watches built in the early 1940s were only exceptionally stamped with a reference number.

From a long-deceased employee comes the following information. After the introduction of its own calibers, lists of reference numbers were handwritten and matched the number in an accompanying photograph (i.e., a picture of the watch in question). In the Patek Philippe archives this registration still exists today, even though all data have long since been computerized.

On some watches listed in this book, we can prove that only from the year 1948 were reference numbers listed on the case back according to a plan. An example can be found on reference 96 watches, which can be seen with and without the reference number (see p. 153). Reference 96 was followed by three-digit reference numbers with 100s, 400s, and 500s regulatory groups. Four-digit reference numbers were then introduced. In the course of the following years these were assigned the regulatory groups 1400, 1500, 2400, 2500, 3400, and 3500. As an example, see pages 162 and 191, where you can clearly see reference numbers 2481 and 3444 on the case backs. On the other hand, one looks in vain for the reference number 592 on the steel watch from the production year 1939 presented on page 157.

Only on complicated watches were reference numbers stamped into the case back, sometimes even before 1948. The following table shows some examples that were produced with and without reference numbers.

On page 192, one can see the complete numbering and hallmarking of an inner case back by looking at the opened watch. In the case of reference 3445-6, the 6 indicates the sixth and last variant of this model.

Let's summarize: During the production of the wristwatches, they were listed internally under the respective reference number; except for the mentioned exceptions, the factory did not begin marking the reference number on the case back until 1948. The reference number itself is always stamped on the inside back and never on the outside.

Illustration 247
At Patek Philippe, all models were listed under reference numbers, but these were not stamped on the inner case backs as planned until 1948. For complicated watches, this marking of the watches had been practiced in part earlier, but without comprehensible logic. This table shows some interesting examples of the sometimes contradictory handling of the reference numbers.

Reference	Complication	Year Made	Movement No.	Case No.	Reference Entry
1436	Double chronograph	1942	862,420	630,552	No
1436	Double chronograph	1943	863,057	646,703	Yes
1463	Chronograph (Walser Wald)	1946	863,724	640,561	No
130	Chronograph	1948	867,454	655,062	Yes
1527	Chronograph with perpetual calendar (also see page 195)	1943	863,247	634,687	No
1518	Chronograph with perpetual calendar	1943	863,195	633,139	No
1518	Chronograph with perpetual calendar	1947	867,243	646,599	No
1518	Chronograph with perpetual calendar	1951	868,004	663,680	Yes
1526	Perpetual calendar with moon phases	1942	864,271	652,371	Yes
1526	Perpetual calendar with moon phases	1948	963,263	657,232	Yes
1415	World time watch	1945	929,694	647,130	Yes

From time to time it happened with old watches that a case back was damaged and had to be remade. Even if the watch in question was from before 1948, in this case it received the reference number noted on the case back.

5.5. Other Identification Markings—Characters, Hallmarks, and Stamps

If one wants to acquire a watch not only as a presentable piece of jewelry but also as a potential investment, one should take a thorough look at all the documentation on the inner case back. In addition to the important statements by the manufacturer, the stamped hallmarks (i.e., official markings) confirm the fineness of the precious metal used. In Switzerland, thirteen assay offices under the directorate in Bern monitor the authenticity of the material specifications. The minimum fineness of 750/1000 gold is represented by the Helvetia, a woman's head with hair ornaments facing left. The Helvetia is the female personification of the Swiss Confederation, comparable to the German Germania. A distinction is made between the small and large Helvetia. The small one is used mostly on the outside of a watchcase, the large one in its inner case back. To demonstrate the many variations of the markings in the inner back, the inner backs of the watches are also illustrated in this book. On page 175 we can see the relevant information on the inner back of the rose-gold automatic Ref. 2540 (production year 1957). In the upper part of the case there are three rows with the letters PATEK-PHILIPPE & Co, –GENEVE–, and SWISS. There is no *accent grave* on the "E" of "GENEVE," but only a dash on the right and left. The SWISS stamp has no dot at the end. In the lower third of the case back, one sees the hallmark of the case maker on the left, the official fineness stamp 18K/0.750 in the center, and the large Helvetia on the right. Below that is the case number 699 844 and finally the reference designation. The latter is stamped RÉF. 2540. The *E* has an *accent aigu* and the *F* is followed by a period.

The inscription of the inner back of the automatic Ref. 3445 (production year 1970) on page 192 shows another variant. At the very top is the large Helvetia, below it the official marking for 18K/0.750. Below that, in three rows, follow the lettering PATEK, PHILIPPE & Co, ≡GENEVE≡, and SWISS. Below that, in three rows, are the case number 328637, the reference designation 3445, and the hallmark of the case maker. The following differences can be found: Between the names PATEK and PHILIPPE, there is no hyphen as before, but a comma. The C° present in the first case has been replaced by Co. The place reference ≡GENEVE≡ now has three dashes on each side. In this case the SWISS proof of origin has been given a closing dot. Finally, the reference designation is done without the previous RÉF. While you are reading this book, we will find and explain quite a few more variants in this field.

Watches with 950 platinum cases have a hallmark in the form of a chamois head looking to the right, with stamped oval framing. As with the Helvetia, two different sizes are used here. The smaller hallmark is usually struck on the outside, the larger one only on the inside.

The Hallmark of Geneva, also known as the Geneva Seal, is today considered a guarantee of origin, and the highest precision, longevity, and watchmaking expertise. Only a few watch manufactories belong to the select circle of those who receive this seal of quality. The three basic requirements for obtaining this seal are as follows:

- The caliber must first be approved and recognized by eight sworn members of the Geneva State Board of Examiners.
- The exclusively mechanical movements must have been assembled and regulated in the canton of Geneva.
- Each caliber must meet the twelve technical criteria of the regulations in terms of its development, characteristics, manufacturing quality, and finishing.

To facilitate the trade and export of articles made of precious metals, the then EFTA countries concluded an international agreement on the testing and marking of precious metals in 1972 (Vienna Convention). This convention also established a common hallmark for gold and platinum. For 18-karat gold, this consists

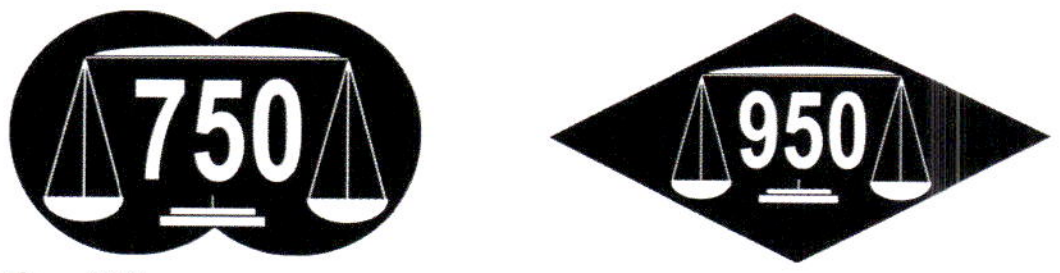

Illustration 248
Hallmarks for gold (*left*) and for platinum (*right*) as per the Vienna Convention of 1972

of a symbolized scale that displays the number 750 in the center and is framed by a double circle. For 950 platinum the scale is in a diamond and is marked with the number 950. On Patek Philippe's newer watches, these hallmarks are stamped on the inner case back and on the outside.

With the Precious Metals Control Act (EMKG), Switzerland introduced new regulations for the circulation of precious metals and precious metal articles as of August 1, 1995. The new official stamp—St. Bernard's Head, affectionately known as "Bello"—hereby replaces the Helvetia, which was valid until then.

To summarize: The watch brand Patek Philippe uses only two precious metals for its cases, 750 gold and 950 platinum. For steel cases this marking is omitted, but the other indications are largely identical. When purchasing an old or new watch of this brand, all markings should be carefully inspected and checked. If in doubt, the German branch in Munich or the head office in Geneva can help.

Material	Specific Gravity γ kg/mm³	Modulus of Elasticity E kg/mm²	Melting Point °C
Gold (Au)	19.32	7,900	1,063
Copper (Cu)	8.96	12,500	1,083
Platinum (Pt)	21.45	17,320	1,774
Silver (Ag)	10.49	8,160	960
Steel × 10 Cr 13	7.75	22,000	1,500

Illustration 249
Table of the main characteristics of the metals used

5.6. Precious Metal or Stainless Steel—a Little Materials Science

When it comes to 750 gold, we are familiar with yellow, red, and white gold. In addition to the proportion of 750/1000 gold, yellow gold is created by admixing silver (125/1000) and copper (125/1000). In red gold, silver (45/1000) and copper (205/1000) are added in a different distribution. In white gold, silver (200/1000) and nickel (50/1000) are used. Gold is often alloyed with palladium, especially because of the skin sensitivity caused by nickel. In very special cases, Patek Philippe also uses cases made of platinum. Platinum may be stamped only with a minimum fineness of 950/1000. No more than 50/1000 parts of copper, gold, rhodium, iridium, and palladium may be fused with pure platinum. The main physical properties of all the metals described are listed in the table.

In addition to platinum, stainless steel, which is alloyed with chromium and is therefore rustproof, has the best prerequisites for the production of watchcases. It is not for nothing that steel watches are particularly popular nowadays. However, the respective processing times for machining operations such as turning, milling, and grinding are around eight times higher for steel than for gold. Like stainless steel, platinum is also difficult to process. In an overall calculation it must still be taken into account that platinum costs around 40% more than gold! If vanity or budget are not the primary consideration when purchasing a new Patek Philippe, watches with stainless-steel or platinum cases would be the best choice.

5.7. The Watches

There are few investments that one can approach with double vigor. On the one hand, this is because one can always enjoy the timeless design and perfect technology of a wristwatch of this brand. On the other hand, because one usually acquires an item that is not dependent on the stock market and that will increase in value in the medium and long term. To enable a newcomer to enter this hitherto unfamiliar field with necessary expertise, we will take a closer look at all the wristwatches listed in this book. As already pointed out in the introduction, the author was granted access to numerous rarities by their respective owners for this documentation.

If you leaf through the now-out-of-print work *Patek Philippe Genève* by Martin Huber and Alain Banbery, you will find well more than two hundred different wristwatches illustrated. The multitude of case variations not only reflects the tastes of the time but also indicates the small quantities in which the models of that time were made. Because less can often mean more, we have concentrated our selection of watches presented here on just under eighty watches. However, they contain interesting technical aspects, largely correspond to today's modern taste, and are often still affordable. In addition to in the previous literature, these timepieces have been described in such detail that even a previously uncertain but well-heeled investor can now dare to purchase a valuable Patek Philippe.

The descriptions of the following watches usually follow the same format. First, the numbers of the movement and case are listed. Then follows the description of the dial, case, and movement. Great importance is given to the inscriptions and hallmarking of the case, as well as the inner case back.

Afterward, the history of the respective watch as well as the prices of acquisition and overhaul is described. Finally, results or prices that these watches have achieved at sales or auctions are often listed.

Because the reference numbers of the watches are always highlighted as headings, the reader can easily orient themselves and quickly find the information they are looking for.

Certainly the following technical texts are not always easy to read. But when a lot of money is at stake, the reader is served neither by flowery texts nor by excitingly formulated style samples.

Technical descriptions and stock market reports have a lot in common. What is needed here are not rhetoric exercises, but precise and directional factual statements, according to which an investor or speculator can safely plan and successfully carry out the purchases they have in mind. The following descriptions were also written with this objective in mind.

A Special Note That Is Present Only in the Patek Philippe Chapter

So that newcomers, watch collectors, and capital investors can receive a recommendation that is legally nonbinding but in line with the trend of the times, the watches presented here have been marked according to the three criteria "To wear," " To collect," and "As an investment." The three grades were given as "good (+)," "very good (++)," or "excellent (+++)."

Of course one can have a different opinion of all three criteria, but it is important that each reader makes their own evaluation and on this basis makes their investment decision. Despite all the opinions and discussions, one thing is certain: a wristwatch from Patek Philippe is usually a safe investment. This is true for most of the older models and many of the current ones. In the case of watches as a capital investment, the question of the taste of the next prospective buyer who wants to purchase the beautiful watch is also decisive. Here the author can serve with his more than thirty years of experience and contribute to the right decision.

The Patek Philippe Classics
The Author's Secret Favorites

Reference 96 Calatrava

The first men's wristwatch sold in large numbers by Patek Philippe reached the market in 1932 with the designation Calatrava. It received the first assigned reference number 96 and was initially equipped with a 12‴ manually wound Jaeger-LeCoultre movement. Only toward the end of 1934 (officially from 1935) were these models equipped with the 12‴-120 caliber movement, which is easily distinguishable from the former. The Jaeger-LeCoultre movement lacks the fine adjustment by means of a micrometer screw, a gooseneck spring on the regulator arm, and the ruby bearing on the center wheel. Furthermore, the escapement wheel bearing lacks the cover plate and the Geneva stripe finish on all bridges and cocks. Finally, the 12‴ caliber from Jaeger-LeCoultre had only fifteen jewels, while the 12‴-120 from Patek Philippe had eighteen.

The reference 96 was produced in many dial variants. More than twenty-six different versions with small seconds (i.e., with the caliber 12‴-120) are known to the author.

Three years later, in 1938, the caliber 12‴-120 was also marketed in a version with center seconds. This version with indirectly driven center second hand has an intermediate gear designed by Victorin Piguet that connects the second wheel via an intermediate wheel with the newly installed second wheel in the center (see also p. 143). Comparing this caliber with its successor 27 SC, the technically trained eye recognizes this not-so-elegant-looking makeshift design that remained in production for more than eleven years. But this version of the Ref. 96 with center seconds also found its share of fans. The author has seen more than eight different dial versions. Despite its rather small case diameter of only 1.20 inches (30.5 mm) (earlier versions even had to make do with 1.18 inches or 30.0 mm), the reference 96 remained in production for almost thirty years and was replaced by the successor model Ref. 3796. Today this classic Calatrava still has many admirers, even if it is now listed under reference 5196 and the case diameter has grown to a stately 1.45 inches (37 mm). Under the reference number 5296, you can even get it with automatic winding!

A small number of further variants of the reference 96 appeared on the market in 1930. These had a full calendar with moon phase at twelve o'clock, or a perpetual calendar with retrograde date display, small seconds, and moon phase at six o'clock. A version with a platinum case can be seen in the Patek Philippe Museum in Geneva.

Illustration 250
The Ref. 96 men's wristwatch shown was unveiled at the end of 1934 (officially from 1935). It was the first Calatrava with the manufacture manually wound caliber 12'''-120 movement. The upper version with rose-gold case and black dial dates from the production year 1942 (movement no. 922,307, case no. 997,660). The reference indication is missing from the inner case back. The yellow-gold model was produced in 1951, and the reference no. 96 is stamped on the inner case back (movement no. 968,921, case no. 303,990).

But now to the jewels of this book: two versions of the reference 96. According to the Patek Philippe pedigree book excerpts, the first timepiece dates from 1942, the second from 1951. The case of the reference 96 is in three parts and consists of a sprung bezel, case middle, and press-fit case back. The case middle is stamped from a solid piece of metal and with the four lugs forms a single unit. The holes for the spring bars are through-holes and thus visible from the outside.

Many people consider the dial to be the face of a watch, and rightly so! That the dial of the rose-gold watch is refurbished can be said only if you know the original imprint used in the manufacture of the watches. This was used until 1952 and thus for both watches reads PATEK, PHILIPPE & Co. Only in the course of 1952 was the abbreviated form introduced and the previous reference to "& Co" was dropped. The *accent grave* on the second "È" of the designation of origin GENÈVE and the SWISS label below six o'clock were retained. With this time indication it must still be kept in mind that Patek Philippe produced only four hundred to five hundred watches per month at that time, and the dials were always planned early. So it could happen that the dials in stock for a model, which was produced in a smaller number of pieces, were sufficient for several years. The year of production of a watch and that of a dial could thus indeed be one or more years apart.

During the overhaul of both watches, the dial of the rose-gold-plated Ref. 96 was refurbished at the Causemann company in Frechen and printed like the version from 1952. The dial of the yellow-gold Ref. 96 still corresponds to the original design but has clearly suffered the effects of time. After their overhauls, both caliber 12'''-120 movements are in almost new condition. The respective movement numbers 922,307 and 968,921 point us to the production periods 1942 and 1951. On the rose-gold Ref. 96, the "small Helvetia" can be found on the outside of the case middle at four o'clock, and on the underside of the lug at eleven o'clock on the yellow-gold watch. On the inside back we notice remarkable differences. On the rose-gold version (i.e., the older version with case no. 997,660), there is still the material thickness indication 4/10 for 0.4 mm gold plate thickness as well as the framed company hallmark "PP.C°." The reference number is missing. The yellow-gold version with case no. 297,660 has the same company details at the bottom, but additionally the number 96 for the reference. On the "–GENÈVE" proof of origin, the corresponding right dash was omitted, but the SWISS proof-of-origin label has both periods. Looking at the opened calibers, one can see another small difference. On the rose-gold version, the Geneva stripe cut is on the axis of the crown center wheel at an angle of 60°, while on the yellow-gold version at an angle of 30°. Was the polisher able to act at his own discretion here due to the lack of precise specifications?

Conclusion: With its case diameter of 1.20 inches (30.5 mm), the reference 96 does not quite correspond to our modern tastes. Nevertheless, the high level of the movement and case still sets standards. Even on the wrist of a pretty lady, this piece of jewelry can still outshine the current models of many competitors.

Reference 96 watches in yellow gold with Breguet numerals (see also p. 157) rarely come on the market and are not offered for less than $10,900. Models with steel cases are even more sought after and are offered between $10,900 and $33,000. The yellow-gold version presented here was purchased by a collector in 2008 for $4,650 (approx. 2,982.87 €) in the USA, and he invested another $1,800 to have the watch serviced at Patek Philippe. The separate refurbishment of the case added another $164. Patek Philippe in Munich charged $82.50 for the installation of new hands. Despite its dire state, the red-gold model changed hands for $4,900 at the Munich Watch Fair in spring 2009. Specialists in Pfrozheim charged a total of $525 to service the movement and refurbish the case. The German company Causemann in Frechen refurbished the dial according to exact specifications and charged $221 for this.

In all, both watch owners can be satisfied with the total price they paid. Only a few watches of the reference 96 model are likely to be found on the market in this excellent condition. Thus, in both cases, one can speak of a safe investment. Even in less good economic times, an excellently preserved Ref. 96 will always find a new admirer.

For wearing: + For collecting: + As an investment: +

Between the References

Of the numerous models that line up between the reference 96 and the following 565, some should still be mentioned. First of all, there is the chronograph reference 130, with a case diameter of just 1.29 inches (33 mm) and a three-piece case including a press-fit case back, which is no longer in line with today's taste. The movement was contributed by the manufacture Valjoux with the caliber designation 23, and the chronograph was operated by two flat rectangular pushers. The case was issued in steel, yellow gold, and red gold and in the combination steel and red gold. The author is aware of more than thirty-three different versions. This watch, launched in 1938, was still in the program in 1961. The selling price at that time for the yellow-gold version was exactly 1,710 DM. The successor model Ref. 530, which was also launched in 1938, is still considered to be up to date today. With a case diameter of 1.45 inches (37 mm) and a width of 0.79 inches (20 mm) between the lugs and the flat pushers, this model is almost identical in appearance to the new chronograph Ref. 5170, which was launched in 2010. This chronograph was also available in steel, yellow-gold, and red-gold cases. The author has encountered six different versions.

Reference 565

In 1938, the noble watchmakers of Geneva launched the house's first waterproof watch. It was assigned reference number 565. A then-opulent case diameter of almost 1.38 inches (35 mm) (precisely 1.37 inches, or 34.8 mm), an inner dustcover, and a solid screw-down case back helped this new release to achieve a sporty, masculine appearance. Even at more than eighty years old, this classic has not lost its attractive appearance and is still considered a real eye-catcher.

Illustration 251
In 1938, Patek Philippe unveiled its first waterproof watch. Our photo composition shows this sought-after collector's item from the front and opened from the back. The watch shown was made in 1952. The encased caliber 12'''-120 has two Geneva hallmarks: one on the main plate near the balance wheel, and one on the mainspring barrel bridge near the movement-retaining ring, with the inscription "METAL" (movement no. 938,562, case no. 667,674).

The watch described here is one of the very rare versions with an original dial in black. The label reads "PATEK PHILIPPE & Co," and at six o'clock there is the "SWISS" proof-of-origin label. According to the pedigree book extract, movement no. 938,562 and case no. 667,674 indicate that this timepiece was made in 1952 and sold on December 2 of the same year. The two-piece case consists of the case middle with integrated bezel and the screw-down case back. The center section was stamped from solid metal, deep drawn, milled, and turned. Most of the surfaces are diamond-finished. A dustcover guarantees additional protection. The screw-down case back closes with a fine thread and decagon. The small Helvetia was stamped at four o'clock on the case middle and on the strap lug at the bottom at eleven o'clock. Beside the patent notice, the entries "PATEK-PHILIPPE & Co," "–GENÈVE–," and "SWISS," here without a period, can be found one below the other. Besides the gold fineness hallmark, large Helvetia, and the case maker's hallmarks (collective-responsibility mark no. 5, key no. 11, Manufacture Taubert, Genève, canceled in 1974), the reference and case numbers can also be seen. The "REF." still has a period but no *accent circonflexe*. The inner case back has a circular-grained surface. However, the dustcover has left its traces there from it being screwed down.

The caliber 12'''-120 movement is in almost top condition. The balance staff has no shock protection. The Geneva hallmark is stamped on the main plate near the balance wheel and on the mainspring bridge. The rarity shown here was called at the auction of Sotheby's in New York on October 17, 2003, as lot no. 316. The estimated price in the catalog was $7,000–$8,000.

An interested German placed a bid by telephone and won the auction at $7,500. Including the premium of 20%, the total price was $9,000. Adding transport and import sales tax, the final price for this purchase was exactly $8,634.

This rare watch was entrusted to Patek Philippe for servicing. In the total costs, which amounted to $2,377.16 (invoice from June 8, 2004), the restoration was the largest item at $1,572. The customer was charged $72 for two gold spring bars and $94.64 for the pedigree extract.

As a "cult watch," the Ref. 565 is one of the most sought-after collectibles. With a yellow-gold case they sell for between $15,000 and $25,000 at auction, and with a steel case for often double the price. A rose-gold model with a black dial comparable to our watch was offered at Christie's in Geneva on November 16, 2009, and sold for 62,500 CHF (approx. $62,202; sold by Freccero, movement no. 938,728, case no. 667,705). At the same auction, a steel version with Breguet numerals, center seconds, and steel bracelet was offered and sold at 93,000 CHF (approx. $92,557; movement no. 700,276, case no. 662,276). Another steel version with two-tone dial and seller's label "GÜBELIN" was offered at Antiquorum New York on June 14, 2006, and sold for $82,000!

For the reopening of the rebuilt headquarters at the Quai du Rhône in Geneva, a reissue of the Ref. 565 in steel was produced in 2006. It was given the reference number 5565 and was limited to three hundred pieces. These collector's items, which immediately sold out, appeared at auctions in the following years and achieved astonishingly high prices. At Antiquorum in New York on June 14, 2007, an example of reference 5565 was called and sold for $45,600 (movement no. 3,559,510, case no. 4,409,445). Another watch was offered by the same auction house on October 16, 2008, and sold for $49,200.

For wearing: ++ For collecting: ++ As an investment: ++

Reference 570 with Small and Center Seconds

Another cult watch with a high degree of popularity is the reference 570, which was delivered with both small and center seconds. Case diameters vary and can be 1.38, 1.40, or 1.41 inches (35, 35.6, or 35.8 mm). The Patek Philippe catalog lists a diameter of 1.38 inches (35 mm). When we checked the two watches described here, we measured 1.39 inches (35.3 mm) for the rose-gold one and 1.40 inches (35.6 mm) for the one in yellow gold. The rose-gold watch is powered by the 12'''-400 caliber and thus has a small seconds. The balance staff is protected by an early Jaeger-LeCoultre shock protection system. The movement number is 723,856 and the case number is 306,726. The latter has a three-piece design and consists of sprung bezel, case middle made in one piece, and press-fit case back. One can find the "small Helvetia" hallmarks on the case middle at four

Illustration 252
The reference 570 is regarded as a cult watch with a high degree of popularity. The first version with small seconds was introduced in 1938, and the model with center seconds followed a year later. The watch shown has a rose-gold case with a diameter of exactly 1.38 inches (35.30 mm); the original measurement should have been 1.40 inches (35.60 mm). The "small Helvetia" stamped on the case middle at four o'clock is barely visible and proves that the case was reworked later. This watch uses the manually wound caliber 12'''-400 movement (movement no. 723,856, case no. 306,726).

o'clock and at the bottom of the upper left strap lug. The yellow-gold version with caliber 27 SC movement is more recent. The movement number is 708,105, the case number first 708,105 and later 4,178,999. During overhaul in Geneva, the case back with the last-mentioned number was found to be from another manufacturer and was replaced by a newly made press-fit case back. The newly issued case number is also noted in the pedigree book. For the expert, the hallmarking of the two press-fit case

Illustration 253
Reference 570 with center seconds and manually wound caliber 27 SC movement. The reference 570 basically has a case with a press-fit case back, meaning it is only splash-proof. The two press-fit backs deserve a special note. The one in front is made of gold and was sand-cast and is therefore a manipulation (fake). The new case back was made as a turned part and is an official replacement case back (movement no. 708,105, wrong case no. 708,105, corrected case no. 4,178,999).

backs should be interesting. The back of our rose-gold watch was made as a deep-drawn part (i.e., it was formed under high pressure). The criticized back of the yellow-gold version was turned from a piece of cast metal. You can see this in the "surface streaks" that occur with simple sand casting, despite the surface polish on the outside and the beading on the inside. The new replacement base turned from the solid metal has a wall thickness of exactly 0.019 inches (0.5 mm), while that of the one made elsewhere was only 0.015 inches (0.4 mm). The interior lettering of the latter is almost perfectly imitated, but the obligatory "large Helvetia" has been forgotten. The hallmarking of the inner case back of our rose-gold reference 570 turns out to be correct, except that in the word "–GENÈVE–," the right dash was once again forgotten. Also there was the repair entry "Cm 26831 6/61." Finally, we take a look at the two dials. The one from the rose-gold case has enameled lettering and was painted later. The dial of the other watch is silvered and has a ray cut. Here, too, the letters of the lettering are made of enamel. Regarding its manufacture, this is almost a miracle, unless the ray cut was applied with an abrasive brush. The fine steel wires of this grinding brush have little effect on the hard enamel.

The yellow-gold watch was purchased by a German collector on July 15, 2002, for $7,000 (approx. 5.930 €). Including transport and import sales tax, about $9,000 had to be invested for this watch. The overhaul at Patek Philippe in Munich cost another $3,171 (invoice from 29/07/2003). The movement service, which cost $1,505, and the new case back made in Geneva for $1,105 were the most-expensive single items in this invoice.

The seller of the watch, a Korean civil engineer from Marina del Rey, near Los Angeles, California, was not happy about the foreign-made back but accepted the costs for the new case back. This reduced the overhaul costs by $1,282 to $1,189.

The rose-gold reference 570 with small seconds changed hands on April 29, 2009, for $6,766 (approx. 5,100 €). Rocky's Gold & Silver, an antique store in Weyers Cave, Virginia, had mistakenly advertised this watch on eBay as "yellow gold" along with a very mediocre photo. Demand remained correspondingly low, and there were only three bidders. The new owner had to invest only $6,713, including transport and import VAT, for this valuable collector's item!

At auctions, reference 570 watches are always in great demand. A rose-gold version sold for CHF 19,200 at Sotheby's in Geneva, and a white-gold version sold for 34,220 CHF at Antiquorum. At Christie's in New York on April 12, 2000, a rose-gold Ref. 570 with Breguet numerals was called and sold for $14,100. The record, however, was broken by a steel-cased model that was auctioned at Sotheby's in Geneva on May 17, 2005, for 110,400 CHF!

For wearing: ++ For collecting: ++ As an investment: ++

Reference 592

One of the minor sensations of our watch selection is the steel version of the reference 592. With a significantly larger case diameter than the standard model Ref. 96, Breguet numerals, and small seconds, this model is considered very rare, and the author had encountered it only once. The caliber 12‴-120 movement bears the number 829,476, the case 506,760. The reference no. 592 is missing in the case back because, according to the registry extract, the watch was manufactured in 1939 and sold on April 19, 1941. The steel case consists of the case middle, including the bezel and the press-fit case back. The case diameter is exactly 1.27 inches (32.5 mm). The case middle was first made as a turned part. Subsequently, four grooves were milled, in each of which the lugs were soldered. A price evaluation of this rarity is certainly not easy, especially since this version has not been offered at auctions before. Because of its larger case diameter, it might be priced higher than the steel version of the reference 96!

The watch shown was purchased by a European watch collector at antique watch dealer FGW (Feel Good Watches) in Scottsdale, Arizona, for $7,000. It was subsequently given to Patek Philippe in Geneva, and an overhaul was commissioned. The total cost was $1,327, of which the movement overhaul at

Illustration 254
The reference 592 in stainless steel and dial with Breguet numerals is considered a rare collector's item. This model was introduced in 1938. The version shown is from the production year 1939. Compared to its predecessor, the reference 96, the Ref. 592's case had a diameter of 1.27 inches (32.5 mm) instead of 1.20 inches (30.5 mm; caliber 12‴-120 with movement no. 829,476, case no. 506,760).

$420 and the dial restoration at $596 were the most-expensive items (invoice from August 10, 2002).

At Antiquorum in New York on September 17, 2009, a reference 592 in steel was called and sold for $19,200. This watch had a black dial with Arabic numerals (movement no. 921,243, case no. 507,497).

For wearing: + For collecting: ++ As an investment: ++

Reference 1491 with Small Seconds

The author knows a rather interesting story about the next wristwatch. A watch enthusiast purchased this rare collectible Ref. 1491 in rose gold at the London branch of Sotheby's. He had to pay around $5,470 for this beautiful piece. With a caliber 12‴-120, movement no. 968,379, and case no. 664,108, this timepiece was registered and known in Geneva. After the bid was accepted, our watch friend wrote to Patek Philippe in Geneva and asked for a pedigree book extract for this newly acquired collector's item. After a short time the answer was on the table. The watch had been reported stolen, and the owner concerned was informed immediately about the appearance of this noble piece, as well as of the request and address of the new owner. Whether in this case the legal rules of data protection were correctly observed would still have to be checked by a lawyer. In any case, after a short time the presumed rightful owner contacted us by letter in excellent German and confidently demanded the free return of the watch. After a short written

Illustration 255
The reference 1491 is one of those watches with an eye-catching case design. The lugs are made from a single piece of metal and rolled to the outside. The fact that these lugs were also soldered to rose-gold cases with yellow-gold solder probably did not bother quality control at the time. The reference 1491 was made with both small and center seconds. The model shown with a rose-gold case and small seconds can tell a special story that begins on this page (movement no. 968,379, case no. 664,108).

intermezzo, both parties contacted each other by phone and discussed the case. The former owner, according to his statement a high administrative official from Tel Aviv, demanded the watch back because it was an heirloom from his father and had been stolen during a burglary. After a lengthy discussion it turned out that the watch had indeed been stolen in a burglary. There was further testimony that a police report had been made and that this case had been reported to the insurance company in question. An insurance payment of $10,000 was also admitted after a lengthy discussion. The Israeli who was robbed was also informed about English law. This states, among other things, that suspected stolen property can be legally reacquired via a public auction.

After a written offer to buy back this heirloom after payment of the costs incurred, there was no reply. Apparently the initial interest to get back the beloved paternal heirloom had suddenly vanished after failing to obtain a settlement at zero cost.

When the case was overhauled, it was found that the exterior of the press-fit case back had been reworked. About 0.003 inches (0.10 mm) of material had been removed. The case back was then professionally restored. Presumably an inscription had been removed that could have indicated an earlier owner.

The two versions of the reference 1491 with small and center second hands were quite popular and in demand in their day. The two-piece case (diameter 1.34 inches or 34 mm) consists of the case middle, including bezel and a press-fit case back. Particularly striking are the lugs, which are made of one piece and rolled in a snail shape. On the version presented, the lettering, the strokes of the small seconds, and the "SWISS" label are enameled, while the dial itself is lacquered. Interesting clues are also revealed by the inscriptions on the case back. At the top they start with the lettering "PATEK, PHILIPPE & Co," the "Co" now without the two dashes previously present below the "o." The ≡GENÈVE≡" proof of origin is framed on the right and left by three dashes. At the very end follows the type reference "REF: 1491," here without an *accent circonflex* but with a colon. The noble timepieces with reference no. 1491 are not often found at auctions. In recent years, prices ranged from 10,000 to 28,000 CHF, depending on the model, condition, and history. Now that ladies also wear men's watches, a Ref. 1491 on the wrist of a pretty lady would certainly make a very special piece of jewelry! At Antiquorum in Geneva, a model in rose gold was offered on October 15, 1989, and sold for 10,800 CHF. Eleven years later a yellow-gold version was offered at Antiquorum in New York on October 16, 2008, and sold for $12,000. At Christie's in Geneva, a yellow-gold watch was offered on May 14, 2007, and was knocked down for 16,800 CHF ($13,835).

For wearing: ++ For collecting: ++ As an investment: ++

Reference 1491 with Central Seconds

As a complement to the reference 1491 with small seconds, here is another version with center seconds. This watch also has a rose-gold case. Inside this collectible is a caliber 12‴ SC movement, a design based on the caliber 12‴-120. This modification did not come from Patek Philippe but was worked out by the Victorin Piguet company. The movement no. is 864,993, the case no. 645,221. The dial is black lacquered and has gold indices and an imprinted minute scale. The case shape corresponds to the watch previously described. The press-fit case back has a thickness of 0.019 inches (0.5 mm). The Victorin Piguet movement design, which needs some getting used to, can be seen particularly well on page 143. Except for the case number, the hallmarking of the inner case back corresponds to the previously described reference 1491. The watch shown was called on November 14, 2009, at Dr. Crott Auctions in Frankfurt/Main but was not sold. The call price of $16,400 to $21,882 was probably set a bit too high, and an underbid of $14,770 was not accepted. At Antiquorum in Geneva, on November 16, 2002, a yellow-gold version was offered and sold for 12,500 CHF.

For wearing: + For collecting: + As an investment: ++

Illustration 256
It is always a question of personal taste whether watches with identical cases look better with small or center seconds. In this case, the Ref. 1491 with center seconds stands out, perhaps because of its black dial (movement no. 864,993, case no. 645,221).

sold on May 12, 1959. The numbers of the two caliber 27SC movements are only 233 apart. Also both watches were delivered with leather straps. Using a stereo microscope, you can see that the former watch still has a dial in its original condition, while the latter has a properly refurbished one.

Now a few words about the case design. The impressive and, for that time, oversized bezel diameter of almost 1.45 inches (37 mm) (1.44 inches or 36.6 mm according to the maker) is actually a bit of a cheat. Turning the case around reveals that the underlying cylindrical part of the case has a diameter of just 1.39 inches (35.5 mm).

The caliber 27 SC has already been mentioned several times. In both cases this movement is very well preserved. The Geneva hallmark is stamped on the main plate near the caliber designation and on the gear train bridge. The size of the movement-retaining ring, which securely protects the movement, is striking. The hallmarking of both watches is the same, even if the case numbers are far apart. The hallmarking starts at the top of the case back and has "PATEK, PHILIPPE & Co," "≡GENÈVE≡," and "SWISS" stamped in three rows. This is followed by the "large Helvetia," then the gold fineness hallmark "18K/0.750," the reference no. "2481," the case number, and the case maker's hallmark. The hammerhead with the number 171 is manufacturer's mark no. 1, behind which is the Bernard Dubois company, La Chaux-de-Fonds, which was canceled in 1987. The indication of the material thickness was omitted, but we have determined it to be 0.015 inches (0.4 mm).

The first-mentioned Ref. 2481 comes from an American collector. He purchased it for $7,563 (approx. 5,957.47 €) and passed it on to a European buyer in exchange. The latter paid another $930 for the import tax. A professional overhaul of the case and movement was discussed, but the costs were still too high for the owner.

The second specimen was offered several times by a Stuttgart antique-watch dealer but found no buyer despite a reduced price! A watch enthusiast from Munich found out at least the one reason for the low demand: the existing lugs did not match the original design. Only when the dealer was asked about this could more details be found out. According to the pedigree book, this Ref. 2481 had a crocodile leather strap when delivered. A later owner had the lugs removed and a solid-gold band soldered in place. The next owner did not like this brazed band. He had it removed and the case rebuilt with four lugs. Unfortunately, the case maker who was entrusted with this task did not have a sample of the Ref. 2481 at hand, so he attached fancy lugs, which look similar to those of the Patek Philippe Ref. 2482 or those of the chronograph Ref. 1579. The Munich watch enthusiast finally bought this watch for $4,900. After its receipt and authenticity were quickly checked, this Ref. 2481, together with an original version, was sent to a case maker in Pforzheim, who was asked for a price for a correct reconstruction. How this story continued we learn from the caption (illus. 263).

The Patek Philippe Ref. 2481 models were also very popular with princes and royalty in their time. For example, Saudi Arabia's King Saud (ruler from 1953 to 1958), successor to famous King Ibn Saud (born November 24, 1880, in Riyadh; died November 09, 1953, in Taif) liked to give away special versions of the reference 2481. With the likeness of his predecessor and ruby indices on the dial, these watches were given to foreign-state guests, deserving subjects, or leaders of the numerous nomadic Bedouin tribes in the country. In 1958, the heavily indebted King Saud abdicated and handed over power to his half brother Faisal. As a result, business with Patek Philippe, which was good at the time, also declined.

Illustration 263
"Two make three" could be the motto of this image. Montages such as this have become possible thanks to modern image-processing techniques. The Ref. 2481 manually wound watch on the left was shown on the previous page. The middle watch of the same reference has lugs, but the wrong ones. These were later removed by a case maker, and new ones were made and soldered in place according to the present pattern. We can see the result of this work on the watch on the right. This example shows us that even a watch with the wrong lugs does not have to be a hopeless case (Ref. 2481, *left*: movement no. 707,889, case no. 2,905,524; Ref. 2481, *right*: movement no. 707,656, case no. 2,607,790).

The Ref. 2481 is still listed in the Patek Philippe catalog from 1961. The case diameter is listed as 1.44 inches (36.6 mm), and the price is 1,295 DM.

In addition to the special versions for royalty and royal houses, the Ref. 2481 was also marketed with special enamel dials. The watches equipped with *cloisonné* enamel dials were already considered valuable and unique pieces at that time and were produced only in small numbers. These *cloisonné* dials were modeled on palm tree landscapes, continents, and similar motifs. During the production of the dials, the outlines and shapes of the motifs to be depicted were first drawn and soldered using thin gold wire. Subsequently the delineated fields and shapes were filled with different-colored enamel powder and fired. The label "PATEK PHILIPPE" was placed in the upper arch below twelve o'clock, and "GENÈVE" in the lower arch above six o'clock, and fired with enamel. Even the *accent grave* on the second "E" of GENÈVE is present as it should be!

A Patek Philippe Ref. 2481 with a *cloisonné* enamel dial came up for auction at Antiquorum in Tokyo on December 16, 1990. The watch, with a palm motif dial, was sold for 10,160,000 CHF (movement no. 702,041, case no. 668,415). At Christie's in New York another reference 2481 with *cloisonné* enamel dial was offered on June 23, 1998, and sold for $178,500 (approx. 319,515 DM). The dial of this watch also featured the well-known palm motif (movement no. 703,043, case no. 678,621). A Ref. 2481 with the likeness of King Ibn Saud in yellow gold and a yellow-gold bracelet ("Klötzli bracelet") was auctioned at Antiquorum on November 16, 2002, and sold for 54,000 CHF (movement no. 704,978, case no. 689,414). From 2007 to 2009, about a dozen normal examples of the Ref. 2481 came up for auction and sold for between $9,600 and $12,000.

In addition to its inherent values, a Ref. 2481 may also be considered a good-value investment in the medium and long term.

For wearing: ++ For collecting: ++ As an investment: ++

Reference 2483

One of the oldest concessionaires in South America is Freccero SA of Montevideo, Uruguay. It was to this company that the Patek Philippe Ref. 2483 shown was delivered. The caliber 12‴-120 movement bears the number 968,697, and the case number is 661,631. Feccero is also one of the selected authorized dealers to have its name present on the dial. Rose gold seems to have been very popular among wealthy South Americans at the time. Thus the dial, indices, and hands of our Ref. 2483 all are made of 18-karat rose gold. Integrating them with the case of the same color gave the watch a very distinguished and elegant look. On the dial is the label "PATEK PHILIPPE, GENÈVE" in keeping with the regulations from 1952. Of course the "SWISS" label below six o'clock is also present, even if it is hidden by the overlapping bezel.

The solid, two-piece case has a screw-down case back and is very elaborately manufactured. The case middle with integrated bezel was stamped from a solid piece of metal, deep-drawn, turned, and milled in one process. The case diameter is 1.28 inches (32.5 mm). Bezel and lugs at the top are diamond-finished; the remaining surfaces are satin-finished and polished. The caliber 12‴-120 movement is in near-mint condition, but the Geneva hallmark is not present. The balance has no shock protection. The hallmarking is interesting and slightly nonstandard. This is perhaps due to the case maker, identified by manufacturer's mark no. 1 (hammerhead with the number 10, Cédex, Charles Dubois SA, Le Locle, canceled in 1982). The "small Helvetia" is stamped on the bottom of the lug at two o'clock, and on the lug at eight o'clock is stamped the number 631. These are the last three numbers of the six-digit case number. This special marking prevented confusion, ensuring that the matching case was always screwed to the correct case back. From the point of view of production technology, this marking had another reason. The finishing of the case and the screw-down case back were

Illustration 264
In this picture the rose-gold Ref. 2483 manually wound watch reveals all its details. The solid rose-gold dial bears the signature FRECCERO. This is the name of the company Freccero SA, Montevideo, Uruguay, which is still a Patek Philippe concessionaire. Only selected authorized dealers had their names printed this way. On the left, one can see the opened screw-down case back with the caliber 12'''-120 movement. The numbers 631 are stamped on the lower-right strap lug. They represent the last three digits of the case number 661,631 and allow a safe allocation of case and screw-down case back. This watch was produced in 1952, and the dial corresponds to the new regulations introduced that year (movement no. 968,697, case no. 661,631).

carried out separately. This also applied to the turning of the fine thread (internal thread) in the housing and the turning of the external thread on the screw-down case back. Subsequently, the various cases were tried out with the different screw-down case backs, and the smoothest running pairings were selected. To avoid confusion, the last three case numbers were stamped on each case.

The heavy case back is turned and deep-drawn, including the six grooves. It is exactly 0.019 inches (0.5 mm) thick, and the inner surface is beaded. In the upper third, the inscription begins with "PATEK, PHILIPPE & Co," "–GENEVE–," and "SWISS." Below that, in a row, are the hallmark of the case maker, the fineness hallmark "18K/0.750," and the "large Helvetia." After that, the numbers for reference 2483 and case 661,631 follow one below the other. The watch shown originates from the USA and was sold on August 12, 2002, for $5,750. Including transport and import tax, the buyer had to invest about $7,360. The complex servicing at Patek Philippe cost another $756. The new owner did not know that the reference 2483 was always delivered with dust protection. On this watch the additional protection was not present.

The caliber 27 SC movement with center seconds was also installed in reference 2483 watches. The models with small seconds are very rare and hard to find. A red-gold version with a caliber 12‴-120 came up for auction at Antiquorum in Tokyo on December 16, 1990, and was knocked down at 10,800 CHF. From 2006 to 2009 only a few of these models were offered in yellow gold and sold between 7,800 and 15,000 CHF. This waterproof model would also make a beautiful ladies watch!

For wearing: ++ For collecting: ++ As an investment: ++

Reference 2508 in Stainless Steel

Another watch that is in high demand today is the reference 2508. Launched in 1950, this manually wound watch was still listed in Patek Philippe's 1961 German catalog and was priced at 1,335 DM in a yellow-gold case. This model was water resistant and had a dustcover and a screw-down case back with decagon. The cases could be ordered in yellow and red gold (case diameter 1.37 inches or 34.8 mm). The "value for money" version at that time was even available with a steel case.

First, let's take a look at the steel version. The watch, equipped with a caliber 27 SC movement, has movement number 708,505 and case number 2,614,125. Besides the screw-down case back, which protects against water, there is also a dustcover. The dial is silvered and printed and has twelve indices. At twelve, three, six, and nine o'clock, these are wedge shaped, and at the remaining hours they are set in a tiny pyramid shape. The company name "PATEK PHILIPPE, GENÈVE" is in capital letters. The dauphine hands go harmoniously with the dial and case. The case consists of a case middle with integrated bezel. It is stamped from a solid piece of metal, turned, milled, and surface-finished. In addition to the screw-down case back, which ensures water resistance, there is also a dustcover. The inner case back is beaded, just like the gold case. The case back is closed by means of a decagon. On older watches a lead seal is still used, while newer ones have a plastic seal. The caliber 27 SC movement is visually in fine condition, but an overhaul by Patek Philippe would have done it good. Stamped on the main plate next to the caliber designation and on the gear train bridge, we find the Geneva hallmark. The inscription on the beaded inner back is a bit different compared to the watch described previously. Listed one below the other, there is the patent certificate, then the inscriptions "VACUUM," "PATEK, PHILIPPE Co=," "–GENEVE–," "ACIER INOXYDABLE," and "REF. 2508," and the case number 2,614,125.

Illustration 265
The Ref. 2508 manually wound watch with stainless-steel case is among the most-sought-after collector's items. The gold-cased version is listed in the 1961 German sales catalog with a price of 1,355 DM. The steel version, launched as a "value for money" watch, could be had for 900 DM. Through the five individually taken pictures, one can see all the details of the watch, including its screw-down case back and dustcover. Also interesting is the lead seal around the movement-retaining ring, which was later replaced by a plastic seal. All the details of the caliber 27 SC hand wind movement can be seen. This is also true of the movement number 708,505 on the mainspring barrel bridge in the upper right, and the case number 2,614,125 on the inner back in the lower right. This watch was sold in spring 2009 for $31,400. At the auction house Antiquorum in Geneva, a reference 2508 in stainless steel was offered on May 13, 2007, and sold for $29,977.19 (27,198.83 €; movement no. 710,519, case no. 2,614,135, production year 1963).

The watch first belonged to a German watch enthusiast who purchased it from a Swedish collector for $8,700 in spring 2007. Two years later the German sold the beautiful piece to a US citizen. Though it had not been overhauled, the American paid all of $31,045 for this steel watch!

For wearing: ++ For collecting: ++ As an investment: +++

Illustration 266
This photo shows the dial of the reference 2508 with steel case and, in comparison with illustration 267, the different ways in which the dials were manufactured. This dial was first painted, then the minute markers and company name were printed, and then the steel indices were applied and riveted. This dial is still in its original condition. Since steel watches were almost a third cheaper than the same reference in precious metal (gold), savings had to be made on the dial as well.

Illustration 267
The dial of the yellow-gold reference 2508 was produced more elaborately than its counterpart shown on the left. The company signature and minute markers are baked enamel. Only then was the dial painted. When the dial was level, the paint flowed off the raised letters and minute markers due to the acceleration caused by gravity. The indices were then placed on the dial and riveted in place. When the dial was refurbished, only the indices had to be taken off and the paint removed. Then the dial was repainted and the indices were put back on. In the case of the dial shown, only an expert can usually tell that it has been refurbished. However, the almost fourfold magnification allows even the layman to recognize minor deviations (e.g., in the letters of the company signature and the minute markers). Had this dial not been refurbished in the USA, but by a specialized company in Switzerland, the rework would probably not have been noticed (movement no. 703,890, case no. 680,970). This watch was produced in 1954.

Reference 2508 in Yellow Gold

The second PATEK PHILIPPE Ref. 2508—the previous example's counterpart in 750 yellow gold—also has a caliber 27 SC movement with the number 703,890. The case number is 680,970. The dial was refurbished (i.e., in this case, newly lacquered). Since the company name and minute scale are in enamel, these were preserved in their original form. In all, this update would hardly be noticeable had the dial maker, who was in charge of it, not forgotten the "SWISS" below six o'clock and put on two of the twelve indices with a less-than-perfect fit!

The design of the gold case corresponds to its counterpart in steel. This also applies to the pierced spring bar holes, the dustcover, and the decagon on the screw-down case back. The thickness of the screw-down case back is exactly 0.019 inches (0.5 mm). The caliber 27 SC is also identical; only its movement number is 4,615, behind that of the steel watch, which is about eight years older. The inscription on the mainspring barrel bridge is the same as on the previous watch. The only difference is the second Geneva hallmark, which is not stamped on the gear train bridge but on the mainspring barrel bridge.

On the case middle, we find the "little Helvetia" on the underside of the upper left strap lug, as well as on the side at four o'clock. The inscription on the beaded inner back also differs in detail from the previous watch. In rows one below the other there are the patent reference with Swiss cross, "PATEK, PHILIPPE & Co=" with symbolized comma, "–GENEVE–," and "SWISS." In the center are stamped the fineness hallmark "18K/0.750," the "large Helvetia," and the hallmark of the case maker (collective-responsibility mark no. 5, key no. 11, Manufacture Taubert, Genève, canceled 1974). Then follow "REF. 2508" and case number "689,970."

This watch was purchased by a collector for $8,065.78 (approx. 6,077.06 €) in the USA. He also paid the import sales tax of about $1,258. The watch was not serviced at Patek Philippe, even though this could have been included in the budget, given this low purchase price!

Reference 2508 watches are in great demand. At the Antiquorum auction in Geneva on March 16, 2008, two gold models were offered. A yellow-gold model with black dial and fitted Milanaise gold bracelet was sold for 38,400 CHF (movement no. 706,435, case no. 696,830). A rose-gold version with silvered dial and leather strap brought 23,400 CHF (movement no. 702,866, case no. 677,478).

For wearing: ++ For collecting: ++ As an investment: ++

Reference 2509 in Rose and Yellow Gold

Among the most-sought-after and most-popular models are the reference 2509 watches. This reference was launched in 1950 and remained in the program for more than fifteen years. With its small seconds, it looks almost more elegant than the previously described Ref. 2508 model. The 1961 German catalog lists the case diameter as 1.37 inches (34.8 mm) and the price as 1,245 DM. Due to the popularity of this model, we have chosen a yellow-gold watch and a rose-gold watch for this chapter.

Illustration 268
Reference 2509 watches are also among the most-popular models. With their small seconds they look almost more elegant than their predecessor, the 2508. In this photo are two different watches, one with a yellow-gold and the other a rose-gold case.

Installed in the first one, with yellow-gold case, is the caliber 12‴-120 movement with the number 938,783 and the case is numbered 575,731. This watch was thus made about three years earlier than its rose-gold counterpart. The case shows that this watch was also worn. But the hallmarks, still positioned low on the case middle at four o'clock, as well as on the bottom of the upper left strap lug, suggest that this case was probably very rarely reworked. This is also revealed by the red coloration of the gold, called gold oxide, which can be seen on the underside.

The refurbishment of the dial can be recognized only by a trained eye. The enameled lettering "PATEK PHILIPPE, GENÈVE," as well as the crosshair for the small seconds, likewise enameled, was preserved. After the golden indices were removed, the paint was removed from the dial and the underlying silver layer was brushed in a longitudinal direction. Thus the "SWISS" marking below six o'clock, which was only printed, was also lost.

Why the crosshair was kept a bit shorter toward six o'clock could not be ascertained. On another model of the Ref. 2509 that was delivered to FRECCERO, Montevideo, the crosshair for the small seconds was made in a similar shape (movement no. 720,057, case no. 672,126, case in rose gold). After the dial was refurbished, the gold indices could have been applied a bit more cleanly and evenly.

The case of the reference 2509 consists of two parts and still has a dust protection cover inside. The lugs have through-holes. The case middle was stamped, deep-drawn, turned, milled, and finished with the desired surface features in one go. The screw-down case back is turned and deep-drawn and has a decagon on the outside with a key width of 1.14 inches (29 mm). The thickness of the case back was measured at 0.019 inches (0.5 mm).

Illustration 269
These different views show all the relevant details of the rose-gold model. This includes the caliber 12'''-120 movement, a rear view of the case, the inner case back with inscription and hallmarks, and part of the dustcover. The caliber 12'''-120 has two stamped Geneva hallmarks: one in the main plate near the balance, and one on the mainspring barrel bridge, but here underlaid with gold bronze. The balance wheel has no shock protection. The inner back is finely beaded, and all inscriptions and hallmarks are clearly visible. This rose-gold watch is quite striking and once belonged to famous American entertainer Chuck Barris.

Two different types of movements were used in the reference 2509. The well-known caliber 12‴-120 was initially used, then from 1950 onward the successor caliber 12‴-400 was also used. The caliber 12‴-120 number 938,783, installed here, was produced in 1953, still without shock protection for the balance staff. The Geneva hallmark is stamped on the main plate near the balance wheel, as well as on the mainspring barrel bridge. Despite its age of about sixty years, this manually wound movement looks almost like new. The fact that there are no watchmaker's marks on the inner case back underlines this observation. The inscriptions on the beaded case back are the same as on the described yellow-gold Ref. 2508, which was also made by Manufacture Taubert, Genève. In contrast to the rose-gold version, which is still to be described, the name "VACUUM" is missing here.

This beautiful collector's item was bought by a German watch enthusiast at an auction in the USA for $7,000 (5,000 €), plus import sales tax of about $870. It could not be determined whether, having obtained the watch at such a low price, the buyer then decided to spend the money to have it overhauled by Patek Philippe.

The rose-gold counterpart was formerly owned by famous American entertainer Chuck Barris. Barris was active in radio and television, and in the 1970s the TV series *The Gong Show* was one of the most popular shows in the USA. This watch was auctioned for $6,037 (approx. 10,806.23 DM) at Christie's New York on June 22, 1998. In this gem is the caliber 12‴-400 with the movement number 726,820, while the case is numbered 693,707. From the pedigree book extract we learn that this rose-gold watch was produced in 1955 and sold on November 18, 1957. Rose-gold versions of the reference 2509 with luminescent hands and indices are rarely found. As a Patek Philippe insider revealed to the author, this version was produced only fifty times.

Our particular interest starts with the dial. The pedigree book extract tells us that this watch was delivered with a black dial. The dial mounted today has the former luminous hands and indices in rose gold but has a silvered finish. As well, the imprint "0 SWISS 0" reveals its later production date after 1968. Because the watch had a rather tightly screwed-down case back after an overhaul at Patek Philippe, the employee of the Ulm concessionaire could not open it for the photo studio. Thus, the description and photographing of the inner workings had to be dispensed with as well. Photos and a description of this watch appeared in an article in the March 2003 issue of *Chronos*. The beaded inner case back also bears the inscription "VACUUM," and the collective mark 5 with key no. 11 tells us the case maker was Manufacture Taubert, Genève.

Reference 2509 watches have always been popular. At Antiquorum in Geneva, a yellow-gold version was offered on February 25, 1990, and sold for 8,400 CHF. On November 14, 1999, a reference 2509 with a steel case was offered at the same auction house and sold for 14,400 CHF. A rose-gold model was also offered at Antiquorum in Geneva on March 16, 2008, and sold for 24,000 CHF (movement no. 725,448, case no. 689,652). A steel reference 2509 was called at Christie's in New York on October 26, 1998, and sold for $8,970 (approx. 14,710.80 DM; movement no. 962,863, case no. 2,605,631).

For wearing: ++ For collecting: ++ As an investment: ++

Reference 2526 Calatrava with Cream-Colored- and Black-Enamel Dial

The 12‴-600 AT automatic caliber gives this watch a level rarely achieved by other brands. On the right we see a particularly beautiful example in 750 yellow gold that was acquired by the heir of the first owner from Portland, Oregon. The reference 2526 was delivered with cases in 750 yellow, rose, and white gold. In addition there were special versions in platinum. These watches should not be thought of as having been prohibitively expensive at the time, as can be seen from Patek Philippe's 1961 catalog, where the successor to the reference 2526, the reference 3428, is advertised at the retail price of 3,880 DM. Because of the special importance of the reference 2526, we present four different variants. The first model has a cream dial and yellow-gold case. This is the watch from Portland, Oregon. According to the pedigree book extract, this automatic with caliber 12‴-600 AT, movement no. 763,637, and case no. 696,013, was made in 1956 and sold on June 28, 1957. This statement is also consistent with the information and documents provided by the first owner, who personally purchased this watch on the aforementioned date at Patek Philippe's headquarters on the Quai du Rhône. Ten years later he traveled back to Geneva and had his watch serviced for the first time. The dial, the real focal point of this watch, gleams in immaculate enamel and has survived the last sixty-three years well. Neither its dauphine hands nor its indices have lost any of their luster.

According to the manufacturer's catalog, the case diameter is 1.40 inches (35.6 mm). The case consists of a case middle with integrated bezel and a screw-down case back. The case middle was stamped from a single piece of metal, deep-drawn, turned, and milled, and the surfaces were finished. The screw-down case back was first manufactured as a turned part and then deep-drawn. The case back has a steel ring pressed into the inside. This accepts a spring clip, which in turn presses on the movement retaining ring when it is screwed down. It must be ensured that the two ends of the spring clip are always inside the steel ring. Otherwise the sharp ends will act like a lathe tool during screwing and can damage the movement-retaining ring. The case middle undergoes the same processing steps.

The automatic movement is in pristine condition and has been overhauled at Patek Philippe. Both technically and aesthetically it is one of the best and most beautiful movements ever made there.

The hallmarking of the case (i.e., the striking of the "small Helvetia") was different on reference 2526 watches. On the first of the three versions described, it was found between the two lugs at the bottom (i.e., at six o'clock). On another watch of this model, the "small Helvetia" was stamped twice, between the lugs as before and on the underside of the upper left lug. On a third watch, again only the last-mentioned place was hallmarked.

Now let's take a look at the inner case back. First of all, the bead-blasted surface can be seen. The inscriptions again appear one below the other with "PATEK, PHILIPPE & Co.," "–GENEVE–," and "SWISS." Then follow in a row, from left to right, the hallmark of the case maker, the gold fineness hallmark, and the "large Helvetia." Stamped below that are the case and reference numbers. The case maker, F. Baumgartner SA., Genève, was registered with the collective-responsibility mark no. 5 under the key no. 2. This trademark was canceled in 1973.

The watch presented was sold over the internet on January 24, 2001, for $12,100, and including transport and import sales tax the buyer paid about $16,198. This watch was in such a good condition that the overhaul performed at Patek Philippe in Munich cost only $620 (invoice dated February 21, 2003).

Among the most-beautiful versions of the reference 2526 are those models that have a black-enamel dial instead of the cream-colored one.

Beginning two years ago, the international auction catalogs put out by Antiquorum, Sotheby's, and Christie's have cited production figures for the reference 2526. According to these,

Illustration 270
Many watch enthusiasts would love to have this watch, the Patek Philippe Ref. 2526 Calatrava, also called "the one with the enamel dial." The timepiece shown comes from its first owner and was almost always kept in a safe (movement no. 763,637, case no. 696,013).

a total of 580 pieces were made: 480 in yellow gold, fifty in rose gold, thirty in white gold, and twenty in platinum. Of this total number, ten watches each with yellow- and rose-gold cases are said to have had black-enamel dials. Only François Cart, the production director at the time and today the administrator of the Patek Philippe archives, could tell us whether this information is correct. But François Cart is no longer among the living, and the guardians of the Patek Philippe archives maintain absolute silence!

Many years ago an employee of Patek Philippe, now deceased, entrusted the author with the figures given. This information was quoted only with reservation, because it could not be verified by personal research on the part of the author.

Dials with an enamel coating can keep their appearance for hundreds of years. Only shock-like temperature changes and mechanical stresses cannot be survived without damage.

A very important point with regard to this reference is the question of replacement dials, but this question is always answered negatively by the Patek Philippe service department. In the early 1960s a small series of replacement dials was issued. Here the number 100 was mentioned, although not documented. However, you can recognize these dials by their color. They do not match their cream-colored predecessors, instead turning out a more clinical white.

When you are buying a reference 2526 watch, the dial is of primary importance. Is it an original dial in good condition, or an original replacement dial, or was a lacquered dial installed

Illustration 271
In contrast to the Calatrava in the photo on the left, here is a very rare example with enamel dial and Breguet numbers. It deserves to be on display in the Patek Philippe museum in Geneva.

as a replacement? The latter can be recognized not only by its weaker shine, but also by the missing constriction at the hole for the small seconds. This feature is typical for the enamel coating. Detailed information on this topic can be found in chapter 12, "The Secret of the Dials."

For reference 2526 watches in good condition, one can estimate auction prices for the yellow-gold version of $43,000 to $55,000, and for the rose-gold model $55,000 to $65,000. A rare version in rose gold with a black-enamel dial was even sold at an auction in Geneva on October 11, 2003, for 157,500 CHF (approx. $174,000; movement no. 761,735, case no. 687,422).

At Antiquorum in New York, a yellow-gold Ref. 2526 with a black-enamel dial was offered on June 18, 2008, and sold for $61,200 (movement no. 762,329, case no. 689,066). On September 22, 2004, a rose-gold Ref. 2526 including gold bracelet (Klötzli strap) was also offered there and sold for $60,000 (movement

Illustration 272
The first self-winding watch from Patek Philippe from 1953 is the dream of every watch collector. The "necking" of the enamel layer can be seen around the hole for the small second hand. The case was made with a diameter of exactly 1.40 inches (35.60 mm). On the right is the cast-iron crown with the PP signet, which was present on all self-winding watches until 1964 (movement no. 763,637, case no. 696, 013).

Illustration 273
The reference 2526 looks even better with a black enameled dial. According to a Patek Philippe employee, exactly 580 examples of the reference 2526 were made: 480 in yellow gold, fifty in rose gold, thirty in white gold, and twenty in platinum. Ten pieces each in yellow and rose gold are said to have had black enameled dials. The model shown is probably one of them.

no. 762,624, case no. 691,288). A record price was achieved by a white-gold model with a fitted "Klötzli strap." It was offered at Christie's in Geneva on November 13, 2006, and sold for 156,000 CHF (movement no. 762,700, case no. 688,740).

Watches with platinum cases achieve even-higher sale prices. At Antiquorum in New York, one such watch was offered on March 24, 2004, and sold for $78,000 (movement no. 761,518, case no. 687,558). At Antiquorum in New York, another reference 2526 with platinum case was offered on June 14, 2007, and sold for $138,000 (movement no. 761,412, case no. 687,561).

An eBay bidder from Texas had also recognized how high the estimated value of the enamel dial of a Ref. 2526 could be. He listed an intact, cream-colored dial first for $20,000, and later for $15,750 (eBay item 290412971446, Providence Golf 1157). Bidding ended on April 11, 2010, without a buyer coming forward. More details can be found in chapter 12, "The Secret of the Dials."

For wearing: +++ For collecting: +++ As an investment: +++

Reference 2533

The 2500 reference series has quite a few outstanding watches, including the references 2532 and 2533, which have identical cases. The case middle made with integrated bezel and screw-down case back make this watch water resistant. Inside the reference 2532 is the 12‴-120 caliber movement with small seconds, while the reference 2533 has the 27 SC caliber with center seconds. The latter variant is presented.

There is a pedigree book extract for our yellow-gold Patek Philippe; there the caliber 27 SC is registered with movement no. 705,194 and case no. 691,533. The watch was made in 1956 and sold on August 28 of the same year. It was delivered with a white dial, gold indices, and leather strap.

The creamy-white dial corresponds to the typical Patek Philippe design. The dashes for the minute scale are divided with four more dashes for twelve seconds each. There are dauphine hands to indicate hours and minutes in 750 yellow gold like the indices. The company label corresponds to the regulations valid from 1952 and reads "PATEK PHILIPPE, GENEVE," the latter without the *accent grave*. The case middle forms a single unit with the bezel. It was stamped from a single piece of metal, deep-drawn, turned, and milled, and the surfaces were given the desired finish. The screw-down case back, which has a decagon, is also stamped from a single piece of metal, deep-drawn, turned, milled, and surface-finished. The thickness of the case back is again exactly 0.019 inches (0.5 mm). The inner case back is bead-blasted. Overall the 1.35-inch-diameter (34.5 mm) case has not received a new polish in decades and has already taken on a reddish surface color due to the gold oxide.

The manually wound caliber 27 SC movement, protected by a dustcover, deserves the "as new" rating. The shock protection is made by Jaeger-LeCoultre. The Geneva hallmark can be found on the main plate and the mainspring barrel bridge. Surprisingly

Illustration 274
The water-resistant Patek Philippe reference 2533A should be a watch for every day and every time of day. The references 2532 and 2533 have the same case but different movements. The former contains the caliber 12‴-120 with small seconds, the latter the caliber 27 SC.

the former is also gold colored (i.e., painted with gold bronze). The "small Helvetia" can be found on the underside of the upper right lug, and the last three numbers of the case number (533) on the underside of the lower left lug.

The inscription on the inner case back is very inconspicuously done and reads in sequence "PATEK, PHILIPPE & Co," "–GENEVE–," and "SWISS." Then, in a row are stamped the case maker's hallmark, the gold fineness hallmark "18K/0.750," and the "large Helvetia." Below that, the case and reference numbers can be found. The presence of two watchmaker repair marks suggests—as does the good case condition—that this watch was rarely used.

The hallmark, manufacturer's mark no. 1—a hammerhead with the number 170—identifies the case maker. It was the Cédex company, Charles Dubois SA, Le Locle. It was canceled in 1982. This same case maker was responsible for the case of the reference 2483 on page 164.

This Patek Philippe was sold at Christie's in New York on April 12, 2000, for $5,875, and another $2,500 (2,262.07 €) had to be paid for the overhaul in Geneva. The two most-expensive single items included in this general overhaul were the movement overhaul, which cost $1,415 and the dial restoration for $628 (invoice from January 25, 2002).

Illustration 275
View of the movement of the reference 2533. The caliber 27 SC is deserving of the "as new" rating.

In recent years, reference 2533 watches have only rarely been offered for sale. At Antiquorum in Geneva a yellow-gold version was sold for 13,800 CHF on October 16, 2005.

For wearing: ++ For collecting: ++ As an investment: ++

Reference 2540

Patek Philippe's second self-winding wristwatch was given reference number 2540 and was unveiled in 1954. It was nicknamed the "TV Patek Philippe." The version presented here has a rose-gold case and the caliber 12‴-600 AT movement and is in absolutely mint condition. The movement number can be read as 764,811, the case number as 699,844. According to the pedigree book extract, this watch was made in 1957 and sold on May 28, 1958.

The dial is silver in color and has baton hands and dash indices in rose gold. The square case back has a base dimension of 1.24 × 1.24 inches (31.5 × 31.5 mm) and soldered lugs. The square case back is deep-drawn and pressed and has a material thickness of 0.015 inches (0.4 mm) at its thinnest point. The case surface is polished all around; only the back is satin-brushed. The caliber 12‴-600 AT movement is mounted in an externally square, very elaborately designed, triple-stepped movement retaining ring (*see next page*). It is attached to the case middle by four cylinder head screws. The movement itself is secured in the movement-retaining ring by three screws.

The "small Helvetia" can be found on the bottom of the upper left lug. The company markings inside the case back consist of "PATEK-PHILIPPE & Co=," "–GENEVE–," and "SWISS." On the lower third of the case back there is a row of markings consisting of the case maker's hallmark, the gold fineness hallmark "18K/0.750," and the "large Helvetia." Below this are the case number and the reference number with the prefix "RÉF." The case maker's hallmark corresponds to collective-responsibility mark no. 5, and the key bears the number 26. It belonged to Ponti, Gennari & Cie., Genève, which was canceled in 1969.

At Antiquorum in Geneva a yellow-gold watch was auctioned for 16,800 CHF on November 13, 2005, and a rose-gold one for 22,800 CHF.

The model shown came from the USA and was purchased there for $12,500. A service at Patek Philippe in Geneva added another $2,045 to the bill. In all a good deal, which shows that such exceptional watches will always be in demand in the future.

To avoid any irritation for the attentive prospective buyer and dedicated collector, we also mention the references 2540/1 and 2540/2. Both models have a manually wound movement. The reference 2540/1 is equipped with the caliber 12‴-400 with small seconds. Inside the reference 2540/2 is the manually wound 27 SC movement with center seconds.

Both variants are offered very rarely. At Christie's in Geneva, a yellow-gold reference 2540/1 was offered on November 17, 2008, and sold for 20,000 CHF ($16,771; movement no. 727,470, case no. 694,485). At Antiquorum in Hong Kong, a yellow-gold version of reference 2540/2 was called on May 27, 1991, and sold for 38,400 CHF (movement no. 706,642, case no. 694,540). An auction of Sotheby's in New York enabled another estimation of value to be made. A yellow-gold reference 2540/2 was offered there, and on April 14, 2005, it sold for $12,000 (caliber 27 SC movement, no. 705,799).

For wearing: ++ For collecting: +++ As an investment: ++

Reference 2551 in Two Different Conditions

The reference 2551 watches are undoubtedly among the most-beautiful creations by Patek Philippe. This model was unveiled in 1954. Here we present two examples of this watch. One can be considered almost virgin, while the other has been serviced several times. In the following we examine both watches side by side and mention possible differences. For simplicity's sake, we will call the watch that is almost still brand new the first, the other the second. Both watches are equipped with the caliber 12‴-600 AT movement. The first has the movement with the number 766,071 and the case no. 2,608,560. According to the registry extract, it was produced in 1958 and sold on October 9, 1959. The second watch has the movement no. 766,216 and

Illustration 276
Anyone who is passionate about watches should have one from Patek Philippe. The reference 2540 is the second model that received a self-winding movement at the Geneva manufacture. Also beautiful is the cast crown, with which all automatic watches were fitted until 1964. All further information about the reference 2540 can be read in the accompanying text.

Illustration 277
The opened reference 2540 reveals the caliber 12'''-600 AT movement as well as the inner case back. The movement is held by an elaborately designed movement-retaining ring, which in turn is attached to the case by four screws (movement no. 764,811, case no. 699,844).

the case no. 2,608,524. Thus, this movement followed 145 pieces later, while the case is thirty-six pieces ahead. Which principle of order Patek Philippe used at that time for the installation of dials and cases can no longer be traced. In any case, both watches have the same year of manufacture, 1958. The differences in the dials are interesting. Baton-shape indices and hands were selected for the first one. The dial is also the original. This can be recognized by the company label "PATEK PHILIPPE, GENÈVE" and "SWISS" below six o'clock. The second watch has wedge-shaped indices and dauphine hands. The dial has been refurbished. The silvered surface is finely brushed in a vertical direction, and the company name, crosshairs for the small seconds, and the proof of origin are partly printed too large or too bold. Because this work was, from a purely technical point of view, neatly executed, the differences should be noticeable only to the expert or by direct comparison.

The reference 2551's two-piece case is very elaborately designed. The case middle with integrated bezel was made as a turned part, while the lugs, which are made as die-cast parts, are profiled and soldered on (i.e., precisely adapted to the outer shape of the case). The screw-down case back was turned, deep-drawn, turned again, and diamond-polished. The case diameter of the first watch is 1.41 inches (36.0 mm), the second one 1.40 inches (35.8 mm). The material thickness of the screw-down case back measures 0.019 inches (0.5 mm) for the first watch and 0.017 inches (0.45 mm) for the second. The case back surface, which was originally diamond-coated, was re-turned on the second watch. This is also the reason for the difference in the thickness of the material. In general, the sharp edges of the first watch show that the case has retained its original shape. The movements of both watches were serviced in Geneva and restored to nearly new condition. Their movement values are excellent.

The case hallmarking is present only on the first watch. The "small Helvetia" is stamped on the outside of the lower right lug. On the second watch this hallmark has been polished off. The surface of the inner back was bead-blasted during production. Why both backs were reworked with emery paper could be determined only for the first watch. Did they want to remove any repair marks from the second one? The markings on the inner case backs of both watches are identical, since the cases were produced by the same maker. Listed from top to bottom are the "large Helvetia," the gold fineness hallmark "0,750," "PATEK-PHILIPPE & Co=," "–GENEVE–," and "SWISS." Below that follow the reference no. "RÉF 2551," the case number, and the case maker's hallmark. The collective-responsibility mark no. 5 with key no. 1 identifies Ed. Wenger SA, Genève, canceled in 1992. There is also a watchmaker's repair mark "W 8215439" scratched into the first watch. According to the owner of this watch, its movement was overhauled in March 1994 at the Munich specialty store Antike Uhren Eder. About ten years later it was overhauled again at Patek Philippe in Geneva.

The second Ref. 2551 comes from another collector who bought his watch in the 1990s for $6,977 and ten years later had it overhauled at Patek Philippe in Geneva for $1,116.

Illustration 278
With a case diameter of 1.41 inches (36 mm), the reference 2551 was one of the largest watches of the 1950s from Patek Philippe. The elaborately profiled case was created as a turned part; the lugs are soldered. The model shown on the right is still in the best original condition and has probably never been refurbished. The watch on the left has a refurbished dial. In comparison with the one on the right, one can clearly see that the company name on the watch on the left was printed too large and too bold. The dial surface is silvered and vertically brushed (watch on the right: movement no. 766,071, case no. 2,608,560; watch on the left: movement no. 766,216, case no. 2,608,524). Interestingly, the movements of these two watches are separated by 145 numbers, and the cases are just thirty-six numbers apart.

At auctions, reference 2551 watches are very much in demand. At Antiquorum in New York, a version with a platinum case was offered on December 7, 2006, and sold for $66,000. A month earlier, on November 12, 2006, a model with a rose-gold case was offered at Antiquorum in Geneva and sold for 36,000 CHF. On later occasions, yellow-gold watches could be purchased for between 20,000 and 30,000 CHF.

For wearing: ++ For collecting: +++ As an investment: ++

Reference 2552 in Two Different Versions

In 1955 Patek Philippe unveiled another self-winding watch, the reference 2552. The wide, double-stepped bezel makes this watch look very large, although it has the same case diameter of 1.41 inches (36 mm) as the previously described reference 2551. Because of the importance of this automatic, we again present two models. The first watch has the movement no. 762,058, case no. 688,312, and was produced in 1955. The second watch has the movement no. 762,428 and the case no. 689,767. According to the logbook extract, it was produced in 1955 and sold on May 14, 1956. Thus, the movement numbers are 370 pieces apart and the cases 1,455 pieces apart. The reference 2552 was issued with cases in yellow, red, and white gold, as well as in platinum.

The dial is made of gold and was galvanically silvered and lacquered. The company name, the graduation marks, and the crosshairs for the small seconds are enameled. The "SWISS" proof of origin below six o'clock is printed. On the first watch presented, the dial is still in original condition. The indices are made of 18K gold and riveted. The dauphine hands complete the elegant look of this watch.

On the second watch the dial has been refurbished, but not authentically. Instead of the seconds scale with crosshairs and eight graduations, the dial has an uneven crosshair with a too-short second hand. The service department of Patek Philippe in Geneva could be convinced of the correct execution of this dial only on the second attempt. The case with integrated bezel is made as a turning part. The lugs are soldered in place. The bezel is double stepped and the cast crown is semirecessed. The best distinguishing feature of this model is its double-stepped bezel. The screw-down case back was turned from a solid piece of metal and deep-drawn. The six grooves ensure secure closure. The whole case is diamond cut; only the flat bottom surface is turned. The material thickness of the case back is only 0.015 inches (0.40 mm) instead of the 0.019 inches (0.50 mm) used on the first watch. Either a dedication engraving was removed here or the back was reworked for other reasons.

The caliber 12‴-600 AT movement is in very good condition. The watch received a complete overhaul at Patek Philippe in Munich. The "small Helvetia" was stamped on the outside of the lower right and upper left lugs. Unfortunately both hallmarks are recognizable only as relics. This case was polished too often and perhaps unprofessionally. The inscriptions on the inner backs of both watches are identical except for the case number, since both cases were made by the same case maker. At the top of the case back, the "large Helvetia" and the gold fineness hallmark "0.750" are stamped one below the other. Then follow the company name "PATEK-PHILIPPE & Co=," "–GENEVE–," and "SWISS." On the lower third of the case back are the reference no. "RÉF. 2552" and the case number. At the very bottom is the case maker's hallmark. In both cases this belongs to the collective-responsibility mark no. 5 with key no. 1, belonging to the company Ed. Wenger SA, Genève, which was canceled in 1992. The inner case back of the first watch was most likely bead-blasted again, probably after the previously described repair. There are a total of three watchmaker repair marks incised. The inner case back of the second clock was sanded out so that neither the bead-blasted surface nor possible watchmaker repair marks are visible. The material thickness of the back is a measured 0.019 inches (0.50 mm). The first reference 2552 was purchased by a German watch enthusiast for $8,377. The second came from the USA and cost $5,569.99 (4,655.24 €). It was subsequently passed on to Patek Philippe in Munich and a complete overhaul was commissioned. Including refurbishment of the dial, $2,500 was charged for this. The most-expensive items in this total were the movement service at $1,242 and the dial restoration at $577.

Illustration 279
The reference 2552 is also one of Patek Philippe's big watches of its day. Its case diameter is exactly 1.41 inches (36 mm). The case was created as a turned part; the lugs are soldered. For protection the crown is semirecessed. The dial of the watch on the left is still original; the one of the right watch is refurbished. The small seconds graduation marks were mistakenly not executed according to type. A "crosshair" like this one is present on the dial of the reference 2551. The small seconds hand is also not true to type (watch on the left: movement no. 762,058, case no. 688,312; watch on the right: movement no. 762,428, case no. 689,767).

Reference 2552 watches are very popular among watch enthusiasts and are also often sold at international auctions. This watch was also called the Disco Volante, or Flying Saucer. At Antiquorum in Geneva, a model with white-gold case was sold on October 15, 2006, and changed hands for 50,400 CHF (movement no. 762,442, case no. 690,970). On November 11, 2007, a watch with a yellow-gold case and dial signature "TIFFANY & CO" was offered by the same auction house and sold for 22,200 CHF (movement no. 762,586, case no. 689,840). At Antiquorum in New York, a rare version in platinum was called for $60,000 on October 16, 2008, but found no buyer (movement no. 761,840, case no. 688,392). Perhaps the economic crisis, which was just beginning at that time, caused potential buyers to be cautious.

For wearing: ++ For collecting: ++ As an investment: ++

Reference 2554

Among the extravagant men's watches was the reference 2554. The special-design distinguishing characteristics are its rectangular dial and the curved sides of its case. It is no coincidence that this design was rediscovered in time and used for the anniversary model celebrating the second millennium of our era. As the reference 5100 "10 Jours," this manually wound watch has a power reserve of a full ten days.

The Ref. 2554 men's wristwatch with caliber 9'''-90 movement was unveiled in 1955. The model presented has movement no. 976,999 and case no. 696,927. According to the logbook extract, this watch was produced in 1956 and sold on September 16, 1957. This watch was available with cases in yellow, red, and white gold, and also in platinum. In the 1961 German Patek Philippe catalog it was listed in yellow gold with a price of 1,285 DM. The white-gold model was priced at 1,475 DM, and the platinum model at 2,740 DM.

The dial is silver lacquered and has baton-shaped indices with matching baton-shaped hands. The case is made in two parts and is very elaborate. The case back is also quite complex and was soldered together from several parts. The half-shell bearing, which is in the press-fit case back, provides guidance and stability to the winding stem. The movement is in mint condition. It is housed in a special movement bearer.

There are no hallmarks on the case. The number 27, the last two digits of the case number, is stamped in the middle of the case at eight o'clock. The markings on the inner bottom differ somewhat from the previous designs. Applied transversely, from top to bottom, are the gold fineness hallmark "0.750," the "large Helvetia," and the case maker's hallmark. They are followed by "PATEK, PHILIPPE & Co=," "–GENEVE–," and "SWISS." Underneath are the case number 696,927 and "REF 2554." The case back surface is bead-blasted. The case maker's hallmark belongs to collective no. 5 and bears the no. 7 key. The supplier was the company Fabior, Olivier Judas, Gimel, which is still in business today.

Figures 280 and 281
Among Patek Philippe's more extravagant watches is the reference 2554, also called the "Manta-Ray" among enthusiasts. The imposing design was also used in the reference 5100 "10 Jours" anniversary model. The occasion was the beginning of the second millennium. This watch has not only a special design, but also a very elaborate case. Impressive is the rectangular 9'''-90 movement, built since 1934 and mounted in a special movement bearer. Also good to see is the Gyromax balance wheel with the adjustable *masselotte* weights. The press-fit case back is closed in such a way that the "nose" visible at the front serves as a half bearing for the balance staff (movement no. 976,999, case no. 696,927).

The watch shown was purchased by a collector for $9,586 at an art auction house in Hamburg. He paid another $1,521 to have the watch serviced. In all a good deal, as well as a decent investment!

Reference 2554 watches often appear at auctions. At Antiquorum in Geneva, a red-gold watch was sold for 40,800 CHF on April 2, 2000. The same auction house in New York sold a yellow-gold version on December 7, 2007, for $25,200.

For wearing: + For collecting: ++ As an investment: ++

Reference 2583

In the 2500 reference series, after references 2551 and 2552 there are only three more models with automatic winding. Of these, references 2583 and 2584 have case diameters of 1.41 inches (36 mm) and press-fit case backs. The reference 2585 is available only with a steel case with a diameter of 1.37 inches (35 mm) and a screw-down case back. Reference 2583 watches are hard to find. The watch presented has the caliber 12‴-600 AT movement, number 763,399, while the case bears the number 697,241.

The dial, which is quite large, with a diameter of about 1.26 inches (32 mm), is silvered and finely brushed in a vertical direction and has applied gold baton indices. Only the "12" is displayed in Arabic numerals. Also present are matching baton hands.

The company signature "PATEK PHILIPPE, GENEVE" and the graduation marks for the small seconds are done in enamel. Only the *accent grave* over the second "E" in "GENEVE" did not survive the dial refurbishment without damage. A lot of effort went into the case of this reference 2583. The bezel is made as a turned part and installed on the case middle. The latter is also made as a turned part. The solid lugs are profiled and soldered. The press-fit case back is deep-drawn; the material thickness is 0.019 inches (0.5 mm). A small lug facilitates the opening of this waterproof case. The latter virtue is achieved by a clever design and a rubber gasket. The lug-to-lug measurement is 0.80 inches (20.2 mm), which is unusual for Patek.

The "small Helvetia" is stamped on the outside of the lower right lug. The information on the inner case back tells us more. At the very top, the markings begin with "PATEK, PHILIPPE & Co=," "–GENEVE–," and "SWISS." Then, in a row, are the case maker's hallmark, the gold fineness hallmark "18K/0.750," and the "large Helvetia." Then, far away, the case and reference numbers are stamped. The surface of the inner back is bead-blasted. A total of seven watchmaker's marks are incised, testifying to the lively use of this watch. The signature "W. E. WARDMAN, 1958" on the outer case back probably identifies the first owner of this watch.

This collector's item was completely overhauled at Patek Philippe; consequently the caliber 12‴-600 AT movement is in mint condition.

Illustration 282
The references 2583 (*left*) and 2584 (*right*) are self-winding watches with a case diameter of 1.41 inches (36 mm). At the top right you can see the reference 2583 from the back and the press-fit case back with its small "nose." Due to a clever design and a special rubber seal, both watches are water resistant (reference 2583 on the left: movement no. 763,399, case no. 697,241; reference 2584 on the right: movement no. 765,654, case no. 2,606,355). The watch on the right has the signature KOCH printed on the dial below the center. This was the name of the concessionaire in Frankfurt am Main at that time.

In the past years the author has seen only two examples of this reference. On October 15, 2006, such a watch came up for auction at Antiquorum in Geneva and sold for 18,000 CHF.

The watch shown was bought by a collector for $5,609 (4,739.53 €) in the USA. Import sales tax and an overhaul at Patek Philippe added another $826 to the bill. Despite all this, this is a rare and sought-after watch that can be expected to increase in value in the future.

For wearing: ++ For collecting: ++ As an investment: ++

Reference 2584

The reference 2584 is another of Patek Philippe's big watches. The one seen here was sold by former concessionaire Juwelier R. Koch, Kaiserstraße 25, Frankfurt am Main, as shown by the small signature "KOCH" below the hands. Today there are only the wholesalers Wempe and Bucherer in the metropolis on the Main. From the Institute for City History of Frankfurt am Main, Company Archives W1, we learned more about the history of the later Patek Philippe concessionaire: "In 1879, Robert Koch opened a jewelry store in which his brother Louis became a partner. They soon earned international fame. A branch was opened in Baden-Baden. Among their customers were numerous royal houses, including the House of Hohenzollern, so that at the turn of the century the Koch brothers acquired the title of court jewelers. Even after the death of the Koch brothers, the store retained its exclusive orientation. The business continued until 1986."

The caliber 12‴-600 AT movement is numbered 765,654 and the case 2,606,355. According to the pedigree book extract, this watch was produced in 1958 and sold on March 12, 1959. The lacquered dial is still original and shines in its old glory despite its age of about fifty-six years. The company signature and the graduation marks of the second scale are in enamel, and the *accent grave* over the second "È" in the proof of origin also looks like new.

Even the case with the cast crown appears almost flawless. The bezel and case middle form one unit. The press-fit case back is deep-drawn, turned, and diamond-finished. Its design can be described as very elaborate. A screw-down case back would probably not have turned out to be much more expensive. The lug-to-lug measurement is 0.767 inches (19.5 mm).

The condition of the movement can be described only as "like new." After a complete overhaul at Patek Philippe in Munich, another assessment would have been out of the question. The "small Helvetia" hallmark can be found on the outside of the upper left lug. The same hallmark is also stamped on the case middle at six o'clock. Stamped on the inner case back, from top to bottom, are "PATEK, PHILIPPE & Co=," "–GENEVE–," and "SWISS." In the center there is another row of markings consisting of the case maker's hallmark, the gold

Illustration 283
Reference 2584 with KOCH, the name of the concessionaire in Frankfurt am Main, printed on the dial

fineness hallmark "18K/0.750," and the "large Helvetia." Below that are the case and reference numbers. The surface of the inner back has not been refinished. The hallmark of the case maker is from collective no. 5 and has the key no. 2. This number stands for the Firma F. Baumgartner SA, Genève, which was canceled in 1973.

The watch presented belongs to a German collector who purchased it on the internet for $7,200. He paid another $3,267 for the complete overhaul. Restoration of the movement at $1,242 and the refurbishment of the dial at $577 were the most-expensive items in this sum (invoice from October 12, 2008).

Reference 2584 watches are often found in international auctions. At Antiquorum in Geneva, a yellow-gold version of the reference 2584 was sold for 20,400 CHF on May 13, 2007. However, it is impossible to say whether this watch was in a comparable condition to our watch shown.

For wearing: ++ For collecting: ++ As an investment: ++

Reference 2585 in Stainless Steel

Here we present another and very special rarity from Patek Philippe, the reference 2585 in stainless steel. Even the author initially had seen only the specimen shown, even though he had already encountered it twice at auctions.

According to the pedigree book extract, this rare piece was produced in 1958 and sold on April 11, 1960. The caliber 12‴-600 AT movement has the number 765,932, and the steel screw-down case has the number 694,414.

The dial is silvered and has steel indices and dauphine hands made of the same material. The case is kept in two parts and consists of the case middle with integrated bezel and a screw-down case back. The case middle is stamped from a single piece, deep-drawn, milled, turned, and finished with the respective surfaces. The case diameter is exactly 1.37 inches (35.0 mm), and the case design is almost identical to that of the reference 2577 R-SCI, one of the few watches with an enamel dial. This similarity can also be seen in the small grooves cut between the tops of the strap lugs and the cylindrical part of the case. The caliber 12‴-600 AT is held in place by three screws. Except for the reference and case numbers, the inscription on the inner back corresponds to that of the reference 3417 (see p. 186).

For both the reader and the author, this rare watch has a particularly interesting history. In the 1980s and the early '90s, the author used to collect wristwatches. As a goal he had planned to acquire only those with automatic winding and screw-down case backs, specifically those made by the noble brands Patek Philippe and Vacheron & Constantin.

The author was present at an auction at Sotheby's on Monday, October 28, 1996, in New York. The reference 2585 shown here was called under lot no. 256. The author bid up to $9,000, but

Illustration 284
One of the very rare watches in this book is this reference 2585 in stainless steel. Apart from Patek Philippe, no one can say how many examples of this rarity were produced. The author estimates that there were probably no more than four or five pieces. The extremely interesting history of this watch can be found in the accompanying text (movement no. 765,932, case no. 694,414). At Patek Philippe there used to be only a few self-winding watches with a water-resistant steel case, which included the reference 2585, the reference 3466 watches, and the Nautilus reference 3700/1. *Christie's photo*

the watch went up in price and was finally sold at $10,000 ($11,500 including a 15% buyer's premium). This was unfortunate for the author, as we will later learn, because apart from the Patek Philippe reference 3454 (see p. 194), this model was the only one missing from his collection at that time!

Almost nine years later, the same watch reappeared at Christie's in Geneva and was called for CHF 40,000 at the auction on May 16, 2005. After a fierce bidding war, this watch finally went to its new owner for exactly 104,400 CHF (approx. $85,194)! In just under nine years, the value of this watch had increased by $73,695, or 740%! There was great surprise when this watch reappeared at a Phillips auction in Geneva on May 10, 2015, and was offered with a call price of 80,000 to 120,000 CHF. Strangely enough, the movement and case numbers were not mentioned in the description of this watch, but the author was discreetly informed that it was the watch that was auctioned at Christie's Geneva on May 16, 2005. The author also had additional proof of the identity of this watch. After a long, fierce bidding war, the auction hammer finally fell for this watch at an incredible CHF 785,000!

That the previously described purchase price was probably an isolated case is evidenced by the offer of yet another Patek Philippe Ref. 2585 at a Phillips auction on November 11–12, 2017. With a call price of 50,000 to 100,000 CHF, this watch went to its new owner for 100,000 CHF (movement no. 762,967, case no. 694,408). With the case numbers 694.408 and 694.414, these watches are only six numbers apart. Does this lead to the conclusion that only six of this reference were made?

Certainly these cases represent special examples of increases in value. However, even in less spectacular cases, considerable returns can be earned, as we will learn later in this chapter.

Finally, a few words about the number of reference 2585 watches made. The author now knows of three, and watch expert Dr. H. Crott four of these rare watches. One can be curious when one of these rare specimens will again come up for auction.

For wearing: ++ For collecting: ++ As an investment: +++

Reference 3403

The 3400 reference range includes many interesting wristwatches. The first one worthy of mention is the reference 3403, a waterproof gold watch with automatic winding. The logbook extract shows that this watch was made in 1956 and sold on August 30, 1958. The caliber 12‴-600 AT movement is numbered 764,061, and the case 309,891. The year of manufacture listed in the pedigree extract as 1956 is somewhat surprising, since according to Patek Philippe, the reference 3403 was not unveiled until 1958. The dial of the watch presented is painted a cream color and is still in its original condition; the light patina does not bother one. The company signature "PATEK PHILIPPE, GENÈVE," the concessionaire signature "TIFFANY & CO.," and the graduations of the small seconds are enameled. The concessionaire signature in enamel is a novelty. Normally the names of the preferred concessionaires were printed on the dials after they were finished.

Strong, profiled indices and dauphine hands convey the exact time to the wearer of this watch. The harmonious case measures exactly 1.28 inches (32.5 mm) in diameter. This value, measured with calipers, contradicts some literature references, which often mention 1.26 or 1.29 inches (32 or 33 mm). The case consists of a solid case middle with an integrated bezel. It is stamped from a single piece of metal, turned, deep-drawn, milled, and surface-finished. The sturdy screw-down case back is also deep-drawn and turned. It is stepped in several stages and has a screw cap with a decagon. Also worth mentioning is the elaborately designed gift engraving on the screw-down case back. The company Forest Oil gave manager Ralph E. Warner this watch in 1958. Forest Oil Corporation was founded in 1916 in the USA and is today one of the largest and still-independent companies for the exploration, production, and marketing of oil and gas. After a complete overhaul at Patek Philippe in Geneva, the movement also shines like new. The Geneva hallmark was placed on the main plate as well as on the changer bridge, in both cases backed with gold paint. For many watch enthusiasts the caliber 12‴-600 AT is also considered the most beautiful automatic movement ever made.

The "small Helvetia" is engraved on the side of the case at four o'clock. Viewed under a stereomicroscope, the letter "G" for the Geneva Examination Office can be seen under Helvetia's head. Further, one gets the impression that this case has not been refinished. The markings on the inner bottom differ somewhat from the practice described so far. In the upper part, the "large Helvetia," the gold fineness hallmark "0.750," and the case maker's hallmark can be found one beneath the other. "PATEK, PHILIPPE & Co=," "–GENEVE–," and "SWISS" are stamped on the lower part of the case back. The "GENEVE" lacks the *accent grave* and a dash on the right, and the "SWISS" is again followed by a period. Underneath is the reference no. 3403 without prefix. The surface of the inner case back seems to be bead-blasted a little bit because the turning marks can still be seen. The collective-responsibility mark no. 5 with case key no. 4 tells us that the case maker was Antoine Gerlach SA, Genève. This company was deleted in 1977.

The watch shown came from Christie's in New York and was purchased on October 24, 2001, for $14,100. The watch was serviced by Patek Philippe in Munich at a cost of $1,146, including an extract from the pedigree book. The two most-expensive items were the movement service at $485 and the gold rotor for $283 (invoice from November 11, 2002).

Reference 3403 watches are rarely offered. At Antiquorum in Geneva, a yellow-gold version was offered on October 12, 2003, and sold for 16,800 CHF. At Antiquorum in New York, a yellow-gold model was called on December 1, 2004, and sold for $16,800.

Another reference 3403 with the dial signature "TIFFANY & CO" and a gift engraving by the Forest Oil Corporation was called and sold for $14,400 (movement no. 764,444, case no. 309,897) at the same auction. Our reference 3403 shown is only six numbers ahead, with case no. 309,891. The watch auctioned at Antiquorum had been sold on the internet three years earlier. The antique dealer Hess Fine Art in Tampa–St. Petersburg, Florida, had offered exactly this collector's item on eBay and sold it on April 28, 2001, for $13,355. If you deduct the consignment fees and the buyer's premium from the auction result of $14,400, the seller lost about $2,500!

For wearing: + For collecting: +++ As an investment: ++

Reference 3415

In the case of the reference 3415 described, we have again presented two examples on page 184. Besides the different cases in rose and yellow gold, these two watches also differ in other details. And they both contain exciting and, for the collector, instructive details that are worth hard cash when buying such valuable collectibles. Even for experienced professionals and dealers, the details and facts described here are of great value—both for buying and selling valuable wristwatches of this brand.

The automatic caliber 12‴-600 AT is found in both watches. The movement number of the rose-gold watch is 766,091, the other 766,095. They are therefore only four numbers apart and were probably produced on the same day. The case number of the rose-gold watch is 2,608,806; that of the yellow-gold watch is 2,608,696. The case numbers are 110 pieces apart. Both cases were supplied by the same case maker. We have a pedigree book extract for the yellow-gold reference 3415. According to it, the watch was produced in 1958 and sold on September 29, 1959. Let's start with a description of the dials. In both cases they are still in their original condition. As usual with Patek Philippe, these two dials are made of 750 gold plate, but

Illustration 285
Comparison of the reference 3403 in yellow gold (*right*) and the 3415 in rose gold (*left*). Both watches are powered by the caliber 12'''-600 AT automatic movement and feature a water-resistant case with a screw-down case back. The only difference between the two watches is the case diameter: the one on the right has a diameter of 1.27 inches (32.5 mm), the one on the left 1.37 inches (35 mm). Both watches still have their original dials. The one on the right bears the vendor's signature TIFFANY & CO. The dial of the reference 3415 has a number of special features. The hour indications for 12, 3, 6, and 9 are applied indices. The remaining hour indications are faceted. In faceting, a diamond steel is used to cut a wedge-shaped indentation into the dial, revealing the gold hidden beneath the varnish. More details about this rarely used finishing technique can be found on this page.

rose gold was used on the first and yellow gold on the second. We can see the importance of this distinction in the design of the two dials. The company signature "PATEK PHILIPPE, GENÈVE" and the graduation marks of the small seconds are enameled, and the surface is painted with light rose or yellow-gold-tone cream. The hour indications 3, 6, and 9 are provided with one index each, the 12 with two indices. The remaining hours are faceted. In faceting, a diamond steel (acute angle of 60° or 90°) is used to draw a wedge-shaped indentation into the dial by machining. This rarely used technique is carried out after the dial has been lacquered. For this purpose the dial is picked up by a dividing head and faceted at the correct angle and the desired length. To prevent the surface of this wedge-shaped depression from oxidizing, the dial must be made of 18-karat gold, as mentioned earlier. The "SWISS" label below six o'clock is not visible. Whether it is present at all can be clarified only when the watch is next disassembled. The yellow-gold watch bears the name of the concessionaire, "WILH. SCHAAFF, HEIDELBERG." The jeweler Wilh. Schaaff, Hauptstraße 34, Heidelberg, is listed in Patek Philippe's German catalog for 1961. At that time, he was one of a total of fifteen concessionaires in Germany. According to a relative of this long-deceased jeweler, the business existed until the 1980s. Today there is no longer a concessionaire for our brand in Heidelberg. The case of the reference 3415 consists of two parts: the case middle with integrated bezel, and the screw-down case back. The case middle is made as a turned part. The lugs are soldered in place. The solid screw-down case back is stamped, deep-drawn, and turned, and the round parts are diamond-polished. The flat section of the screw-down case back has a measured material thickness of 0.019 inches (0.5 mm). The case measures 1.37 inches (35.0 mm) in diameter. In the bezel area the case middle is twice slightly tapered. The outside is kept cylindrical and has a small step and a conical finish at the bottom. A case contour like this can be achieved only by turning.

Illustration 286
Fine-adjustment mechanism of the caliber 12'''-600 AT movement in about 20× magnification. One does not have to be a watchmaker to recognize all the relevant components. First, a note about the Gyromax balance wheel. On the first design, the inertia weights were positioned in recesses in the balance ring. One can also see the Breguet hairspring, the hairspring stud, the regulator arm, and the micrometer screw. Several other details can also be seen (e.g., the fact that the gooseneck spring is double-screwed from below). To adjust the rate, one can use either a regulator arm, which limits the oscillation length of the hairspring, or a Gyromax balance wheel, which makes it possible to control the GD^2 moment of oscillation. Here both options are installed, so the system is doubled, as they say in the vernacular. On later Patek Philippe movements, only a Gyromax balance wheel was installed.

Illustration 287
At first glance the references 3415 and 3429 look confusingly similar. The case middles with integral bezels are identical; only the case backs differ in minor details. The dial serves as a reliable distinguishing feature. That of the reference 3429 has applied indices, while the dial of the reference 3415 has the faceting explained in the caption for illustration 288. On the two watches below, one can clearly see the faceting on the one on the right. This reference 3415 in yellow gold bears the concessionaire's signature WILH. SCHAAFF, HEIDELBERG. This outlet closed in the early 1980s. In the upper row, the rose-gold reference 3415 is on the left, the reference 3429 on the right. One can see there are different calibers ticking in the two references after opening the back: in the reference 3415, the 12'''-600 AT; in the 3429, the successor caliber 27-460.

Illustration 288
The lower right lug of a yellow-gold reference 3415 has the "large Helvetia" stamped on it. If one looks closely, one can still see the *G* for the Geneva Examination Office on the left. The various scratches and the material distortions on the groove of the screw-down case back are evidence of the heavy use this watch endured. The depth of the hallmark also allows us to assume that this case has probably never been refinished.

Illustration 289
This lug also belongs to a reference 3415, but in this case to a rose-gold one. It is no longer possible to determine who did the work improperly on this watch made in 1958. The fact is that 0.2–0.3 millimeters of valuable gold were senselessly removed, but the original surface structure was ignored because, as delivered from the factory, the entire case of the reference 3415 was only diamond-polished. The overhauling of a watchcase should always be carefully considered.

The statements about the surfaces originally present here are of great value. The case of the yellow-gold specimen has clear signs of use, but it still has the factory-made surfaces. All surfaces of the case middle are diamond-finished without exception. The "small Helvetia" stamped on the outside of the lower right and upper left lugs is still deeply recessed, and the "G" for the Geneva Examination Office is also still visible. The screw-down back, which has six grooves, has been diamond-cut in its round section, and the flat part of the case back has been turned. In the case of the rose-gold specimen, the responsible *poliseur* did not have these exact specifications and proceeded according to his own discretion. Thus, the cylindrical part of the case was satin-finished in the longitudinal direction, while the graduated bezel and the lower conical part of the case were correctly diamond-polished. The outer surfaces of the lugs received a satin finish in the vertical direction, but the "small Helvetia" stamped there got quite a bit of it and is barely recognizable (see fig. 289). The screw-down case back received the correct finishes.

The rose-gold reference 3415 was overhauled at Patek Philippe in Munich, and the condition of the movement is as one would expect. Apart from the lack of lubrication, the movement of the second watch is also in good condition. The comparative view of the two inner case backs is also very interesting. The one from the rose-gold specimen still has its original surface. After all the information was stamped on it, this inner case back was very finely bead-blasted. The only additional information is the abbreviation "L.F." noted by a watchmaker with a fine felt-tip pen.

The inner back of the yellow-gold case looks a little bit different. Here, the back was clamped into the chuck of a lathe, and the inner surface was reworked with sandpaper. The information stamped on the case back suffered as a result of this improper treatment. As a rule, one must assume that such reworking was intended to remove earlier watchmaker's marks. A watchmaker's signature with "9674 RPU+" was engraved only after the mentioned treatment with emery paper. Except for the case numbers, the information present on the inner case backs of these two watches matches. Some details show that the same case maker was active here. Stamped from top to bottom are the "Helvetia," the gold fineness hallmark "0,750," "PATEK, PHILIPPE & Co=," "–GENEVE–," and "SWISS." At a distance below follow "REF 3415," the case number, and the case maker's hallmark. The case maker had collective-responsibility mark no. 5 with case key no. 1, identifying the company Ed. Wenger SA, Genève, which was canceled in 1992.

The rose-gold reference 3415 came from the Munich antique-watch dealer Bachmann & Scher and was priced at $11,359. The author has known Thomas Bachmann for more than twenty years and appreciates him as a serious and competent professional. The overhaul, which was done at Patek Philippe in Munich, cost $544. The yellow-gold version was bought at Henry's auction house in Mutterstadt for $7,862, but the watch had not been serviced.

At Antiquorum in Geneva, such a model was auctioned on April 23, 1995, for 6,840 CHF. Ten years later a rose-gold version was sold there on October 16, 2005, for 15,000 CHF (movement no. 766,165, case no. 2,608,800). The year of production listed in the auction catalog as 1960 might not have been quite correct, since this watch is from the same year as our two specimens.

In summary, it should be emphasized that watchcases that are still in their original condition should be left as they are. Small traces of use on the still-original case are much less disturbing than a watch whose exterior was incorrectly refurbished, as demonstrated by the two reference 3415 watches examined.

For wearing: ++ For collecting: ++ As an investment: ++

Reference 3417 Amagnetic

Among the most-sought-after watches from our Geneva manufacturer is the reference 3417 Amagnetic, which could be had only with a steel case. In Patek Philippe's German catalog for 1961, it was listed on page 11. The price at that time was 900 DM. Today, watches of this reference are offered for fifty to eighty times that price. Our collector's piece houses the caliber 27-AM 400 movement no. 732,829. The case number is 2,633,182. The watch was made in 1963. The dial is painted silver, while the maker's signature and the graduation marks for minutes are printed. Not every dial that is still original has "SWISS" printed below six o'clock. Except for the 12, which has Arabic numerals, the hours are marked with indices in bar form, except for the 6. The baton-shaped hands go perfectly with this. The case is of a very elaborate design, so that possible magnetic fields cannot disturb the accurate running of the watch. This is ensured by a special dial, a movement-retaining ring, and a special inner case back, all of which are made of soft iron of the permalium type. The outer case consists of a case middle with integrated bezel and a very strong screw-down case back, both made of stainless steel. The case middle is stamped from a single piece of metal, turned, milled, and surface-finished. The diameter is 1.37 inches (35.0 mm). The base was created as a turned part and has six deep-drawn grooves. The material thickness is measured at 0.019 inches (0.5 mm). The inner case back is turned from a single piece and engages cartouche-like in the soft-iron ring of the movement. Two small grooves provide free passage of the winding stem and the fixing pin for the exact positioning of the movement. The caliber 27-AM 400 movement differs from the caliber 12‴-400 in many details. The components of the oscillating system are made of antimagnetic material (armature of gold-plated beryllium bronze, lever disc of brass, escapement wheel of Durochrome). On the caliber 27-AM 400, the shaft for the second hand as well as the minute tube are also longer than on the caliber 12‴-400 to compensate for the greater thickness of the soft-iron dial.

The reference 3417 was issued in two series. In the second series, the Amagnetic label was omitted and the seconds scale was made simpler. Why some dials bear the "SWISS" label and others do not could not be clarified at Patek Philippe in Geneva.

In contrast to cases made of precious metal, the inscription on the inner case back is much simpler. Stamped from top to bottom are the company signature "PATEK, PHILIPPE & Co," "≡GENEVE≡," and "SWISS." This is followed by case and reference numbers without additional designation. The surface is circular grained, and a total of three watchmakers' marks have been scratched into it. The watch presented was purchased by a collector for $10,700 in Los Angeles, California. Another $1,189 had to be paid for the professional overhaul at Patek Philippe in Geneva. All in all, a good investment.

Illustration 290
One of the most sought-after watches from our Geneva manufacture is the reference 3417 Amagnetic, which comes only with a steel case. In 1961 this watch, which is protected against electromagnetic fields, could be purchased for 900 DM. Today one must pay fifty to eighty times that amount. The reference 3417 was produced in two series. Only the watches of the first series are marked Amagnetic, and they often fetch higher prices at auctions. Watches that are protected against electromagnetic fields are required by only a few professions. Investment bankers are certainly not among them. Rather, it is drivers of electric locomotives or heavy-current electricians in power plants who indeed need such watches. More information about this watch can be found on this page (movement no. 732,829, case no. 2,633,182).

At Antiquorum in New York, a reference 3417 watch was knocked down at $18,400 on March 24, 2004. At Sotheby's in Geneva, a reference 3417 of the second series (without the "Amagnetic" label) changed hands for 32,400 CHF on May 16, 2006.

At Christie's in Geneva on November 16, 2009, a reference 3417 with a fitted Milanaise steel bracelet was called and sold for 42,500 CHF (movement no. 728,934, case no. 2,610,407, year of manufacture 1960).

For wearing: ++ For collecting: ++ As an investment: +++

Reference 3419

The development engineers at Patek Philippe in Geneva came up with another ingenious design in the form of the reference 3419 manually wound watch, which was made in steel only. Like the previous model Ref. 3418, which also was available only with a steel case, an integrated interchangeable system for the use of leather or steel braided straps was also to be implemented for the reference 3419. At first glance the watches of these two references are indistinguishable; only the expert recognizes the small difference. The case of the reference 3419 is slightly countersunk on the right side for the crown (i.e., the crown sits slightly recessed in the case). For the reference 3419 presented, we have a pedigree extract. The caliber 27-AM 400 has the movement number 730,715, and the case number is 2,619,877. The watch was produced in 1962 and sold on April 13 of the same year. As you can see, it was delivered with a leather strap. The dial is silvered. The company signature, the graduation marks for the minutes and the small seconds, and the "SWISS" proof of origin are printed on it. Except for the 6, all hour indices are faceted, but for the 12 they are faceted twice. The watch has matching baton hands. The case is made of stainless steel and has a diameter of 1.35 inches (34.4 mm) according to the brochure. The design of this case is quite elaborate. The case middle with integrated bezel consists of two turned parts set into each other. The outer ring has the special cutouts on either side of the 6 and the 12 to accommodate the connectors for the leather strap or the steel braided strap, depending on the application. The inner part is cylindrical and has the fine thread for the ring nut screwed on the outside. The inner case holds the movement and is closed by a screw-down case back with six grooves. By inserting the previously described band connectors into the outer case, the ring nut can be screwed on and tightened. A special ring wrench is required for tightening the round nut, the driving pins of which engage in the two radial grooves of the ring nut. The caliber 27-AM 400 is mounted in a soft-iron ring that corresponds in its design to the one in the reference 3417. Since this watch was overhauled in Geneva, the movement is almost in new condition. The Geneva hallmark is stamped on the mainspring barrel bridge and on the main plate near the balance. A flat gasket ensures guaranteed water resistance.

Illustration 291
Like the reference 3417, the reference 3419 was made exclusively with a steel case. It is not easy to distinguish it from the 3418 model, which preceded it. The watch presented has the strap-changing system shown in the photo above right. The best way to distinguish these two references is by the crown, which is slightly recessed only on the reference 3419. The model shown is still original in all respects, including the dial with faceted hour bars. In this case the dial was silvered only after faceting, so that the recesses could not oxidize (movement no. 730,715, case no. 2,619,877).

Illustration 292
The reference 3419 has an integrated interchangeable bracelet system. In the area of the two lugs, the middle part of the case has special recesses that accommodate the connections for a leather strap or a steel braided strap, depending on the version. Once the two fitting pieces are inserted, they are held in place by a ring nut, which is screwed to the case via a fine thread. In this picture the case back is screwed shut again.

The finish and inscription of the inner back are basically the same as on the reference 3417. The reference 3419 described came from the USA and was purchased by a German collector for $4,950. Another $620 had to be paid for the complete overhaul (invoice from June 25, 2001).

These models are rarely offered at auctions. At the auction house Antiquorum in Geneva, a Ref. 3419 was sold on April 13, 1997, for 5,040 CHF. Seven years later, on March 24, 2004, the same auction house sold one for $6,960 in New York. At the Heritage auction house in Dallas, Texas, USA, an overhauled specimen was called on May 10, 2010, and sold for $15,535 (movement no. 729,198, case no. 2,606,134).

For wearing: + For collecting: ++ As an investment: +

Reference 3425 in Two Educational Examples

The third model from Patek Philippe equipped with the then-new caliber 27-460 movement is the reference 3425. With a case diameter of 1.31 inches (33.5 mm), purely outwardly it is somewhat modest in appearance but shows its true strengths after a brief acquaintance.

We again present two specimens of this model, because the first one is almost as good as new and has been preserved original in all respects, while the second specimen had been badly battered and quite defaced. In addition to this comparison, we show with all the details how the second watch, despite all its handicaps, was extensively overhauled and restored to almost new condition.

This comparison is particularly exciting due to the fact that Patek Philippe in Geneva does not normally offer such an extensive and elaborate restoration. The first example of the reference 3425 contains the caliber 27-460 with movement number 1,114,454 and has the case number 2,639,270. As the dial signature tells us, this watch was delivered to concessionaire Hausmann & Co. in Rome. The second specimen has the movement number 1,114,987 and the case number 2,666,201. Thus, the automatic movements are 533 numbers apart, the cases 26,931 numbers apart.

Our first automatic is still in mint condition. An Italian watch enthusiast bought it on August 12, 2002, from a dealer in New York for $4,500. For a subsequent service at Patek Philippe he invested another $1,897 net (invoice from August 26, 2003). The watch was originally purchased in 1966 in Rome from jeweler Hausmann & Co. Hausmann & Co. is one of the oldest Patek Philippe concessionaires in Italy and celebrated its two hundredth anniversary in 1994. The company name tells us that this jewelry store has German roots. The second watch comes from the USA and was purchased by a German watch lover via the internet. This Patek Philippe had to endure a great deal in the last forty years, as we will now learn in the comparative description. First of all, it should be noted that the reference 3425 was listed in the 1961 German Patek Philippe catalog and was available in yellow gold for 1,755 DM.

The dial of this reference is painted silver. The company signature "PATEK PHILIPPE, GENÈVE," the graduation marks for the minutes, the small seconds, the concessionaire signature "HAUSMANN & Co.," and the "SWISS" proof of origin are printed in one pass. The indices are bar shaped and riveted in place. The baton hands match this design exactly.

On the second watch the dial has been refurbished, but with some flaws. The company signature turned out a bit too big, and the *accent grave* over the second "È" of "GENÈVE" was forgotten. The arrangement of the graduation marks of the seconds scale does not match the original, neither in diameter nor in stroke width. Also the "SWISS" label below six o'clock has been omitted.

This somewhat out-of-line dial was submitted to the dial factory Bethge & Söhne, near Pforzheim, and was professionally reworked using the original dial from the first watch as a pattern; $305 was charged for this work. After the new dial was installed, the two watches were indistinguishable to the

Illustration 293
These two reference 3425 wristwatches tell us a particularly exciting story. This model is an elegant men's wristwatch with automatic winding and 1.29-inch (33 mm) case diameter. Unveiled in 1960, some watches of the first production series were still equipped with the caliber 12'''-600 AT movement, while all those that followed had the successor caliber 27-460. Here, it is about a comparison of a like-new version with a badly battered one, as well as how to restore the latter to near-new condition. This concerns both the dial and the case. Along with the mint-condition watch, the dial, which had been refurbished in the USA, was handed over to the Bethge & Söhne watch factory in Ispringen, and a replacement dial was ordered. The result was so successful that even an expert can hardly distinguish the two dials. Thus, on the left the watch is shown with the incorrectly refurbished dial, and on the right with the dial correctly printed by Bethge & Söhne company.

Illustration 294
The watch at the bottom in this picture is the reference 3425 that originated in the USA. The extensive inscription was certainly well intentioned but makes the watch hard to sell. Following the example of the mint version, the back was removed and reworked by a Pforzheim specialist. All the inscriptions were filled using a laser and gold wire and in the second step were turned over and diamond-polished. The watch at the top shows the result—an almost-mint screw-down case back (movement no. 1,114,987, case no. 2,666,201).

layman. An expert would have noticed in direct comparison at most the slightly lighter color of the replacement. The case of the reference 3425 is in two parts and closed with a screw-down case back. The case middle with integrated bezel was made as a turned part, and the four lugs are soldered in place. The bezel is set off in a fillet shape; the middle part of the case is cylindrical. The screw-down case back is deep-drawn, turned, and diamond-polished and has six grooves. It has a double taper on the outside and is flat in the middle. The metal thickness there is 0.019 inches (0.50 mm).

The case of the second specimen is basically identical, except that the screw-down case back has an extensive inscription. "50 YEARS OUTSTANDING SERVICE, PRESENTED TO C. S. WHITE-SPUNNER, DECEMBER 1969 BY TURNER SUPPLY CO." reads the complete engraving. Turner Supply Co. is based in Mobile, Alabama, and was founded in 1905. As a distributor, Turner now sells industrial products from leading manufacturers such as 3M, Norton, Sandvik, and many others.

Since a Patek Philippe with such an "embellished" case can be sold only at a significant discount, the case was sent to a Pforzheim case maker for reconditioning, and the first watch was submitted as a sample. The screw-down case back was placed in a special receptacle, and all the inscriptions were filled with gold by laser. Subsequently the back was turned, diamond-polished, and restored to its original condition. The comparative illustration on page 188 confirms the successful effort by the case maker; $234 was paid for this work.

Unlike on other watches, there was no gold hallmark on the outside of the case of the reference 3425. The inscription on the inner case back showed that both cases came from the same supplier. From top to bottom are the "large Helvetia," the gold fineness hallmark "18K/0,750," the company signature "PATEK-PHILIPPE & Co=," "–GENEVE–," "SWISS," the marking "RÉF. 3425," the case number, and the case maker's hallmark. The latter consists of the collective-responsibility mark no. 5 with case key no. 1, which is the company Ed. Wenger SA, Genève, canceled in 1992. The surface of the inner case back of the first watch is finely bead-blasted and has no watchmaker repair marks. The inner case back of the refurbished specimen was also bead-blasted again after the screw-down case back was overhauled, albeit a bit more coarse grained than originally.

Reference 3425 watches are always in demand. Whether for everyday use or as a collector's item, they always retain their value. At Antiquorum in Geneva, a watch was sold for 7,200 CHF on October 20, 1991. At the same auction house in New York, a reference 3425 in yellow gold was knocked down for $6,348 on June 20, 1998. Finally, again at Antiquorum in Geneva, an example in white gold came up for bid on March 29, 2009, and brought it to 9,600 CHF (movement no. 1,113,274, case no. 2,636,947).

For wearing: ++ For collecting: ++ As an investment: +

Reference 3429

The reference 3429 is another popular automatic. It was produced with yellow- and rose-gold cases. The case middle with integrated bezel is similar to that of the reference 3415, only the screw-down case back is not round or diamond-polished, but two-stage tapered. At first glance it could easily be confused with the reference 3415, if the latter did not have faceted indices on the dial. Inside this watch is the caliber 27-460 with movement no. 1,113,443 and the case number is 2,644,656.

The reference 3429's dial is made of yellow gold, and the company signature and the graduations for the minutes and the small seconds are enameled. Following this preliminary work, the dial received a galvanic silver coating and was given a sunburst finish. The gold indices in bar form are riveted on, with two side by side at twelve o'clock. The watch has matching baton hands. The "SWISS" proof of origin is imprinted below six o'clock.

According to the 1961 sales catalog, the case of the reference 3429 has a diameter of exactly 1.37 inches (35 mm). The case of the first watch has a "small Helvetia" stamped on the case side at four o'clock. On the movement side, all case surfaces except for the turned case back are diamond-polished!

The watch with a caliber 27-460 movement was overhauled at Patek Philippe in Munich and optically restored to new condition. The model with movement no. 1,113,443 was in such a good condition that the overhaul costs were kept within limits, and the owner was charged only $818 (invoice from December 13, 2002) by Patek Philippe. After this work, the watch achieved chronometer rates again.

The reference 3429 described was purchased on August 22, 2002, from an antique-watch dealer in New York for exactly $4,785 USD (approx. 4,800 €). The gold tang buckle, the gold

Illustration 295
The reference 3429 is one of the most-pleasing wristwatches from our Geneva manufacture. Except for the partially faceted indices, it looks very similar to the reference 3415. The screw-down case back of the reference 3429 has a double conical offset, while that of the 3415 is diamond-polished round or spherical. Instead of the caliber 12'''-600 AT, it uses the successor caliber 27-460 (movement no. 1,113,443, case no. 1,644,656).

spring bars, and the original crocodile leather strap were still present on this watch. According to the 1961 German Patek Philippe catalog, a yellow-gold reference 3429 cost exactly 1,755 DM and a white-gold one cost 1,930 DM. At the Antiquorum auction house in New York, a yellow-gold version was sold for $9,000 on June 14, 2006, and a white-gold one for $26,400 on May 28, 2007 (movement no. 1,115,296, case no. 2,644,740).

For wearing: ++ For collecting: ++ As an investment: ++

Illustration 296
At Patek Philippe, attention is paid to even the smallest detail. Here we see the reference 3429 with alligator strap removed (exactly: Louisiana Alligator Strap). This strap is hand-stitched and is commonly called the Umbugg strap. Next to it are the tang buckle and spring bar—both made of 18K yellow gold. If one wants to detach the strap from the case, one takes a suitable tool from Patek Philippe or even a toothpick to push the pin in and remove the strap. The case of the reference 3429 was made as a turned part. The soldered lugs are clearly visible. The bridge width is 0.70 inches (18 mm).

Reference 3433

Among the watches of the 3400 reference range with a case diameter of 1.41 inches (36 mm) is the reference 3433. In the German Patek Philippe catalog from 1961, it is listed on page 14, and the stated case diameter is 1.41 inches (36 mm). The yellow-gold version is listed at 1,855 DM. However, this reference was also made with platinum cases but perhaps was not offered in Germany.

The automatic shown is considered to be the successor of the reference 2551 and looks very similar to it but is powered by the caliber 27-460 movement. The movement number is 1,112,899, the case number 2,626,193. The dial was replaced when the watch was overhauled at Patek Philippe in Munich. In the current version it is finished in cream-colored lacquer and has rounded bar indices that are riveted in place. The upper parts of these indices are painted black. At six o'clock the index is shortened because of the seconds crosshairs, and three indices are riveted on at twelve o'clock. The watch has matching baton hands.

The case consists of two parts. The bezel is integrated into the case middle, and both are made as turned parts. The four lugs are profiled and soldered (i.e., exactly matched to the case profile). The case back is deep-drawn, turned, and surface-finished. Like the screw-down case back of the reference 3429, it is two-step tapered, and the flat case back area is turned. Why the six deep-drawn grooves are much smaller than on the reference 3425 and 3429 is not clear.

This watch received a complete overhaul at Patek Philippe in Munich. This overhaul cost $3,305, including $1,526 for the restoration, $692 for the replacement dial, $144 for the minute wheel with minute tube, and $133 for the ball bearing for the gold rotor (invoice from May 17, 2006). After this work the movement looks almost new again. The rate values have also been restored.

The inscription on the inner base again reveals the handwriting of the case supplier. Stamped from top to bottom are the "large Helvetia," the gold fineness hallmark "0.750,"

Illustration 297
The reference 3433 is only rarely offered for sale. It is considered the successor to the reference 2551 and also has a case diameter of 1.41 inches (36 mm). From the outside it can be distinguished only by its screw-down case back. On the reference 2551, the case back is diamond-polished round or spherical, while on the reference 3433 it is double conical. I was unable to discover why the deep-drawn grooves are much smaller than usual.

and the company signature "PATEK PHILIPPE," "–GENEVE–," and "SWISS." This is followed by "REF 3433," the case number and the hallmark of the case maker. With the collective-responsibility mark no. 5 and case key no. 1, this again turns out to be the company Ed. Wenger SA, Genève. Unfortunately this company and thus its hallmark were canceled in 1992.

The automatic presented belongs to a German collector. The watch, which originated from Italy, was purchased at the Watch Fair in Munich for exactly $5,234.

Reference 3433 watches do not appear too often on the market. At an auction of Christie's in Geneva, a yellow-gold specimen was sold for 16,800 CHF on November 13, 2006. A year before, on July 10, 2005, Antiquorum in Hong Kong called a model with a platinum case and sold it for 600,000 HKD (about $65,423; movement no. 1,111,112, case no. 2,616,375). Another version with a platinum case was offered by Christie's in Geneva on May 12, 2008, and sold for 85,000 CHF ($81,284; movement no. 1,232,822, case no. 2,626,268).

For wearing: ++ For collecting: ++ As an investment: ++

Reference 3444

Before describing our next watch, the automatic reference 3444, we should mention the Ref. 3440 automatic model. It has a case with a diameter of 35.0 mm and a screw-down case back. It is powered by the caliber 27-460 movement, but the design of this model takes some getting used to for most watch collectors, because the case middle is profiled in a double-V shape.

The reference 3444 we present is one of the quite rarely offered automatics. From the logbook extract we can see that this watch was produced in 1961 and sold on October 2 of the same year. The caliber 27-460 has the movement number 1,111,024; the case number is 2,619,254. The dial is still in original condition and a feast for the eyes. The company signature "PATEK PHILIPPE, GENÈVE," the graduations for the minutes, and the small seconds are enameled. After this operation the gold dial was galvanically silvered and sunburst. Subsequently the indices were riveted on. At twelve, three, six, and nine o'clock positions, they are slightly wedge shaped. The remaining eight hours have indices that taper toward both ends. This form of hour marker has not been found on any of the watches presented so far.

Below six o'clock, one can still faintly see the "SWISS" proof of origin. The case of the reference 3444 is very elaborately designed. The case middle, with integrated bezel, is multistepped, rounded, and made as a turned part from a single piece of metal. The strikingly strong lugs are soldered on in a profiled manner. The screw-down case back is deep-drawn and conically stepped and has six grooves. All surfaces are diamond-brushed.

This watch had received a complete overhaul at Patek Philippe in Munich, and thus the movement also shines in nearly new condition. The gold rotor was also replaced. There is a small circular abrasion on the inside back. These marks were caused by one of three cylinder head screws used to attach the rotor. It was not tightened and was left sticking out too far, leaving its marks on the case back.

Probably due to a lack of suitable surfaces, the case does not have a small Helvetia stamped on the outside. However, the inscription on the inner case back is interesting. In the upper part the maker's signature is stamped, with "PATEK, PHILIPPE & Co," "≡GENEVE≡," and "SWISS." In the center there are the case maker's hallmark, the gold fineness hallmark "18K/0.750," and the "large Helvetia." Differing from previous versions, the case maker's hallmark extends vertically upward. In the lower part of the inner case back, the case number and reference number are placed one above the other, the latter without addition. The inner case back is fine bead-blasted, and there are two watchmaker repair marks scratched into its surface. The case maker's hallmark is collective-responsibility mark no. 5 with case key no. 4. This hallmark belongs to the company Antoine Gerlach SA, Genève, which was canceled in 1977.

Illustration 298
The reference 3444 is a very rare find. The case and lugs are very elaborately made. The screw-down case back also does not correspond to any of the designs seen so far. Also interesting is a small detail in the screw-down case back. One of the three cylinder head screws with which the rotor is screwed down has left its mark in the center. Since it was not screwed all the way in, it brushed against the base and caused damage. If it had been screwed out a little farther, the rotor would probably have been jammed. More information about this watch can be found on this page (movement no. 1,111,024, case no. 2,619,254).

The watch shown was purchased by a collector from the jeweler TNS Diamonds Inc. in Philadelphia, Pennsylvania, for $7,000. The price advertised in the store was initially $12,950 and was reduced to $7,000 after a lengthy discussion. A further $848 had to be paid to have the watch serviced (invoice from October 12, 2008).

At auctions, representatives of the reference 3444 are hard to find. At Antiquorum in New York, such a model was offered on March 28, 2007, and sold for $12,500 (movement no. 1,111,096). At Christie's in Geneva a reference 3444 was offered on November 13, 2006, and sold for 22,800 CHF (yellow-gold case, movement no. 1,112,096, case no. 2,628,014).

For wearing: ++ For collecting: ++ As an investment: +++

Reference 3445 in Different Versions

In 1961 Patek Philippe released its first automatic with date display. It was given the reference number 3445 and was initially powered by the caliber 27-460 M movement, while the final series had the caliber 27-460 M PM (here "M" stands for "Monodate," and "PM" for *piton mobile*," or mobile stud carrier). The latter caliber is easy to recognize, since the seven-digit movement number always begins with the numbers 123. The company signature, consisting of "PATEK, PHILIPPE & Co," "≡GENEVE≡," and "SWISS," is stamped. In the center there are the case maker's hallmark, the gold fineness hallmark "18K/0.750," and the "large Helvetia." Differing from previous versions, the case maker's mark is applied vertically. In the lower part of the inner case back, the case number and reference number are noted one below the other, the latter without addition. The inner case back is fine bead-blasted, and there are two watchmaker repair marks scratched into its surface. The case maker's hallmark belongs to collective-responsibility mark no. 5 and bears the key no. 4. This hallmark identifies the case maker Antoine Gerlach SA, Genève, which was canceled in 1977. The reference 3445 was the model most in demand in the 1960s. It was fitted with cases in yellow, rose, and white gold, as well as platinum. Due to the high quantities on the one hand and the numerous dial variants on the other hand, we present three different models.

Before we go into the individual variants, we will describe the design of this automatic wristwatch. The dial is made of yellow or white gold, depending on the case. If indices are planned, the necessary holes are drilled and then the dial is silver-plated. In the next step the sunburst finish and any lacquer finish were applied. The next step was to print the company signature and the graduation marks for the small seconds, followed by the protective lacquering.

The case is made up of two parts and consists of the case middle and screw-down case back. The bezel is an integral part of the case middle and is stepped in two stages. The case middle

Illustration 299
In 1961, Patek Philippe released its first watch with a date display. It was given the reference number 3445. The watch presented was purchased by a watch enthusiast from Leihhaus Bodenhagen GmbH in Mönchengladbach for $4,089. At that time, the watch still had the dial shown at below right. A check of the dial code number, 93Y3514A, revealed that this dial did not belong to this reference 3445, but to a reference 3514, a later model. But the reference 3514 also uses the caliber 27-460 M and has the same dial dimensions. On the misplaced dial, further damage to the Roman numerals IV, X, and XI had been repaired manually, but not professionally. While being serviced at Patek Philippe, a new but this time correct dial was fitted. This service cost was $1,441.50 (movement no. 1,118,030, case no. 328,637).

Illustration 300
Two particularly impressive examples of the reference 3445 in white gold. The dials of both watches are made of white-gold plate and have a sunburst finish. The model on the right later received a blue finish and white-gold indices. Even the applied frame for the date window is in white gold. The watch on the left received a gray lacquer finish with Roman numerals printed on the dial. This special lacquer almost looks like enamel. The white-gold hands are lacquered in the same color. The bezel of this watch is chased (i.e., has decorative patterns applied with a steel engraver). Both watches originate from Japan and were later serviced at Patek Philippe. The case diameter is exactly 1.37 inches (35.0 mm). As a protective measure, the crowns are semirecessed (watch on the left: movement no. 1,230,402, case no. 329,656; watch on the right: movement no. 1,230,663, case no. 318,158).

is initially cylindrical and then tapers toward the bottom. It was created as a turned part from a single piece of metal. The plain lugs are soldered in place. All surfaces are diamond-polished. The screw-down case back is also turned from a single piece of metal and milled. A stout decagon ensures secure closure. The attachment surfaces of the decagon, as well as the remaining surface up to the fine thread, are bead-blasted. The flat surface of the case back is turned, and the rounded transition to the decagon is diamond-cut. The case diameter is 1.37 inches (35.0 mm), the overall length with lugs is 1.63 inches (41.5 mm), and the height is around 0.39 inches (10 mm).

It was previously mentioned that the caliber 27-460 M movement was used until the end of the 1960s, followed by the modified 27-460 M PM version. Besides the signature on the main plate, one can also see the differences on the balance cock. There you find a slotted hole radially to the balance staff. With its cylinder head screw, the hairspring stud can be loosened and adjusted. This allows the "drop" (i.e., the symmetry of the oscillating system) to be precisely adjusted.

The signature and the surfaces of the inner case backs are identical on all four watches presented. In fact, they originate from the same case maker. Like the previously described reference 3444, this is listed under the collective-responsibility mark no. 5, case key no. 4. Behind this is the company Antoine Gerlach SA, Genève, which was canceled in 1977. We thus describe only one of the three inner case backs (also see fig. 299, p. 192). The "large Helvetia" is stamped at the very top. Below that follow the gold fineness hallmark "18K/0.750" and the company signature "PATEK, PHILIPPE & Co," "≡GENEVE≡," and "SWISS." Below that are the case number, reference number, and the case maker's hallmark.

The first reference 3445 presented has the movement no. 1,118,030 and case no. 328,637. According to the pedigree book, which we have, this yellow-gold watch was produced in 1970 and sold on May 5, 1971. After forty-one years of wandering, the stages of which we do not know, it finally ended up at Leihhaus Bodenhagen GmbH in Mönchengladbach in 2002. The automatic, which was offered on the internet, was finally sold and went to a German watch enthusiast for $4,045 on June 12, 2002 (see p. 192). During the initial inspection of the freshly purchased watch, the defects already mentioned by the seller were checked and confirmed. On the dial with Roman numerals, which was still in its original condition, damage to the numerals IV, X, and XI had been repaired by hand, although not masterfully. During an overhaul at Patek Philippe in Munich, the dial was also exchanged for a much-nicer one. The new one is silvered and sunburst. Also the larger Roman numerals in combination with the silvered dial give a much more pleasing overall impression. The cost of the service, including the creation of a pedigree book, was calculated as $1,426. The two most-expensive items in this total were the movement service at $485 and the replacement of the dial with an old original at $322 (invoice from July 8, 2002).

The next Ref. 3445 automatic watches presented both have a white-gold case. The first one has the movement no. 1,230,663 and case no. 318,158. The dial is impressive. It has a sunburst finish and was subsequently painted dark blue. The proof-of-origin "0 SWISS 0" below six tells us that this dial has gold—in our case white-gold—indices riveted in place. As already described in chapter 12, "The Secret of Dials," this regulation came into effect only in 1968. This watch originates from Japan and was purchased by a German watch enthusiast on August 31, 2001, for $4,550 (approx. 5,056 €). Another $1,553 was paid for a complete overhaul by Patek Philippe in Munich. In addition to $436 for the movement overhaul, the gold rotor in exchange at $214 and the replacement dial at $296 were the most-important items (invoice from January 22, 2003). The second white-gold Ref. 3445 has the movement no. 1,230,402 and case no. 329,656. According to the pedigree book extract, which we have, this watch was produced in 1972 and sold on June 10 of the same year. The sunburst finish was painted gray and printed with creamy-white Roman numerals. At first glance one may come to the conclusion that the Roman numerals are enameled. The company signature and the graduation marks for small seconds are printed the same way.

What differs is the chiseled bezel, which may have been executed later. This watch also comes from Japan and was purchased by a European watch enthusiast for $7,000 (approx. 5,000 €). The service performed by Patek Philippe in Munich cost another $1,800. Despite everything, this was a good deal, because at auctions this variant of the reference 3445 rarely appears and is probably not to be had for less than $13,000 to $16,400! The inner case backs of both white-gold watches are still in their original condition and show no repair marks.

Models of the reference 3445 are offered at auctions quite often. At Antiquorum in New York, a version with a white-gold case and silvered dial was offered on June 18, 2008, and sold for $19,200. Surely this price was influenced by the overhaul receipt from Patek Philippe. On October 5, 2009, a yellow-gold model was offered at Antiquorum in Geneva and sold for 15,600 CHF. Interesting for the collector are also the watches for sale, which can be found on the internet via www.chrono24.com. Often ten or more examples of the reference 3445 are offered there. A spot check by the author revealed that none of the watches offered between $9,200 and $14,700 still had all their original or factory case surfaces.

For wearing: ++ For collecting: ++ As an investment: ++

Reference 3454, an Almost Unknown Watch

One of the sensations of this book is the Ref. 3454 automatic presented. This model has not been mentioned in literature before and cannot be found in the Patek Philippe museum in Geneva. This is also only the second time that the author has come across this timepiece.

We have a pedigree extract for this watch. The caliber 27-460 has movement number 1,112,614, and the case number is 312,558. The watch was produced in 1962 and sold on October 16, 1963. The dial is painted a cream color and is still in original condition. The company signature, the graduation marks for the minutes, the crosshairs for the small seconds, and the "SWISS" proof of origin are printed. Specially shaped bar indices indicate the hours, with those for twelve, three, six, and nine o'clock being double wide. The watch has matching baton hands.

Illustration 301
The reference 3454 shown was another sensation. Fact is, only four or five examples of this model were made, and there is not one in the Patek Philippe museum in Geneva. Even Oswaldo Patrizzi did not include this watch in his diverse and comprehensive books. As well, the example shown had a varied history. The entire story is told in the accompanying text (movement no. 1,112,614, case no. 312,558).

Illustration 302
When auctioned at Antiquorum, the reference 3454 was fitted with the sapphire crystal case back shown in the photo on the right, but with two major flaws. First, this additional case back was made of rose instead of yellow gold, and second, when it was properly screwed down it blocked the rotor. The "case specialist" had most likely erred by 0.011 inches (0.3 mm) when they measured the inner height.

Illustration 303
The reference 3454 with the new glass case back made in Pforzheim. In the photo on the left is the case back that was mounted *ex manufacture*. It showed clear traces of machining. By checking the thickness of the material, it was possible to prove that the case back was only 0.015 inches (0.4 mm) thick instead of the original's 0.019 inches (0.5 mm). The new glass case back was supplied by a case maker in Pforzheim. Instead of making the deep-drawn grooves, they created six grooves made with an end mill. The original finely bead-blasted inner base was subsequently sanded, probably to cover up unwelcome traces.

The case measures 1.36 inches (34.6 mm) in diameter and has a two-piece design. The narrow, partially fluted bezel is integral, and the four lugs are profile-soldered. The case middle, turned from a single piece of metal, has only a short cylindrical height and is conically stepped at the bottom. The double-conical, stepped, screw-down case back fits well to this shape. It was turned from a single piece of metal and was deep-drawn and has diamond-coated and turned surfaces.

The screw-down back was included with the watch only because it was sealed with a sapphire crystal case back that was made later (this will be discussed).

The case middle is still in original condition. The case has no hallmarks on its outer surfaces. Like the cases of the four previously described Ref. 3445 automatics, this one was supplied by Antoine Gerlach SA, Genève. Except for the reference number, the entire inscription on the inner case back is identical. The original fine-bead-blasted surface of the inner back has been sanded improperly. Furthermore, two watchmaker repair marks are still visible. This rare piece was auctioned on February 4, 2006, at Antiquorum in Geneva and sold for 21,600 CHF.

After receiving the watch, the new owner sought out the author and asked for his opinion. The first item of business was to inspect the watch itself, then the two screw-down case backs. The first thing I noticed about the watch was the incorrectly refurbished dial, and its rate accuracy also left something to be desired. Further findings were provided by the two screw-down case backs. The original version, which was loosely enclosed, showed small dents and traces of refinishing on its inner surface. The outer surface had been re-turned. In such cases this may infer the repair of damage or the removal of an inscription. Measurement of the case back thickness, which it turned out had been turned down from the original 0.019 to 0.015 inches (0.5 to 0.4 mm), provided confirmation.

The sapphire crystal case back, which was manufactured as an add-on, also showed signs of use and blemishes. Its inner height was too small, so that the sapphire crystal touched the gold rotor when it was screwed down and ultimately blocked it. The choice of material may be claimed as a flaw, because instead of yellow gold the case back is made of rose gold. This rare automatic was first entrusted to Patek Philippe in Munich and an overhaul was commissioned; $1,808 was paid for this service (invoice from February 3, 2007). In this, the complete overhaul at $611 and the replacement dial at $724 were the largest single items.

Later a new sapphire crystal case back was ordered from a case maker in Pforzheim, and $1,123 was paid for this functionally correct piece.

Including import tax, a new strap, and a gold pin buckle, as well as all related value transport costs, the total investment for this watch added up to about $19.410. All in all, a safe and valuable investment, especially because this watch is not in the Patek Philippe museum in Geneva.

Because the reference 3454 is considered extremely rare, the author has searched intensively for this watch and came across some strange happenings. On eBay, a yellow-gold automatic reference 3454 was offered for $10,000 in December 2001, but the seller, from West Hollywood, California, failed to find a buyer. On November 15, 2004, Christie's in Geneva offered a reference 3454 and sold it for 14,340 CHF (about $12,276). The author concentrated his further research on this watch, because despite its different appearance, the numbers of the movement and case were identical to those of the reference 3454 presented in this book. What had happened here?

The description of this watch in the Christie's catalog brought the first insights. As to the dial, it could be read that it has baton indices and the concessionaire's signature "B. D. HOWES & SON." A lengthy dedication was engraved on the screw-down case back ("Los Angeles Realty Board, Service Watch Presented to George H. Shellenberger, February 16, 1967, Los Angeles, Most Useful Citizen in 1966"). In the Christie's description, in conclusion it was noted that only four examples of this reference are known to exist.

What happened to this watch in the period from the auction at Christie's on November 15, 2004, for 14,340 CHF to the sale at Antiquorum on February 4, 2006, for 21,260 CHF? First, the dial of this watch was refurbished, but not extensively! Then the inscription on the back was removed, but with visible traces! The previously made assumption had thus been confirmed.

Using the example of our reference 3425 automatic (see p. 188), you can see that all this can be done much better—namely, correctly—and the refinished case back retains its original dimensions.

What did the seller wish to cover up with all these measures? Perhaps the fact that this watch had already been auctioned at Christie's two years earlier? Was the unprofessionally made screw-down case back part of this game? Was this meant to be another red herring, or to feign an additional appreciation in value? Because after deducting the premium and the consignment fee of 20% each, the seller would have been left with only about 14,160 CHF, and the suspicion is warranted that in this case a collector or speculator has taken action and initiated the dial correction and the creation of the new sapphire crystal case back. However, this assumption cannot be proven.

But this story teaches us one thing: if the checklist in this book is followed when buying a watch, such stories can certainly be avoided (see p. 422).

For wearing: ++ For collecting: +++ As an investment: +++

Reference 3466

One of the highly prized watches from Patek Philippe is the reference 3466, which is extremely rare. This automatic was unveiled in 1965 and was made only with a steel case (diameter 1.37 inches or 35.0 mm) and in small numbers. It has a screw-down case back. This timepiece's simple beauty is probably hard to beat. The watch is powered by the caliber 27-460 movement, in this case movement number 1,116,171. The case number is 2,664,436. The dial is silvered. The company signature and the graduation marks for the small seconds are printed. The indices are riveted on—two in the case of the 12. The bezel is integral with the case middle. However, it could not be determined beyond doubt how the lugs were made. The transition points on the case were so neatly reworked that they cannot be recognized as soldered joints. The sturdy screw-down case back was preturned from a single piece of metal, deep-drawn, turned, and diamond-finished. The six large grooves are used to securely close the case.

Illustration 304
After the reference 2585 (see fig. 284, p. 180), the reference 3466 was Patek Philippe's second watch with automatic winding and a steel case. Here we see this rare watch from the front and back. The bezel is integral with the case middle. The screw-down case back is deep-drawn, turned, and diamond-finished.

The inside surface of the screw-down case back is circular grained. As for markings, we find the company signature "PATEK, PHILIPPE & Co," "≡GENEVE≡," and "SWISS," the latter much larger than in gold watches. Below that follow the case and reference numbers. The Ref. 3466 presented was bought by a collector at Sotheby's in Geneva on November 19, 2003, for 12,266.40 CHF. The watch, which is in very good condition, was passed on to Patek Philippe to be serviced; $511 was charged for this work (invoice from February 11, 2004).

On May 15, 2006, another Patek Philippe Ref. 3466 was offered at Christie's in Geneva. This time, however, the buyer had to dig much deeper into his pocket, since the final price was 33,600 CHF (movement no. 1,116,692, case no. 2,664,370). Today these rare models are offered on the internet for $32,000 to 44,000.

For wearing: ++ For collecting: ++ As an investment: +++

Reference 3473

One of the watches that the author has seen only twice is the automatic reference 3473. With its fluted bezel and screw-down case back, it is one of the water-resistant watches from the reference 3400 series and is the second-to-last automatic of this series before the Ref. 3485 with square case. The model shown is powered by the caliber 27-460, which has the movement number 1,112,437. The case number is 2,629,723. The dial is cream colored and has the same indexes riveted on it as the dial

Illustration 305
This extremely rare wristwatch with the reference number 3473 has been seen by the author only twice. This is also the only watch from Patek Philippe with a finely fluted bezel. Inside this model is the caliber 27-460 movement with automatic winding. The case is of three-piece design and has a screw-down back. Everything on this watch is still original, and it was probably rarely worn. The current owner got this watch by a lucky coincidence. The purchase price, import fees, and the service at Patek Philippe added up to exactly $5,328. Today this watch is available only for twice the price—if you should get to see one (movement no. 1,112,437, case no. 2,629,723).

of the reference 3454 on page 194. The watch has matching baton hands. The company signature, the crosshairs for the small seconds, and the "SWISS" proof of origin are printed on the dial. The case consists of three parts. The bezel is finely fluted. The case middle is turned from a single piece of metal. The base diameter is 1.37 inches (35 mm). The lugs are soldered. The strong screw-down case back ensures water resistance.

The watch presented, which is a very special one indeed, has an interesting history. First it was offered on October 3, 1998, at an auction at Christie's in New York under lot no. 30, with a suggested price of $4,500–$5,000, but did not find a buyer. Perhaps the poor photo in the auction catalog, which shows the watch with the crown pulled out, played a part in this.

The owner of this watch turned the unsold piece over to rival auction house Sotheby's of New York. At the auction on June 22, 1998, this Patek Philippe came under lot no. 121 and was sold for $4,025.

A German watch collector had bid by phone and won the auction. Due to the exchange rate at that time of 1.79 DM to the dollar, the price for this watch, including transport and import tax, was exactly $4,749.

The watch was overhauled at Patek Philippe in Munich, and the movement was restored to almost new condition. The owner had to pay $522 for the overhaul, a new crocodile leather strap, and a new gold tang buckle. The total investment for this Ref. 3473 was exactly $5,271. The reference 3473 appears only very rarely at auctions. At Antiquorum in Geneva, this model was offered on April 14, 2002, and sold for 22,800 CHF (movement no. 1,112,289, case no. 2,629,719). Another example was offered by Christie's in Geneva on November 13, 2006, and sold for 21,600 CHF (movement no. 1,112,496, case no. 2,629,734). So far the author has not come across a third specimen.

For wearing: ++ For collecting: ++ As an investment: ++

Reference 3483

The next watch presented corresponds to a modern trend, because steel watches from the noble brands are again strongly on the rise. One of the very rare models from our Geneva brand is the reference 3483. This waterproof, manually wound watch was produced only with a steel case (diameter 1.37 inches or 35 mm) and in small numbers. We have a pedigree book extract for this watch. According to this, it was produced in 1963 and sold on September 19 of the same year. It was delivered with a leather strap and a steel tang buckle. The caliber 27 SC has the movement number 710,826, and the watchcase number is 2,632,453. The dial is silvered, the indices are in bar form, and the baton hands are steel. The company signature and "SWISS" proof of origin were printed on the dial.

The steel case is in two parts and has a sturdy screw-down case back. The case middle, with integral bezel, was created as a turned part, and the four lugs are soldered in place. The screw-down case back was deep-drawn as well as turned and has six grooves for secure closure. The case is almost identical to that

Illustration 306
Wearing a Patek Philippe wristwatch in stainless steel is considered particularly chic today. The reference 3483 not only represents a special rarity, but with its 1.37-inch (35 mm) case diameter, it is still an up-to-date size! This model was unveiled in 1963 but was made in small numbers. The case is similar in design to the reference 3466 (see p. 196, fig. 304). Due to the lower height of the caliber 27 SC, the screw-down case back has a different shape.

of the reference 3466 and differs only in the height of the case middle and the design of the screw-down case back (see p. 196). The caliber 27 SC is supported by a sturdy movement-retaining ring. The Geneva hallmark is stamped on the main plate near the screw balance wheel, as well as on the gear train bridge, the latter underlaid with gold paint. After an overhaul at Patek Philippe in Munich, this beautiful movement shines again, almost like new. On the inner case back, all surfaces of which are beaded, we find engraved the company signature "PATEK, PHILIPPE & Co," "≡GENEVE≡," and "SWISS." This is followed by the case and reference numbers. The documented watch came up for bid at Antiquorum in New York on March 24, 2004, and sold for $6,480. With the addition of transport costs and import sales tax, the purchase came to $7,115 for the German watch enthusiast. Having the watch serviced cost another $959, bringing the total investment to exactly $8,074.56.

As mentioned, reference 3483 manually wound watches are very rare. The last examples known to the author were put up for auction at the end of the 1990s. In June 2010, a reference 3483 was up for sale at the Parisian antique watch dealer ROMAIN REA and was listed at $18,693. How this Patek Philippe was finally sold is not known.

For wearing: ++ For collecting: ++ As an investment: ++

Reference 3514 in Different Versions

The second wristwatch with date display to be unveiled by Patek Philippe was given the reference number 3514. It was also the first automatic in the reference 3500 range. Along with its predecessor Ref. 3445, the reference 3514 is one of the most popular and bestselling men's wristwatches from the Geneva-based watchmaker. With a diameter of 1.32 inches (33.6 mm), the cases were made in yellow, rose, and white gold, as well as platinum. In addition to numerous dial variations, this model could also be purchased with a Milanese gold bracelet attached or soldered on. Because of the special significance of this watch, we present three different versions. The basic description of this model, which was launched in 1965, is based on a watch from the first year of production. According to the available logbook extract, this watch was produced in 1965 and sold on March 31, 1966. The caliber 27-460 M movement is numbered 1,123,252, and the case is numbered 319,417. The dial, made of gold, was first electroplated silver and then given a sunburst finish. The company signature, the graduations for the small seconds, and the "SWISS" proof of origin are printed in black. The watch has baton hands matching the riveted bar indices for the hours.

The case middle with integral bezel, including the lugs, is stamped from a single piece of metal, turned, and finished with the desired surfaces. The bezel and lugs are diamond-coated at the top, and the sides are satin-finished. The cone that closes the middle part of the case at the bottom is again diamond-finished. The construction diameter of the case is exactly 1.32 inches (33.6 mm) and not 1.29 or 1.33 inches (33 or 34 mm), as it is often listed in literature or in auction catalogs. The solid screw-down case back is turned from a single piece of metal, milled, and diamond-finished. The decagon is milled and has a width across flats of 1.14 inches (29 mm). The material thickness of the base is 0.019 inches (0.5 mm).

Having been serviced by Patek Philippe in Munich, the caliber 27-460 M movement shines almost like new again. For this service, including the preparation of a pedigree book, the

Illustration 307
The reference 3514 was the second model from Patek Philippe with date display (after the reference 3445) and was unveiled in 1965. This watch was produced in large numbers and in numerous versions. Despite its case diameter of 1.32 inches (33.6 mm), the Ref. 3514 was very popular and was issued in yellow, red, and white gold, and platinum. Because of the special significance of this watch, we present three variants. On the far left is a version with white-gold case, silvered white-gold dial and indices, and 750 white-gold hands. Like the others of its kind, this watch has the automatic caliber 27-460 M (movement no. 1,126,585, case no. 322,943). This watch must have been delivered with a white-gold bracelet, because on the lugs, one can still see the small roundings for the lug counter bearing. The second version from the left has a yellow-gold case. The dial of this watch is made of sheet metal that was silver-electroplated and painted silver (movement no. 1,123,956, case no. 320,650). The watch on the far right represents the standard version of this reference, a simple beauty for everyday use (movement no. 1,123,252, case no. 319,417). This model was produced in 1965 and sold on March 31, 1966.

owner had to pay exactly $1,901. The major items were the movement overhaul at $485, the two automatic bridges for $165, and the gold rotor in exchange for $284 (invoice from November 11, 2002). On the two versions presented with yellow-gold case and crocodile leather strap, the "small Helvetia" was struck on the bottom of the upper right and the lower left strap lug. The white-gold case has no such hallmark because it was shipped with a white-gold band attached. One can see this by the slight indentations that are present on the underside of the lugs to accommodate the band's counter bearings. The inscriptions on the inner backs are basically identical; they were supplied by the same case maker. The inner surface is finely bead-blasted. *From top to bottom*, one below the other are stamped the "large Helvetia," the gold fineness hallmark "18K/0.750," and the company signature "PATEK, PHILIPPE & Co," "≡GENEVE≡," and "SWISS." Case and reference numbers follow below, as well as the case maker's hallmark. This belongs to collective-responsibility mark no. 5 with case key no. 4, indicating the company Antoine Gerlach SA, Genève, which was deleted from the commercial register in 1977.

The second model with gold case has a silver-lacquered dial with black Roman numerals. To match this, the yellow-gold baton hands have been painted black. The dial is made of steel plate and galvanically silvered. The identifier 93Y3514A is stamped on the back. The number 93 stands for Patek Philippe, the 3514 for the reference. It is almost identical to the dial on page 192 (illus. 299). This watch, listed at Patek Philippe as a reference 3514 J, has movement no. 1,123,956 and case no. 320,650. A Munich watch enthusiast discovered this watch on the internet and purchased it on November 30, 2003, for $6,301 (approx. 5,252.80 €). With transport costs and import sales tax, his total investment came to about $6,723. An overhaul commissioned from Patek Philippe in Munich cost $1,189 (invoice dated April 06, 2004).

While the two listed watches with yellow-gold cases were serviced by Patek Philippe in Munich, the white-gold version went to Geneva for *service-aprés-vente* (after-sale service). It had been called at Antiquorum in Geneva on October 17, 1993, and sold at auction for 7,000 CHF, plus buyer's premium. After inspection by Christophe Harsch, then service manager of Patek Philippe in Geneva, the new owner of this watch was not pleased; he had to pay 1,120 CHF for a thorough overhaul (movement no. 1,126,585, case no. 322,943).

Judging by the auction prices, reference 3514 automatic watches seem to be in great demand today. At Antiquorum in Geneva, a yellow-gold reference 3514 was called on April 2, 2006, and sold for 18,000 CHF. At Antiquorum in New York, a yellow-gold reference 3514 was called on March 28, 2007, and sold for $13,200. At the same auction house, on September 26 of the same year, another yellow-gold model 3514 was offered and sold at auction for $18,800. At Live Auctioneers, a reference 3514 with platinum case was called and knocked down at $30,000 (movement no. 1,124,043, case no. 320,385).

Since the trend is toward more-"modest"-looking white-gold watches, the automatic with white-gold case shown and purchased years ago should also be a good investment.

For wearing: + For collecting: ++ As an investment: ++

Reference 3541

Patek Philippe's third men's wristwatch with date display was launched in 1967 and was given the reference number 3541. At first glance this model can easily be confused with its predecessor, the reference 3445 (see p. 192). The only obvious differences are the partially recessed crown and the larger diameter. The reference 3541 was produced with yellow- and white-gold cases. The version presented has the movement no. 1,124,874 and case no. 320,933. The dial was electroplated-silvered and given a sunburst finish. The company lettering, the Roman numerals, and the graduations for the small seconds are printed on it.

The case could possibly have been stamped from a single piece of metal, turned, and diamond-finished, since no soldering or lug marks are visible on the lugs. Maybe the case was also made as a turned part and the lugs were soldered in. The crown is countersunk to two-thirds of its height and thus well protected. The bezel is integrated into the case middle. The latter is kept cylindrical for the most part, then stepped down and turned to taper toward the screw-down case back. A very elaborate design! The case diameter is exactly 1.40 inches (35.60 mm), while 1.41 inches (36 mm) is often cited in literature, as well as in auction catalogs. The solid screw-down case back is turned from a single piece of metal and has a decagon for

Illustration 308
The reference 3541 is considered to be the successor to the reference 3445 and is the third Patek Philippe model to have a date display. The almost completely recessed winding crown serves as a reliable distinguishing feature. The case diameter is exactly 1.40 inches (35.60 mm), which is 0.02 inches (0.6 mm) more than the reference 3445. The watch shown looks very elegant. From the back you can see the screw-down case back with turned surface, as well as the decagon with 1.14-inch (29 mm) width across flats.

closing; the width across flats is 1.14 inches (29 mm). The back of this watch is turned over the surface and outer bulge. The caliber 27-460 M was overhauled at Patek Philippe in Munich, but the case was left as it was because of its mint condition. On the underside of the upper right and lower left lugs, we find the "small Helvetia" stamp. The inscription on the finely bead-blasted and still-untouched inner case back corresponds, except for case and reference number, exactly to that in the backs of the three reference 3514s. Here again, the same case maker with case key no. 4, Antoine Gerlach SA, Geneva, was the supplier for Patek Philippe.

The model shown was bought by a European watch enthusiast at Sotheby's in New York on April 6, 2004, by telephone for $7,200. Transport and import sales tax totaled about $1,020, with another $840 to have the watch serviced by Patek Philippe in Munich. Even if the total investment adds up to $9,162, the new owner can be satisfied with it. At Christie's in Geneva, a yellow-gold reference 3541 came up for auction on May 12, 2008, and was knocked down for 13,125 CHF (approx. $12,531). On April 2, 2006, a yellow-gold version was sold at Antiquorum in Geneva for 17,400 CHF (movement no. 1,124,855, case no. 320,952). At the same auction house, a watch with a white-gold case came up for auction on October 15, 2006, and went to its new owner for 25,200 CHF (movement no. 1,128,794, case no. 321,093).

For wearing: +++ For collecting: ++ As an investment: ++

Reference 3542

Patek Philippe's next automatic was the reference 3542, albeit with some differences compared to the previously discussed reference 3541. Due to the small number of these watches made, this model is also considered a rarity. The case diameter is a verified 1.41 inches (36.0 mm), which is exactly 0.015 inches (0.4 mm) more than the reference 3541.

The caliber 27-460 has the movement number 1,115,027, while the case number is 321,169. The dial is silver-lacquered. The Roman numerals, the company signature, the dashes for the small seconds, and the "SWISS" proof of origin below six o'clock are printed on the dial.

The case of the reference 3542 might have been designed a bit differently than that of the reference 3541. The case middle with integral bezel seems to have been created as a turned part, after which the four lugs were soldered in place. The fact that they were so heavily undercut on their underside when turned could further confirm the author's assumption. The stout screw-down decagon case back has a width across flats of 1.14 inches (29 mm). Except for the case and reference number the inscription on the finely bead-blasted inner case back is identical to the one described. Once again, Patek Philippe turned to the same case maker.

Reference 3542 watches are hard to find. On April 2, 2006, a watch of this type was called at Antiquorum in Geneva and sold for 15,000 CHF (movement no. 1,116,234, case no. 321,245). On September 13, 2004, a collector purchased the watch shown for $6,000 USD (approx. 4,894.14 €). This watch was equipped

Illustration 309
The reference 3542 was produced only in small numbers. Its case diameter is exactly 1.41 inches (36.0 mm). The version presented has a silver-lacquered dial and black-lacquered hands. The solid case is closed by a screw-down back. The winding crown is almost completely recessed for safety reasons.

with a Milanaise gold mesh bracelet. The service carried out at Patek Philippe in Munich cost another $1,073. In view of the rarity of this watch, on the one hand, and the previously mentioned auction price, on the other hand, the collector may have already made a safe investment. It is not known if the watch sold at Antiquorum for 15,000 CHF had been previously overhauled at Patek Philippe.

In Germany, a yellow-gold reference 3542 came up for bid on eBay and was sold for $9,000 on May 14, 2003. The seller had indicated in his listing that this watch had been overhauled at Patek Philippe for $1,581.

For wearing: +++ For collecting: ++ As an investment: ++

Reference 3558 in Two Different Versions

Of all the watches with the caliber 27-460 or 27-460 M movement presented so far, the reference 3558 holds a special place. Its gold case has the greatest material volume and thus the highest weight in gold. This model was very popular in the USA and came with a yellow- or white-gold case. To present several of the variants of this watch, we have listed two different models. The first watch has the caliber 27-460 M PM movement no. 1,230,449, and the case no. is 2,695,315. This watch was produced in 1971 and was fitted with a leather strap from the factory, which was later replaced by a Milanaise gold mesh bracelet

Illustration 310
This photo illustrates how different watches of the same reference can look. Perhaps the horizontal reference 3558 with the Milanaise strap is the right one for a birthday party, while the one with the blue dial together with the cuff links is the right one for a visit to the opera. The old rule of wearing only white-gold watches on such occasions has become outmoded. A beautiful Patek Philippe wristwatch combined with matching cuff links always betrays style. Erica Pappritz, the federal chancellor's wife, sometimes referred to as the countess, would have been delighted with it.

supplied by Patek Philippe. This pleasing bracelet also fits the references 3428, 3429, 3433, 3435, and 3444 and cost exactly 970 DM extra in 1961. The second version of the reference 3558, with blue dial, still uses the earlier caliber 27-460 M movement, here with the number 1,118,342. The case number is 2,689,535. This watch was produced and shipped to the USA in 1970. Both models had the same case supplier, Ed. Wenger SA of Geneva, which was deleted from the commercial register in 1992.

The reference 3558's dial is made of 18K yellow gold and has a sunburst finish. The dial of the second watch was also painted blue. The indexes are made of 18K yellow gold, and on the first watch they are lacquered in black, including the hands. The company signature, the dashes for the small seconds, and the "SWISS" proof of origin are printed on the dial.

The case is of very elaborate design. The inclined bezel is cylindrical and conically turned and diamond-finished. The case middle was stamped from a single piece of metal, turned, milled, and finished with various surfaces. The case diameter is exactly 1.33 inches (33.8 mm), even though the watch appears much larger. The solid screw-down case back is also turned from a single piece of metal, deep-drawn and double-tapered. Six grooves ensure secure closure. The bezel, the 45° break of the lugs, and the two conical surfaces of the screwed back are diamond-finished; the remaining surfaces are satin-finished or turned.

Both watches, and thus their movements as well, received a complete overhaul at Patek Philippe in Munich, so that the movements also present themselves in almost new condition and exhibit excellent rate values.

The "small Helvetia" is stamped on the underside of the upper right lug. Although the cases of both watches are from the same case maker, there is a slight difference in the inscriptions on the inner case back. On the older watch, the "large Helvetia" and the gold fineness hallmark "18K/0.750" are stamped next to each other in the upper area, while on the newer watch they are stamped one below the other. Then follow the company signature "PATEK-PHILIPPE & Co=," "–GENEVE–," and "SWISS." Below that, in three rows, are the reference designation "REF. 3558," case number, and the case maker's hallmark. The latter belongs to collective-responsibility mark no. 5 with key no. 1, indicating Ed. Wenger SA, Geneva. Both watches originate from the USA. Including transport and import VAT, about $6,379 was paid for the first one and about $6,100 for the second one. The cost of the complete overhaul was $1,353 for the first watch with the gold band and $1,745 for the second.

The reference 3558 is not offered very often. At Antiquorum in Geneva on November 19, 2000, a yellow-gold version was called and sold at 12,000 CHF. On May 13, 2007, a white-gold model with blue dial was offered at the same auction house and sold for 30,000 CHF (movement no. 1,230,245, case no. 2,695,410).

For wearing: ++ For collecting: ++ As an investment: ++

Reference 3561, a Special Rarity

The reference 3561 automatic can be numbered among the rarest collector's items from Patek Philippe. After the reference 3439/1, it is the second wristwatch from Patek Philippe fitted with a screw-down case back with sapphire crystal at the factory. This eliminated the need for an inner case back to hold all relevant key data. We have a pedigree extract for our watch. The caliber 27-460 has the movement number 1,116,766. The case number is 2,681,327. This watch was produced in 1970 and sold on October 6, 1971. We also know the subsequent life story of this timepiece. The warranty certificate, which is available to the author in the original, shows that this rare piece was sold by concessionaire Swiss Stores Ltd. in Jamaica on March 15, 1975.

The dial is made of 18K yellow gold, which was galvanically silvered and then satin-finished lengthwise. The company signature, the graduation marks for the minute scale and the small seconds, and the "SWISS" proof of origin are printed on the dial. There are matching baton hands installed.

Illustration 311
The reference 3561 was unveiled in 1968 and is considered the successor to the reference 3439/1 from 1965. Both watches have screw-down case backs with sapphire crystal inserts and were the world's first watches with a transparent back. Although other watch manufacturers may claim this, today these claims should not be taken seriously. As in other areas, Patek Philippe was again a pioneer. Since the inner surface of the screw-down case back was not available for the usual markings, the gold fineness hallmark "0.750," the "small Helvetia," and the case maker's hallmark were stamped between the lugs below six o'clock from left to right.

Illustration 312
Our reference 3561 seen from behind with the sapphire screw-down case back removed. The screw-down back ring consists of a turned part that was given six grooves in the deep-drawing process. In the 1960s, sapphire crystals of this size could be produced only with considerable effort and were very expensive. It was not until twenty years later, when new technologies made synthetic glasses easier and cheaper to produce, that watches with transparent backs became widespread.

The case is made in two parts and has a screw-down case back. The diameter is 1.31 inches (33.5 mm). The case middle with integral bezel was created as a turned part, and the four lugs are soldered in. In addition to the cylindrical center part, the bezel has a throaty design, and the connection for the screwed case back is conically stepped. All surfaces are diamond-brushed. The screw-down case back was created as a turned part and was deep-drawn and diamond-finished. The sapphire crystal case back is glued in place.

The movement of this watch was overhauled at Patek Philippe in Munich and is now in "like new" condition. This is also revealed by the view through the glass bottom. Since this watch had probably spent most of its time in a safe, a refurbishment of the case was not necessary.

The basic differences from the watches presented so far should be emphasized. The gold fineness hallmark "0.750," the "small Helvetia," and the case maker's hallmark are stamped between the lower lugs, from left to right! Such an arrangement is a novelty. "18K" is stamped on the underside of the lower right lug, presumably by the customs authorities of the island state of Jamaica (member of the Commonwealth, same customs requirements as in the UK). The differences become even more obvious after opening the screw-down case back. Engraved (i.e., made by machining) on the inside of the outer ring of the case back is "PATEK PHILIPPE GENEVE REF. 3561 No. 2681327." If one removes the movement, one can see that the reference and case numbers are also manually engraved (scribed) with a steel chisel at the twelve and six o'clock positions.

The consequences of this realization are also important. Watches from Patek Philippe with glass backs have often appeared at auctions. Mostly these were reference 3445 automatic models that were subsequently given an additional screw-down case back with sapphire crystal. If in such cases the original case back is also included, one can go into this business with peace of mind. If not, you should keep your hands off it!

The watch presented comes from an Indian engineer who had emigrated from India to Canada and purchased this watch on March 15, 1975, in Jamaica as its first owner. After his retirement, he sold the watch to a German watch enthusiast for $6,195 USD. Including transportation and import sales tax, the purchase price was exactly $8,098. The overhaul performed at Patek Philippe in Munich cost another $491 (invoice dated July 11, 2001). Due to the mint condition of this watch, only the crown, winding stem, mainspring, and various small parts had to be replaced during the overhaul.

Reference 3561 watches are hard to find. At Antiquorum in Geneva on April 2, 2006, this rare Patek Philippe was offered and sold for 21,500 CHF (movement no. 1,116,781, case no. 2,681,339). Seven months later the identical watch was again offered at Antiquorum in Geneva and this time sold for just 18,000 CHF. A bitter loss for the consignor! With the numbers mentioned, the watch auctioned twice in Geneva was fifteen numbers ahead of the movement and twelve numbers ahead of the case shown. Thus, they both were probably produced the same week.

For wearing: + For collecting: ++ As an investment: +++

Reference 3563, Also Called the Backwinder

No other watch at Patek Philippe gave as little pleasure as the Backwinder series. The reason for this was the new caliber 350 movement, which was designed in a completely new way. Details were already explained in the presentation of the automatic calibers at the beginning of this chapter. The instability of the regulating system, the low power reserve, and the vulnerability of the outer rotor led to such massive complaints that the Patek Philippe branch in New York had to take back large numbers of watches. The detail problems with the ball bearing of the rotating ring, as well as the changer for the bilateral winding mechanism, in particular, could never be solved satisfactorily and made further development of this caliber necessary. In 1979 the new version, designated 1-350, appeared with an 18-karat gold segment replacing the heavy metal weight. A new ball bearing, the elimination of the changer, and improvement of the manual winding and setting mechanism brought advantages not only for the user of the watch, but also for its service conditions. However, Patek Philippe was never completely satisfied with this caliber, and production ended in 1985. The reference 3563/2 presented in 18-karat white gold with the same type of integral gold bracelet was provided to us by a Japanese watch collector. He had previously purchased it on the internet for $2,300. It was professionally overhauled by Patek Philippe in Geneva and subsequently achieved passable movement performance. The cost of the service was $1,261.70 (invoice from August 10, 2002). The largest items were the movement service at $484 and the replacement of the dial, which cost $528. When this

Illustration 313
Reference 3563 with soldered strap in 750 white gold. This model was unveiled in 1969 and was available both with crocodile and gold bracelets. In the USA and in the Far East, this watch was considered particularly chic due to its new design without a crown and initially found many enthusiasts. Only when the problems with the new 350 automatic movement became more and more obvious did interest in this watch collapse. The version shown was produced in 1974 and sold in Japan on March 14 of the same year (movement no. 1,185,655, case no. 2,727,909).

Illustration 314
This reference 3563 Backwinder looks chic, with its bezel and dial in nail pattern decor (Clous de Paris). Although the case diameter is only 1.35 inches (34.4 mm), the lug-to-lug measurement for the alligator strap is 0.78 inches (20 mm). Therefore, the lugs are a bit narrow.

Illustration 315
This photo shows what a Backwinder looks like when it is open. The outer circumferential rotor has a golden flywheel weight. One can also see some of the ball bearings under the note NO OIL. On the case back, a small groove can be seen both on the top and bottom edges. When the case back is placed on the case, these two grooves must precisely engage their counterparts on the case. This positions the winding stem, which protrudes from the mainspring barrel bridge, precisely in the center of the through-hole. Only when everything fits together may the ring nut be put on and tightened. A small detail on the mainspring barrel bridge: a lengthy recess has been milled there, using an end mill. In it may be found the caliber designation 1-350.

watch accidentally fell from a height of about 12 inches (30 cm) onto a wooden table during photo work, it left the recently serviced watch totally out of adjustment again. It was serviced again at Patek Philippe in Geneva, but as a gesture of goodwill without charge.

We have selected two of these watches, on the one hand to allow us to describe the calibers 350 and 1-350, and on the other hand to compare the models with and without gold bracelet. The first one has already been described. It has a white-gold case with a soldered strap and the caliber 350 movement. The second model has a yellow-gold case with a leather strap and has the successor caliber 1-350. For the first model we have a pedigree extract. The reference 3563/2 watch was produced in 1974 and sold on March 14 of the same year. The caliber 350 has the movement number 1,185,655, and the case number is 2,727,909. The yellow-gold counterpart is from the production year 1984. The caliber 1-350 movement has the number 1,449,198, and the case number is 2,698,334. Different dial variants were installed on the reference 3563. Blue-lacquered dials were popular on the models with a soldered gold bracelet. The indices for the hours, as well as the matching baton hands, are made of gold. Guilloché dials made of 18-karat gold, like the one on the second Backwinder with crocodile leather strap shown, were also very popular. They had gold graduated indices with the upper surface painted black. The baton hands were executed in a matching design.

The reference 3563 has a case diameter of 1.35 inches (34.4 mm) and is quite elaborately designed. The case middle, with integrated bezel, is made as a turned part; the thin lugs are soldered on. The width between the lugs is 0.78 inches (20 mm), which is not a common measurement for normal Patek Philippe watches. A pot-shaped, deep-drawn and turned case back, which has keyways at six and twelve o'clock, is screwed in place by a ring nut that has six grooves. The very precise positioning of the case back is important for fixing the setting crown, which is screwed tight after the back has been put in place. The material thickness of the gold case back is 0.019 inches (0.50 mm). Overall, the case design gives the impression that the product planning for this reference had called for the economical use of the precious metal gold. Otherwise the material thicknesses on the case and lugs would not have been so sparsely thin in places!

The caliber 350 and 1-350 were already presented and the most important differences were described (see pp. 147–148). Besides the difference between the rotor with heavy metal segment on the caliber 350 and the golden rotor weight on the caliber 1-350, the mainspring barrel bridge reveals an interesting detail. During servicing of the caliber 350, no changes had to be made in the area of the mainspring barrel bridge. Thus, this component could continue in use without any problems. But because a large number of this movement part was already in stock, and originally nobody thought about a modification of this movement, they simply milled out the existing caliber designation 350 with an end mill and engraved the new caliber designation 1-350 into this free space. Even the movement of the watch with yellow-gold case presented was done in the described way, although it was number 9,198 of 10,000 produced. Thus, the movement shown was made in the last production year of this caliber type.

The "small Helvetia" is stamped on the outside of the case back. The inscription on the bead-blasted inner back tells us more. Engraved there, from top to bottom, are the gold fineness hallmark "18K/0.750," followed by the case maker's hallmark and the "large Helvetia." The company signature "PATEK-PHILIPPE & C°," "–GENEVE–," and "SWISS" follow. Below that can be found the reference number 3563 3 and the case number. The case maker responsible belongs to collective-responsibility mark no. 5 with case key no. 1, which stands for Ed. Wenger SA, Genève. This company was removed from the commercial register in 1992.

A German watch enthusiast purchased the yellow-gold reference 3563 3 via the internet in the USA for $4,000 (approx. 3,150.85 €). With transport and import sales tax, this total added up to $4,153. Noteworthy about this purchase was the fact that neither the PPC pin buckle nor an original PPC crocodile leather strap were present, but the two gold spring bars on the case were.

Reference 3563 watches, especially the version with the soldered gold bracelet, are not considered to be in high demand among collectors. They are still sought after in the Far East. On December 3, 2008, a white-gold reference 3563 3 with white-gold bracelet came up for auction at Christie's auction house in Hong Kong and was sold for $9,721. As of June 20, 2010, a white-gold 3563 3 was offered for $11,900 and a yellow-gold 3563 2 for $10,950 at internet auctions in the USA. Whether or not they were also sold for this price is not known.

For wearing: + For collecting: + As an investment:

Reference 3585

Here we present another version from the Backwater series that has quite an elegant appearance. The blue dial and the white-gold case give this watch an elegance you can expect only from a Patek Philippe. There were a total of five references in the Backwinder series: 3563 (p. 204), 3569, 3573, 3585 (p. 206), and 3586. Submodels of these five references also exist in multiple versions. The reference 3585 was unveiled in 1970 and was sold primarily in the American and Far Eastern markets. The model shown was also chosen because of some special features, and its pedigree was reviewed. This shows that it has the automatic caliber 350. The movement number is 1,188,464, and the case number 2,728,813. The extract also shows that the original automatic movement was replaced by a new one. The caliber that still exists today was installed during a service performed in Geneva in 1977. If a watch was built in 1970, sold on April 20, 1974, and needed a replacement movement after just three years of use, this does not correspond to the high standard of quality normally expected from Patek Philippe.

Illustration 316
The Backwinder series includes five references: 3563, 3569, 3573, 3585, and 3586, and here we present the penultimate model. Not only does the case have a very elegant shape, but it is also the first one to receive a sapphire crystal instead of the Plexiglas previously used. The white gold of the case is well matched by the blue dial and white-gold hands and indexes. The case middle, with a size of 1.41 × 1.65 inches (36.0 × 42.0 mm), is made from a single piece of metal, and the case back is screwed in place with a ring nut as in illustration 315.

The dial of this watch is painted blue. The white-gold indices and the baton hands still have a blue background. This version was the first to have a sapphire crystal installed.

The case middle, measuring 1.41 × 1.65 inches (36.0 × 42.0 mm), is made in one piece, and the case back is attached and screwed with a ring nut. The strap connection width measures exactly 0.79 inches (20.2 mm).

Because this watch could not be opened at the Ulm concessionaire, neither the movement nor the inscription on the inner case back could be seen.

Now, some remarks about the setting crown of this backwinding watch. In position 0, the movement can be wound. If one pulls out the crown in position 1, one can set the time. But the movement does not have a stop-second mechanism.

Opening the case back is done in two steps. First, the small cylinder head screw that holds the crown must be loosened and removed together with the crown. Then the ring nut is loosened with a special wrench, and only then can the back, which is positioned by two grooves at twelve and six o'clock, be carefully removed.

This watch was bought by a watch lover for $4,338 on the internet, but because of the considerable costs, he could not decide for the necessary overhaul at Patek Philippe.

Reference 3585 are rarely offered. At Antiquorum in Geneva, a reference 3585 came up for auction on November 17, 2002, and was sold for 6,720 CHF. At Christie's in Rome, an identical model was offered on June 1, 2005, and sold for $6,760.

For wearing: + For collecting: + As an investment:

Reference 3588, the "Pope's Watch"

Here we describe a special watch with a somewhat mysterious history. The descriptions given are those that we have received from the owner of this rare piece. In 1970, the company Hausmann & Co, Via-del-Corso 406, Rome, received an order from the Vatican to supply twelve wristwatches of special design from Patek Philippe. The model chosen was the new, superflat automatic Ref. 3588. Instead of the normal version, the dial was to be in the Vatican color, a special purple. Instead of Roman numerals, the hours were to have faceted indices. The profiled faceting allows the gold of which the dial is made to reappear, giving it an almost mystical appearance. Hausmann delivered the dozen watches to the Vatican in 1971. This was during the time of Pope Paul VI, who sat on the "Chair of Peter" in Rome from 1963 to 1978 as the 261st pope. Subsequently, as a special honor, deserving church leaders were presented with one of the watches from the hand of His Holiness. One of the recipients was a cardinal from the USA who was the granduncle of the current owner.

On page 208 we see this watch together with a complimentary ticket to the pope's mass in Munich on October 9, 2006. The colors match exactly. On that Sunday, Pope Benedict XVI celebrated a service with more than 250,000 faithful on the open-air grounds of the New Munich Trade Fair Center, which went down in church history as the "Pope's Mass in Munich."

Readers will surely ask themselves about the fates of the other eleven watches. Unfortunately, only the Vatican archives can provide an answer.

Our Ref. 3588 "Pope's Watch" is powered by the then-new automatic caliber 28-255. The movement bears the number 1,280,422, while the case number is 2,714,748. This watch was produced in 1970 and sold the following year. The dial is made of yellow gold. Its surface was electroplated silver, sunburst, and fine-lacquered. The next step was to imprint the company signature and SWISS proof of origin. Subsequently the dial was clamped onto a dividing head, and the twelve-hour indices were faceted. In contrast to the dial of the reference 3415 (pp. 182 and 184), the indices are not faceted in a V shape but in a W shape. From a manufacturing point of view, a dial with small seconds is easier to facet, because the hole for the small seconds serves as a position fixation. The current dial does not have a fixation point and must be clamped in the center position by using the two dial feet and negative pressure. Only then can the diamond make the engraving into the gold by chip removal.

Our "Pope's Watch" was also analyzed under a stereomicroscope. This led to the conclusion that the Vatican purple was applied as a second coat of paint and that the first color was probably the familiar blue. However, the fact that the company signature and proof of origin were printed only after this second color application again speaks for the authenticity of the dial. It can be assumed that for the small number of twelve dials, stock parts were used, which were then repainted in the desired Vatican color.

The case measures exactly 1.37 inches (35.0 mm) in diameter and is only 0.240 inches (6.1 mm) thick, despite the sapphire crystal. It consists of three parts: bezel, case middle, and press-fit case back. The case middle, including the lugs, was presumably stamped from a single piece of metal, turned, milled, and surface-finished. The crown has been slightly recessed for its protection. The press-fit case back has a material thickness of 0.019 inches (0.5 mm).

The caliber 28-255 measures exactly 0.10 inches (2.55 mm) in height and occupies only 42% of the watch's total thickness. The rotor has a 21-karat gold oscillating weight and winds in both directions.

A "small Helvetia" is stamped on the press-fit case back. The inscription in the bead-blasted inside of the case back is neatly arranged over the entire height. Stamped one below the other are the "large Helvetia," the gold fineness hallmark "18K/0.750," and the company signature "PATEK-PHILIPPE & Co=," "–GENEVE–," and "SWISS."

Below this follow the reference and case numbers and the case maker's hallmark. This belongs to collective-responsibility mark no. 5 with key no. 1, indicating Ed. Wenger SA, Genève, which was deleted from the commercial registry in 1992.

Illustration 317
This picture shows not only the "Pope's Watch," but also an admission ticket to the "Pope's Mass" in Munich. According to the owner of this watch, his great uncle, an American cardinal, received this watch as a special award from the hands of Pope Paul VI in Rome. The ticket for the guests of honor at the Pope's Mass celebrated in Munich on September 10, 2006, bears the same color as the dial of the Pope's Watch, the special purple of the Vatican. Further details about this unique clock can be found in the accompanying text.

Our "Pope's Watch" was sent to Geneva via Patek Philippe in Munich and was serviced there. The charge for this work was $815 (invoice from September 26, 2003). A pedigree book extract was requested, but this was denied for the time being for the reason that the movement and case numbers of this watch would not match. Had the Holy See in this case also ensured the discretion, which is a matter of course in upper church circles?

For wearing: For collecting: +++ As an investment: ++

Reference 3700/1 Nautilus

One of the most spectacular watches from Patek Philippe is the Nautilus model. This sports watch was unveiled in 1976 and is still one of the brand's most popular models. Originally this steel watch was not planned but was the result of third-party initiatives. One was the resounding success of the Royal Oak sports model from competitor Audemars Piguet of Le Brassus, and the other was the offer from Gerald Genta, designer of the Royal Oak, to create a noble sports watch for Patek Philippe as well. Although both designs are half-shell cases, the Royal Oak model was initially better received by the public because of its convincing design. Surely this was also due to its clean lines and the typical porthole face of the Royal Oak. Even diehard friends of our Geneva brand found it difficult to get used to the appearance of the Nautilus and often opted for the rival product from Le Brassus. Today the tide has turned in favor of Patek Philippe. For a Nautilus Ref. 3700/1 from the 1970s and '80s, also affectionately called a "Jumbo," you often have to dig deeper into your pocket than for a Royal Oak from the same era.

Illustration 318
The Nautilus was launched in 1976 as a sports watch and remains one of Patek Philippe's most popular models. In spring 2021, Thierry Stern, president of Patek Philippe, initiated the discontinuation of the steel version of this watch. The reason for this was the excessive demand, which had led to astronomical price increases.

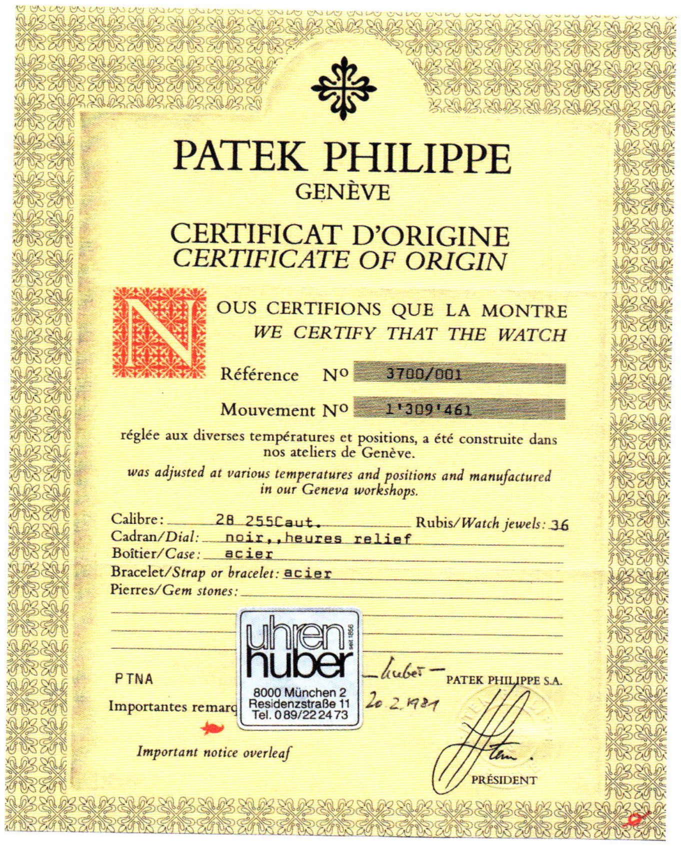

PATEK PHILIPPE
GENÈVE

CERTIFICAT D'ORIGINE
CERTIFICATE OF ORIGIN

NOUS CERTIFIONS QUE LA MONTRE
WE CERTIFY THAT THE WATCH

Référence N° 3700/001

Mouvement N° 1'309'461

réglée aux diverses températures et positions, a été construite dans nos ateliers de Genève.

was adjusted at various temperatures and positions and manufactured in our Geneva workshops.

Calibre: 28 255Caut. Rubis/*Watch jewels*: 36
Cadran/*Dial*: noir,,heures relief
Boîtier/*Case*: acier
Bracelet/*Strap or bracelet*: acier
Pierres/*Gem stones*:

uhren huber
8000 München 2
Residenzstraße 11
Tel. 089/22 24 73

PTNA

Importantes remarq

Important notice overleaf

PATEK PHILIPPE S.A.

PRÉSIDENT

Illustration 319
Sales certificate for the reference 3700/001 Nautilus shown on the left, issued by Uhren Huber in Munich on February 20, 1981

At least with the Nautilus model series, Patek Philippe was able to appeal to those customers who were looking for a new noble sports watch with a famous name.

The Nautilus Ref. 3700/1 was offered in steel, steel and gold, and yellow and white gold, and in platinum in small numbers. Besides its futuristic look, it was also the unusual case size that immediately caught the eye of every observer. With an overall width of a full 1.65 inches (42 mm without crown), the Nautilus was far ahead of its time and an early pioneer for the lush dimensions of today's watch models.

The steel version shown has movement no. 1,309,461 and case no. 540,862. It was bought on February 20, 1981, at Uhren-Huber in Munich, Residenzstraße 11, for exactly $2,789.

The dial has horizontal stripes and is anthracite colored. The indices and the two hands are luminescent, which is applied in milled grooves. The date display is at three o'clock. The case, with its width of 1.65 inches (42 mm), looks almost like a porthole—some say like a TV. It is made of stainless steel and features a very special design. On page 212 you can find an exploded view of this case and a disassembled watch. From the point of view of technical mechanics, as well as some watchmakers, this case is not optimally designed. A proper screw-down case back, like the ones usually found on Patek Philippe watches, would have been a better solution. From a mechanical point of view, the bezel of this single case is a bridge. After the movement is mounted, the bezel is placed on the case and connected on both sides with two set screws each. The truncated cone attached to the tip of each grub screw fits into a recess in the case, thus fixing the position of the bezel. In conjunction with a specially shaped gasket, this guarantees water resistance to a depth of 393 feet (120 m). Bracing over four cones that are precisely aligned does not allow any vertical forces. Thus, the water resistance has to be taken over by the flexibility of the special gasket alone. Such a solution can occur only to a designer who is not a specialist in the field and who is only vaguely familiar with technical mechanics.

With increasing external pressure (i.e., at greater water depths), the mentioned design weaknesses become equalized because the entire housing is compressed from the outside. The forces that the dial would have to withstand at a depth of 393 feet (120 m) are astonishing. At this depth the water column exerts a force of 132 tons or 264,554 pounds (120 tonnes, i.e., 120,000 kg) on a surface of 10.76 square feet (1 square meter). In relation to the surface area of the sapphire glass of around 1.03 in^2 (6.7 cm^2), 1.03 × 172 = 177.25 lbs. (80.4 kg)!

28-255

Illustration 320
This nearly 15× magnification shows a section of the caliber 28-255 movement. In the foreground is the angular profiled rotor ring, which runs on four vertically arranged ruby bearings. The balance cock with KIF Ultraflex shock protection, Gyromax balance wheel, and flat hairspring are also clearly visible. The movement number 1,292,432 can be seen at the bottom left, and to the right of it the caliber 28-255 and, quite modestly, the Geneva Seal. The very thinly made lugs are not exactly neatly soldered.

Illustration 321
The Nautilus reference 3700/1 has a monoblock case with a suspended metal bracelet. The movement is inserted from above, with the dial mounted and closed with the bezel. On the inverted bezel you can see the gasket already inserted (movement no. 1,309,461, case no. 540,862).

The movement bears the caliber number 28-255 C and has been used in many Patek Philippe watches.

The inscription on the finely bead-blasted inner back of the half-shell case is quite simple. On the upper part is the company signature "PATEK PHILIPPE" and "GENEVE," in the middle is the case number, and at the very bottom is the reference number 3700/1. Both case parts are made from a single piece of metal and must have been very labor intensive and expensive to produce. This is also true for the upper part, which is well visible here.

Illustration 322
This exploded view of the Nautilus clearly shows the entire design. The load that the watch glass would have had to withstand at a depth of 393 feet (120 m) can be read here in the text.

Today, more than ever, the Nautilus is one of Patek Philippe's successful models. Over the years the case design was revised somewhat, and the two side case lugs were slightly rounded. Additional models expanded the range, so much so that there are a full sixteen pages dedicated to this line in the Patek Philippe catalog. The watch presented here belongs to a friend of the author. After fourteen years he had brought his beautiful piece back to concessionaire Andreas Huber in Munich to have it serviced. He paid exactly $534 for this service according to service card no. 40467, dated October 5, 1995. The warranty covered a period of six months.

Illustration 323
Gyromax balance wheel with flat hairspring of the caliber 28-255 C
By turning the six mass balance weights, the moment of oscillation GD^2 and thus the rate can be adjusted. If the slots are aligned to the center (i.e., inward), the moment of inertia increases and the watch runs more slowly. If the slots are turned outward, the moment of oscillation is reduced and the watch runs faster. Due to the low height of this caliber, a Breguet hairspring could not be installed.

The Nautilus Ref. 3700/1 is often offered at auctions. At Antiquorum in Geneva, a white-gold version was sold on November 15, 2005, for 78,000 CHF. At the same auction house in Geneva, a stainless-steel model was called on November 11, 2007, and sold for 36,999 CHF. Furthermore, a yellow-gold version was offered by this auction house on October 14, 2009, and went to the new owner for 33,600 CHF. You can see here that "Jumbos" made of steel are traded at a similar price to models in yellow gold.

Because the Nautilus is the only watch line at Patek Philippe that has been in the program for about thirty-five years and has not changed its outward appearance, the successor models should be briefly discussed. The "Jumbo" Ref. 3700/1 was followed by the somewhat smaller Ref. 3800/1. Now again equipped with the in-house automatic movement caliber 335 with date display and center second hand, it was one of the house's most famous "long runners." The following model Ref. 3900/1 was first equipped with a quartz movement and later with the automatic caliber 330 SC. In the 1991 price list, the Nautilus Ref. 3800/1 was listed in steel at 9,750 DM, in yellow gold at 35,300 DM, in white gold at 37,550 DM, and in platinum at 74,800 DM.

The Nautilus Ref. no. 5711/1 cost exactly $15,362 in stainless steel in 2009–2010. This price comparison also shows that a thirty-year-old Nautilus Ref. 3700/1 often costs several times as much as a new model. More about the technology and the sensational price development can be found in chapter 14, "The Noble Sports Watches," on page 345.

For wearing: +++ For collecting: +++ As an investment: +++

This description of the Nautilus concludes our series of collectible wristwatches from our manufacturer in Geneva.

5.8. Concluding Potpourri

So that the reader can also see some current models of the Geneva noble brand at the end of this chapter, a potpourri of newer models can be found on the following two-page spread. Three of these watches were purchased almost anonymously by the author from retailers and put through their paces as test watches. Accuracy, functional reliability, and usability were in demand here, and all three watches were able to meet high expectations.

In our potpourri, readers will also learn how the watches presented are traded on the secondary market.

Illustration 324
Thierry Stern, the current president of Patek Philippe, and his father, Philippe Stern, the honorary president, in conversation with the author

Illustration 325

Illustration 330

The Noblest of the Noble
(Image descriptions clockwise from top left)

Illustration 325
The reference 5940G-001 automatic with perpetual calendar, white-gold case, and Breguet numerals begins our series of photos. This watch is quite commonly found on the secondhand market, and the average asking price is between $57,000 and $61,000.

Illustration 326
The Nautilus Ref. 5990/1A-001 in stainless steel is ideal for frequent travelers. With a chronograph movement, second time zone, date, and day/night display, this watch provides all the information one needs. The fact that the average price for this hard-to-find model on the secondhand market is between about $130,000 and $140,000 proves that it is in high demand.

Illustration 327
This white-gold Ref. 5227G-001 with ivory dial and 1.53-inch (39 mm) case diameter is representative of the classic Calatrava. This model was one of our test watches. On the secondhand market this fine timepiece is priced between $30,000 and $35,000.

Illustration 328
The classic Ref. 3970EP perpetual-calendar chronograph with platinum case. Unveiled in 1985, this direct successor to the reference 2499 commands prices of between $130,000 and $150,000 on the secondhand market.

Illustration 329
This Ref. 5159G-001, with perpetual calendar and white-gold case, is an elegant watch suitable for an evening reception. With a case diameter of 1.49 inches (38 mm) and height of 0.48 inches (12.40 mm), it fits well under the left shirt cuff. Personally, however, the author has never encountered this watch.

Illustration 330
Despite an outward similarity, these two handsome models are miles apart in age and value. The model on the right is representative of the automatic drive, annual calendar, and rose-gold case of the current reference 5396R-011, while the one on the left is the reference 3448J from 1977, with perpetual calendar and automatic caliber 27-460. On the secondhand market the watch on the right sells for between $35,000 and $55,000. The one on the left was sold in the past years with prices ranging from $200,000 to $500,000. It is easy to understand that wristwatches from the house of Patek Philippe are not only among the most noble but also among the most valuable products of the Swiss watch industry.

Illustration 326

Illustration 327

Illustration 329

Illustration 328

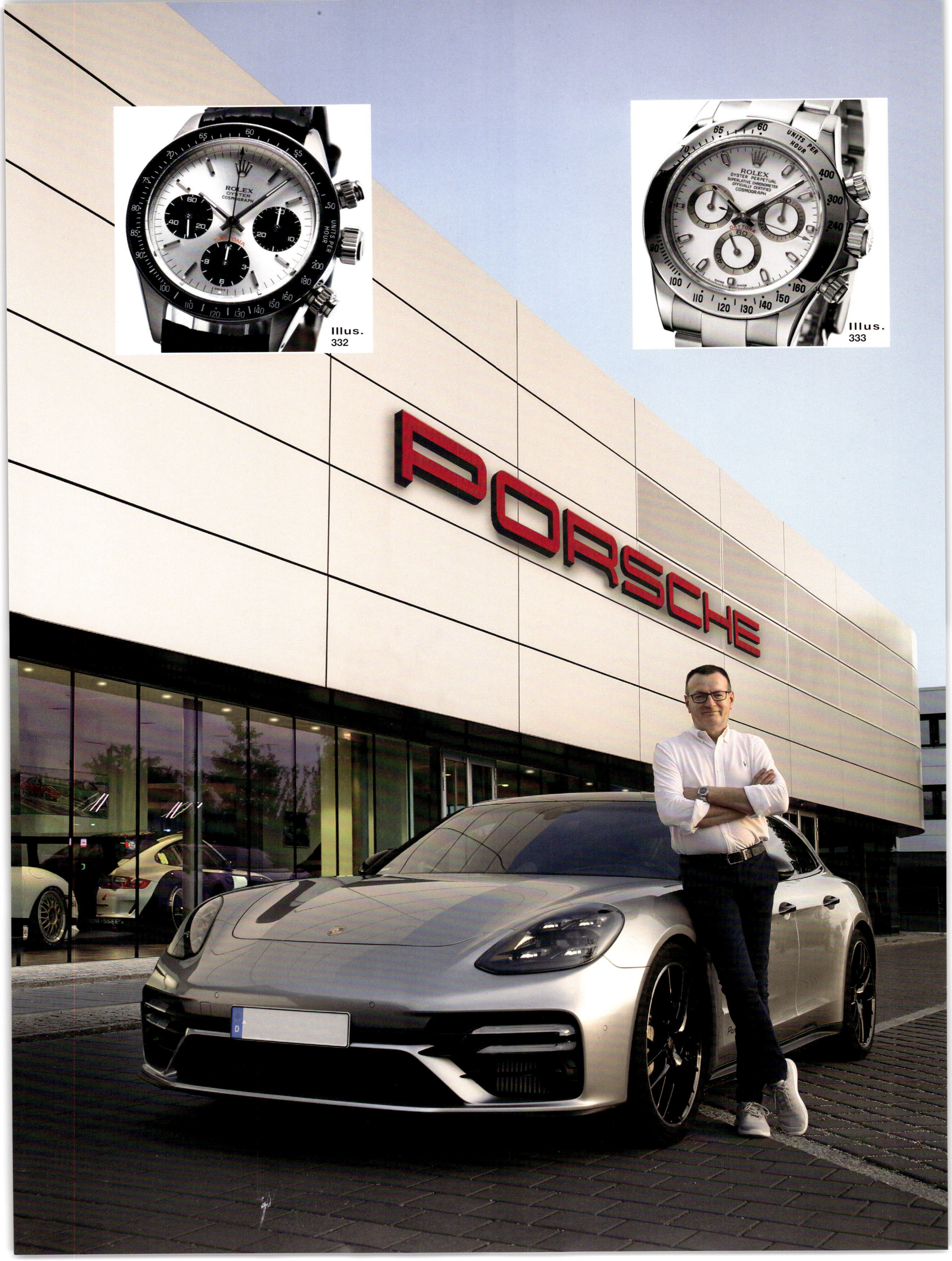

Illus. 332

Illus. 333

6. Rolex

6.1. Introduction by Gisbert L. Brunner

Rolex: Precision Watches under the Sign of the Crown:

A Merchant Discovers the Wristwatch

Anyone who has the good fortune to be able to rummage through the archives of one of the world's most important watch manufacturers to their heart's content will not find a person named Rolex, despite a meticulous search. They simply do not exist. But with a differentiated analysis of the history of the chronometric empire, which is more than a hundred years old, one finally finds a man who would very well deserve this name. The first retrospective encounter with him takes place in a location that actually has nothing to do with watches and time measurement. The northern Bavarian town of Kulmbach, which is the subject of this article, is primarily known for its excellent beers.

On March 22, 1881, a boy first saw the light of day in this small town, and his parents baptized him Hans Wilhelm. The three children of Anna and Ferdinand Wilsdorf did not grow up under a particularly favorable star, because in 1893 they became orphans. Thanks to financial reserves—the mother came from a well-known brewery dynasty—and the sale of a flourishing hardware and household goods store on Kulmbach's market square founded by their grandfather Carl Wilsdorf, they nevertheless did not miss out on schooling and vocational training. After graduating from high school and completing a commercial apprenticeship in nearby Bayreuth, Hans Wilsdorf moved to Switzerland at age nineteen. In Geneva, where he later lived, he worked briefly for a pearl dealer as a trainee. He then moved to the Jura Mountains in western Switzerland, which had some scenic similarities to his native Franconia. In the industrial metropolis of La Chaux de Fonds, an interesting job beckoned. A successful watch importer paid the language-savvy young professional eighty francs a month to handle correspondence in English and do general office work. In addition, Hans Wilsdorf had to wind up several hundred watches a day and check their running. With such products Cuno Korten earned more than one million Swiss francs annually. This remarkable turnover with watches from German, Swiss, and French manufacturers awakened in Hans Wilsdorf the desire to go into business for himself. The ambitious young businessman ordered and sold three gold pocket watches with official certificates from the Neuchâtel Observatory on his own account. His superiors were not at all annoyed by his initiative. On the contrary, they gave their enterprising employee new, more-responsible tasks. But his suspicion was aroused and, after an involuntary military service intermezzo in Germany, he wanted more, much more.

Long Live Independence

Hans Wilsdorf first demonstrated his entrepreneurial talent in 1905. A total of 33,000 German gold marks had been stolen from him while on a voyage to England. After losing his inheritance, he had to ask his two siblings for a loan to start his own business. Like it or not, the victim of robbery also took his much-older partner Alfred James Davis on board. The latter would later turn out to be a solid *filou* (friend).

At Hatton Gardens 83, London, the start-up took its course. The purpose of Wilsdorf & Davis Ltd. was to sell watches "*en gros*." Because of the large number of competing companies, Wilsdorf had to come up not only with particularly attractive products, but also with innovative marketing and sales concepts. In this context the new timepieces for wear on the wrist came just in time. After moving to Holborn Viaduct 44, Hans Wilsdorf activated his existing contacts with the Swiss watch industry. In particular he had fond memories of Aegler SA, founded in 1878. "The Aegler company," he wrote in his memoirs, "was

Illustration 331
Franz Wörle, the managing director of Porsche dealership "Porsche 5 Seen" in Gilching, near Munich, with his Porsche Panamera. Wörle is known as a lover of fast cars and precision wristwatches.

Illustration 332
Rolex Ref. 6263 Daytona with manual winding, also called Big Red

Illustration 333
Rolex Daytona Ref. 116.520, the current model

Illustration 334
Rolex automatic caliber 1520 in about 14× magnification. At the front left is the balance wheel with Breguet hairspring. On the balance staff, one can see one of the two *masselotte* weights used for fine adjustment of the rate. The red-and-gold-colored gears of the automatic device module are described in more detail in the text.

SWISS
1520

the first watch factory to start the production of a watch with an eleven-lever movement. Since the success of this innovation, which was subsequently to prove of the greatest significance for the entire watch industry, the future of the wristwatch could be regarded as ensured." So Hans Wilsdorf traveled to Biel in 1905. There, his offer to sell wristwatches made by Aegler in Great Britain met with a positive response, resulting in a first large order. It was valued at several hundred thousand Swiss francs, which corresponded to five times the operating capital of Wilsdorf & Davis. His courage quickly paid off, however, since the imported goods sold like hotcakes. Further orders, new product ideas, and suggestions for improvement followed in quick succession. Thus Hans Wilsdorf, who firmly established himself in England and took on British citizenship, made a not-inconsiderable contribution to the triumphant advance of the then-new star in the watchmaking firmament.

A Name for the World

There was only one shadow over the joy of success: important specialty stores such as the Goldsmiths Company and Asprey wanted to see only their own names on the watch dials. Manufacturers and suppliers had no business there, or on the movement and case. Wilsdorf wanted to break with this tradition at all costs. But this could be achieved only if he could offer unique products with a high level of desirability. And a distinctive signature was also needed. Hans Wilsdorf had the latter protected in 1908. There are many stories about the background of the name Rolex. In any case, it is undisputed that it can be used with equal ease in all languages. Even Americans cannot succeed in mispronouncing the name. In addition, the two-syllable term took up so little space on the dial that the dealer's lettering could also be printed on it without any problem. Finally concise, five-letter words were very popular. This is evidenced by Kodak, another legendary name of the era.

Rolex, some claim, is derived from *horlogerie exquise*, meaning outstanding watchmaking. Hans Wilsdorf's widow, on the other hand, has come up with a different version. According to her, Rolex is nothing more than an abbreviation of Rolling Export. This refers to the almost uninterrupted flow of watches from Aegler to Wilsdorf & Davis. Be that as it may, the creation was and is unsurpassed in terms of genius.

Precision Writ Large

Even when the company had just been named, the young company boss refused to accept that wristwatches should be less accurate than their chain-worn counterparts. On March 22, 1910, the official watch inspection office of the Biel Watchmaking School awarded a Rolex movement with a diameter of 0.98 inches (25 mm) a first-class certificate after fourteen days of detailed testing. In 1914, after a forty-five-day check, the observatory in Kew, England, awarded the first A Class certificate for a wristwatch. After that, the buck stopped with Aegler in Biel; he had to regulate all movements according to chronometer criteria. In 1932 the observatory in Besançon, France, wrote: "Rolex, as the only one in the world, has obtained a first[-]class certificate for a wrist chronometer of small format at each of the four observatories in Kew, Geneva, Neuchâtel, and Besançon." In 1945 Rolex, now the world's largest producer of officially certified chronometers, was already able to show its 50,000th rate certificate. In the year of the company's fortieth anniversary, the childless patron established the Hans Wilsdorf Foundation, which still commands Rolex today.

In the Spirit of Perfection

Because imperfect cases ran counter to his efforts to achieve great accuracy, Hans Wilsdorf next turned his attention to this watch component. The result of his extensive efforts was patented in 1926. Once again, a suitable name was quickly found. Oyster aptly described the characteristics of the new thing. The Oyster A passed its first endurance test on October 7, 1927, on the wrist of the London steno typist Mercedes Gleitze. After fifteen hours and fifteen minutes in the cold waters of the English Channel, at 6:10 p.m. the exhausted and completely hypothermic Amazon climbed onto the British limestone rocks somewhere between South Foreland and St. Margaret's Bay. Her Rolex ticked along as if nothing had happened.

That's exactly what earned Hans Wilsdorf a blanket ad on the front page of London's *Daily Mail*. The one-off appearance on November 24, 1927, in a newspaper with more than two million readers cost the equivalent of $27,260—a gigantic sum by today's standards. But it put the Oyster on everyone's lips and enabled Hans Wilsdorf to increasingly assert the Rolex signature among concessionaires. Anyone who insisted on his own signature had to do without his business.

"The logical consequence of the Rolex Oyster was the creation of the self-winding watch, the movement of which wound itself over and over again and ensured uninterrupted running," Hans Wilsdorf wrote in his autobiography. Only a self-winding movement effectively prevented the daily unscrewing of the crown. This was the hour of technician Emil Borer, who in 1931 developed the unidirectional rotor winding system for the 0.295-inch-high (7.5-millimeter) caliber NA 620 movement. Afterward Rolex's lawyers did a great job, because for fifteen years the company's competitors had to do without this innovation.

Further Milestones by Hans Wilsdorf and Rolex

In 1945, for his company's fortieth birthday, Hans Wilsdorf launched the so-called Datejust, a synthesis of the Oyster, automatic movement with chronometer certificate, and easily readable window date at the three o'clock position on the watch, which moved to the next day "just in time" at midnight.

On May 29, 1953, at 11:30 a.m. sharp, Colonel Sir John Hunt, Edmund P. Hillary, and the Sherpa Tenzing Norgay reached the summit of Mount Everest. In terms of timekeeping, the team had relied on waterproof automatic wristwatches that Rolex later launched under the name Explorer.

In 1953, the management of Pan American Airlines asked Rolex to create a wristwatch that would keep local time as well as home time. A year later the groundbreaking GMT-Master was ready, which Rolex marketed from 1955 under the reference number 6542.

In 1954, visitors to the Basel Watch Fair were able to admire the "Submariner—the Diver's Friend" in one of five Rolex display windows. This was a wristwatch with a matte black dial, oversized luminous numerals and indexes, and luminous hands for hours, minutes, and seconds. In addition, it had a handy rotating bezel with minute divisions. Its water resistance was up to 145 pounds per square inch (10 bar).

The year was 1956 when Rolex launched a new wristwatch with the reference number 6511. Next to the window date with magnifying glass, the name of the day of the week could be read in full in the "north of the dial." However, the Day-Date was available only in 18-karat gold or platinum.

Until 2004—something many may not know—there were two separate Rolex companies: in Biel, Manufacture des Montres Rolex SA, controlled by the Aegler family through their descendants, for the production of watch movements; and in Geneva, Montres Rolex SA, founded in 1920 by Hans Wilsdorf, which was responsible for finished watches. After tough negotiations, a merger took place in 2004 under the umbrella of the Geneva branch of the company. Hans Wilsdorf was able to experience the unstoppable rise to a global corporation only in part; he died on July 6, 1961.

6.2. The Movements

There are hardly any other watch series that can boast such a high degree of recognition as the Oyster models from the house of Rolex. Accuracy, water resistance, and robustness are among the most-important attributes of these legendary timepieces. Only a few authors have dared to tackle the subject of Rolex, especially since the market leader in Geneva has always been very cagey. In addition to including the pictorial representation of these coveted watches, the author tries in this chapter to give the subject of Rolex a little more transparency and to address readers' previously unanswered questions.

Manually Wound Movements

Men's wristwatches with manual winding are no longer part of Rolex's Oyster line. However, in the 1940s and '50s they were still an important part of the company's product range. Today we can find watches with these calibers only at watch fairs or auctions. If repairs are needed, the task can be very difficult, because Rolex has not kept any spare parts for these watches in stock for many years. Two of these manually wound movements are presented here.

700 10½''' (illus. 335)

For this movement Rolex used a caliber from the Gruen company (base caliber 825 there). The case diameter is 0.92 inches (23.40 mm), its height 0.14 inches (3.65 mm). The movement contains eighteen rubies, and the balance runs at 18,000 beats per hour. It was unveiled in 1938. While the 700 version is equipped with small seconds, the 710 variant has a center second hand.

Illustration 335

1210, 1215, and 1220

The caliber 1210, a solid precision movement with manual winding, was unveiled in 1954. The installation diameter is 0.93 inches (23.80 mm), and the movement height is 0.168 inches (4.27 mm). There are eighteen rubies in the movement, and the balance runs at 18,000 A/h. This basic caliber has a center second hand. The 1215 version features a date display (movement height 0.199 inches or 5.07 mm). The calibers 1220 and 1225 largely correspond to the calibers 1210 and 1215, except for improvements in the oscillation system, but operate at 19,800 A/h.

Chronograph Caliber Valjoux 13''' VZ (illus. 336)

This chronograph movement, which was already in wide use in the 1930s, was also used by other Swiss watch manufacturers,

Illustration 336

such as Audemars Piguet, Patek Philippe, and Vacheron & Constantin, as an *ébauche* and installed with the company's own finish. Movement diameter is 1.16 inches (29.50 mm or 13'''), the height is 0.230 inches (5.85 mm), and the beat rate is 18,000 vibrations per hour. The standard version, like the one used by Rolex, has seventeen jewels. Patek Philippe's first water-resistant chronograph, the reference 1463, unveiled in 1940, has twenty-three jewels. The movement has a small seconds and a thirty-minute counter. The power reserve is about forty hours.

Valjoux 727 (illus. 337)

The chronograph caliber 727 was a caliber 726 modified for Rolex. The fine-adjustment mechanism has a Breguet hairspring, the balance wheel has four Microstella screws, and the frequency of the balance is 21,600 vibrations per hour. Movement diameter is 1.16 inches (29.50 mm or 13'''), height is 0.27 inches (6.95 mm), and the movement has seventeen jewels. Shock protection for the balance staff is provided by KIF. This caliber remained in production until 1987, when it was replaced by the self-winding caliber 4030 made by Zenith. This Valjoux caliber was also used by numerous other Swiss watch manufacturers, but not in the version seen here, which was refined for Rolex.

Illustration 337

Automatic Movements

Compared to the other six brands presented in this book, automatic winding found its way into Rolex watches very early. It began offering the first self-winding timepieces in the early 1930s. Further development of automatics was slow but consistent. Some examples are provided here.

655 (illus. 338)

The design of the 645 11¾''' caliber goes back to Rolex's first automatic movement. However, with the same movement diameter of 1.03 inches (26.40 mm), the height of the movement was reduced by 1 millimeter to 0.255 inches (6.50 mm). Additionally, a cutout was added at the formerly almost closed automatic bridge so that the balance wheel became visible. The

Illustration 338

movement contains twenty jewels, and the balance with Breguet hairspring runs at 18,000 A/h. The rotor winds in only one direction. For this caliber too, spare parts are no longer available.

765 (illus. 339)

With the caliber 740, Rolex introduced the date display at three o'clock in 1945. This gave birth to the designation Datejust, which is still in use today. The 765 version described here was unveiled in 1950.

The diameter of the movement is 1.16 inches (29.50 mm), and the movement height is 0.255 inches (6.50 mm). Here, too, winding is in one direction only. This movement is easily recognizable by the double, free-milled, self-winding bridge positioned directly above the balance staff. The movement contains twenty rubies, and the balance wheel has a Breguet hairspring and runs at 18,000 A/h. The basic movement is based on the manually wound caliber 700 (see also p. 221).

Illustration 339

1030, 1035, 1055, and 1065 (illus. 340)

The base caliber 1030 was introduced in 1950. The movement has a diameter of 1.12 inches (28.50 mm) and a height of 0.230 inches (5.85 mm). This caliber broke completely new ground in terms of design, with its rotor now winding in both directions. The rotor is supported by a steel axle and carbide bearings. The rotor plate is only 0.011 inches (0.3 mm) thick and has two slots. On the outside is a heavy metal flywheel weight that is triple bolted. In conjunction with the spring steel spring ring (according to the parts list "Wedge for oscillating mass axis," part no. 7016), which fixes the rotor axis, the thin rotor plate acts as a spring element, which can absorb shocks acting on the rotor. The oscillating system consists of a screw balance wheel equipped with regulating discs, a Breguet hairspring, a shock protection system, and a combi-chuck for the escape wheel bearing (Duofix). The balance runs at 18,000 A/h. There are twenty-five rubies installed in the movement. New in the automatic bridge is the changer, which works via two springless twin ratchet wheels. For the first time, gears made of anodized aluminum are used here—an almost ingenious solution, as we will read later. In the

Illustration 340

version 1035 this caliber features a date indication at three o'clock. For this reason the overall height of the movement increased to 0.262 inches (6.66 mm). The 1055 version of this caliber was used in the Day-Date model. Besides the date indication, a weekday indication was added. The movement height increased to 0.275 inches (7.00 mm) because of these additional installations. The version 1065 is a revision of the caliber 1030. The automatic bridge, the date indication, the balance cock with oscillating system, and many other things were changed. Diameter, movement height, and beat rate remained the same.

1520, 1530, 1555, 1556, 1565, and 1575 (illus. 341)

The new basic caliber 1530 was unveiled in 1957. It was so well designed that it was able to stay in the program for several years. From the outside it was easily distinguishable from its predecessors. The rotor, automatic bridge, fine-adjustment mechanism, and plate were all completely new designs.

The basic caliber had only a center seconds hand. The movement diameter was 1.12 inches (28.50 mm), the height 5.75 mm. The balance with Breguet hairspring runs at 18,000 A/h. There are twenty-six jewels in the movement.

The 1556 version appeared in 1965, was intended for the DAY-DATE model, and had weekday and date switching. With the same diameter, the movement height of this caliber is 0.27

Illustration 341
Caliber 1520

inches (7.03 mm), and the balance runs at 19,800 A/h. The caliber 1565, with date indication, results from the basic caliber 1530 and was used in numerous models from 1959 onward. The movement height is 0.248 inches (6.30 mm), and the movement diameter is 1.12 inches (28.50 mm). However, the beat rate is only 18,000 A/h. The successor caliber 1575 was launched in 1965. The dimensions were identical to caliber 1565, but the beat rate was increased to 19,800 A/h. The movement diameter was 1.10 inches (28 mm).

Caliber 3035 (illus. 342)

Ten years after the launch of the automatic caliber 1530, its successor, designated 3035, entered service in 1977. This also represented a giant leap forward for Rolex. The entire movement had been redesigned, and only the best solutions had been accepted in the choice of manufacturing methods and materials. Everything had also been thought of in terms of serviceability.

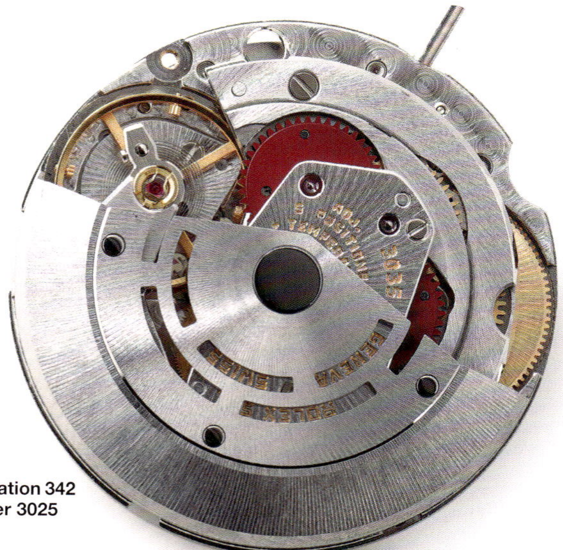

Illustration 342
Caliber 3025

Besides the balance stop, the caliber variant 3055 for the Day-Date model Ref. 18038 also had a date quickset installed, also called Quickset.

With the same diameter of 1.106 inches (28.10 mm or 12½‴) and a height of 0.25 inches (6.35 mm), its rate was now 28,800 vibrations per hour, and twenty-seven jewels are installed. The power reserve is fifty hours. Now the four-spoke balance wheel with four Microstella regulating screws, an adjustable hairspring stud, and the escapement wheel bearing as combined Duofix combi-chuck also ensure a smooth rate. KIF-Ultraflex is used as shock protection for the balance staff.

Caliber 3135 (illus. 343)

Eleven years after the introduction of the caliber 3035, its successor, with the caliber designation 3135, was unveiled in 1998. Again, various novelties were installed, the most obvious of which was the new balance bridge. From the mechanical

Illustration 343
Caliber 3135

engineer's point of view, here all technical details were designed with a mix of previous experience, on the one hand, and innovative ideas, on the other. The caliber 3135 can be described as the Porsche 911 of the watch industry. In chapter 9, "New Automatic Movements from Switzerland," this automatic movement is used as a reference, and its strengths are once again described in detail.

Until the appearance of the new Rolex caliber 3235, the automatic movement 3135 could be considered the undisputed market leader.

Caliber 3235 (illus. 344)

Even the first glance at this automatic movement launched in 2015 reveals to a production engineer who has previously taken a close look at the predecessor caliber 3135 the successful

Illustration 344
The new automatic caliber 3235 was shown for the first time at Basel World 2015 and declared to be an absolutely new development. Most of this work focused on the inner workings (i.e., the automatic device module, the mainspring, the gear train, and the fine-adjustment mechanism). Visible in this image are the new rotor, now consisting of one piece, with ball bearings instead of the previously used synthetic rubies, as well as the newly designed balance bridge.

Illustration 345
The balance bridge of the caliber 3135 has integrated knurled screws on both sides to adjust the height of the balance staff. The clamping device on the right is used for locking the hairspring stud. The outside of the balance is covered by a small bridge.

Illustration 346
From the point of view of a development engineer, the innovations in this control system might be considered quite banal but smart ideas. After all, you first have to come up with such seemingly simple ideas. The prebridge is omitted and the balance bridge is designed as an integrated component. The height adjustment for the balance staff is done only on the right side. Thus, the bridge itself acts as a beam clamped on one side. The height adjustment for the balance staff is now done only with a knurled screw on the right side. The locking of the hairspring stud is elegantly done in the adjustment mechanism itself via an integrated screw. Also new is the shock protection, which no longer comes from KIF-Ultraflex but is an in-house development by Rolex under the name Paraflex-Antishock-System. Also new is the balance wheel with blue Parachrom hairspring and—not visible—the Chronergy escapement with optimized efficiency, which was also a new development.

implementation of various demands placed on this new movement. In addition to the desire to improve performance and increase the power reserve, various potentials were also exploited in terms of production costs. A comparison of the balance bridges of the caliber 3135 and 3235 alone shows the new movement's impressive design simplifications. Thanks to almost banal but nevertheless clever ideas, it was possible to save more than three parts in this area alone and to significantly reduce assembly and adjustment times. A further example is the new rotor, which now consists of just one part, thus reducing production costs. Such mass-produced parts can be manufactured cost effectively with a machining center. Material reinforcement on the outer edge of the rotor largely compensates for the absence of the heavy metal hoop. It was therefore possible to dispense with the rotor, which in the case of the 3135 caliber still consists of two parts. As a further potential saving, a uniform surface structure was achieved on the various movement components without detracting from the good overall impression of this movement. A 911 in plain silver gray looks good because its design alone radiates sufficient harmony. Why should it be any different with Rolex?

Chronograph Caliber 4130 (illus. 347)

With the chronograph caliber 4130 Daytona, in 2000 Rolex launched its first manufacture chronograph movement with automatic winding. The high expectations that were associated with the appearance of this new movement were also completely fulfilled.

Illustration 347
Rolex Chronograph caliber 4130, the first movement of this kind with automatic winding. *Rolex photo*

With a movement diameter of 1.20 inches (30.50 mm), height of 0.25 inches (6.50 mm), forty-four jewels, 28,800 vibrations per hour, and a power reserve of seventy-two hours, not only had the internal development goal been achieved, but the number of fans of the Rolex Daytona had also increased significantly.

6.3. The Reference Numbers

Because of the multitude of Rolex models, on the one hand, and the space available here, on the other hand, only a few examples can be listed on this topic. With few exceptions, the reference numbers in the period under consideration are made up of four digits. To get an idea of the multitude of models, let's take a look at the caliber 700 10½''' shown on page 221. This widely used manually wound movement, which was built from 1938 to 1951, was installed in more than 120 different model variants. Here, the four-digit reference numbers started with the numbers 2, 3, 4, 5, 6, and 7, with the exception of the references 978, 1504, and 1873 (all Oyster models). With the end of the 1950s, automatics were given reference numbers that began with a 1. For subvariants, a slash and another number (e.g., 1803/9 for the Rolex Oyster Perpetual Day-Date in 750 white gold) were added. From 1977 on, five-digit reference numbers followed. Here, the first four digits were retained for the respective model, and a fifth digit was added at the end on the right. Thus, in 1977 a model of the new Day-Date with the caliber 3055 (from base caliber 3035) received the reference number 18038. For comparison, Rolex today assigns six-digit reference numbers. Thus, an Oyster Perpetual Day-Date in 750 white gold with the corresponding President bracelet carries the reference number 118239. In the time period described here, the reference numbers are usually listed twice on Rolex men's wristwatches. They are engraved inside the case back and then a second time between the lugs at the twelve o'clock position. In the 1950s and '60s, the inner case back also listed references to the production period. For example, the engraving II.58 means that this watch was produced in the second quarter of 1958. Rolex was unable to explain why these indications are found only in steel cases. Some concrete examples will be listed in the following text.

6.4. The Case Numbers

Rolex case numbers consist of six or seven digits. We always find them engraved between the lugs at six o'clock. With many years of use, the use of metal bracelets, and improper refurbishment of the case, it can happen that the engraved number becomes only partially legible or disappears completely. This happened on a Datejust reference 1601 in 750 red gold, which we can admire on page 177. If the case number is no longer recognizable, the only way to identify the watch is to send it to Rolex.

6.5. The Watches

A survey conducted many years ago showed that buyers of timepieces with the "crown on the dial" are primarily guided by the brand name, its high degree of recognition, and the eye-catching appearance. The excellent workmanship of the case and dial, on the one hand, and the high level of design and manufacturing technology of the automatic caliber, on the other, usually play only a secondary role. But it is precisely these two criteria that give a Rolex wristwatch the outstanding reputation that it undoubtedly deserves.

Approximately thirty watches from the house of Rolex are presented in this chapter. For this purpose they not only represent half a century of watch history but also convey the diversity of models from earlier years. In addition, Rolex segmented its worldwide markets and supplied them individually. If a new model was launched, this happened only after extensive testing, as well as varied trials. One example is worth mentioning. When the Oyster Datejust (Ref. 17000) and Day-Date (Ref. 19000) models with quartz movements were launched in the 1970s, they were given a new design for the strap and case. To be able to put this new design through its paces, automatic movements were encased in about nine hundred pieces (Ref. 1530 with sapphire crystal) of these new cases and exported as a special model for the South American market. Only when proof of the reliability and functionality of this new design was available was production of the new generation of quartz watches approved.

Manually Wound Watches

Reference Numbers 6082 and 6426, Chronograph 3525 and Chronograph Daytona 6263

Our chapter begins with two already very old Oyster models with manual winding. The first one, with the caliber 700 movement and small seconds, was found by a watch collector in the display of a Munich pawn shop. After a lengthy discussion the rare piece changed hands for $697. The timepiece with reference number 6082 has a very special dial and can be assigned to the production

Illustration 348
When the 1950s began, Rolex did not yet offer the extensive range of men's wristwatches that it does today. Therefore, the interested customer was still offered quite inexpensive models with manual winding. The Rolex Oyster reference 6082 presented has a manually wound caliber 700 10½''' movement with small seconds. This movement was unveiled in 1938. The dial is deep-drawn and painted in two colors. The 12 and the indices are set in recesses. On this two-piece steel case, the back and crown are screw-down. The reference number is engraved between the lugs at twelve o'clock, the case no. 814,358 between the lugs at six o'clock. The movement no. 53,497 can be found on the plate near the balance wheel. The caliber designation can be read only after removing the balance cock.

year 1950. Such an elaborate dial can be found only on a version of the Rolex chronograph Ref. 6034 with Valjoux caliber 72 from the early 1950s. The caliber 700 was in production from 1938 to 1951. The three gold *chatons* visible on the gear train bridge also deserve a note, the right one being the escapement wheel bearing. The second Rolex Oyster, with manual winding Ref. 6426 Oyster Precision in a steel case with rose-gold plating, was purchased by a watch enthusiast in spring 2001 from a dealer in Fountain Hills, Arizona. He did not want to discuss whether the $1,125 paid at the time was in line with the market value. He liked the watch so much that he gladly paid another $290 to have the watch overhauled and the case refurbished. The USA was one of Rolex's most important export markets in the 1950s and '60s. The respective Oyster cases were available in different material combinations. In addition to the steel cases with gold-plated bezel and case middle, there was also the combination of steel and gold. In the latter case, people preferred only the fineness of 585 for the *mixte* (mixed) models, but Oyster models with full gold cases in 14-karat alloy were also very popular. Only the Rolex products that went on sale in the Old World usually had cases made of 750/1000 gold.

Illustration 349
This reference 6426 Oyster Precision with rose-gold-plated steel case is regarded as another example of an affordable Rolex from the 1950s. Here, too, the watch has a screw-down case back and crown. The thickness of the gold layer is 40 μ (0.0015 inches or 0.040 mm), and in the steel screw-down case back you can find, among other things, the indication *CAPRURÉ LUNETTE PLAQUÉ OR LAMINÉ 40 MICRONS* (CAPRURÉ GOLD-PLATED BEZEL LAMINATED 40 MICRONS). In contrast to the manually wound caliber 700 movement shown on the left, the caliber 1210, which is not visible here, uses a combi-chuck for the escapement wheel bearing (movement no. 55,334, case no. 61,473). All high-end watch brands use so-called combi-chucks for the escapement wheel bearing of their movements. These not only absorb unwanted shocks but also improve oil retention. This is particularly important in the case of extreme shock loads.

Illustration 350
One could call this picture Fascination Rolex pur (Pure Rolex Fascination). The chronograph reference 3525 was launched in 1940 and came with different dials and cases. The case was offered in 750 yellow and rose gold, as well as stainless steel. In this model presented, both the case and the dial are made of 750 yellow gold. Working inside is the Valjoux-sourced caliber 13''' VZ with Breguet hairspring and seventeen jewels. It has no movement number engraved. The case number 47,255 is found not between the lugs, but on the screw-down case back.

Chronographs from Rolex have always had their admirers. In addition, it was precisely in the 1940s and '50s that watches with a stop mechanism had become particularly fashionable. Despite the multitude of brand names, there were only a manageable number of around two dozen manufacturers producing the coveted chronograph calibers. Rolex did not have its own movement and sourced its chronograph calibers from the supplier Valjoux. Since Rolex watches have created their own circle of enthusiasts, only one chronograph is listed in this book. The reference 3525 shown is privately owned in Munich and was

Illustration 351
Rolex Ref. 6263 Daytona in stainless steel, also called Big Red (caliber 727, case no. 6,393,352)

received by the current owner more than two decades ago as compensation for unpaid debts. From a neutral point of view, the Rolex chronograph presented is an inexpensive watch to produce. The case consists of just two parts: the case middle with integral bezel and the screw-down case back. An inner case back was omitted. The Valjoux caliber has no special finish and was probably encased as delivered. The movement was not numbered, but the case was. The number is not placed between the lugs but instead is stamped on the outer case back (see fig. 350, p. 227).

Chronographs from Rolex have always been in the fashion trend of the time. The manually wound Daytona models of the 1970s and '80s are in great demand today. Due to this permanent demand, they are constantly increasing in price, as we can explain here using a reference 6263, case no. 6,056,787—also called "Big Red"—as an example.

The very well-preserved Daytona Ref. 6263/0 was auctioned on April 18, 2010, at KEUP Kunstauktionen in Regensburg for $22,097. This Rolex Oyster Cosmograph Daytona came with papers, box, and the original jubilee metal bracelet.

For an authenticity check and servicing, the beautiful piece was passed on to Rolex Cologne. After about five months this watch came back from Geneva almost like new. During this overhaul, the crown, dial, hands, and case back had been replaced. For the overhaul, Rolex charged exactly $3,600 and at the same time pointed out the great exception they would have made with the overhaul for this book.

Today, some ten years after the acquisition of this watch, reference 6263 "Big Red" Daytonas are sold for double to triple the purchase price paid here. On October 27, 2018, a Daytona Ref. 6263 with case no. 6,393,352 was hammered in Hong Kong for around $76,327. Today other sales are also in this price range.

Reference 6090 (2), 6305, 6309, 6605, and 6609
Automatic Watches

Even in earlier times, elegant ladies of distinction discovered their love for certain men's wristwatches from the house of Rolex. These included the various Oyster models, such as those shown on page 229. We have explained in the caption why these Oyster models were also given the suffix Bombé or Bombay. Both Oyster models were purchased by a collector in spring 2001. He bought the older model with guilloché dial at the Munich Watch Fair for $2,705. The newer one came from an amateur in San Diego, California, and changed hands for $2,233. The usual import duties had to be added. Same model, same reference, same price could be the credo for these two Bombés. On direct comparison and close inspection, both calibers were in good condition corresponding to their age. Also the number of watchmaker marks was identical at three. Even the cases were in similar condition. So only the different appearance of the dials remained. Here, too, the Bombé with guilloché dial finally got the higher score.

The first Rolex Datejust was unveiled in 1945 under the reference number 4467 and represented a watchmaking masterpiece at that time. The successor model with the reference number 6305 followed about seven years later.

Soon after the end of WWII, the founder of the Rolex company, Hans Wilsdorf, came up with the marketing idea of presenting prominent persons from modern history with a watch made by his company Rolex. The first president of the United States of America to receive a timepiece with the crown from Geneva in this way was Dwight D. Eisenhower (thirty-fourth president of the United States, term 1953–1961). This Rolex Datejust in 18K yellow gold was a reference 6305 with special significance. This was also the 150,000th watch made by Rolex. Eisenhower is said to have worn this watch until the end of his life. The case and dial of the Datejust Ref. 6305 presented here

Illustration 352
Among the most popular models from the house of Rolex are the Oyster-Perpetual Bombé. The term *bombé* refers to the curved lugs and comes from the French verb *bomber*, to bulge. Due to a confusion of terms, this model series is often referred to as "Bombay" in publications and at auctions. We have also seen this type of lug from watches made by IWC and Universal Genève, and in one case also from the Patek Philippe Ref. 2537. The Bombé models, including the Ref. 6090 shown, included a total of ten different versions. These were powered by three different calibers over time: 645, 1030, and 1560 (*left*: movement no. 20,587 [F 4723], case no. 967,914; *right*: movement no. 97,621 [F 24,945], case no. 729,953). The Bombé on the left is the older one and has a guilloché dial with the "SWISS" label below six o'clock. The one on the right has luminous hands, a lacquered dial with minute divisions, and the label "SWISS MADE" below six o'clock.

correspond exactly to the watch presented at that time, only the president wore his Rolex not with a crocodile leather strap, but with a gold Jubilee strap.

The Rolex Oyster Perpetual Datejust Ref. 6305 in 750 GG shown also has an interesting history. It comes from a watch enthusiast in Offenburg. He had this watch overhauled at the Rolex concessionaire in Freiburg. This service included refurbishment of the movement and case, as well as refreshing of the dial. After the return of this watch, the rate was poor and the rotor grazed the screw-down case back. The owner did not notice all this and sold the watch for $2,510. Only the new owner noticed these serious defects. The beautiful piece was immediately passed on to Rolex Cologne and professionally restored under the watchful eye of Georg Ressel, technical manager.

The pricing of this work was also very accommodating. The entire process was brought to the attention of the Rolex concessionaire in Freiburg, and a request was made for reimbursement of the costs incurred by Rolex. However, the desired reaction was not forthcoming, and the aggrieved party is still waiting for payment.

The Thunderbird presented here has several special features. It belongs to the first models of the reference 6309 and still has

Illustration 353
This Rolex Oyster Perpetual Datejust in 750 yellow gold looks to be in almost-new condition. This automatic, a reference 6305, was serviced by Rolex, which included professional refurbishment of the dial. The first Oyster Perpetual model with date was unveiled in 1945.

Illustration 354
View of the caliber 765 movement in the reference 6305 illustrated above. On the caliber version shown, one can see that on the automatic bridge there is no oblong hole, but instead a groove milled in its side for setting the regulator arm (movement no. 83,777, case no. 114,630).

the caliber 765 movement. From 1955 onward this movement was replaced by the new caliber 1065. The Thunderbird shown is an heirloom donated by an American. The watch had belonged to his father, who had served as an airman and later as leader of the Thunderbirds air display team. Retired with the rank of colonel, the officer had served in three wars (WWII, Korea, and Vietnam) and was considered an aviator par excellence. Besides the history of this timepiece, the bezel deserves special mention. Instead of fluting, the bezel was decorated with a fine nail pattern, also called Clous de Paris.

Illustration 355
This reference 6309 in 750 yellow gold is a Rolex Turn-O-Graph (or Thunderbird) of a special kind. It once belonged to a US Air Force pilot in Phoenix, Arizona. The officer, who retired with the rank of colonel, once served in the famous American air display squadron the Thunderbirds. The first Rolex Turn-O-Graph with the typical rotating bezel was assigned the reference number 6202 and was unveiled for the first time at the Basel Watch Fair in 1954. It had neither a date display nor a watch rate certificate. Both could be found only on the subsequent model shown. This model had another special feature. Neither its predecessor nor any of its successors used the bezel seen here. It is not fluted but has a fine guilloché finish called Clous de Paris (Paris hobnail) (movement no. 70,946 [H 94.848], case no. 989,125). Also typical for that time was the red date display.

In the 1950s the German owner of the House of Rolex is said to have been very busy. It is thanks to him that various versions of the Oyster Perpetual Datejust were given dial inscriptions that deviated from the standard design. The Oyster Perpetual Datejust with steel case and fluted bezel in yellow gold also belongs to this group of watches. Instead of a dial inscription, in this case the ROLEX lettering is executed in raised gold letters. Also the reference to the preserved chronometer certifications reads here in German "Superlativer Chronometer,

Illustration 356
Caliber 765 of the Thunderbird shown in the photo on the left

amtlich geprüft" (superlative chronometer, officially tested). Although the dial has quite significant corrosion damage, this watch still has a special charisma. This Datejust belongs to the reference 6605, and inside it is the caliber 1065 movement. The reference III.56 tells us the manufacturing period; namely, the third quarter of 1956. Also for the statistician, the two watches shown in figs. 357 and 358 are of special interest. The reference 6605 was manufactured in the third quarter of 1956, the reference 6609 in the second quarter of 1958. If one now calculates the

Illustration 357
This Rolex Oyster Perpetual Datejust chronometer with the interesting dial inscription *SUPERLATIVER CHRONOMETER, AMTLICH GEPRÜFT* (Superlative Chronometer, Officially Certified) is clearly getting on in years. The signs of age can be seen not only on the dial, but also on the caliber 1065 movement. The rotor carrier plate is double slotted and can therefore absorb shocks from the outside. The additional information III.56 in the screw-down back shows that this Ref. 6605 watch was made in the third quarter of 1956 (movement no. N 697,472, case no. 220,977).

difference between the two case numbers 388,325 and 220,977, the result is 167,348 cases. Dividing this number by seven—the number of quarters in between—gives the quarterly number of cases produced, in this instance precisely 23,906. Since the case numbers were assigned consecutively, this also gives the number of watches produced in each quarter. If one multiplies the quarterly production by four, the result is an approximate annual production figure, which was about 95,000 (exactly 95,627) watches at that time. If one makes a cross reference and takes a look at the lists of the BO's (Bureau Officiel, official office for chronometer testing), one finds some interesting statements. According to a list by Dr.-Ing. Fritz von Osterhausen (†), the Rolex company submitted 90,868 watches for chronometer testing in 1958. But only about 40% of them, exactly 36,347 movements, received the coveted certificate. The rest failed. In the following year (1959), 97,095 movements were submitted for chronometer testing, of which 42% (i.e., 40,780 calibers) were allowed to bear the title Superlative Chronometer—officially certified. Compared to the documented numbers for the years 1958 and 1959, we are not wrong with our previously made theoretical projection for an annual production figure. With the caliber 3135 used today, in the opinion of the author one of the best automatic movements ever, the success rate is certainly 100%.

On the reference 6609 Turn-O-Graph in stainless steel presented, the name ROLEX OYSTER PERPETUAL DATEJUST can be found on the still-readable original dial below twelve o'clock. References to the chronometer status of this watch are not recorded, although the caliber 1065 has the usual indications for rate regulation. Here, it remains to be clarified whether customs reasons were responsible or—quite banally—the dial was improperly replaced at a later date. A note about the color of the date indication must be made. Since the introduction of the Oyster Datejust in 1945, the date numerals were printed in two different colors: red for even date indications, black for odd (see pp. 229 and 230).

Illustration 358
Among the most popular collectibles are the various models of Turn-O-Graph (also called Thunderbird). Here we see an example from the 1950s, a Ref. 6609 in stainless steel. The bezel can be turned in both directions. The caliber 1065 has twenty jewels and is regulated in five positions and temperatures. It belongs to the 1030 caliber family, with a diameter of 1.12 inches (12½''' or 28.50 mm) and a height of 0.23 inches (5.85 mm). Due to the date mechanism, the movement height of the caliber 1065 increased to 0.246 inches (6.26 mm; movement no. N 866,939, case no. 388,325). This caliber family introduced the bidirectional winding rotor. The numbers II. 1958 engraved in the screw-down case back mean that this Rolex was produced in the second quarter of 1958.

Reference 1500, 1501, 1601 (3), 1625 (3), and 1803 Automatic Watches

Sometimes human destinies are connected through watches. Such an example has also become known to the author. The Rolex Oyster Perpetual Date shown was sold in the USA on April 30, 1967. The owner of this watch in turn gave it to his son for his twenty-first birthday. As was learned from the son, his father had left his mother when he was just a year old, so he grew up without a father and did not meet him until his twenty-first birthday. With the comment "My father did not care for me all his life; now there is no need for him to give me his watch for my twenty-first birthday!," the son sold the watch to an acquaintance of the author, far away from any feelings. For this reason it also came to Rolex and there received a professional servicing of the movement, case, and Oyster band. For the statisticians among the Rolex collectors, we bring another interesting note. The caliber 1565 used in this watch was unveiled in 1959. It operates at 18,000 vibrations per hour.

The successor caliber 1575 came on the market in 1965 and increased the beat rate to 19,800 A/h. From the movement certificate we can see that the steel Ref. 1500 received its chronometer dedications on August 30, 1966.

Another beauty (*right*) powered by the aforementioned caliber 1575 is a reference 1501 and comes from Athens, Greece. A watch enthusiast purchased this golden piece from a jewelry dealer named Stavros Mouflouselis in September 2000 for exactly $3,030. The reference 1501 Date is quite a well-known model, especially due to its shapely bezel, and the steel version has sold in large numbers. The version in 750 rose gold can be considered quite rare, especially with the matching Oyster bracelet. The aforementioned watch enthusiast soon became

Illustration 359
Indeed, everything about this classic Rolex Oyster Perpetual Date Ref. 1500 is known to the author. This applies not only to the technical details, but also to the complete history of the watch. We see here not only this Rolex reference 1500 in stainless steel with Oyster strap, which was typical and popular at that time, but also the official chronometer certificate and the warranty card. Until the end of the 1960s, all chronometers from Rolex were accompanied by the chronometer certificate. Today they save this effort and deposit the official certificate on discs. The strap bears the number 57, which reveals the connection width of 0.74 inches (19 mm). The second number, 7205, is the reference of this riveted Oyster strap. You can read why this beautiful collectible did not have a happy ending in the text of this chapter (caliber 1565, movement no. 2,113,580, case no. 1,459,862). The case number is stamped in the warranty certificate. The seller was PACEX RUN, and the date of sale is April 30, 1967. According to the movement certificate, the chronometer testing was done in the official testing office in Biel/Bienne and was signed there on August 30, 1966. The average daily rate in different positions was +2.2 seconds at that time.

disillusioned about this beautiful Rolex when he took it to a Munich watchmaker to have it serviced. The latter informed him that the case and movement were still in a passable condition, but that the rose-gold oyster band was worn out and irreparable. Thus, his only option was a leather strap. The collector subsequently contacted the author and asked him for advice. Due to his promise to make the watch available for this publication, the beautiful piece was turned over to Rolex. After more than half a year, the watch came back almost like new, except for a tiny quirk on the bezel at four o'clock. The Oyster strap would have deserved the description "almost brand new." However, Rolex made no secret of the fact that this service was again a major exception and that the Oyster bracelet had been made with the last available spare links available in Geneva. They also asked that we not publish the agreed-upon service price.

Among the best-known and most-popular men's wrist watches in the world are the different versions of the Rolex Oyster Perpetual Datejust Ref. 1601. This model was unveiled in steel, steel and gold, and gold versions. Early on, your author experienced the "dress code" in the USA. The young manager wore the steel version, the head of the department wore the steel-and-gold combination, and the director the Datejust wore the version in full gold. The chairman of the board wore the reference 1803 Day-Date with a President gold bracelet (see p. 240). The three different versions of the Ref. 1601 illustrated on these pages are also of interest to the statisticians among Rolex collectors. The steel-and-gold version was, according to the engraving IV.68 on the inner back, produced in the fourth quarter of 1968 and has the case number 2,005,554. The Ref. 1500 presented on page 232 was produced in the third quarter of 1966 and has the case number 1,459,862. If one again calculates the difference between these two case numbers, the result is 545,692. If one divides this again by 9 (nine quarters), the result is a quarterly production volume of 60,632 watches and thus an annual value of about 242,530 pieces. Not all Rolex timepieces were sold as certified chronometers, such as the Air King and

Illustration 360
The motto "almost too good to wear" could be the credo for this watch. The Rolex reference 1501 was offered in various versions. Quite well known was the steel version with Oyster strap. The models in yellow and rose gold were the most-beautiful versions in the program. The version shown in rose gold with silvered dial resembles a little star, which is probably best worn on the left wrist of a sporty and elegant lady. The case has a diameter of 1.37 inches (35 mm), and the Oyster strap shines in near-mint condition (caliber 1575, movement no. D 335,227, case no. 1,796,266).

Illustration 361
Among the world's most popular men's wristwatches are the various versions of the Rolex Oyster Perpetual Datejust Ref. 1601. This version in 750 yellow gold with silver-plated dial may already be called a noble watch. The standard drive of the reference 1601 is the caliber 1575, with 19,600 vibrations per hour (movement no. D 14,144, case no. 1,433,778). This timepiece received a professional overhaul at Rolex that included the complete refurbishment of the case.

Illustration 362
The steel-and-gold version of the Ref. 1601 shown must be among the most-widespread watches from the house of Rolex. This model has a gold-plated dial and is fitted with a Jubilé bracelet (movement no. D 302,435, case no. 2,005,554). The Jubilé strap bears the numbers J 55 and 6251 H. The former is the identifier for the 0.78-inch-wide (20 mm) connector; the latter is the strap reference number. The engraving IV.68 in the inner back tells us that the watch was made in the fourth quarter of 1968.

Illustration 363
Here is the back of the Ref. 1601 in 750 yellow gold shown on the left. The case finish reveals to us the exact surface finish. The case back is radially satin-brushed, and on the lug at the top left you can still see the "small Helvetia" indicating 750/1,000 gold fineness. The double-set screw-down back has the bevel diamond-finished and the back surface satin-finished. The back can be opened professionally only with a special Rolex tool.

Illustration 364
Close-up of our Datejust shown on the left. The fluted bezel and the screw-down crown are in 750 yellow gold, the case stainless steel. Also not to be overlooked is the date magnifier on the Plexiglas. This watch was also overhauled by Rolex.

Air King Date models. This left Rolex with two options if a movement submitted for testing did not meet the high chronometer requirements. First, it could be installed in models that did not require chronometer movements. Second, the calibers could be carefully checked, regulated again, and resubmitted for testing. For this reason alone, one should not equate the number of movements submitted annually to the testing offices with the annual production figure.

Among the most-popular watches from Rolex are the various versions of the Turn-O-Graph model. This designation results from the rotating bezel, which can be used to control or measure time segments. A ten-minute speech or even a five-minute egg are tasks that can be easily and safely accomplished with this watch. In addition to the official name Rolex Oyster Perpetual Datejust Turn-O-Graph, this watch also had an unofficial name, under which it became very popular, especially in the USA: Thunderbird. It comes from the name of the US Air Force air display team, which is known for its performances in the USA and elsewhere. This unit named the Thunderbirds (officially the United States Air Force Air Demonstration Squadron) was created on May 25, 1953. Luke Air Force Base, Arizona, USA, was selected as its home base. Its first aircraft was the Republic F-84G Thunderjet, a subsonic single-seater fighter with straight wings. Just two years later it converted to the F-84F Thunderstreak with swept wings. This aircraft type was also used by the German air force as a photoreconnaissance aircraft. As early as 1956, another conversion took place, but this time to the USAF's first supersonic single-seat fighter, the North American F-100 Super Sabre. At the same time, this unit was transferred to Nellis Air Force Base, Nevada. Today the Thunderbirds fly the General Dynamics F-16C Fighting Falcon and enjoy great popularity (see www.thunderbirds.acc.af.mil/ for event dates).

Illustration 365
One could rate this Ref. 1601 in 750 red gold as "almost brand new." The dial, plus indices and hands, is also in 750 red gold. Also impressive is the watch's caliber 1575 movement. To be precise, one can see the rotor with automatic assembly and the screw balance wheel of the basic movement. This view of the inside case back reveals the complete inscription and hallmarking (movement no. D 191,091, case no. 186 . . . [remaining numbers not legible]). This watch was produced in 1969 and was delivered with a Jubilé bracelet in 750 red gold.

Illustration 366
Even if the position of the bezel is not exactly adjusted to twelve o'clock, it should not affect the charisma of this Rolex. The reference 1625 in steel and gold with matching Oyster strap is considered a superlative chronometer even for those watch lovers who are not so much inspired by its exterior, but by the inner values of this watch. Inside this almost-mint-condition watch is a caliber 1575 movement that runs at 19,800 vibrations per hour. The fact that this watch has already served reliably and precisely for more than thirty years is not apparent at first sight (movement no. D 245,320, case no. 1,470,626).

The Thunderbird, which was first launched under the reference numbers 6202, 6309, and 6609, was followed in 1959 by a fourth version with the reference 1625. This one featured the caliber 1565 movement. The Thunderbird presented also has a personal history that is well worth reading. It came from a then-eighty-two-year-old watchmaker named Phil Harmage from Hollywood, California. It was purchased in January 2001 via the internet by a German watch enthusiast. The purchase price was $3,051 plus transport costs and the usual import fees, so that the total price ended up at $4,462. Judging by the appealing appearance of this watch, the purchase price did not seem excessive, but fair. Only after measuring the rate and opening the Oyster case did disillusion come. On the main plate and all the bridges, including the automatic assembly, the protective rhodium plating (*rhodinage*) was largely missing, so that almost everywhere the bare brass surface was revealed (see fig. 368). Among other striking signs of wear, the rotor bearing was also badly worn. Only a radical cure involving replacement of the complete movement helped. As a major exception, Rolex Cologne replaced the movement and professionally refurbished the case.

For this work, too, the watch lover had to dig deep into his pocket again, although Rolex had been very accommodating with regard to this publication.

The next case also shows us that expertise and caution are always appropriate when acquiring an older Rolex. Sometimes one comes across watches that have been given away but rarely worn. This includes a Turn-O-Graph in stainless steel that the author discovered in the display of a watch dealer in Los Angeles, California. Unfortunately the associated papers were no longer available, but the original box was. But how can one find out the production year and thus the age of a watch in such a case? The easiest way is to have it serviced by Rolex, but the necessary information can be obtained only on special request. Another possibility is to analyze the caliber and its movement number. Inside the Turn-O-Graph shown (illus. 370), there is a caliber 1575 with movement number D 002.768. The parts list reveals that this caliber was introduced in 1965. From this Rolex document one can also see all the references in which the caliber 1575 was used. The list starts with the reference 1500 and ends with the reference 1680—thirty-one references in total. If all the information is summarized and extrapolated to the annual production number, this means that the watch shown can be assigned to the production year 1965.

Finally, some information about the "leader model," the Oyster Perpetual Day-Date, which for decades has been a status symbol par excellence. The first Day-Date, or President as it is often called in the USA, came on the market in 1955 under the reference number 6511. The watch was powered by the caliber 1055 movement, unveiled in 1954, and later by the 1055 B. The difference between the two versions consisted of the improved date switching for day and date locking. Thus, the overall height had increased from 0.275 to 0.279 inches (7.00 to 7.10 mm). Five years later (i.e., in 1960), the new model series with the reference range 18XY appeared. Basically the Day-Date was offered only in precious metal. Under the reference 1802, the cases were offered with a smooth bezel and were manufactured in yellow gold, white gold, and platinum. The reference 1803 had the fluted bezel and was produced in yellow and white gold.

Illustration 367
Another version of the Thunderbird is the reference 1625 in 750 yellow gold and a dark-brown dial. After the references 6202, 6309, and 6609, this was the fourth version of this popular watch.

Illustration 368
The original caliber 1560 movement exhibiting strong signs of wear. Both on the winding bridge, with the movement designation 1560, and on other components, surface-protection coating, also called rhodium plating, is missing. Of course the heavy wear on the gear train is not visible here.

Illustration 369
As part of a comprehensive overhaul at Rolex Cologne, this watch's entire movement was also replaced as a one-time exception for this book. In direct comparison between old and new, one can see the drastic difference. The caliber 1560, a version of the 1565, has a bidirectional rotor and runs at 18,000 A/h. The original caliber was numbered D 97.691. The new movement is numbered D 000.970. The case number is no longer legible, except for the final two digits 16. You can find out why in the text of this chapter.

Illustration 370
Among the most-sought-after variants of the Rolex Oyster Perpetual Datejust Turn-O-Graph reference 1625 is the one with a yellow-gold case and black dial. A lined black Louisiana alligator strap further emphasizes the elegant appearance of this timepiece. The fact that this watch still shines with excellent chronometer values is due to the overhaul carried out at Rolex (movement no. D 890,944, case no. 2,793,197). A look at the automatic caliber 1575 makes it clear that this one looks almost like new. Also interesting is the close-up of the tang buckle with soldered Rolex crown.

The reference 1804 had a diamond bezel and was produced in yellow gold, white gold, and platinum. The diamond bezel was also always in platinum on the first-mentioned versions. The reference 1807 received a bezel with bark engraving and was available in yellow-gold and white-gold versions. According to the Rolex model list, there were also the references 1805, 1806, 1808, and 1810. Except for the last version, the author had never knowingly seen these references. In addition there were some exotics, such as the reference 1816, which had a bezel with twenty-four baguette-shaped gemstones. These baguettes could be ordered in diamond, ruby, or sapphire versions. In the identification of precious metals, Rolex assigned the number 6 for platinum, the number 8 for yellow gold, and the number 9 for white gold. The different shapes of the bezel also need to be addressed. The first Ref. 6511 Day-Date had a fluted bezel. As the Ref. 6612 it was delivered with a smooth bezel.

The caliber 1555 movement was installed from 1960 onward, beginning with the 18XY series. In 1965 it was replaced by the improved caliber 1570, with 19,800 vibrations per hour. In 1972 the caliber 1570 received a further improvement in the form of a second-stop mechanism, which allows the watch to be set to the second by pulling the crown. A disadvantage of this caliber was often the date setting, where each day's date had to be set individually. Only with the introduction of the fundamentally redesigned caliber 3055, which was launched in 1977, was the Day-Date equipped with a date quickset.

Illustration 371
In almost mint condition is this Rolex Oyster Perpetual Datejust Ref. 1625 in stainless steel with Oyster strap and original box. The rotating bezel is made of 18K white gold; the six indices are not flat but are inclined slightly like a roof. Inside is the caliber 1575 movement, which in contrast to the earlier caliber 1565 runs at 19,800 A/h (movement no. D 002,768, case no. 1,503,243). The Oyster strap has a safety clasp. The numbering 501 B indicates the connection width of 0.78 inches (20 mm). The number 78,790 is the reference number of this steel bracelet. Why the screw-down case back is engraved with 1016 rather than the reference number 1625 is easily explained. Presumably, the original back was so damaged that a replacement became necessary. In this case the involved watchmaker made it easy for themselves and took the screw-down back from a Rolex Explorer I and installed it on this Turn-O-Graph.

Illustration 372
The renown and résumé of the Rolex Oyster Perpetual Day-Date are unsurpassed. The two Ref. 1803 masterpieces presented in 750 yellow and white gold were always among the top models from the house of Rolex. As a rule, these watches were delivered with President- or Jubilé-type gold bracelets. Only in Austria were these timepieces also offered with leather straps. The reason for this was that country's special tax on luxury goods, which not everyone was willing to pay. In the foreground is the gold bracelet with folding clasp, which belongs to the Day-Date on the left (watch on left: movement no. DD 279,043, case no. 2,092,352; watch on right: movement no. unknown, since watch was not opened after overhaul, case no. 1,746,861). The watch on the left has an amusing story that you can read in chapter 8, "Buying and Overhauling Special Wristwatches," on page 294. The one on the right had been purchased by a watch enthusiast from a Viennese watch dealer. Only during the overhaul carried out by Rolex did it become apparent how much wear had been caused by decades of use. But also in this case, the Rolex master watchmakers showed how a watch can be restored to almost-new condition even in difficult cases.

Whoever sees this missing date quickset as a deficiency can remedy it with a watch winder. This is offered for $100 to $200 on the internet. Because why should you let the joy of a beautiful old Rolex Day-Date be spoiled only because of small things?

Closing Chord for Rolex

Because the Rolex brand always emphasizes its special quality (i.e., accuracy, robustness, temperature independence) and much more, we wanted to know whether these high demands are also permanently fulfilled.

Thus, over a longer period of time, more than half a dozen different models of this manufacturer were purchased in the normal way from specialized retailers and subjected to intensive testing. Extensive wearing tests carried out by representatives of various professions allowed a representative statement to be made about the precision and serviceability of these timepieces. Witschi's test equipment (time scales) was used to determine the rate values of these watches before and after the wear tests and to compare them with the practical records of the test wearers.

Without exception, every one of the watches examined here, from the simple Datejust Turn-O-Graph to the sinfully expensive Day-Date with Ice-Blue dial in platinum, achieved the precision proclaimed by Rolex. Astonishing was the significant but not-too-great difference in the performance between the calibers 3135 and 3235.

The Rolex Sea Dweller in stainless steel, with the new automatic movement 3235 seen on this two-page spread, may almost be understood as a work of art, and it is followed on the next double page by further timepieces from Geneva that were part of the aforementioned test purchase. The expert description of these impressive wristwatches has been kindly done by Peter Braun, editor in chief of *Armbanduhren*.

Surely this watch parade can also serve as a suggestion for one's next watch purchase—surreptitious advertising excluded.

Illustration 373
The Rolex Daytona (Ref. 116523) is one of the most successful models from Rolex. Five-year waits for delivery of the steel version are not uncommon. In 2000, Rolex replaced the base caliber previously supplied by Zenith with the in-house-developed caliber 4130, making the Daytona even more desirable. Just like the El Primero movement, the 4130 controls the chronograph functions by means of a wheel. However, like all Rolex movements it was consistently developed for serviceability, robustness, and longevity. This includes the stable balance bridge, which prevents vibrations from being transmitted to the balance wheel, and the significantly increased height of the rotor transmission wheels. In addition to the zero setting, the power transmission to the chronograph has also been improved: the patented vertical clutch works backlash-free and friction-locked, thus reliably preventing jerking of the stop-seconds hand. At seventy-two hours, the movement also manages significantly longer without the energy of the rotor. And of course the movement features a Breguet hairspring, a Glucydur balance wheel with Microstella regulating screws, and an official chronometer certificate from the COSC. But it is not just the movement that's been improved: Rolex also took the opportunity to address the weak points of the Oysterlock bracelet and folding clasp. The bracelet is now much more solid, and parts of the clasp are made of milled metal rather than thin, stamped sheet metal. In addition, the bracelet can easily be lengthened for increased comfort in hot temperatures.

Comparison of the Rolex Calibers 3035, 3135, and 3235

[1] Blue Parachrom from 2006 (Note: Parachrom is the way it is spelled on the Rolex website) [2] KIF Ultraflex [3] Rotor mounted on ball bearings.

Caliber	3035	3135	3235
Outer diameter (mm)	28.50 12½'''	28.50	28.50
Height (mm)	6.30	6.00	5.37 [3]
Hairspring	Breguet	Breguet [1]	Blaue Parachrom
Shock Protection	KIF	KIF [2]	Paraflex
Vibration Frequency	28,800	28,800	28,800
Power Reserve hrs	48	50	70
Introduction	1977	1988	2015
Jewels	27	31	31

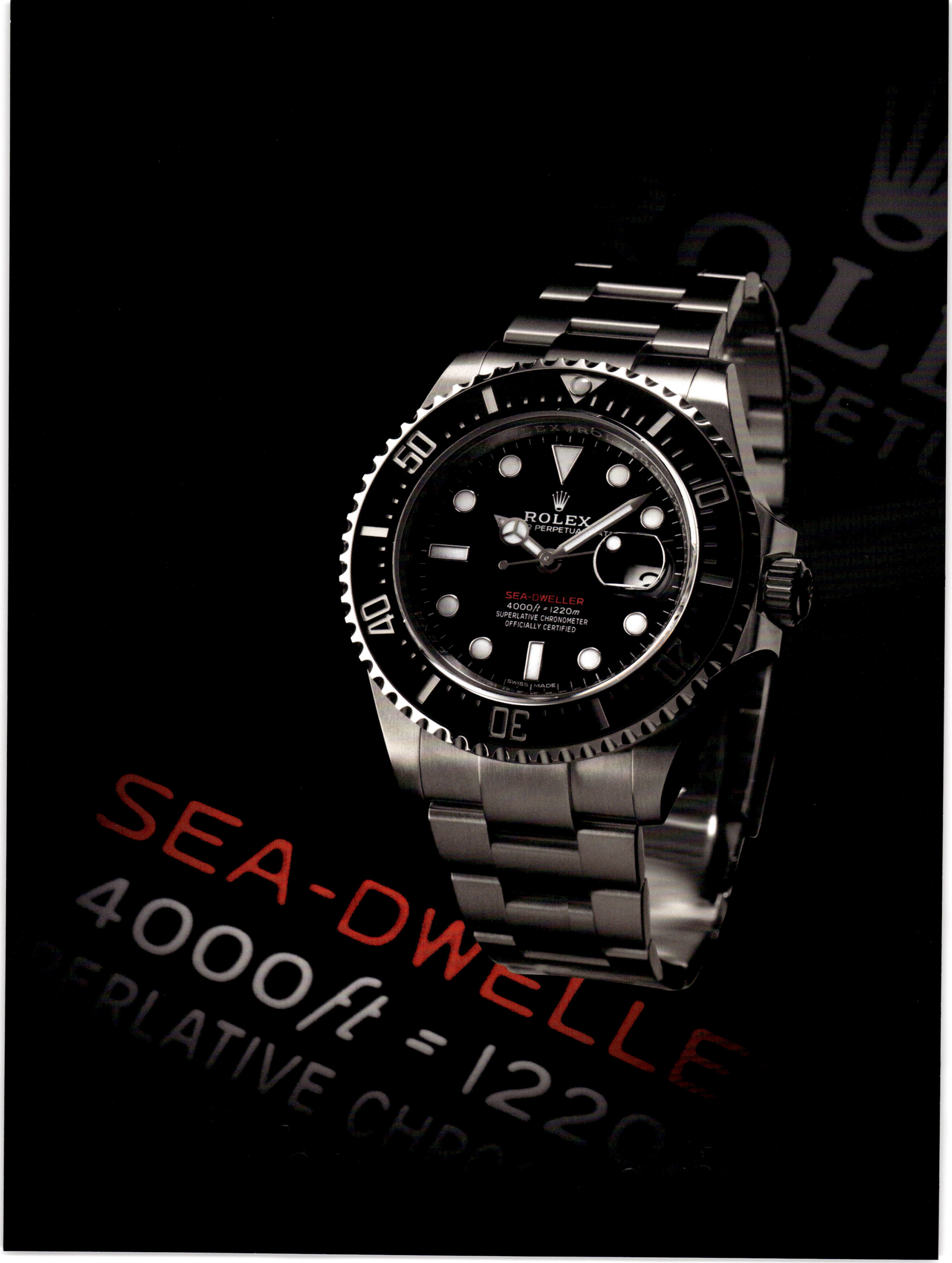
ROLEX
SEA-DWELLER
4000ft = 1220m
SUPERLATIVE CHRONOMETER
OFFICIALLY CERTIFIED
SWISS MADE

Illustration 374

With a case diameter of 1.73 inches (44 mm), this new Sea Dweller diver's watch in stainless steel probably fits only the wrists of strong men. This reference no. 126600 watch is powered by the new Rolex caliber 3235. Its guaranteed water resistance to a depth of 4,002 feet (1,220 m) will hardly be used by any wearer of this watch. The photos on the right offer a look into the inner workings of this watch. While the lower picture shows the rotor, in the upper picture we can see the gear train—not shown in the Rolex advertisement—which belongs mostly to the automatic device module. Here you can again see that the caliber 3235 is a completely new design.

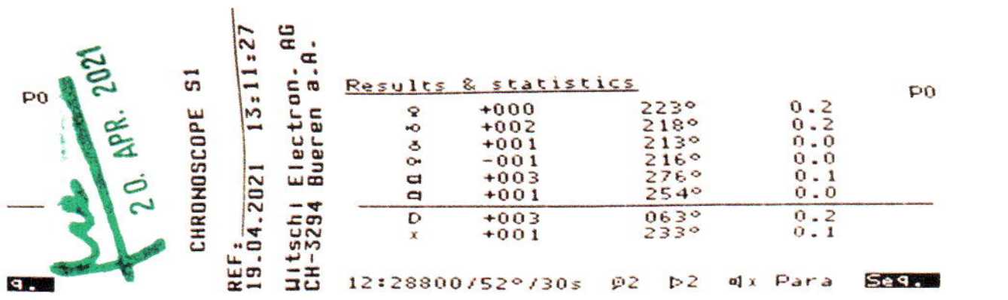

20. APR. 2021

CHRONOSCOPE S1

REF:
19.04.2021 13:11:27

Witschi Electron. AG
CH-3294 Bueren a.A.

Results & statistics

	+000	223°	0.2
	+002	218°	0.2
	+001	213°	0.0
	-001	216°	0.0
	+003	276°	0.1
	+001	254°	0.0
D	+003	063°	0.2
x	+001	233°	0.1

12:28800/52°/30s

Illustration 375

Measurement record of the accuracy of the watch shown. It advances exactly one second per day. The rate difference D in the six positions is 3 sec. / 24 hrs.

Illustration 376

Illustration 377

Illustration 378

Certificate of Quality

All the models listed here were purchased from specialized dealers over a long period of time. After purchase they underwent extensive functional, accuracy, and usage testing. The high expectations, which are appropriate for market leaders, were fully met. All watches met the specifications of the Swiss testing offices on the Witschi rate-measuring device. An acquisition of one of these watches can certainly not be considered a bad purchase.

Illustration 376
In 2000 the legendary "Daytona" became a genuine manufacture chronograph. The Rolex caliber 4130, developed in-house, replaced the modified Zenith automatic movement used in the Oyster Perpetual Cosmograph since 1988. The picture shows a reference 116520 in stainless steel from 2008.

Illustration 377
The so-called Tool Watches from Rolex enjoy particular popularity among collectors. As early as the 1950s, various robust models were created for special occupational groups, such as the GMT-Master with an additional twenty-four-hour display for airline personnel. The GMT-Master II model, introduced in 1980, featured not only the twenty-four-hour arrow hand but also, for the first time, a main hour hand that could be gradually adjusted by means of the crown for rapid adaptation to a new local time. The Rolex GMT-Master II Ref. 126710 BLRO is especially appreciated by the younger public and nicknamed Pepsi. Only mint-condition watches with Oyster strap are traded in the secondhand market, at an average of $24,000 to $28,000.

Illustration 378
The archetype of the Rolex Tool Watch is the Submariner diver's watch, which has set the industry standard since its debut in 1953. This applies not only to the key technical data, but also to the appearance of a diving watch for professional use, reduced to the essential and absolutely necessary. This Rolex Submariner without date, Ref. 124060, corresponds to the original model. Inside is the new caliber 3230 movement, with a power reserve of approximately seventy hours.

Illustration 379

Illustration 380

Illustration 381

Illustration 379

Introduced in 1969, the Submariner Date made the first—and so far only—concession to everyday use with its date display. In terms of readability, Rolex made no compromises and equipped the sapphire crystal with a ground-in magnifying glass to make the rather small date display more visible. In this Ref. 126610 watch is the new caliber 3235 movement. Prices on the used watch market are between $12,000 and $14,500.

Illustration 380

A small leap for the date, but a giant leap for the watch world: the Rolex Datejust, with its date jumping exactly at midnight, laid the foundation for the Day-Date unveiled in 1956, which made the previously maintained full calendar superfluous by adding a weekday display. The Day-Date became the archetype of a new generation of practical wristwatches for businessmen. This Ref. 118206 Rolex Day-Date with Ice-Blue dial and 1.41-inch (36 mm) platinum case is offered on the used-watch market at prices ranging from $31,000 to $36,000. The price of a new watch is around $59,000. The watch is considered very rare because people often shy away from the high new price.

Illustration 381

With diamond-cut bezel and the combination of stainless steel and rose gold, called Rolésor at Rolex, the Rolex Datejust has high recognition value. This Rolex Oyster Datejust Ref. 116261, also called the Turn-O-Graph, has a long tradition. This model dates to the 2005 production year. The combination of steel and rose gold, called Everose by Rolex, and the black dial give the watch a high degree of elegance. On the used-watch market, one can find this model for $11,000 to $14,250 on the internet.

Illus. 382

Illus. 384

7. Vacheron & Constantin

7.1. Introduction by Gisbert L. Brunner

Tradition is the foundation on which the present and future are built. This insight also applies in a special way to the watchmaking industry. But here, too, there are small but immensely significant differences. On the basis of current knowledge, there is no Swiss maker of the absolute luxury class of watch that has been in the service of time measurement longer, without interruption, than Vacheron (&) Constantin. Moreover, no other company of this kind and its high-quality products have borne the same name longer than this one. The history books tell us that it all began in 1751, when the twenty-year-old, multitalented, and interested Jean-Marc Vacheron was admitted to the elite circle of about eight hundred Geneva *cabinotiers*. They worked in bright "cabinets" directly under the roofs of the Saint-Gervais district. There they filled orders from the established *etablisseurs* (i.e., the makers of watches). Compared to ordinary industrial workers, these craftsmen enjoyed some remarkable privileges; for example, twice the normal income. In addition they were allowed to enjoy their free time from as early as three o'clock in the afternoon or to spend time furthering their education.

Fruitful Partnership

Under these pleasant auspices of a flourishing epoch, Jean-Marc Vacheron would have been able to enjoy his watchmaking life to the fullest. But that is precisely what he did not want. He had higher aspirations. That is why in 1755 he opened his own watchmaking workshop in Geneva's old town, the Cité, where he employed one or even two apprentices. Logically his ticking products bore the name Jean-Marc Vacheron. As it turned out, the second half of the eighteenth century was not a simple time for the young entrepreneur as a result of economic and political upheavals.

In 1810, Jacques-Barthélemy (1787–1864) followed in the footsteps of his father. To secure his future, in 1819 he allied himself with François Constantin, the son of a wealthy fabric and grain merchant. This alliance led to the founding of Vacheron & Constantin. Through extravagant and at times quite unorthodox methods, François Constantin succeeded in reviving the company, which had been ailing after the French Revolution. In 1839 the pair hired Georges-Auguste Leschot. The ingenious watchmaker and inventor had quietly developed various machines for the accurate manufacture of watch movements. Their improved accuracy was accompanied by better interchangeability of watch parts. Vacheron & Constantin started the production of new precision calibers. This enabled the up-and-coming company to be considered a true manufacture. After all, all the necessary components were created in the ateliers on the island in the Rhône in the middle of Geneva. Since then, there have also been exact records about every watch made.

After the surprising resignation of partner Jean-François Constantin, a nephew of François Constantin who left the company in 1849, in 1867 the company name changed to César Vacheron & Co. After the tragic death of Charles Vacheron, the last scion of the watchmaking dynasty, the manufacture was renamed Vve. César Vacheron & Cie. To revive the widely known name of Vacheron & Constantin, Jean-François Constantin reconsidered his decision and returned in 1875. In 1880 he had the famous Maltese cross legally protected as the company logo. And in 1881 a watch from the house of Vacheron & Constantin set a record in a competition at the Yale College Observatory. Other achievements of this kind followed. The first wristwatches entered the collection in 1910; very different models for men and women, with and without complications.

In the years between the two world wars, Vacheron & Constantin also felt the effects of the economic downturn. In 1931, after the start of the Great Depression, short-time work was the order of the day. Through outstanding commitment, the company's management and employees ensured its survival. In the wake of the severe financial crisis, in 1940 Charles Constantin sold the majority of shares to Bernese industrialist Georges Ketterer. As a result of close economic and financial ties with the *ébauche* manufacturer LeCoultre in Le Sentier, from then on the latter's raw movements ticked in the majority of timepieces bearing the Vacheron & Constantin signature. The manufacture mutated into a luxury establishment without its own movement production.

In 1955, the year of the company's two hundredth anniversary, Messrs. Bulganin, Eden, Eisenhower, and Faure met on July 18

Illustration 382
Michelin-starred chef Bobby Bräuer from the Esszimmer restaurant in BMW World in Munich. Behind him is the BMW i4, a real premiere—BMW's new all-electric model.

Illustration 383
According to experts, this chronograph, which has resurfaced after some fifty years, may be one of the two prototypes of the reference 6087 shown on the right.

Illustration 384
Vacheron & Constantin chronograph Ref. 6087 in 18K yellow gold, also called Cornes de Vache (cow horns), with waterproof case

Illustration 385
Chronographs have always been considered particularly impressive wristwatches, also called "small complications" in professional parlance. The stainless-steel "time recorder" on the left is probably one of the two prototypes made during development of the Ref. 6087 water-resistant chronograph, probably in 1953. On the right is a current model Ref. 47192/000B-9352 from 2012. The screw-down case is made of 18K white gold, while the minute hand, second hand, pushers, and crown are 18K red gold. The case of the watch on the left has a diameter of 1.45 inches (37 mm), the one on the right 1.65 inches (42 mm). The prototype uses the caliber 492 movement with seventeen jewels; the new watch, the caliber 1141, with twenty-one jewels. The fact that the two subdials are so close together on the newer model was dictated by the designer of this chronograph caliber. The new price of the chronograph on the right was about $48,000 in spring 2021, while on the secondhand market this watch in mint condition could be had for about $25,000 to $27,000. The prototype in stainless steel on the left should probably be almost priceless. Actually, it belongs in the Vacheron Constantin museum in Geneva.

in Geneva's Palais des Nations to answer questions concerning world peace. At the end of the conference, twenty Geneva citizens presented each of these statesmen with a Vacheron & Constantin watch with the message "May this watch always show happy hours—for yourself, for your people, and the peace of the world."

Toward the end of the 1970s, the Geneva workshops produced the Kallista, then the most expensive wristwatch in the world and therefore immortalized in the Guinness Book of Records. This gem, consisting of 4.93 ounces (140 grams) of gold set with 130 carats of precious stones, cost five million dollars. Soon after, the serious quartz watch crisis once again called for a financially strong partner. This was found in 1987 in the person of Sheik Yamani, a friend of the Ketterer family. The former oil minister of Saudi Arabia ultimately took over 95% and thus the overwhelming majority of Vacheron Constantin. However, the interest of the millionaire investor from the Middle East and his willingness to provide the traditional brand with the necessary finances for further international development were limited. Consequently, the scope for a global presence of the traditional brand was relatively narrow. The real moment of happiness came in 1996, when the Vendôme Group (today Richemont SA) of South African billionaire Johann Rupert rounded off its brand portfolio with the House of Vacheron Constantin. The subsequent history under the umbrella of the luxury multinational fully confirms this statement. Especially in the Far East, Vacheron Constantin again enjoys a very high status. In the Geneva suburb of Plan-les-Ouates, a company building that is both functional and futuristic in appearance announces the elevated status of the soon-to-be-260-year-old luxury watch manufacturer. And in 2013 Vacheron Constantin was able to inaugurate a brand-new manufacturing facility for watch movements and complex mechanisms in the Vallée de Joux.

Illustration 386
One of the most famous models from Vacheron & Constantin was given the name Chronomètre Royal. The chronometer caliber P 1008/BS is encased in this watch. The case, made of 18K yellow gold, has a diameter of 1.37 inches (35 mm) and a sturdy screw-down case back. In addition to the name Chronomètre Royal, the case back is also engraved with the reference number 6105 and case number 352,541. The latter is an exception.

Back to the Manufacture

Since the turn of the millennium, Vacheron Constantin has once again made manufacturing a top priority. In this sense the production and timing of its own calibers of more or less greater complexity required not only qualified personnel, but also space. Because there was an increasing lack of both in Geneva, years ago Vacheron Constantin sought salvation in the Vallée de Joux at an elevation of 3,280 feet (1,000 m). When the rented makeshift branch near Le Sentier threatened to burst at the seams, there was no way around constructing a spacious new building in the middle of a greenfield site. The opening of the 75,347-square-foot (7,000 square meter) branch in the neighborhood of Le Brassus took place on August 19, 2013. Around two hundred employees of mostly French nationality are currently on the payroll there. In keeping with the signs of the times, the construction of the new building was dominated by energy efficiency. Management has designed the ultramodern production facility for the production of 15,000 to 20,000 movements per year. The production spectrum, which comprises eight to ten exclusively mechanical caliber families for up to forty models, also includes modules for various additional functions. In addition to raw-movement production, the employees are also responsible for component finishing and the preassembly of some subassemblies. Incidentally, all products—where required—already meet the strict specifications of the Poinçon de Genève introduced in 1868 at the design stage. After Vacheron Constantin could no longer identify with the catalog of criteria, which was tightened in 1957 but nevertheless remained lax, it pushed for a new, more modern version starting in 2009. The conflict-ridden task fell to the Timelab. It was settled under the umbrella of a foundation of the Geneva Laboratory of Watchmaking and Micromechanics as the "Geneva Seal" department. On November 9, 2011, the partially retreaded, partially adapted, but in any case significantly stricter regulations were unveiled, and they came into force on June 1, 2012. As always, ticking movements and additional mechanisms such as calendar movements had to be assembled, adjusted, and encased in the canton of Geneva. From then on, the movement and case formed an inseparable unit. Supplied components or assemblies were subject to the same strict guidelines as those manufactured in-house. In contrast to modern silicon components, which traditionalist Vacheron Constantin currently categorically rejects, plastic components are definitely excluded. Regarding technical and manual execution, similarly strict standards apply as before. In the end, everything revolves around the finished watch. What is meant here is water resistance, functional reliability, rate autonomy, and, in particular, precision for seven consecutive days. Only when the deviation determined by incorruptible computers does not exceed one minute do watches bearing the Geneva Seal find their way to customers all over the world. Anyone who straps on a mechanical Vacheron Constantin watch from current production—for example, the ultrathin minute repeater Patrimony Contemporaire with caliber 1731 movement—can be sure that it has passed every test without any problems.

This is currently guaranteed by around a thousand employees around the globe; 480 of them do their jobs at the company's Geneva headquarters. At present, this also calls for significant expansion, because success does come at a price.

7.2. The Movements

Watches from Vacheron & Constantin have always been among the best made in Switzerland. Especially in the 1950s and '60s, the models of this long-established Geneva manufacturer were in no way inferior to those of Patek Philippe and were on the same level in terms of design, technology, and diversity. The annual production figures of these two Geneva watch manufacturers were also of the same order of magnitude. By the early 1970s it was already apparent that this noble brand was visibly losing its reputation and that the Geneva competitors had already achieved better market positions. In addition to the emergence of quartz technology and the associated changes in the market, the company's own lack of willingness to innovate also led to market losses that threatened the company's existence. Due to differences of opinion, there were changes in ownership in 1968 that finally led to the separation from sister company Jaeger-LeCoultre. The Le Sentier–based company had served for decades as a supplier of raw movements (*ébauches*) and had brought about many an innovation. Neither this parting of the ways nor its background has ever been well known. Nor is anything to be found on this subject in the relevant literature. Some of these secrets will be revealed in this chapter.

Manually Wound Movements

K 453/3B 12½''' and P 453/3C 12½'''

Like all other Vacheron & Constantin watch movements, these two calibers were also procured from Jaeger-LeCoultre in Le Sentier as raw movements (*ébauches*). The caliber 543 is regarded as a base watch movement for different variants. Installation diameter is 1.13 inches (28.8 mm), and the movement height is 0.157 inches (4 mm). The double-spoke screw balance wheel has a Breguet hairspring and runs at 18,000 A/h. A micrometer screw and a gooseneck spring are used for fine adjustment of the regulator arm. The movement has seventeen jewels. All surfaces are circular grained or Geneva striped.

Illustration 387
Caliber K 453/3B

Illustration 388
Caliber P 453/3C

The differences between the otherwise identical calibers are, among others, that the caliber K 453/3B (illus. 387) not only has eighteen jewels (rubies) but is also regulated for heat, cold, and isochronism, and in five positions. Furthermore, it bears the coveted Geneva Seal. Thus, this is a Geneva product of the highest quality, which at that time could be found only at Patek Philippe. According to the inscription on the mainspring barrel bridge, the caliber P 453/3C (illus. 388) has the usual seventeen jewels and is regulated in temperature only. These differences were also related to the destination countries to which such watches were exported. The Geneva Seal and chronometer accuracy were of interest only to customers who knew what to make of the quality terms described. Thus, an Argentinean cattle baron was quite indifferent to chronometer values or Geneva Seals. The main thing was that the watch had a nice-sounding brand name and looked good.

P 454/5B 12½'''

The caliber 454 is very similar to the 453. Instead of the small seconds at six o'clock, this caliber has a center second hand. With the same installation diameter of 1.13 inches (28.80 mm), the movement height is 0.196 inches (5.00 mm) because of the fourth wheel. There are seventeen rubies installed, and the movement is adjusted to temperature. There are other variants of the basic caliber 453. The caliber 485 displays day, month, and date. The 495 version also has a moon phase installed.

P 1007/BS and P 1008/BS

These two chronometer movement variants were created from the basic calibers 453 and 454. Here, too, the external difference consists of small and center seconds. The internal differences concern the escape wheel and its lower bearing. More obvious is the stop-seconds mechanism. When the crown is pulled, the balance wheel is stopped by a clearly visible bail as well as via a steel spring (*tirette*) so that an exact time setting is possible. These two chronometer movements were produced between 1953 and 1962 in a quantity of three hundred pieces for the P 1007/BS (illus. 389) and a thousand pieces for the P 1008/BS (illus. 390), but not all of them were encased. According to Rudolf Büll, fewer than one thousand movements were used, and they can be found only in models of the Chronomètre Royale (French) or Chronometer Royal (English) series. A comparable device for stop-seconds can also be found in the Jaeger-LeCoultre Geophysic chronometer (see p. 102) and the Pilot's Watch Mk 11 (see p. 105).

Illustration 389
Caliber P 1007/BS

Illustration 390
Caliber P 1008/BS

The Automatic Movements

477 and 477/1

With the installation of these calibers, Vacheron & Constantin began the age of the self-winding wristwatch. The design of the caliber 477 corresponded to Jaeger-LeCoultre's own automatic caliber 476, which had been unveiled in 1948. However, in terms of content, the Vacheron & Constantin version featured some significant improvements. For example, this automaton is adjusted to temperature and has a Breguet hairspring, and its surfaces are finished with the Côtes de Genève (Geneva stripes) and *perlage* (circular graining). One such caliber is shown on page 252. A counterpart from Jaeger-LeCoultre can be found in the caliber 481 variant on page 100. The caliber 477 has an installation diameter of 1.16 inches (29.56 mm) and an overall height of 0.242 inches (6.15 mm). It has seventeen jewels and operates at 18,000 vibrations per hour. Jaeger-LeCoultre's proprietary shock protection is also standard on this caliber. Automatic winding is handled by a limited oscillating weight popularly known as the "automatic hammer." However, the

Illustration 391

winding is in one direction only. The 447/1 version (illus. 391) received several improvements: one wheel of the winding mechanism received a bridge, the pivot of the hammer was strengthened, and the decoration of the movement was changed. Not all profiled watchmakers were satisfied with this caliber, because the gear section is very difficult to access. Also a specially sophisticated balance cock was required for assembly to, for example, check the air pressure at high elevations.

Illustration 392
Approximately 7× detail enlargement of the caliber 477/1 with oscillating weight bridge (movement part 1341). On the upper left you can see the winding stem, and on the edge of the mainspring barrel bridge the cylinder head slotted screw for the bell crank. Typically the original cylindrical recess for this screw is damaged, probably because the watchmaker working on this movement was too comfortable to use a smaller screwdriver to loosen this screw.

498 and 499 plus P 1019, P 1019/1, and P 1019/2

Just four years after the introduction of the first automatic movement with limited oscillating weight, the next generation followed in 1953 with self-winding, but this time with a 360° rotating rotor. The caliber 498 has a small seconds at six o'clock, while the caliber 499 has center seconds. Thus the caliber designations determined by Jaeger-LeCoultre were initially

Illustration 393
Caliber P 1019

adopted by Vacheron & Constantin. These movements have a case diameter of 1.14 inches (29.00 mm) and a case height of 0.212 inches (5.40 mm). A Breguet hairspring and Jaeger-LeCoultre shock protection are also used here.

The movement operates at 18,000 vibrations per hour and has a power reserve of approximately forty hours. The year the movement was introduced, the caliber designation 12‴-499 was changed at Vacheron & Constantin to P 1019 (illus. 393) and subsequently supplemented with the suffixes /1 and /2. Compared to the caliber 499, here are the main changes:

1. the automatic winding bridge at the top and bottom (not interchangeable)
2. the oscillating weight (with reinforced axle)
3. the winding wheel (first nine, then eight teeth; crown wheel first twenty-four, then thirty-two teeth)
4. driving wheel for crown wheel
5. regulator arm with gap

This caliber series remained in production until 1958.

K 1070 and K 1071

These two successor calibers were unveiled in 1959. They again came as raw movements (*ébauches*) from Jaeger-LeCoultre and were largely identical to the calibers K 2070 and K 2071 from Audemars Piguet (see p. 31). The caliber P 1019 also served as a starting point. While the basic movement remained almost the

Illustration 394
Caliber K 1071

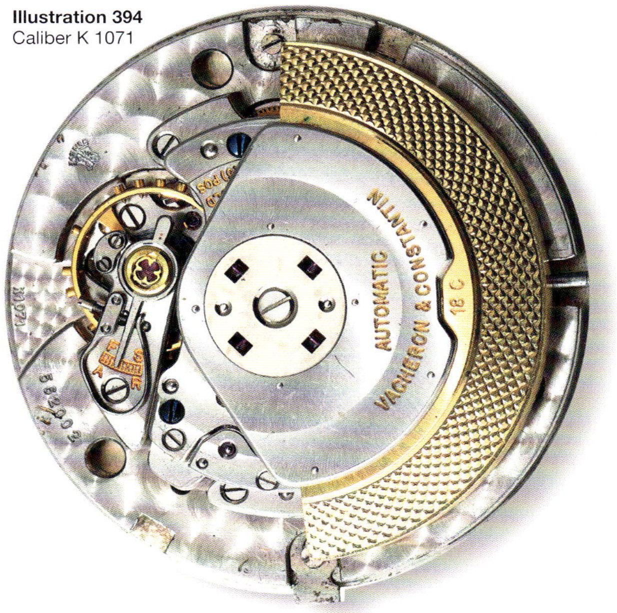

Illustration 396
Caliber K 1072/1

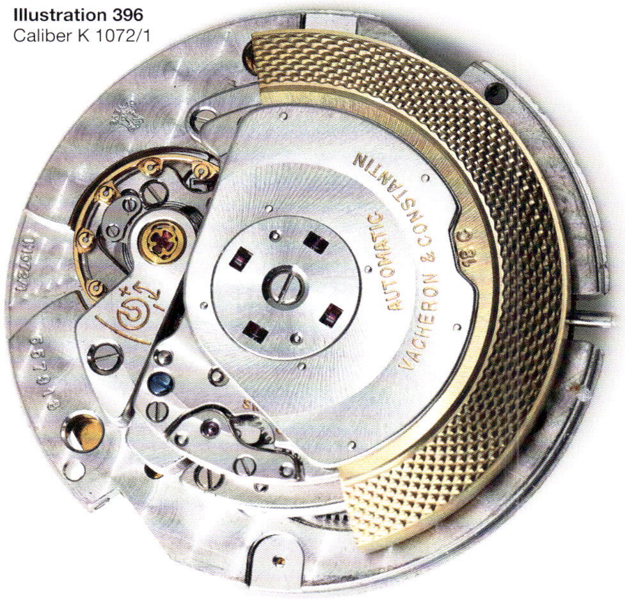

same, the automatic device module was extensively revised. The differences are clearly visible in the design of the rotor, now supported by four ruby bearings.

The rotor bearer plate features a sunburst finish. The 18-karat oscillating weight is quadruple-screwed and has a beautiful guilloché finish. The pawl has been changed and is now on a separate bridge, which in turn is screwed to the mainspring barrel bridge from below. The shock protection of the balance staff is now done by Parechoc "KIF" Réf. Flector. The calibers 1070 and 1071 (illus. 394) have twenty-nine rubies and identical dimensions: installation diameter 1.14 inches (29.00 mm), overall height 0.212 inches (5.50 mm).

The screw balance wheel uses a Breguet hairspring and runs at 18,000 A/h. Fine adjustment is done by a micrometer screw with gooseneck spring. All surfaces are sunburst or circular grained. These calibers bear the Geneva Seal and are regulated in temperatures and five positions.

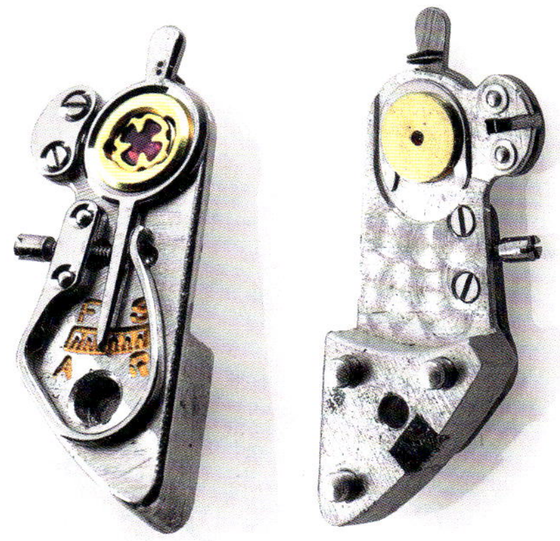

Illustration 395
Balance cock of a caliber K 1071 by Vacheron & Constantin (see fig. 394), the top side on the left and the bottom side on the right. The regulator arm, micrometer screw, and gooseneck spring are clearly visible. The balance staff shock protection system is from Incabloc. If the micrometer screw moves the regulator to the left, the watch runs faster (F). A shift to the right makes the watch run slower (S). This balance cock came from the American used-parts dealer francis.antictac in New York and cost $250 plus $20.68 shipping, for a total of $270.68. This part is no longer available from Vacheron Constantin in Geneva.

K 1072 and K 1072/1

The version of the caliber K 1071 with date display at three o'clock and the new caliber designation K 1072 was launched in 1961. Almost identical in design, the movement height increased from 0.212 inches (5.50 mm) to 0.248 inches (6.30 mm). From 1966 onward the caliber K 1072 was equipped with a Gyromax balance wheel and the supplementary designation K 1072/1 (illus. 396). In contrast to Patek Philippe's oscillating systems, this Gyromax balance wheel was equipped with ten rather than eight mass balance weights (also see p. 254).

The movement numbers have six digits and were assigned consecutively. For example, if one compares the chronograph Ref. 321 from 1938 with movement no. 420,637 (case no. 264,454) with the automatic Ref. 6782 with movement no. 598,456 (case no. 433,445) from the production year 1968, the difference is 598,456 minus 420,637, or 177,819. If one then divides this number by 30, one can calculate an average of 5,927 movements per year.

Caliber K 1072/1

Illustration 397
At the beginning of the 1960s, the caliber K 1072 was one of the best automatic movements in Switzerland. In terms of design and execution, its technical standard largely corresponded to the caliber 27-460 M from Patek Philippe, and in terms of rate performance to the chronometer caliber 564 from OMEGA, which followed somewhat later. From 1966 on, a Gyromax balance wheel with ten mass balance weights was used, and the movement was now called the K 1072/1. A Breguet hairspring with a bent-up end curve ensures good rates. The installation diameter is 1.14 inches (29 mm) and the overall height 0.24 inches (6.30 mm). The rotor, with an 18K yellow-gold oscillating weight, runs elegantly on four artificial ruby bearings with almost no wear. A changer allows the rotor to wind in both directions. All in all, this automatic movement uses everything that was "state of the art" at that time. After a professional overhaul, a watch with this caliber still achieves chronometer values.

Illustration 398
This 18K yellow-gold Gyromax balance wheel with ten *masselotte* weights and Breguet hairspring testifies to the highest level of watchmaking (approximately 8× magnification). At the very bottom right, you can still see the stamped Geneva Seal.

Caliber 2455

Illustration 399
The design of the caliber 2455 is about forty years younger than that of its predecessor, the K 1072/1 shown above. The movement diameter was reduced from 1.14 to 1.03 inches (29 to 26.20 mm), the overall height from 0.248 to 0.141 inches (6.30 to 3.60 mm). The dogma of low movement height found its way into almost all Swiss watch manufacturers in the last quarter of the last century and was practically considered an important survival factor. Only Rolex remained true to the rules of technical mechanics and gave its new automatic caliber 3135, launched in 1988, a full 0.236-inch (6.00 mm) overall height. If one omits certain technically high-grade components from an automatic movement, such as a Breguet hairspring, Gyromax balance wheel, changer, double-offset rotor bearing, etc., and replaces them with simple components, some of which are mass-produced, such as a flat hairspring, Triovis regulator, and simple balance wheel, this leads to considerable manufacturing cost savings. Whether the high expectations of a buyer who is prepared to spend more than $20,000 for a particularly valuable wristwatch are also met here remains an open question. Nevertheless, for the consumer this caliber 2455 is nice to look at, especially since all watches equipped with this movement have a case with a glass back. By the way, the balance shafts of both movements presented have a shock protection system by KIF: on the old design on the left, the KIF Flector; on the right, the new version, the KIF Ultraflex.

Illustration 400
It is hard to play in the upper class of high-quality movements with a flat hairspring, Triovis regulator, and a simple balance wheel (approximately 8× magnification).

7.3. The Reference Numbers

The assignment of four-digit reference numbers in the case backs was not introduced until the beginning of the 1950s, but the watches manufactured before then also had reference numbers for internal identification. As a retired employee confided to the author, the four-digit number on the photo of each was immediately used as a reference number. This classification system had been started in 1940 with the number 4000. Before that time, three-digit reference numbers were common. However, the system behind this could not be determined. For example, shown on page 272 is an automatic with the reference number 4871. It was made in 1949, and its reference number was not yet present in the case back.

Illustration 401
Inside the case back of a Vacheron & Constantin Automatic Ref. 6394 Q in 18K yellow gold (also see fig. 442, p. 276). Stamped beneath the company logo is the reference number and then the case number 409,841. Below that, in the form of a key with the number 1, is the responsibility stamp of the case maker Ed. Wenger SA from Geneva, which ceased operations on April 7, 1992.

7.4. The Case Numbers

In the period under consideration, the case numbers were always kept in six digits. The fact that the case numbers do not correspond to those of the movement numbers, mostly differing by an order of magnitude of 160,000, results, among other things, from the history and logistics of the suppliers. The movements were delivered in raw form (*ébauches*) and finished and reassembled only at Vacheron & Constantin. This is also when the various markings and engravings, including the Geneva Seal, were applied. In contrast, the case manufacturers delivered their products in a ready-to-install condition. In addition to the complete hallmarking, this also included the reference and case numbers.

7.5. The Watches

Manually Wound Watches

Watches from Vacheron & Constantin were always small beauties. Case, dial, and movement always formed a harmonious whole, even if the latter remained closed to the eye of the beholder. Obviously, the imaginatively designed dials as well as the different case variations always impressed potential buyers.

The first example in our picture gallery is an absolute rarity. The manually wound movement with the number 433,783 is probably an early version of the caliber 453 from 1941. The balance shaft has no shock protection, nor does the escapement wheel shaft have a cover plate. The regulator arm lacks the later self-evident fine adjustment. The watch with small seconds has

Illustration 402
Watches with stainless-steel cases and screw-down case backs were always a rarity at the noble brands Vacheron & Constantin and Patek Philippe. This model has no reference number but can be assigned to the production year 1941 by way of its movement and case numbers.

a steel case with screw-down back and dustcover and is therefore water resistant. This steel case measures 1.37 inches (35 mm) in diameter and 0.393 inches (10 mm) in height. The case number is 292,877. The small elevation in the center of this cover ensures that when the screw-down back is closed, this cover is elastically pressed against the movement ring. Also interesting is the twelve-edged screw-down case back, which we already know from watches made by IWC. The impressive dial with the applied Arabic numerals in white gold completes the noble image of this watch. The value of this collector's item should probably not be less than that of a comparable watch with gold case.

Illustration 403
When it's opened, one can see every detail of this watch shown on the previous page. Below is the dustcover, which is placed on the movement before the screw-down case back is closed. The circular-grained finish on the inner surface of the screw-down case back shows the effort that used to be put into finishing components.

Another rarity is the red-gold watch with caliber V 454 movement produced for Cartier. The caliber V 454 is numbered 459,958, and the case is numbered 302,399. This three-tier case consists of a bezel, case middle, and press-fit case back. Although the diameter is only 1.31 inches (33.50 mm), the lug-to-lug measurement is 0.78 inches (20 mm).

The rosé dial features six Arabic numerals as well as six pyramid-shaped indexes. This rare watch comes from an English used-watch dealer and was sold on May 2, 2013, for 2,250 pounds ($3,520 at that time). According to today's modern taste, this beautiful piece should also cut a fine figure on the arm of a fine lady.

Illustration 404
Produced for Cartier Paris, this manually wound watch with rose-gold case and dial would also look good on the wrist of a fine lady. The case has a diameter of 1.31 inches (33.5 mm) and a simple press-fit case back.

Illustration 405
Vacheron & Constantin's first chronograph with a water-resistant case was given reference number 6087. The two preceding prototypes had two goals. An existing case from sister company Jaeger-LeCoultre was used for the first prototype, and the caliber V 492 movement was installed. This advance model was used to test function and to verify its water resistance. The second prototype, not shown here, received the original case later called Cornes de Vache (cow horns) to confirm this rather unusual design concept. The whereabouts of this second prototype will probably forever remain a mystery.

The chronographs launched by Vacheron & Constantin until 1954 were optical beauties, but they were not equipped with water-resistant cases. Its Geneva-based competitor Patek Philippe had already had such a model with the reference number 1453, on offer since 1940.

As mentioned in the introduction, two different prototypes were produced for the construction of this chronograph. One of them used an existing steel case from sister company Jaeger-LeCoultre for testing watch function and water resistance, and the intended chronograph caliber V 492 was installed in it. For the other, the prototype series case later called Cornes de Vache, or cow horns in English, was used.

Before we go further into this chronograph, we should also take a brief look at the little-known history of Vacheron & Constantin. After Vacheron & Constantin's very successful president and owner Georges Ketterer passed away in 1969, his son Jacques Ketterer took over the company's fortunes. In those rather difficult times for the Swiss watch industry, the emergence of inexpensive but very fashionable quartz watches demanded a radical rethinking and a completely new market strategy.

Jacques Ketterer, who was also said to have a penchant for fast cars, beautiful women, and high-proof drinks from Scotland, was probably overwhelmed by this new market orientation. In 1987, at the youthful age of fifty-seven, he died unexpectedly. He left his family a heavily indebted company without any concept for the future. This circumstance was all the more painful because in the 1950s and 1960s, the Vacheron & Constantin brand had been on a par with its market companion Patek Philippe in terms of quality, technology, and sales. Salvation came from the former Saudi Arabian oil minister Sheik Ahmed Zaki Jamani,

Illustration 406
After removing the screw-down case back, one has a clear view of the chronograph caliber V 492. Some parts were also replaced during the overhaul by Vacheron & Constantin, including the balance cock. Since only one without surface treatment (i.e., without rhodium plating) was available in the spare-parts stores, and subsequent treatment would have taken some time, at the request of the author the balance cock was installed as is. Also, the balance staff lacks shock protection, and the regulator arm lacks fine adjustment via micrometer screw and gooseneck spring.

who was himself a great watch enthusiast. Jacques Ketterer and Sheikh Jamani had also known each other from a young age, because they had not only attended the same English boarding school but also shared a room there.

Because the following story is largely uncovered in the literature so far, we want to take a closer look at what happened in Geneva at that time. In fall 1986 (i.e., during Jacques Ketterer's lifetime), there were rumors in Geneva that Sheik Jamani wanted to buy the oldest watch brand in the world. This story, which can be compared to a thriller, was published in the Swiss Handelszeitung on February 22, 2005. We reproduce it here:

> Vacheron Constantin, the noble manufacture with a fabulously adventurous history, was to fall into foreign hands. To understand the scope of the deal, one must know that Vacheron Constantin was counted among the big three between the end of the Second World War and the beginning of the watch crisis in the 1970s. Together with Patek Philippe and Audemars Piguet, the house embodied Swiss *haute horlogèrie* and stood for the finest mechanical complications in the top league.
>
> It is fitting for the eventful history of the renowned watch manufacturer that the final sale to the sheikh was preceded by an exciting takeover thriller. Because, according to Lex Friedrich, company acquisitions by foreigners were permitted only if the value of any company real estate did not exceed one-third of the company's assets; therefore the former president of Opec was initially denied the purchase.
>
> Jamani had submitted three appraisals that put the market value of the company's real estate at just 8.9 to 11.3 million Swiss francs, which would have made the purchase possible. However, the estimates seemed so suspect to Geneva construction director Jean-Philippe

Maitre that he commissioned a reappraisal. Promptly the presumed price was officially raised to 20 million francs—more than the permissible third of the assets: Sheikh Jamani will not be able to buy Vacheron Constantin, the local press gleefully concluded with lavish exclamation marks.

But the newspapers had hardly been bundled into waste paper when the former oil minister definitely struck with a clever coup. The real estate was now suddenly worth a fantastic 43 million francs, the price for which it was sold to the city's well-known real estate shark Jean-Pierre Magnin. This meant that Vacheron Constantin no longer had any real estate of its own, but Jamani, a passionate watch collector by the way, was finally able to get his hands on it.

If Jamani had not done so, the manufacture would probably have gone under. As had often been the case in the past, Vacheron Constantin was on the brink of ruin. After these financial manipulations Jamani still acquired 95% of the share capital of Vacheron & Constantin in a second attempt in 1987 and at the end of 1988 appointed

Illustration 407
The ratchet wheel is the "heart" of a chronograph caliber, shown here magnified about ten times.

the then-44-year-old Neuchâtel economist Claude-Daniel Proellochs as its new president. At that time Vacheron & Constantin's workforce was only 65 people, and the number of watches produced annually was around 3,400.

But the *Handelszeitung* also has more internal information about Claude-Daniel Proellochs: "The man who was to bring about a turnaround for the better was Claude-Daniel Proellochs, and it should only be mentioned here in passing that in wooing him away from Eterna, a gentleman had a hand in it who had shaped the industry like no other: Nicolas G. Hayek, whose company Hayek Engineering had been brought in by Sheikh Jamani as a consulting firm. Proellochs later compared his first contact with Vacheron Constantin to the discovery of an old Rolls-Royce that had been left to rot in a barn. To make the bodywork shine again, he reasoned, it would probably be enough to wipe off a bit of dust and apply the high-gloss polish. To start the car and steer it through international traffic, on the other hand, there would be a lot more to do. The tires, to stay with the picture, were flat and brittle, the battery was dead, and the exhaust did not meet the new requirements. But the potential, Proellochs recognized immediately, the potential was there." (Pierre-André Schmitt, "Vacheron Constantin: The Crazy Rescue of a Manufactory," *Handelszeitung*, February 22, 2005)

At that time, the development, production, and sales departments of Vacheron & Constantin in Geneva were spread over a total of five locations, so that effective management of this company and the necessary overview were hardly guaranteed. It is easy to understand that not only a lot of know-how but also some material assets left the company in a not entirely legal way during that critical time, combined with an ongoing loss of personnel. The two prototypes of the chronograph reference 6087 mentioned probably also disappeared and took a path into the unknown that is no longer comprehensible today.

It is now impossible to say how the chronograph mentioned in the introduction came to be in Lisbon. But why the seller immediately set such a high price of 10,000 euros for a steel watch is also hard to understand. Before we go into this story further, we must first take a look at the legal situation of an eBay sale. According to German and Swiss law, a stolen item cannot be sold or bought. Only if it is offered on a public auction and everyone has access to the sale can the item be passed on and acquired legally. Since an eBay auction can be equated with a publicly accessible offer of an auction house, an object acquired there can be described as a legally correct transfer of ownership.

Since, on the one hand, the offer of $10,900 seemed too high to the author but, on the other hand, this watch had still not found a buyer after more than four weeks, a sale price of $8,178 was agreed upon, as well as the usual right of return.

After the watch arrived, it was immediately sent to Vacheron Constantin in Geneva for evaluation. Only after some doubts did they declare that it could possibly be one of the two prototypes from 1953. The seller in Lisbon was then informed by email about the positive assessment from Geneva. However, this message was immediately returned for the reason that the specified email address did not exist. A letter subsequently sent to the seller's postal address by registered mail, which was supposed to verify the purchase, was returned after two weeks with the reason "recipient unknown."

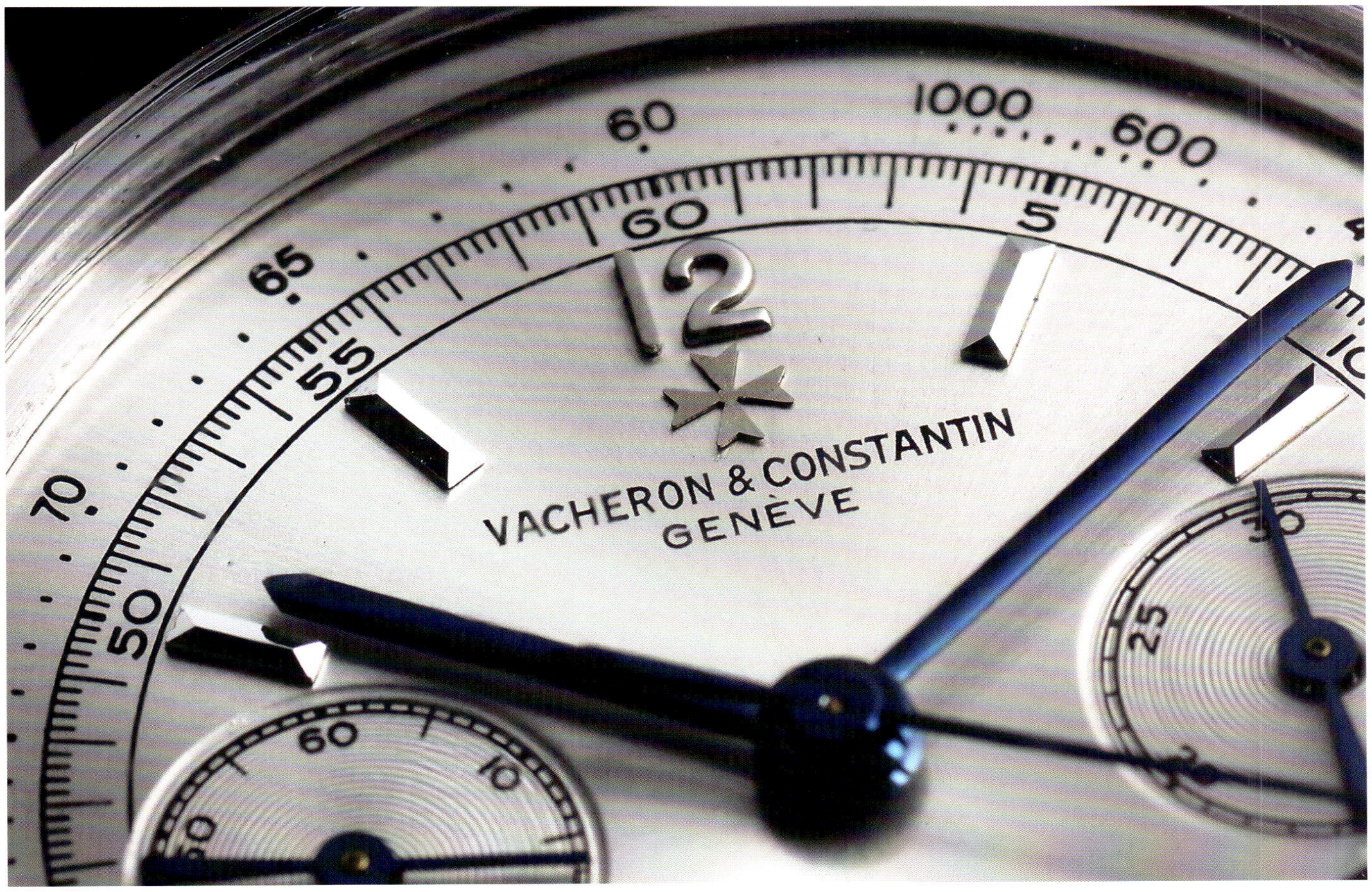

Illustration 408
This detail enlargement of the dial illustrates the high quality that was a matter of course at Vacheron & Constantin. Indices, numerals, and company signet are made of 18K white gold, the hands of blued steel.

Despite this most mysterious situation, an overhaul of this chronograph was ordered from Vacheron Constantin. After almost six months the notification of completion finally arrived from Geneva, as well as the notification of shipment to Munich. This service and overhaul were anything but routine; rather, they were a special project under the technical responsibility of Rudolf Büll, at that time assistant to the company management.

Upon return, the prototype chronograph appeared almost in new condition. This applies not only to the waterproof case with screw-down back, but also to the masterfully refurbished dial. The eight indices and the numbers at twelve and six o'clock were made of white gold. This is impressive, but so too is the successful, precise printing of the dial and the dark-blue burnished hands. The whole thing leaves a professional impression. The case diameter is 1.45 inches (37 mm), and the height is 0.511 (13 mm). As with any previous Valjoux chronograph caliber, in this case the 23 13‴, the lower pusher at four o'clock is farther from the crown than the upper one.

The Jaeger-LeCoultre screw-down case back is made as a decagon (12 sides) with a key width of 1.22 inches (31 mm). The lug-to-lug measurement is 0.74 inches (19 mm).

During the overhaul of the caliber V 492, some components, which were no longer in complete form, had to be replaced. This includes the balance cock, which was available only as a non-rhodium-plated part. At the request of the author, the cock was installed without surface protection and thus represents the compromises that were unavoidable during this overhaul. Nevertheless, one may thank the Vacheron Constantin employees responsible at that time, primarily Mr. Rudolf Büll, for this energetic but unofficial support at this point.

Actually, this rare piece should get a place of honor in the Vacheron Constantin museum, even if the author still has a residue of doubt about the originality of this valuable collector's item despite the confirmation of authenticity from Geneva. In addition, a veteran of Vacheron & Constantin believes that before this watch was illegally taken, the movement would also have been exchanged to make later identification more difficult.

In 1996, Sheikh Jamani's share capital was acquired by the Vendôme Luxury Group, part of the Richemont Group. Claude-Daniel Proellochs remained chairman and in 2005 was privileged to celebrate Vacheron Constantin's much-publicized 250th anniversary in Geneva. At that time the company again had about four hundred employees and produced about 15,000 watches per year. The oldest Swiss watch manufacturer will continue to be a force to be reckoned with in the future.

References 6304, 6305, and 4600

Particularly noteworthy are the case sizes, which are almost 1.45 inches (37 mm) in diameter—unusual for watches made in the 1940s and '50s. Two typical beauties are shown.

The watch shown first comes from the USA and is still in its original condition with all details intact. The second timepiece shown is, contrary to expectations, a *mariage*—a watch that marries parts from different watches. Even the Dutch collector, from whom this otherwise quite noble piece comes, was not aware of this fact. Only an inspection by Vacheron & Constantin in Geneva revealed that the case number 365,977 originally contained a caliber P 454/5 B with center seconds (movement no. 516,147). The caliber P 453/3 C with movement no. 479,153 shown originally belonged to a watch with reference no. 4195 and case no. 311,137. This fact also explains the large numerical distance between the movement numbers of the two watches.

Another watch from Vacheron & Constantin gained great popularity in the early 1950s: the manually wound watch Ref. 4600 Jubilé. Thanks to a harmoniously designed case with a diameter of 1.41 inches (36 mm), case height of just 0.322 inches (8.2 mm), and lug-to-lug measurement of 0.74 inches (19 mm), this noble piece fits discreetly under any shirt cuff. The name JUBILÉ is engraved on the outer cone of the press-fit case back vis-à-vis the crown. This is something you should definitely pay attention to when making a purchase, because if this engraving should be found only faintly or not at all, then the press-fit bottom could have been repolished several times.

Illustration 409
An interesting comparison allows us to see the wristwatches with manual winding and small seconds described on this page. Design of the dial and case are typical of the 1950s. The dial of this Ref. 6304 is gold-plated and has dash indices and baton hands. The caliber K 453/3B 12½''', supplied by Jaeger-LeCoultre as a raw movement (*ébauche*), has eighteen jewels and is adjusted for heat and cold, for isochronism, and in five positions. The Geneva Seal is stamped on the gear train bridge and plate. The case measures 1.43 inches (36.5 mm) in diameter and is made of 750 yellow gold. The case middle was kept quite massive, and the caliber was directly inserted without a separate movement retaining ring (movement no. 480,562, case no. 365,932).

Illustration 410
This timepiece is listed under reference 6305 and is the follow-up model to the Ref. 6304 shown opposite. The dial shines with Roman numerals and elaborately designed minute and second displays. There are also differences in the manually wound movements. Inside this watch is the caliber P 453/3C 12½'''. On the mainspring barrel bridge, one can read that this movement has seventeen jewels and is regulated in temperatures. Unlike the previous caliber, the fine adjustment via micrometer screw and swan-neck spring are missing from the regulator arm. Both components were originally present and must have been removed at some point. Also missing is the Geneva Seal. Compared to the Ref. 6305, the numbers of the caliber are around 1,409 behind, and those of the case around forty-five behind compared to the Ref. 6304 shown on the left. Where these differences come from we learn in the text of this chapter (movement no. 479,153, case no. 365,977). Research by Vacheron & Constantin revealed that movement no. 479,153 originally belonged to case no. 311,137 and was a reference 4195. Case no. 365,977 originally encased the manually wound caliber 454 movement no. 516,147 with center seconds.

This model uses the caliber P 453/3C with seventeen jewels. The fine-adjustment mechanism consists of a double-spoke screw balance wheel, Breguet hairspring, micrometer setting with gooseneck spring, and Jaeger-LeCoultre's own shock protection system. The escape wheel shaft is mounted in cover plates. Even movements from Patek Philippe could not have been more beautiful in those days.

The two watches presented have an interesting history. The beauty in rose-gold case comes from Bogota, the capital of Colombia. In the 1950s, watches in rose gold were very popular and widespread in South America. The father of the current owner had personally purchased this watch from Vacheron & Constantin in Geneva in 1954 and, at a business meeting in Geneva six years later, had the watch refurbished there. Subsequently, this watch spent all its time in a safe, since the father had switched to a robust steel watch of the Rolex brand. Including import sales tax and transport, this almost mint heirloom changed hands for $2,820 to delight those interested in beautiful timepieces from the house of Vacheron Constantin in this chapter (movement no. 512,887, case no. 335,043).

Illustration 411
Among the most-beautiful models of the 1950s was the reference 4600, nicknamed the Jubilé. Here we see a version with a yellow-gold case and black dial. The case diameter (without crown) is 1.41 inches (36 mm).

Illustration 412
This Jubilé model has a rose-gold case and comes from Bogota, Colombia. In contrast to the watch on the left, this version has the typical Maltese cross between twelve o'clock and the company signature. The seconds display has only a dash cross.

The counterpart with black dial and yellow-gold case has a fundamentally different look. Especially as an evening watch or for a visit to the opera, this valuable piece should fit well.

This Vacheron & Constantin Ref. 4600 comes from Italy. Including purchase, transport, and a professional overhaul, it cost about $3,470 (movement no. 487,274, case no. 317,111).

From just these two examples—masterfully made wristwatches with a high standard of quality—the watch lover and collector should take note that they are undoubtedly at the same level as watches from market companion Patek Philippe. The unfounded misconception that this is not so means that Vacheron & Constantin models comparable to those from Patek Philippe often cost only about half as much at present. For example, if you compare a manually wound Patek Philippe Ref. 3420 watch with a Vacheron & Constantin Ref. 4600 Jubilé, the Patek Philippe costs almost twice as much as the Vacheron & Constantin on the secondhand market.

In May 2021, there were six Patek Philippe Ref. 3420 watches offered on the Chrono24.com internet portal. The prices of these watches ranged from $7,500 to $13,000.

At the same time, only two examples of Vacheron & Constantin's Ref. 4600 Jubilé were offered. The two watches were priced at $5,254 and 5,744 euros. The Patek Philippe model has a case diameter of only 1.33 inches (34 mm), while the Vacheron & Constantin model measures 1.36 inches (36 mm). That is why the Vacheron & Constantin watch makes a much more noble impression than its Patek Philippe counterpart. The two manually wound movements 27-AM 400 and P 453/3C can also be considered equivalent, even if the design of the second movement is more than ten years older than that of the Patek Philippe product.

From a purely objective point of view, the Ref. 4600 from Vacheron & Constantin should be the absolutely better buy. I wonder if this well-intentioned signal will finally be received by watch fans?

In the 1940s, '50s, and '60s the case diameter of wristwatches produced by well-known watch brands was usually 35 mm—

Illustration 413
Rear view of the reference 4600 shown above right reveals the flat case shape with a height of 0.322 inches (8.2 mm). The press-fit case back is only splash resistant. Anyone who owns such an elegant watch is unlikely to wear it when washing their car.

often less, but rarely more. The masterpiece shown brings with its rose-gold case a stately 37.5 mm, an unusual value for the time. Only the wristwatches made by IWC for Portugal as late as the 1930s had cases with a diameter of 38 mm (see p. 65, fig. 103), but despite all their efforts, they did not possess the high standard and noble aura of a Vacheron & Constantin.

The fact that this watch's case already shows clear signs of corrosion is due to its great age. An exact identification can be made only on the basis of the movement number 493,682 and case number 323,382, since this watch has no stamped reference number. The manually wound caliber P 453/3B movement has a neatly fitted movement-retaining ring. When looking at the dial, one can still see the imagination that inspired the designers of that time. There are four Arabic numerals for the quarter subdivisions, dash and pyramid indices for the five-minute and hour subdivisions. The small seconds fits stylishly between the hand shafts on the 6, which is slightly shortened at the top. A beautiful watch indeed! The excellent impression is completed by the elegant Baton hands. At auctions this special piece should bring $7,500 to $8,680.

Vacheron & Constantin distinguished itself with one model series in particular and, with it, took a leading position in the Swiss watch industry. These watches were given the pleasing name Chronomètre Royale (later Royal Chronometer in English). Three different models of these chronometers are shown. Either the caliber P 1007/BS or P 1008/BS is encased in these watches. They were produced from 1953 to 1962, and production totaled just 479 pieces.

Our display of rarities continues with a Chronometer Royal reference 6105 in 18K yellow gold. In contrast to the previously mentioned Ref. 4338 and 4907 models, with press-fit case backs, this watch has a water-resistant case and screw-down case back. Engraved on this screw-down back are Chronomètre Royal, the case number 352541, and the reference number 6105. The fact that this watch has the case number and the reference number engraved on the outside is indeed unusual.

Illustration 414
In the middle of the last century, the standard for watchcase diameters was 1.37 inches (35 mm), and often less. The model shown has a case with a diameter of 1.47 inches (37.5 mm), which was unusual at that time. The dial is also deserving of a close look. It has four Arabic numerals for the quarter-hour subdivisions, and dash and pyramid indices for the five-minute and hour subdivisions. The small seconds display fits stylishly between the hand shafts and the 6, the top of which is slightly shortened.

The case itself measures 1.37 inches (35 mm) in diameter and 0.42 inches (10.7 mm) in height. The arrow-shaped lugs, also called Broad Arrow lugs, have a lug-to-lug measurement of 0.70 inches (18 mm).

The chronometer caliber P 1008/PS has the movement number 509,362. The manufacturer's mark in the form of a key with the number 23 and the collective mark 5 identify case supplier Eggly & Cie. from Geneva. This company went out of business on March 22, 1990.

Illustration 415
Among the most-sought-after wristwatches are the different models of the famous Chronomètre Royale line. This version, Ref. 4838, has a small seconds display at six o'clock. The 750 yellow-gold case is a two-piece design and has a press-fit case back.

This rare reference 6105 watch could not be found on the internet or at well-known auction houses, but copies of the comparable references 6107, 6110, 6111, and 6161 were.

Models with screw-down case backs that did sell ranged in price between $12,000 and $14,000.

The rare piece shown next was the inspiration for the current Fiftysix Ref. 51326 model. The company's own advertising from 1956 identifies this watch as a Ref. 6073, but this is not quite true. Since the new version is a three-hand watch with center seconds, the model shown should be a Ref. 6069! The special feature of this beautiful piece is the Maltese cross, which is present on the dial. The latter can also be found on the reference 6073, but on a dial with small seconds. Why did the responsible advertising and marketing staff not pay closer attention?

Illustration 416
Probably the most beautiful version of the Chronomètre Royale line that the author has come across is this Ref. 6161 model in 750 rose gold. The Arrow lugs of the three-piece screw-down case define the appearance of this rare chronometer, and the silvered dial with the highly complex indices is the face of this timepiece. Here, too, the caliber P 1008/BS is encased (movement no. 537,125, case no. 376,145).

The reference 6069 shown uses the caliber 1002. The movement number is 518,798, and the case number is 346,808. The case has a press-fit case back and is 1.25 inches (32 mm) in diameter. The height is only 0.295 inches (7.5 mm). Again, we find the Broad Arrow lugs, and the lug-to-lug measurement is 0.70 inches (18 mm).

This watch comes from a watch collector in Antwerp, Belgium, and was sold on eBay on March 13, 2013, for $2,767 plus shipping. The seller probably could not have guessed that this model would become a great example seven years later.

If a dial were to win a beauty award, the "face" of this watch would be a serious contender. On this dial the circular face has a fine wave pattern, with only the company name on a cutout. In addition to the numerals at three, six, nine, and twelve o'clock, four wedge and four dot indices mark the hour indication. The baton hands fit very well in the overall picture.

Illustration 417
The second version of the Chronomètre Royale has central seconds and has the reference number 4907. "Chronomètre Royale" is engraved only on the press-bit case back.

Illustration 418
One of Vacheron & Constantin's most famous lines was given the name Chronomètre Royale. The chronometer caliber P 1008/BS is encased in this watch. The 18K yellow-gold case has a diameter of 1.37 inches (35 mm) and a sturdy screw-down back. In addition to "Chronomètre Royal," the case back is also engraved with the reference number 6105 and the case number 352,541. The watch shown may be considered a special case. The reference number and case number, which are engraved on the inner case back as standard, are also engraved on the outside of the screw-down case back. The other markings, such as the precious metal hallmark, company logo, and case supplier, are also listed.

Illustration 419
Opened reference 6105 with the excellent chronometer caliber P 1008/BS, movement number 509,030

Illustration 420
In this approximately 45× enlargement, the case number 352,541 and the reference number 6105 are more clearly visible.

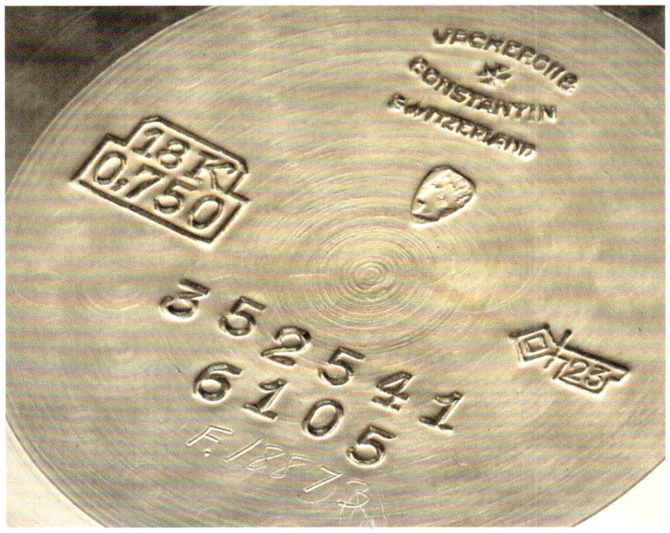

Illustration 421
This additional enlargement makes it possible to read the manufacturing and processing marks on the inside of the case back.

The case has a two-piece design and a screw-down case back with six deep-drawn grooves. The case diameter is 1.39 inches (35.5 mm); the height is 0.38 inches (9.8 mm). The lug-to-lug measurement is 0.71 inches (18.2 mm). Inside the case

Illustration 422
This lovely manually wound reference 6069 with the Maltese cross on the dial from 1956 serves as a template for the current three-hand, self-winding Fiftysix 1326.

Illustration 423
Fiftysix Ref. 1326, a three-hand watch with automatic movement and stainless-steel case. The Ref. 6069 seen above served as a model for this watch. *Vacheron Constantin photo*

there is the caliber P 455/5B. The movement number is 515,508, the case number 348,610. A reference number is not available, but the hammerhead signet with the number 141 identifies the case supplier Gabus Frères, Oracier, Le Locle. This company ceased operations on June 26, 1980.

Another gem from the Chronomètre Royal series is the reference 6340, with center seconds and yellow-gold case. The slightly conical crown, which is found only on this model, is also typical. Inside this watch is the chronometer caliber P 1008/BS; the movement number is 531,122. The 18K yellow-gold case has a press-fit case back with a diameter of 1.40 inches (35.6 mm); the case number is 375,985. The yellow-gold dial has four double and eight small indices. CHRONOMETRE ROYAL is engraved in large letters 2 millimeters high on the press-fit case back in the conical outer area.

The manufacturer's mark in the form of a key with the number 23 and collective mark 5 identifies the case supplier as Eggly & Cie. from Geneva. This type of Chronometer Royal is relatively rarely offered. At auctions a purchase price range of $4,300 and $7,600 can be expected.

The last in our series of photos of beautiful manually wound wristwatches is another reference 6400 Jubilé. In addition to its harmonious rose-gold case, the elaborately designed dial also shines. Three different surfaces, Arabic numerals, and wedge-shaped indices, as well as a framed company name, highlight the special features of this dial (caliber P 453/3B, movement no. 500,254, case no. 329,581).

Among the noblest collectors' items from the Geneva manufacture are the chronographs produced in various versions. Examples are the references 4072, 4075, 4083, 4111, 4177, and 4178, which were produced in the 1940s. The only "time

Illustration 424
This watch, with its special dial, should be considered for entry in a beauty contest.

recorder" equipped with a waterproof case was given the reference 6087, and just thirty-six examples of this watch were made. The last model of this rare chronograph was made in November 1959 and exported in February of the following year (see pp. 270, 271).

Watches with Automatic Winding

Vacheron & Constantin began the era of self-winding wristwatches before its competitor Patek Philippe, but after market leader Rolex. *Ébauches* resupplied by Jaeger-LeCoultre gave rise to the calibers 477 and 477/1. The automatic presented in 18K yellow gold has the caliber 447/1 with movement number 494,732 encased. The two-piece case with the number 322,821 measures 1.37 inches (34.8 mm) in diameter and 0.46 inches (11.7 mm) in height, and it has a conical screw-down case back with the usual twelve corners, and a key width of 1.14 inches (29.2 mm). The curved lugs with a lug-to-lug measurement of 0.669 inches (17 mm) are also elegant looking. The manufacturer's mark no. 1 with the hammerhead and stamped number 136 reveal that the case supplier was C. R. Spillmann SA, Chêne-Bourg. This company ceased operations on April 5, 1988.

References 6038, 6307, 6378, 6394, and 6426

Some watches also write history several times, as one can learn from the automatic reference 6038. According to the story told

Illustration 425
Another variant of the Chronomètre Royale is this three-hand watch in 18K yellow gold. The case has a press-fit back. Why an inwardly tapered crown was used only on this model, no one at Vacheron Constantin could say.

Illustration 426
On this model the Chronomètre Royal logo is engraved on the case back.

Illustration 427
This reference 6400 Jubilé with a very special dial and 18K rose-gold case is also a real eye-catcher. Can anyone still make such beautiful watches today?

by its last owner, this watch, made in 1958, was presented as a gift to the then Romanian ambassador from the hand of the president of the Swiss Confederation. About a quarter of a century later, the by-then-retired diplomat had to undergo a serious, life-sustaining heart operation. Since such an extensive operation was possible only in exchange for hard foreign currency, even in his socialist homeland, he sold the self-winding watch to a merchant from St. Gallen for 6,000 CHF. With this considerable sum the retiree was able to cover not only the costs of the operation but also those of the associated medication. In this way the Vacheron & Constantin that is shown unexpectedly made its mark as a lifesaver.

The second example of this series has the same case but in 18K yellow gold and has the same reference no. 6038. The dial indices and the hands are also identical, only the Maltese cross

Illustration 428
Among the most-beautiful and rarest watches ever made by Vacheron & Constantin is this chronograph Ref. 6087 in 750 yellow gold. It is considered a counterpart of the chronograph Ref. 1463 built by Patek Philippe and, like the latter, is the only one to have a case with a screw-down back, making it water resistant. In the period from 1954 to 1959, just thirty-six examples of this precious "time recorder" were produced: twenty-six in yellow gold, eight in rose gold, and two in platinum. The specimen shown is the last of the thirty-six chronographs to be made. According to the archival record for this watch, it was completed in November 1959; placed in stock on January 1, 1960; and exported to the United States the following February. The retail price at the time was approximately $3,200. Unlike with previous case designs of this chronograph, the bezel of this example is not rounded, but elegantly beveled. The dial shines in a warm silver tone.

Illustration 429
The inner workings of this beautiful chronograph with caliber 492 movement are revealed (movement no. 470,084, case no. 373,268). The raw movement of this chronograph movement with ratchet wheel control is a Valjoux 23 13''' with seventeen jewels, outer diameter of 1.16 inches (29.5 mm), and height of 0.230 inches (5.85 mm). However, the Vacheron & Constantin version had nineteen jewels and was finely regulated. In the last four years, exactly four examples of this rare chronograph appeared at international auctions. The chronograph presented was called for CHF 140,000 at the Antiquorum auction house in Geneva on November 12, 2006, and sold for CHF 141,000, about $98,700 at the time.

below twelve o'clock is missing. Did only the rose-gold version get the Maltese cross back then? The caliber P 1019/1 is reasonably well preserved and bears the movement number 528,200. The inside of the screw-down case back is turned. Stamped there are the reference number 6038, the case number 359,427, and the case maker's key with the number 1. The case diameter is 1.37 inches (35 mm), and the height is 0.44 inches (11.3 mm). This case was delivered by Ed. Wenger SA, Geneva.

Illustration 430
This shapely specimen was one of the first wristwatches with automatic drive. The two-piece 18K yellow-gold case looks elegant thanks to its curved lugs (movement no. 494,732, case no. 233,321).

Illustration 431
At first glance it is difficult to understand the design of a movement with the so-called automatic hammer (i.e., with limited oscillating weight). The angle of rotation is only about 180°, and on each side a hairspring catches this rotating body. All usual hallmarks are stamped in the circular-grained screw-down case back; only the reference number is missing.

This beautiful collectible comes from a Viennese auction house. It is likely that only an expert could tell that the dial has been refurbished. In this case the varnish was stripped and a fine brush finish applied to the silvered dial. The enameled lettering was preserved and only had to be repainted. The SWISS label below six o'clock was probably forgotten. The sixty index markers for the minute timer are also missing.

With the launch of the advanced calibers K 1070 and K 1071 in 1959 and the K 1072 version with date display in 1961, Vacheron & Constantin actually began the great era of self-winding watches. Until the end of the 1960s it produced numerous

Illustration 432
With its rose-gold case and silver-plated dial, according to its owner this Ref. 6038 automatic watch from 1958 was presented as a gift to the Romanian ambassador by the president of the Swiss Confederation. Why this watch acted as a lifesaver decades later you will learn in the text of this chapter.

models with impressively designed dials and cases. The various references used cases in yellow, rose, and white gold. Only a very few were given steel cases, such as the 6307 and 6562 shown (see p. 281).

In addition to the automatic Ref. 6307 in stainless steel shown, many years later the author encountered a second example, this time in 18K rose gold, at a Phillips auction in Geneva on November 10, 2018. The current undervaluation of this brand is evidenced by the hammer price of just CHF 10,000. A comparable automatic from Patek Philippe—for example, a reference 3415 in rose gold—would likely have fetched twice that price. This is also the only Vacheron & Constantin model with a crown that is completely flush with the edge of the case when in the zero position and thus completely protected. For an avid collector this reference should be a real challenge!

Illustration 433
The second automatic caliber unveiled by Vacheron & Constantin also came as a reserved raw movement (*ébauche*) from Jaeger-LeCoultre and was initially given the caliber designation 498 12''' for the small seconds version and 499 12''' for the center seconds version. Later the caliber designations were changed to P 1019 and P 1019/1. This designation is also engraved on the automatic bridge. Unlike in their Audemars Piguet counterparts 2498 and 2499, the rotor was not fitted with a gold oscillating weight, but instead one made of heavy metal (also see p. 36, Audemars Piguet). Otherwise, both caliber versions are almost identical (movement no. 536,251, case no. 362,212).

A special noble piece can be found on the facing page. The reference 6426 was encountered by the author only once in the past years. Its rose-gold case with Arrow lugs, guilloché rose-gold dial, and fancifully designed indices and its dauphine hands give this gem a very special look. The automatic, which appears on the right, should actually be an even-greater rarity, because the hands with small seconds suggest the caliber K 1070, photographs of which have never before been published. However, the research conducted by Vacheron & Constantin in Geneva

Illustration 434
Watches from Vacheron & Constantin have always made an understated, distinguished expression. This is also true of the yellow-gold reference 6038 with its wedge-shaped indices in 18K yellow gold.

concluded that this is in fact a *mariage*. Case and movement thus did not originally belong together. Accordingly, the almost criminalistic finding is as follows: the rose-gold case with screw-down case back belonged to a manually wound K 1002 caliber watch made by Vacheron & Constantin in 1955. The automatic caliber K 1072 (center seconds, date), movement no. 571,133, had originally been encased in an automatic with 750 yellow-gold case. This watch, probably reference 6378, was exported

Illustration 435
Although the caliber P 1019/1 was Vacheron & Constantin's first automatic movement with a rotor, it underwent step-by-step improvement from the beginning. One of its weak points was the rotor mounting over four steel ball bearings. This design's shortcomings were never fully resolved. Also, one should avoid intensive wearing of this automatic movement, since the automatic bridge and the rotor bearing are no longer available as spare parts (movement no. 528.200, case no. 359.427).

Illustration 436
Close-up of the fine-adjustment mechanism of the caliber P 1019/1 seen above. The screw balance, Jaeger-LeCoultre shock protection, and regulator arm with micrometer screw and gooseneck spring are easy to recognize. This is also true for movement no. 528,200.

to Great Britain in 1965. How and when the automatic movement was put into the watch that we have is unknown. The fact is that the calendar mechanism and the drive for the center second hand were removed from this caliber K 1072. Furthermore, a fourth wheel with long pivot was used to drive the small seconds. In this condition the caliber was finally placed in the red-gold

Illustration 437
The reference 6307 is a self-winding watch that is very rarely seen. Here is a version with a rose-gold case. In this picture the crown is pulled out and protrudes over the case. In the pushed-in position, it is flush with the case. *Phillips Geneva photo*

Illustration 438
This image is typical for the then-new automatic caliber K 1071. It has ruby bearings for the rotor with gold segment, circular-grained plates, and screw-down case back. The movement has the number 545,203 and the case the number 365,989. *Phillips Geneva photo*

case, which is about ten years older than the movement. The numbering of the case back revealed further inconsistencies. Here, the first digit of the six-digit case number had been polished out, and a zero had been added to the right. The first digit of the reference designation was also removed and replaced by a two.

Illustration 439
Wristwatches in stainless steel are rarely found among the noble brands. In addition to Patek Philippe, this statement also applies to timepieces from Vacheron & Constantin. The automatic Ref. 6307 shown dates from the production year 1961. The two-piece steel case is screwed together, and the winding crown is completely protected. You can pull it out only from the bottom side to set the time. The watch is driven by the caliber K 1071 movement, which was unveiled in 1959 (movement no. 545,689, case no. 366,067).

Illustration 440
This comparison shows the pleasing appearance of the *mariage* seen here on the right, and that it can also stand up to comparison with a noble automatic from Patek Philippe. Like the *mariage*, the Patek Philippe Ref. 2526 has a 750 rose-gold case (caliber 12'''-600 AT, movement no. 764,093, case no. 698,280).

What moved the person responsible to make the described change will probably remain a mystery forever. After this disappointment, the hunt for a Vacheron & Constantin with the rare caliber K 1070 is also on again.

If one compares wristwatches from Patek Philippe and Vacheron & Constantin from identical eras, nether is in any way inferior in charisma and technology. Juxtaposed on page 274 are a very rare Patek Philippe Ref. 2526 and our previously described *mariage*, both with 750 rose-gold cases. At first glance it would

Illustration 441
"Two special beauties in rose gold"—that's what you might think when you first see these two exceptional timepieces. On the left, the extremely rare Ref. 6426 with guilloché dial and automatic caliber K 1071 with center seconds; on the right, the probably unique version with caliber K 1070 with small seconds at six o'clock. The fact that only the watch on the left passed the authenticity test and the other watch turned out to be an elaborate *mariage* can be read in the text (*left*: movement no. 557,620, case no. 383,130; *right*: movement no. 571,133 but rebuilt; case no. 886,570, unfortunately faked).

be hard to find one of these watches more beautiful than the other. The Vacheron & Constantin reference 6732 Q, which we find on page 277, has a very pleasing case design. This appearance is further emphasized by the beautiful K 1072/1 caliber shown next to it. In addition to the dominant winding rotor with guilloché oscillating weight in 18K gold, we can also see the Gyromax balance wheel with its ten adjustable *masselotte* weights. A look at the inner case back reveals all the hallmarking and numbering. Among the most-beautiful models from the house of Vacheron & Constantin is the automatic Ref. 6782 Q, illustrated here on this double-page spread. The yellow-gold case has a rotating bezel and a special crown used exclusively on this model. This is the only watch from Vacheron & Constantin with these features. Its silvered dial harmonizes well with the dauphine hands and yellow-

Illustration 442
The Ref. 6394 Q was made in different dial and case material variants. Here the version in yellow gold with a silvered dial with vertically positioned line indices.

Illustration 444
The Ref. 7397 Q with the most commonly produced dial version, a very sought-after model

Illustration 443
The Ref. 6394 Q, here opened, and below it the screw-down case back. The automatic caliber K 1072 still has a fine-adjustment mechanism with screw balance and micrometer screw (movement no. 582,417, case no. 409,641).

Illustration 445
The second opened Ref. 6394 Q with its screw-down case back. The automatic caliber K 1072/1 has a fine-adjustment mechanism with Gyromax balance (movement no. 587,293, case no. 418,936).

Illustration 446
The Ref. 6732 Q timepieces are among the most beautifully designed automatics unveiled by Vacheron & Constantin. The three-piece case has a screw-down case back. The dial is silvered, while the indices, hands, and date window are in 750 rose gold, as is the case. The K 1072/1 caliber has a Gyromax balance wheel with ten *masselotte* weights and the Geneva Seal stamped on the plate and automatic bridge. At that time, wristwatches from Vacheron & Constantin were among the best to be found in Switzerland and were in no way inferior to the noble timepieces from Patek Philippe (movement no. 587,210, case no. 419,789).

gold indices. Looking at the caliber K 1072/1 under a magnifying glass, one can see almost every detail of this excellent movement. Underneath the left end of the 18-karat oscillating weight, one can even see one of the two dial pins and the end of the set screw that locks this pin. For better understanding, here is a brief explanation of the dial-locking system. The dials of wrist watches have two pins soldered on their back, usually opposite each other. During assembly, these pins are inserted into two holes drilled into the dial. The dial is locked in place by two threaded pins, which are centered and at right angles to these pins. Even Patek Philippe could not keep up with the model shown, because the competitor from Geneva never had a comparable watch in its program. Just three hundred examples of this Turn-O-Graph were made. The author has seen six different dial variants.

Illustration 447
The only Vacheron & Constantin automatic with a case with rotating bezel is the Ref. 6782 Q. This model was available from 1967 to 1973. Production of this watch totaled about three hundred, with six different dial versions known to the author. The case was made only in 750 yellow gold. The winding crown, which was used only by this reference, is deserving of special mention. The caliber K 1072/1 was considered *non plus ultra* in its time. Breguet hairspring, Gyromax balance, rotor with ruby bearings, rotor bearer with sunburst finish, and guilloché oscillating weight in 750 gold, circular grained, and angled bridges and plate, Geneva Seal: all these are attributes that are found only on masterpieces from Vacheron & Constantin and Patek Philippe (movement no. 598,456, case no. 433,445). According to Vacheron & Constantin's master book excerpt, this reference 6782 Q was manufactured in 1968 and sold the same year.

In the photo we see three reference 6782 Q watches. On the right is the previously illustrated version with the silvered dial and dauphine hands. The watch in the middle has a gold-plated dial with baton hands. The version on the left also has a gold-plated dial and dauphine hands. The caption refers to the special history of this watch. If you find an opportunity to get ahold of an example of this special model for a price around $9,700, you should not miss out on this beautiful piece, because no more of these rare watches will ever be made, but they will become more and more valuable!

In chapter 8, "Buying and Overhauling Special Wristwatches on the Basis of Specific Examples," we wrote about the Royal Chronometer *mariage* (see p. 302). The fact that this watch's case and oscillating weight were made of 18-karat gold sounds almost mysterious. But again, the impression that this watch has hardly lost any of its charisma, even in a plain steel case, is astonishing. However, you will find the comparison of this *mariage* and the Chronometers Royal, which is presented on the next page, interesting.

According to Vacheron & Constantin, exactly 1,227 officially certified examples of the reference 6694 Q and the successor models 7258 Q (case with soldered gold bracelet) and 7375 Q (tonneau case with leather strap) were made in which the calibers K 1072 and K 1072/1 were encased. From today's perspective, it is difficult to describe the case design of these three models as particularly successful. Neither the oversized lugs on the reference 6694 Q nor the tonneau case on the reference 7375 Q, also known as the fried-egg design, were particularly well received by the public. At that time, the brand image, as well as the high-quality automatic movements, could still make up for minor design flaws. Despite repeated presentations, one simply can't get enough of the extremely successful and noble automatic caliber K 1072/1. That's why it was the final piece in our picture gallery of classic beauties with the calibers K 1073 and K 1072.

Illustration 448
This photograph documents three different dial versions of the Ref. 6782 Q. On the left, a model with a gilded dial and dauphine hands. This watch was offered by a Geneva auction house in November 2000 but was not sold. It finally changed hands for 8,500 CHF in a postsale. This jewel, which was in quite moderate condition, underwent a thorough refurbishment of its case and movement and was restored to almost-new condition. In April 2005 this watch came up for auction again at the same auction house, and this time was hammered at 24,150 CHF. The watch in the center also has a gilded dial but black-backed baton hands. It was bought by its current owner at the aforementioned auction, but only 14,950 CHF was paid for it. The model on the right has already been seen on the two previous pages. The dial is silvered but matches the dauphine hands and the yellow-gold case well (*left*: year of manufacture 1972, movement no. 643,782, case no. 472,376; *center*: year of manufacture 1970, movement no. 644,048, case no. 485,867).

Illustration 449
Here the opened stainless-steel case of a Ref. 6562, which contains the caliber K 1072/1 movement with chronometer certificate, including the associated Ref. 6694Q dial (movement no. 630.082, made in 1971; case no. 382.366, made in 1962)

Illustration 450
Both watches have the stainless-steel case of the reference 6562. The one on the left is an original with caliber K 1071 (movement no. 562,030, case no. 435,838), while the one on the right, previously seen on page 302, is a *mariage* with the dial of a Chronometer Royal and a caliber K 1072/1 (movement no. 630,032, case no. 382,366). There never was a Chronometer Royal with stainless-steel case.

Illustration 451
Self-winding watches with white-gold cases were not as much in demand in the 1960s as their counterparts in yellow and rose gold. The watches shown with the reference numbers 7397 Q (*left*) and 6394 Q (*right*) were sold on the island of Okinawa (Japan) in 1969. Both timepieces have the caliber K 1072/1 movement (/1 stands for Gyromax balance). The two movements are only nineteen numbers apart (*left*: movement no. 629,594, case no. 436,515; *right*: movement no. 629,613, case no. 447,522). As you can see, the original sales documents have survived.

Illustration 452
The diversity of designs that can occur with automatics of the same reference is illustrated by this comparison using the Ref. 6592 as an example. On the left is the version with a silvered dial, and on the right the model with a gold-plated dial and date. On the latter watch, too, there is only the reference designation 6592 on the screw-down case back, without the usual Q (*left*: movement no. 561,443, case no. 392,541, year of manufacture 1962; *right*: movement no. 570,833, case no. 408,150, year of manufacture 1963). As Vacheron & Constantin discovered, movement no. 570,833 originally belonged in case no. 402,142 and reference 6594 Q. Movement no. 571,333, caliber K 1072, was originally in case no. 408,150 and had the reference number 6592.

Illustration 453
Among the most-sought-after collectibles from the Chronomètre Royal line are these two officially certified Ref. 6694 Q timepieces with unusual lugs. This model was launched in 1965 and remained in production until 1969. The eye-catching cases were made in yellow gold, white gold, and platinum. On the left is a Chronomètre Royal from 1965. The chronometer certification is not present on the dial. However, *CHRONOMETRE ROYAL* and a Maltese cross are engraved on the outside of the screw-down case back. The watch on the right was taken into stock on September 19, 1969, and sold the following month. It is from the last production series of this model. Our picture now shows the later version of the Ref. 6694 Q, which was kept until the end. The Maltese cross is adorned with a diamond, and the inscription ROYAL CHRONOMETER indicates the special value of this watch. The signature on the screw-down case back has been kept in its earlier form (*left*: movement no. 582,341, case no. 407,816; *right*: movement no. 618,947, case no. 440,465). View on the caliber K 1072/1 movement of the diamond-decorated Chronomètre Royal. Almost every detail of this fantastically beautiful automatic caliber contains artistic accents. The engraving on the balance cock explains the action of the *masselotte* weights, ten of which are positioned on the ring of the Gyromax balance wheel. Turning the *masselotte* weight clockwise by 90° increases the oscillation moment (GD^2) and the watch runs slower, but turning it counterclockwise reduces the oscillation moment and the watch runs faster.

Watches with the Caliber K 1120 and K 1121

The advent of quartz watches, which were not only very inexpensive but also were very flat, forced the traditional watch industry, especially the noble brands, to also provide timepieces with mechanical drives and flat cases. As with Audemars Piguet, the newly developed automatic caliber 920 from movement supplier Jaeger-LeCoultre enabled Vacheron & Constantin to construct very flat wristwatches. The new automatic movement, introduced at V&C under the caliber designation K 1120 without and K 1121 with date, had an installation diameter of 1.02 inches (26.0 mm) and a height of 0.09 inches (2.45 mm) and ran at 21,600 vibrations per hour. The version with date indication had an overall height of only 0.120 inches (3.05 mm).

We would like to present some watches with this new but very flat caliber. The automatic Ref. 7592 has a steel case with a screw-down case back. The case back has eighteen corners, with a width of 1.20 inches (28 mm) across flats. We know this design from some flat automatics from IWC (p. 86). The diameter of the case is 1.29 inches (33 mm), the height a full 0.267 inches (6.80 mm; movement no. 608,033, case no. 438,610). With its elegant simplicity, one could well imagine this watch on a lady's wrist.

Here we present another beauty from the house of Vacheron Constantin with a steel case, but with a press-fit case back. The movement number of the caliber K 1121 remains unknown, as does the reference number, because the case remained unopened.

Illustration 454
Elegant, flat, and made of stainless steel. This is how you could introduce this rather small men's wristwatch with the reference number 7512. In the elaborately designed screw-down case, the ultraflat automatic caliber K 1120 performs its services. The case diameter is 1.29 inches (33.00 mm), height just 0.267 inches (6.80 mm).

Illustration 455
Such an ultraflat case depends not only on the height of the movement but also on the design of the screw-down case back. A special wrench is needed to open this eighteen-edge case back, which has a key width of 1.10 inches (28 mm).

Illustration 456
Opening the screw-down back allows one to look inside this elegant watch. The caliber K 1120 uses a Gyromax balance wheel, the balance shaft shock protection system comes from KIF, and the escape wheel shaft is mounted in removable cover plates (movement no. 608,033, case no. 438,610).

Illustration 457
Enlargement of the caliber K 1120 movement. The fine-adjustment mechanism consists of a flat hairspring, Gyromax balance wheel, and KIF shock protection. The escapement wheel shaft is mounted in removable cover plates, also called combi-chucks.

The diameter of this watch is 1.33 inches (33.80 mm), and the height is 0.244 inches (6.20 mm). The silvered dial has indices and baton hands made of 18K yellow gold, each of which is still painted in black, and the "Ó SWISS Ó" label below six o'clock indicating that it complies with APRIOR regulations.

Illustration 458
Like the previous men's watch, this model is also one of the ultraflat versions in stainless steel fashionable in the 1980s. Because of its waterproof press-fit case back, this watch remained unopened. The case number 465,906 is also engraved on the outside of the press-fit case back.

Illustration 459
A special case design also makes possible a waterproof press-fit case back. Although the caliber K 1121 with date is 0.023 inches (0.6 mm) higher than the version without date, the height of this watch is only 0.244 inches (6.20 mm). The sapphire crystal contributes 0.078 inches (2 mm) to this height, leaving only 0.165 inches (4.20 mm) for the watch itself. On the back one can read the case number 465,906.

Chronomètre Royal

The earlier successful Chronomètre Royal model was reissued in 1975 as the reference 2215. The chronometer caliber 1096 was encased in a rather unusual case design. This movement is

Illustration 460
Chronomètre Royal Ref. 2215 in stainless steel from 1976. This model was intended to replace the previously shown reference 6694 Q in the 1970s, but it was not a resounding success. Neither the case design, which was also conceived for a chronometer from Jaeger-LeCoultre developed at the same time (see p. 115), nor the chronometer caliber 1096, which was based on the Jaeger-LeCoultre caliber 906 (movement no. 673.720, case no. 514.442), were convincing.

Illustration 461
After opening the screw-down case back, one can see the details of the chronometer caliber 1096. What catches the eye is not only the lighter rotor carrier bar, but also the massive rotor segment made of 18K yellow gold. In a comparison with the base caliber 906 from Jaeger-LeCoultre, here the operation of the stop-seconds is not done via a poison-green injection-molded part (see p. 116), but via a metal rotating body with a steel *tirette*. Fine adjustment is accomplished by means of a Triovis regulator, with KIF shock protection for the balance staff.

Illustration 462
The Vacheron Constantin Model 222 sports watch Ref. 44018 in steel and gold was not exactly a crowd favorite. Was it just the design of the case and metal bracelet, or did flashy details dominate the overall impression, such as the bezel designed as a fine-thread ring nut? The elaborate but not innovatively designed case was probably the responsibility of a watchmaker who had risen to the rank of designer. The case consists of a cartouche-shaped inner part inserted into the case middle from below. The upper end of this cartridge has a fine thread on the outside, which is then screwed tight by the bezel with internal thread, which acts as a nut. The case middle and the cartouche must be in a precisely predefined position, since the through-hole for the rip-crown shaft on the case middle must meet exactly with the through-hole in the cartouche. Only a rip crown can be used in this complex but not innovative design, because otherwise the cartouche can neither be kept watertight nor removed downward. Did Jacques Ketterer, the owner of V&C at the time, know what a great effort had been made here for a merely average design? According to a Vacheron & Constantin employee, there was little market demand for this model, and thus it was not a success. Furthermore, there would still be about 340 movements, which were never installed, remaining in stock. The reader should also have a look at the original version of this chronometer movement as used by Jaeger-LeCoultre in their own chronometers. In particular, the execution of the second stop with a bilious-green plastic injection-molded part is worth a look (p. 116, fig. 205).

based on the Jaeger-LeCoultre caliber 906, which we described on page 116. But a metal component was installed for the stop-seconds instead of the venomous green injection-molded part. The rotor consists of a weight-optimized carrier to which a segment-shaped gold weight is screwed. No one at Vacheron Constantin wanted to answer the question as to why, on the one hand, this carrier is weight-optimized and, on the other hand, it has to carry a relatively heavy gold weight so that the whole weight optimization process becomes a farce.

Two versions of this reference were released, one in stainless steel and one in 18K yellow gold. These watches rarely show up at auctions. A version with 18K yellow-gold case was hammered at $5,116 at Antiquorum on December 7, 2005, and a steel version sold for $5,803 at Christie's on June 21, 2021. As a now-retired Vacheron Constantin employee told the author years ago, there were still more than three hundred examples of this chronometer caliber, which was not particularly popular, in stock at the time. One wonders if these movements are still available today?

Illustration 463
Close-up of the lug on the case middle of the model 222 by Vacheron Constantin. The last intermediate piece of the metal bracelet is screwed to the case middle via a set screw. At the top, one can still see part of the ring nut, as well as the Maltese cross, Vacheron Constantin's signet, embedded in the case. The aforementioned intermediate piece is made of steel and has only one gold segment (i.e., gold plate) attached. Whether such technical solutions can also save manufacturing costs remains an open question.

Model 222 Sports Watch

For a designer, keeping up with the public's taste at any given time is like walking a tightrope. Vacheron Constantin was once again last in line after Audemars Piguet and Patek Philippe when it came to launching a sports watch with public appeal. After Audemars Piguet had already launched its completely redesigned Royal Oak Ref. 5402 ST sports watch in stainless steel in 1972, Patek Philippe followed suit in 1976 with the Nautilus Ref. 3700/1. It was not until a year later that Vacheron Constantin responded with the new Model 222 sports watch, represented here by the reference 44018/411 in stainless steel and gold.

However, this model was not granted great and lasting success. The stainless-steel version remained in production only until 1981. The steel-and-gold version shown, as well as the full-gold model, were made only until 1985. At that time, when Jacques Ketterer was at the helm of Vacheron Constantin, the Swiss watch industry was already in turmoil because there were no innovative alternatives left against the overwhelming power of quartz watches.

The 222 model was created without any specifications by watchmaker and designer Gerald Genta. Unfortunately, it was not until too late that the watchmaker realized what he had let himself in for with the elaborately designed case. A watchmaker probably sets completely different standards for the design of a functional object than does a graduate of a technical university.

The diameter of the case is 1.45 inches (37 mm); the inner diameter of the bezel, which serves as a threaded ring and thus as a lock nut, is 1.10 inches (28 mm). The height of the case is only 0.275 inches (7 mm). If the caliber K 1121 (i.e., with date) has a height of only 0.12 inches (3.05 mm), this leaves exactly 0.155 inches (3.95 mm) for the case and sapphire crystal (movement no. 521,495, case no. 549,158).

Perpetual Calendar Ref. 43031/3

After Audemars Piguet unveiled the flattest wristwatch with perpetual calendar and automatic winding on July 1, 1978, Vacheron Constantin followed with its own shortly thereafter. Here the automatic movements K 1120 and 2120 were almost

Illustration 464
Vacheron Constantin with Perpetual Calendar Ref. 43031 in 18K yellow gold. Soon after the launch of the first self-winding perpetual calendar by Audemars Piguet (see p. 46) in 1978, Vacheron Constantin followed with a similar model. Here, too, the K 1120 serves as the base caliber. The difference from the Audemars Piguet version is that here a leap year indicator is integrated into the four-year sequence. The ultrathin automatic caliber 920 developed by Jaeger-LeCoultre, which has never been used even in their own house, allowed with the caliber K 1120 and the perpetual-calendar module a total height of only 0.159 inches (4.05 mm). Thus, the case diameter is 1.41 inches (36.00 mm), but the case height is only 0.275 inches (7.00 mm).

identical, but the module from the Geneva noble forge still had a leap year display with the four-year sequence on the dial. The model presented here, Ref. 43031/3, was purchased via eBay on February 8, 2007, from a Berlin antique-watch dealer for exactly $12,228 (movement no. 689,935, case no. 626,388).

This price was a good one for that time. A comparable model was sold on May 14, 2006, at the auction house Antiquorum in Geneva for 23,600 CHF (movement no. 690,147, case no. 611,733). On the basis of the rate of exchange at that time, that was about $16,000. Since the two movement numbers are only 242 digits apart, both watches are probably from the production year 1988.

The caliber K 1120 QP has thirty-six jewels and operates at 21,600 A/h. It is regulated in five positions and temperatures. The perpetual calendar is set by three auxiliary pushers at two, four, and ten o'clock. In recent years, auction prices for the reference 43031 in 18K yellow gold have ranged between $10,800 and $12,000 and thus have changed little even after more than ten years. The possibly comparable reference 3940 model from Patek Philippe is usually three times as expensive. Even if one would grant an additional bonus of 100% to this competitor's watch because of the Patek Philippe 240 Q automatic movement with micro motor, the price would be only twice as high.

Here, too, the phenomenon of the undervaluation of Vacheron Constantin's classic watches reappears, for which no objective reasons can be found. Actually, only the collector of beautiful vintage watches can be happy about this, because they do not have to dig as deep into their pocket for a watch that technically is almost equal. Even for the author, vintage models from Vacheron & Constantin are still in no way inferior to those from its competitor Patek Phillipe.

Automatic Ref. 46014/47014

With the appearance of the 889/1 automatic movement newly developed by Jaeger-LeCoultre, with a movement diameter of 1.02 inches (26 mm) and height of 0.127 inches (3.25 mm), 28,800 A/h, thirty-three jewels, date display, and forty-hour power reserve, completely new possibilities also arose at Vacheron Constantin. Here we present the reference 460/47014, the first model in which the new caliber 889/1, designated 1126 at Vacheron Constantin, was installed. In the 1990s, men's watches with a case diameter of just 1.33 inches (34 mm) were quite en vogue. The case height reached just 0.255 inches (6.50 mm),

Illustration 465
The automatic reference 47014 also marked a new beginning for Vacheron Constantin, because the new 889/1 automatic caliber from the traditional movement supplier Jaeger-LeCoultre in Le Sentier was used here for the first time. With a case diameter of only 1.33 inches (34.00 mm), this model did not appeal to all potential buyers at the time.

Illustration 466
With the case back removed, one can see Jaeger-LeCoultre's automatic caliber 889/1. A possible steel insert in the case back is conspicuous. For a material expert, the suspicion arises that a steel inner case back was chosen for stability reasons to keep the proportionate gold case back thinner. The hallmarking with the small Helvetia and 18K/750, however, suggests that this inner base was solid but still postgilded in white gold. The case back is fixed to the case middle with four cylinder-head screws. Considering the diameter of the movement of only 1.02 inches (26.00 mm) and that of the case with 1.33 inches (34.00 mm), there is only 0.105 inches (4 mm) left radially from the movement to the edge of the case.

allowing this watch to still fit under any shirt cuff. It is also worth mentioning that Jaeger-LeCoultre launched its new Master Control 1000 Hours series at almost the same time and entered into direct competition with Vacheron Constantin. These much-larger watches with 1.45-inch (37 mm) case diameters made a much-better impression than the smaller ones from Vacheron Constantin and were also much more affordable.

In May 2021, a Vacheron Constantin Automatic Ref. 47014 in 18K yellow gold was listed on eBay USA at $8,500. If it had been purchased in Germany, the addition of shipping, customs duties, and import sales tax would have meant a total price of around $9,500. At the same time, a Jaeger-LeCoultre Master Control 1000 Hours Ref. 140.1.89 in 18K yellow gold was also offered, for which the German seller wanted a Buy Now price of $6,500. In addition to the price difference of about $1,500, the difference in case weight should be noted, which is certainly almost 50% higher for the Jaeger-LeCoultre model. What intentions the marketing department of Vacheron Constantin had at the time with this model series can no longer be understood today. This almost dainty Vacheron Constantin automatic Ref. 46014/47014 was not a resounding success.

Illustration 467
Approximately 5× enlargement of a section of the case back interior. The company logo "VACHERON & CONSTANTIN" is stamped on the top. Although the "&" has been deleted from the company name for many years, it is still found in this case back. The numbers for reference 46014 and case no. 591,153 are manually stamped.

Automatic Ref. 87172

A watch that instantly pleases doesn't have to have spectacular features. The self-winding Vacheron Constantin Traditionelle in 18K white gold, Ref. 87172/000G-9301, has a tidy dial with a small seconds at nine o'clock and a date window at three o'clock. The dauphine hands for hours and minutes are made of 18K white gold, as are the indices and the Maltese cross. The case has a diameter of 1.49 inches (38 mm), and the height is just 0.314 inches (8 mm). The case consists of bezel, case middle, and screw-down case back. The latter has deep-drawn grooves as well as knurling applied to the circumference. Both mean a great effort in terms of manufacturing technology and are to be emphasized as special-quality features. Here is a brief explanation: most watch manufacturers who offer cases with screw-down backs have the six grooves milled into the back by using a dividing head and end mill. This is a fairly simple manufacturing process, which means an additional but simple step in the production of this screw-down back in the machining center.

Illustration 468
Approximately 5× enlargement of the fine-adjustment mechanism of the automatic caliber 2455. Instead of the Gyromax balance wheel and Breguet hairspring formerly used, now a simple flat hairspring is used, as well as a Triovis regulator. The shock protection comes from KIF.

Deep-drawn grooves require a completely different manufacturing step. An appropriate tool is used, and the six grooves are produced under high pressure by shape-forming material displacement (i.e., a noncutting process). In earlier times OMEGA's water-resistant Constellation and Seamaster chronometers had screw-down backs with deep-drawn grooves. The knurling on the screw-down case back is also worthy of mention. Once you have loosened a screw-down case back that has a knurl with the help of a wrench, you can easily and simply unscrew the rest of the threads by hand, thanks to the knurl. But why the designer of this watchcase wanted seven grooves for the screw-down case back instead of the usual and more sensible six, no one could explain to the author. These seven grooves are also not convincing from a design point of view and can be called absolute nonsense because they are of no use to anyone but cost more to manufacture. Concessionaires offer this watch for about $27,000. For this high price, the automatic movement should also be first class. Unfortunately, this is not the case. This watch's regulator arm is quite simple. Instead of the Gyromax balance wheel with the formerly used *masselotte* weights, the rate is set by a simple TRIOVIS regulator. The winding rotor winds in only one direction, so there's no need for a changer. Finally, this automatic movement uses a simple flat hairspring instead of a Breguet hairspring.

Illustration 469
This automatic can be called an understated design for connoisseurs. The uncluttered dial, the clean design, and the skillfully designed case initially suggest a top place on the list of the Swiss watch industry. The screw-down case back with deep-drawn grooves and knurling also supports this assumption. The first disillusionment was the screw-down back with seven instead of six grooves. The automatic movement also put a damper on our initial enthusiasm. It uses a flat hairspring and a Triovis regulator instead of the former Breguet hairspring and Gyromax balance, winds the rotor in only one direction, and doesn't have the stylish movement design that can still be seen on the competition from Patek Philippe. For the list price of about $28,000, you can get a comparable automatic, but with full calendar and moon phase, for example, at competitor Jaeger-LeCoultre. Vacheron Constantin should also think about this thoroughly.

The caliber 2455 has a diameter of 1.03 inches (26.20 mm), a height of 0.141 inches (3.60 mm), and twenty-seven jewels and beats at 28,800 A/h. The movement consists of 194 parts, and power reserve is forty hours.

If one compares this caliber 2455 with the previous and outstanding caliber K 1072/1, the earlier and the new caliber have almost renounced all former virtues. As already mentioned in the presentation of Vacheron & Constantin's automatic calibers, the K 1072/1 has all the good features that such a movement should have: a fine-adjustment mechanism with Breguet hairspring and Gyromax balance wheel, a changer for bidirectional winding, a rotor with gold segment, and an escapement wheel bearing with removable cover plates.

Ceramic ball bearings for the rotor are sometimes useful but by no means innovative and advantageous, as was heard from a large, well-known German ball-bearing manufacturer.

Illustration 470
The manually wound chronograph Ref. 47192/000B-9352 with 18K white-gold case and hands, indices, pushers, and crown in rose gold always leaves a special impression with its 1.65-inch (42 mm) case. The author himself has not yet experienced this impressive "time recorder" on someone else's wrist. Perhaps the list price of around $50,000 was too high for most interested parties. In the secondary market, this version in 18K white gold is already offered from around $30,700 (i.e., with an almost 40% discount from the list price).

Ref. 47192 Manually Wound Chronograph

Chronographs have always had a great tradition at Vacheron & Constantin. Time recorders—as chronographs are also called—with a water-resistant case were first introduced with the reference 6087 in 1955. Vacheron Constantin did not want to miss this trend toward large watches and, at the beginning of the second decade of this century, launched the manual-winding chronograph Ref. 47192 with white-gold case and hands and rose-gold pushers and crown. The case diameter is 1.65 inches (42 mm), and running inside it is the manually wound chronograph caliber 1141. This caliber is based on the Lemania movement 2320 and still beats at the classic 18,000 vibrations per hour. It has a diameter of 1.08 inches (27.50 mm) and a height of 0.220 inches (5.60 mm) and contains twenty-three jewels. The whole movement consists of 164 individual parts; power reserve is forty-eight hours.

Unusually the screw-down case back has seven grooves, in terms of production technique again high-quality deep-drawn with knurling as well. The sapphire crystal back reveals the chronograph caliber 1141. The two rectangular pushers were probably a marketing decision, although there is no

Illustration 471
A look through the sapphire crystal of the screw-down case back of the chronograph Ref. 47192 reveals to us the details of the chronograph caliber 1141, which is based on the Lemania 2320 movement and was released in 2009. While the watchcase comes to 1.65 inches (42 mm), the diameter of the caliber 1141 is only 1.08 inches (27.50 mm or 12'''). Noticeable on the screw-down case back with commendable knurling are again the seven deep-drawn grooves, which make little sense from a watchmaker's point of view. With the well-known case opener for amateurs with three adjustable pins, this screw-back can also be opened without any problems. Interestingly, the story about the seven grooves probably did not come to the attention of any of the editors who wrote about this chronograph.

visual advantage to be seen. From the point of view of technical mechanics, flat pushers have no functional advantages, but in most cases they are much more expensive to manufacture and seal.

Finally, there remains the question of the sense of seven instead of six grooves. The fact that no editor of a watch magazine has noticed this technical nonsense is probably due to their limited knowledge in the field of technical mechanics. But this fact will also seem completely insignificant to most lovers of Swiss watches.

We should take a look at the commercial side of this chronograph. Its retail price at the concessionaire is about $49,000. Several of these watches are on offer on the secondhand market, and the cheapest as an immediate purchase was about 40% below the previously mentioned list price. Such examples cannot be found in watches from Patek Philippe. A Patek Philippe Ref. 5170G-001 model comparable in size and function was offered on the internet starting at a price range of only $58,500, which is close to the original price. The managers responsible for sales and marketing at Vacheron Constantin should soon start thinking about how and when they can reach the quality and price level of the aforementioned competitor from Geneva again. Watch collectors would certainly thank our brand presented here.

Nostalgia in Closing: Automatic Ref. 6441

At the end of our picture gallery we present a special classic model from Vacheron & Constantin that even the author has seen only once, but that is a collector's item worthy of notice. One of the special features is the design of the case, here more precisely the top view. It is modeled after the reference 4600 Jubilé and has a diameter of 1.41 inches (36 mm; see also p. 263). The immediately recognizable difference from the reference 4600 is the center seconds display.

The other difference is the automatic movement and the screw-down case back, which makes this watch water resistant. Since all other automatics listed here have a case diameter of 1.37 inches (35 mm), this reference 6441 with a 1.41-inch (36 mm) case is considered to be a special one.

This watch uses the caliber K 1071 movement.

This rare collector's item may be offered more frequently in Italy. And so the hunt is on!

Illustration 472
Vacheron & Constantin Ref. 6441 Automatic with caliber K 1071 (movement no. 561,495, case no. 390,201). Seen from above, the case, measuring 1.41 inches (36 mm) in diameter, is exactly the same as the Ref. 4600 Jubilé model. Only the back is modified and features a solid screw-down case back. While the Ref. 4600 comes to a case height of only 0.322 inches (8.2 mm), most automatic watches from Vacheron & Constantin are around 0.472 inches (12 mm), as in this case.

Illustration 474
The back of the Ref. 6441 shows us the screw-down case back with a dodecagonal and 1.16-inch (29.5 mm) wrench size. The collective responsibility mark No. 1 with the hammerhead and the number 136 tells us the case supplier C. R. Spillmann SA, Chêne-Bourg, which went out of business on April 5, 1988.

Illustration 473
Juan-Carlos Torres, CEO and head of Vacheron & Constantin until 2017 (*center*), together with Rudolf Büll, his specialist in classic watches (*left*), in the reception hall of the new factory inspecting a guilloché machine from 1884. The author is on the right.

Illustration 475
Potpourri of six watch scenes, some of which only the watchmaker can see

8. Buying and Overhauling Special Wristwatches on the Basis of Concrete Examples

8.1. How It Used to Be

How would a normal contemporary have reacted about half a century ago if he had inherited a valuable gold wristwatch—for example, from the noble house of Patek Philippe in Geneva—from his father? In one case this valuable timepiece would have been stowed away in a safe without much notice. In the other case, when the focus was more on cash, the unemotional son would certainly have targeted the nearest competent watch store and inquired about the sale value of his heirloom. At that time, lack of interest, on the one hand, and widespread ignorance, on the other hand, often still allowed the dedicated collector to find real "bargains" here or there and to acquire a sought-after watch very cheaply. These times belong to the past. Nevertheless, there are still ways and means of finding the coveted collector's item and acquiring it relatively cheaply. The fact that a value-preserving overhaul still has to be taken into account when buying is often forgotten. How this can be done in practice is described with the help of a few concrete examples.

8.2. Proper Preparation

Once one has made up their mind to purchase an old noble wristwatch, one should first consider whether this investment should be regarded as a one-time decision or the beginning of a new and momentous hobby.

Because the latter—as experience teaches us—can rarely be ruled out, the potential buyer should already be clear about what they are getting into in the future. Building up an extensive Swatch collection will certainly not lead to personal ruin. With timepieces from Swiss noble brands, however, it can have the opposite effect, because a single noble watch can cost the equivalent of one or two net salaries of a well-paid manager. Before the desired piece of jewelry is purchased, there are still some questions relevant to one's decision that must be clarified:

- Brand and model (here we want to consider only noble Swiss manufacturers).
- Technique and case material (whether steel or precious metal, automatic or manual winding, whether small or large complication)
- Appearance and size (applies to dial, case, case back design, lugs, etc.)
- Purchase price
- Subsequent investments for preservation of value (overhaul costs, replacement strap and buckle, etc.)
- Value estimation, market position, and further price development

Once one has clarified these questions, discussed them with like-minded people or dealers, and narrowed down the target range of one's wishes, one can get down to business.

8.3. Sources

With watches, the process is not unlike that of buying a used car. Not only the search for the object of desire is in the foreground, but also the comprehensive comparison of different offers. But where can these different offers be found? The most obvious possibilities are as follows:

- Local dealers in old watches (in Munich alone, there are more than five stores selling used noble wristwatches)
- The relevant watch fairs in Germany, Switzerland, and the USA (whereby the Munich Watch Fair is one of the most important events of this kind)
- National and international auction houses (Antiquorum, Christie's, Dr. Crott, Phillips, Sotheby's, etc.)
- Advertisements in the daily newspapers, in specialized magazines, and on the internet
- internet auctions (e.g., market leader eBay)

Whereas the first two sources still provide one the opportunity to personally inspect the piece of jewelry you are looking for and discuss its properties, origin, and price with the seller, with the other three sales channels these possibilities either do not exist at all or are limited. Liability for hidden defects or the right of return are often excluded, so that the unencumbered buyer can only hope for luck. This increased risk is often offset by significantly lower prices, so that a comparable assessment can be made only at the end.

Certainly all theoretical advice, study of the relevant literature, and expert suggestions from third parties are of great benefit for upcoming watch purchases. But we wanted to be more certain and explain to our readers with the help of some practical examples about how one proceeds with these watches and what one experiences.

8.4. The Used Wristwatches Purchased and Discussed Here

In selecting the new acquisitions described here, care was taken not only to ensure that current public taste was met, but also to report on special features and curiosities. Regardless of the description of the watches by the respective buyer, they were subjected to a comprehensive inspection. The dial, case, and movement were thoroughly examined, and this always included a rate test. It is true that most watch collectors have neither the necessary prior knowledge nor the testing equipment required for acceptance tests. After studying this book, our readers will know much more about what to look for when buying a watch and when and where it is better to leave what initially appears to be an inexpensive purchase alone.

8.5. Audemars Piguet Automatic Ref. 5112 BA (BA Stands for 750 Yellow Gold)

To say it in advance: self-winding men's wristwatches that were manufactured by Audemars Piguet in the fifties and sixties are extremely difficult to find. Even a look at the relevant literature cannot help much here. The books *Automatik Armbanduhren* and *Audemars Piguet* (both from Callwey Munich) and the Chronos special *Audemars Piguet* (Ebner Ulm), for lack of other options, include very few of these watches, and for the most part the same ones. Only in the March 2003 issue of *Chronos* (Ebner Ulm) was this topic taken up by its author, who provided some insight into that almost unknown era of the fifties and sixties.

But the presentation of this collector's item is a premiere in itself, because an AP Automatic with the rare caliber K 2070 movement has never before been illustrated and described in a published work. This is somewhat surprising to the expert, because the precious timepieces from Audemars Piguet from the village of Le Brassus in the Jura may be mentioned in the same breath as the other two noble brands, Patek Philippe and Vacheron & Constantin of Geneva. Certainly we want to know why AP watches with the automatic caliber K 2070 are so rare. Beginning in 1953, AP purchased self-winding raw movements (*ébauches*) from Jaeger-LeCoultre in Le Sentier, just 1.8 miles (3 km) away. The first two calibers were designated P 2498 (small seconds at six o'clock) and P 2499 (center seconds). Unlike the *ébauches* supplied to Vacheron & Constantin (caliber P 1019), which had heavy metal oscillating weights, the calibers for AP were equipped from the outset with a rotor whose supporting plate was decorated with Geneva stripes and had an elaborate guilloché 18-karat gold segment. In the subsequent caliber, introduced in 1959, the basic design was largely retained, but the pawl was arranged on a separate bridge and screwed to the barrel bridge from below.

Illustration 476
This timelessly beautiful Audemars Piguet Automatic Ref. 5112 BA from 1959 appears almost like new. The original dial is set off in two colors, with hands and indices in 750 yellow gold. AP watches with the caliber K 2070 with small seconds were produced only in small numbers, and for years they have been impossible to find.

Illustration 477
The caliber K 2070 movement with twenty-nine jewels. The rotor is supported by four ruby rollers, and the rotor carrier plate is decorated with Geneva stripes. In the foreground is the fine-adjustment mechanism with screw balance, micrometer screw, and gooseneck spring.

Illustration 478
Approximately 4× cutaway magnification of the AP caliber K 2070. Engraved on the automatic bridge is TWENTY NINE (29) JEWELS, ADJUSTED TO HEAT COLD ISOCHRONISM AND FIVE (5) POS.

The most obvious innovation concerned the rotor bearings. Instead of four steel ball bearings, which were later replaced by sapphire bearings, four ruby rollers were now installed. This innovation gave the rotor greater stability and reduced wear. The new movement family included the calibers K 2070 (small seconds at six o'clock), K 2071 (center seconds), and K 2072 (center seconds and date at three o'clock). The last version delivered by Jaeger-LeCoultre had a Gyromax balance instead of the previous balance wheel with micrometer screw and gooseneck spring and was designated K 2072/1.

Illustration 479
From the side, one can see the elaborate case with screw-down case back and case number 17 756. In contrast to the watches from Patek Philippe and Vacheron & Constantin, the case number was always stamped on the outside of Audemars Piguet watches, and the reference was always omitted. If the movement and case number of a watch are available, the reference can be determined by the manufacturer.

AP timepieces with the former caliber K 2070 movement could be found neither in the literature nor in auctions held in recent years. One of the few exceptions is the AP Ref. 5112 in 750 white gold and company signature Guebelin, which was sold under lot no. 18 at the forty-seventh auction by Dr. H. Crott on November 13, 1993. The auctioneer was probably not aware of the rarity he offered for sale. One wonders if the lucky buyer knew it?

The AP presented was the result of years of research. American Rick Gilbert, coauthor of *The Complete Price Guide to Watches* and dedicated watch collector, offered this rarity for sale on the internet. However, his initial asking price of $3,850 drew little interest. After a lengthy back and forth, he and a potential buyer agreed on $3,250, so that this rare AP, including transport and import VAT, came into the hands of the new owner for about $3,526. The automatic was in almost-new condition, but the case surfaces already exhibited some reddish tarnish (formation of gold oxide). The new owner ordered an overhaul of the movement and case to preserve the value of the watch. This overhaul cost the new owner about $360. Due to the excellent condition of the dial, hands, and movement, no parts had to be replaced. Only the crocodile leather strap, on which the original gold buckle was still mounted, was replaced. To determine the history of this watch, Audemars Piguet in Le Brassus was contacted and asked for verification. Christoph Guhl, one of the house's distinguished technicians, confirmed the authenticity of this automatic and reported that this Ref. 5112 with caliber K 2070, movement number 74 754 and case number 17 756, had been manufactured on May 16, 1959. He also advised that the reference 5112 had been launched in 1956 and was initially equipped with the predecessor caliber P 2498. The total price of around $3,960 for a timelessly beautiful watch that cannot even be found in Audemars Piguet's museum should certainly be a justifiable investment, and an affordable and promising one at that!

8.6. IWC Automatic Yacht Club Ref. 811 AD in Steel

The timepieces from Schaffhausen on the Upper Rhine, also known for the famous Rhine Falls near Neuhausen, are among the most-popular collectible watches, especially since you can often still obtain high-quality technology at a reasonable price.

Among the most-famous models of this fine manufacture is our Yacht Club. Not only the name of this watch, but also its robust and durable design suggested sportiness and dynamism to its wearer. With a production of more than 45,000 pieces, about 10% of which were gold-cased, the Yacht Club represented IWC's most widely built men's automatic at the time. In addition to the case variants in stainless steel and yellow and white gold, there were also different dial variants. The dials of the steel versions were either silver-plated or coated with blue and black enamel. On the white- and yellow-gold versions, the dials were either silvered or gold-plated (champagne colored). The models in stainless steel and 750 gold were listed under the reference

Illustration 480
The IWC Yacht Club Ref. R 811 AD in steel after its arrival from the USA. The case has been unprofessionally refinished, so even the spring bar holes are almost "in the wind." The dial is not original and came from a gold-cased watch. The crown was not made by IWC.

Illustration 481
The rear of the case with the back unscrewed. The case middle is heavily corroded in the area of the seal. The seven rubber segments for elastic mounting of the movement in the case are missing and have been replaced by a brass section bent to fit.

Illustration 482
In addition to the case number 1929857, the complete reference designation R 811 AD can be found on the inside of the screw-down case back.

Illustration 483
Our IWC Yacht Club could be called brand new after its overhaul at IWC. During this overhaul the dial and hands were replaced, as was the case middle with bezel and crown.

Illustration 484
The movement is also barely recognizable. Many parts were replaced, so that here, too, one can almost speak of like-new condition. Typical for the automatic caliber C.8541 B movement is the fine-adjustment mechanism with eccentric fine adjustment of drop and rate.

Illustration 485
The case back underwent a professional refurbishment, though it was impossible to remove every trace of corrosion.

designation R 811 AD, the *mixte* (mixed) version with case in 585 yellow gold and steel back under R 911 AD. After the reorganization of the company's reference designations in 1971, the Yacht Club line was designated R 1811 AD or R 1911 AD. Our Yacht Club R 811 AD in stainless steel was discovered at the antique-watch dealer Alec Stromberg in Los Angeles, California (www.strombergtime.com). The Yacht Club, which was offered on the internet, did not find a buyer at first, despite its low starting price of $899. Subsequently an agreement was reached on the purchase price so that this timepiece, including freight and insurance, changed hands for $855 (approx. 775 €). Including import tax and customs duties, the purchase price came to about $976, a fair market value. However, the anticipation of this new acquisition was quickly dispelled when the watch was subjected to the usual inspection upon its arrival. The seller's assessment, "overall condition very good with evidence of regular wear," unfortunately did not apply in all relevant areas. The dial, which had gold indices and was quite worn, did not match the steel case. The latter had been refinished several times, although not professionally. The condition of the caliber C.8541 B movement turned out to be a minor disaster. Instead of the seven rubber segments for the elastic and shock-protected mounting of the movement, a small brass profile was used, which was bent accordingly. The self-winding movement itself was in blatantly bad condition. The almost scrap-worthy automatic was personally handed over to the production manager of IWC Schaffhausen, Ronald Jäger, and together with the head of the service department, René Schwarz, the intended overhaul was discussed. As part of this complete overhaul, the dial with hands, as well as the case middle, including the bezel, were replaced, and the movement was restored to almost-new condition. Also the good movement performance has to be mentioned, which again achieved the strict chronometer values from the sixties. For this exemplary work, including pedigree book, 935 SFR (approx. $660) had to be paid.

Adding to these costs the purchase price and the additional outlay for the leather strap and steel buckle, the investment came to a total of around $1,736. On the one hand, this is a lot of money for a wristwatch that is more than thirty-five years old, but on the other hand, it will be a faithful companion for coming decades, since you can certainly no longer find new watches at this price with these technical features.

No description can convey the condition this watch was in before and after its overhaul as clearly as the photo comparison shown.

8.7. Jaeger-LeCoultre Memovox Date E 855 Automatic in Steel

Back in the 1950s and '60s, Jaeger-LeCoultre was already one of the big names in the watch business. The manufacture in Le Sentier was primarily known as a manufacturer of movements, but it also produced numerous wristwatches itself. In addition to the well-known Futurmatic models and the automatics with power reserve display at twelve o'clock, the various Memovox versions were among the most-successful timepieces of that era.

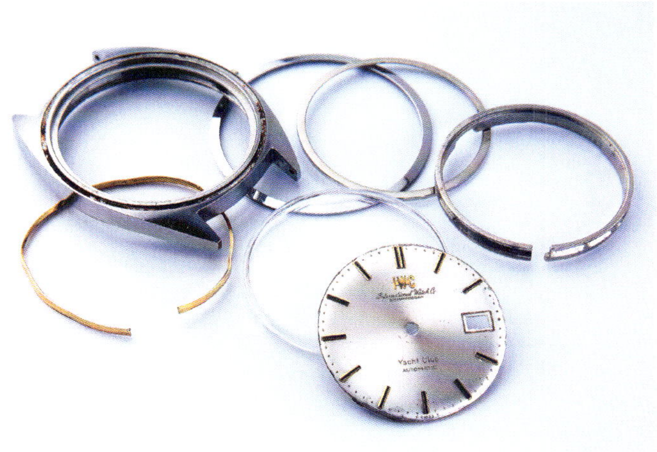

Illustration 486
This photo reveals which parts of the case had to be replaced during the IWC overhaul. At the bottom left is the bent brass section that had replaced the seven rubber shock-absorbing elements.

Here we present a Memovox Date Ref. E 855 from the late fifties. As part of the preparatory work for a documentary on Jaeger-LeCoultre men's wristwatches, we also looked internationally. Among the various search results is this Memovox Date, which has some special features. First, this reference E 855 is a Swiss product. This is also confirmed by the "T Swiss Made T" label on the dial below six o'clock. In this case, "T" stands for tritium, a nonradioactive luminous material. On the other hand, this Memovox has the coveted black dial as well as logo and company signature at twelve o'clock. What did the Memovox have ahead of other watches, and why was it so widely sold? Besides the time, the Memovox had an alarm function, which could be used in many ways. Whether as a reminder of an important appointment, as a parking-time reminder, or simply as a morning alarm clock, the Memovox made itself almost indispensable in everyday life. The Memovox presented originates from the USA and was acquired in Los Angeles during a business trip by a watch collector. Since no suitable case key was at hand, this alarm clock changed hands unopened after no more than a functional and visual inspection. The purchase price after a lengthy discussion was $850. A thorough inspection of this new acquisition could be made only after returning to Munich. The watch, identified as a reference E 855 with the automatic caliber K 825, movement number 1 936 863, and screw-down case back (case number 1 131 186), was examined more closely. In contrast to its visual appearance, the automatic movement with alarm function showed signs of heavy wear. This was not due to constructional or functional reasons but was probably the result of decades of use without regular, value-preserving maintenance. Although the caliber K 825 movement, with limited oscillating weight, colloquially known as a hammer movement, is considered to be particularly easy to maintain and very robust, even the best design cannot withstand decades of use without the prescribed lubricants, various oils, and greases. Accordingly, the overhaul carried out by JLC Germany, at that time still in Nuremberg, was unsuccessful. One of our photographs of this watch documents the many individual parts and components

Illustration 487
Jaeger-LeCoultre Memovox Automatic Date Ref. E 855 with the sought-after black dial and Jaeger-LeCoultre logo and signature at twelve o'clock. Whether as a parking-time reminder or morning alarm clock, this Memovox was always needed.

that had to be replaced during the factory overhaul. Many items had to be requested from the head office in Le Sentier, causing the work to drag on for almost four months.

The basic overhaul cost was $564, the various replacement components $507, and the crocodile leather strap with steel buckle $143. In all, $1,215.56, which added up to an impressive $1,409 after adding VAT. After the Memovox was returned, we took a closer look at it and checked the mechanics with a stereo microscope, and the rate with the Witschi Mikrometre Professional rate tester. The results showed once again the high watchmaking level achieved by Andreas Böhm, who supervised this overhaul,

Illustration 488
The automatic caliber 825 is still wound by means of a limited oscillating weight, also called an automatic hammer in colloquial language. Next to it you can see the case back and the five-grooved ring nut. On the pin, which is welded to the inner base, the hammer of the ringing mechanism strikes and produces the alarm tone via the base as a resonance body (movement no. 1,936,863, case no. 1,131,186).

and his colleagues at JLC Germany. But no evaluation without a final total. The overhaul costs must be added to the purchase price of about $770, so that the total investment comes to $3,038. For some watch lovers, this already raises the question of whether an overhaul, whose cost is almost 1.8 times the purchase price, still seems justified. Certainly a local watchmaker will not offer such horrendous prices and may charge only $160 to $220 for an overhaul. In most cases this should also be enough, if the movement is only cleaned and supplied with the necessary lubricants. After all, collectible watches spend most of their time in the display case or in a safe and are not worn on the wrist every day.

8.8. Patek Philippe Automatic Date Ref. 3445 in 750 Yellow Gold

When talking about noble old wristwatches, one can confidently give a top place to the valuable timepieces from the house of Patek Philippe. Many of the models launched in the 1950s and 1960s are now sought-after collectors' items of ever-increasing value.

The era of self-winding wristwatches with date display began at PPC in 1961, with the introduction of the reference 3445. A total of four different models were subsequently equipped with the automatic caliber 27-640 M (M for Monodate; references 3445, 3517, 3541, and 3558, not counting the IOS reference 3565). While only 750 yellow and white gold, as well as platinum, were used for the cases, there were a variety of dial versions. The watches came with silvered, gold-plated, or lacquered (blue, black, gray, etc.) surface finishes, and indices or Roman numerals, and in a few cases also Arabic

Illustration 489
The reverse side of our Memovox. Around the case, one can see all the movement parts that were replaced during the overhaul (or perhaps more accurately, restoration). Among them are bridges, barrels, and a lot of other components that can be found only in the spare-parts store at JLC in Le Sentier.

numerals provided the hour markers. One of the most beautiful "faces" is probably the silvered-dial version with black Roman numerals presented here.

Searching for this special variant, one could not find it on watch exchanges or auctions. At a Munich watch fair, a Viennese antique-clock dealer had a version in white gold with a black dial and light-gray Roman numerals on offer, but the $10,300 he was asking was well above my intended budget. An internet search finally led to the desired success. A pawn shop in Mönchengladbach was offering a yellow-gold Ref. 3445 with Roman numerals. The handicap of this watch was the dial, which had been damaged at six o'clock and only amateurishly repaired. For only $4,025—a real bargain—the piece changed hands and was immediately passed on to the German branch of Patek Philippe in Munich. Under the watchful eye of Birte Zweiniger, the head of the atelier there, this Ref. 3445 received a comprehensive overhaul. As part of this work, various parts were replaced, most notably the dial, in this case by an original from the 1960s. Including the preparation of a pedigree book extract, the overhaul costs amounted to $1,260 plus the usual 16% VAT. With a new band, Patek Philippe gold clasp, and box, the total cost came to exactly $6,203. A truly good price for a Patek Philippe Ref. 3445 in yellow gold, especially for one in mint condition with such a rare dial.

Illustration 491
Among the most-beautiful automatic movements ever made in Switzerland is the Patek Philippe caliber 27-460 M (M for Monodate). Its gold rotor, very efficient winding via eccentric wheel and pawl lever, Gyromax balance, and power reserve of more than fifty hours are only some attributes of this excellent design. On this movement number 1 118 030, notice the balance cock with adjustable hairspring stud. This design was used only in the last production series, with the movement number 123 XXXX. On this occasion the caliber was given the designation 27-460 M PM (PM stands for *piton mobil*, French for "adjustable hairspring stud carrier"). With this design, the drop (i.e., the symmetry between balance, pallet, and escape wheel) can be established without loosening the Breguet hairspring.

8.9. Vacheron & Constantin Automatic Date Chronometer Royal, Special Version

While searching the internet for special wristwatches from this noble Geneva watch manufacturer, I came across an offer that initially forced me to think. It was a Chronometer Royal in a simple stainless-steel case. In addition, the automatic caliber K 1072/1 lacked the golden oscillating weight! So the automatic winding mechanism was not working anyway. Despite this handicap, which was clearly described in the offer, the watch sold for about $3,800. Obviously the buyer had not understood the reference to the missing oscillating weight correctly and later canceled the purchase because the watch's winding mechanism did not work. To get to the bottom of this matter and the history of this watch, after lengthy discussions this curious piece was acquired for $2,300, but not without having been granted a right of return. Including freight, customs, and import sales tax, there was about $2,495 to be invested. The

Illustration 490
The Patek Philippe Automatic reference 3445 was unveiled in 1961 and remained in the program until the early 1970s. The version with the silvered dial and Roman numerals is relatively rarely found. The same is true of the visual and technical condition of this noble timepiece.

Illustration 492
Reverse of the reference 3445, with ten-sided screw-down case back and Patek Philippe international service emblem

usual inspection quickly revealed that this new acquisition was a *mariage*, albeit a rather strange one! A query to Vacheron & Constantin in Geneva quickly clarified the matter. The automatic movement caliber K 1072/1 with movement number 630 082 originally belonged to a yellow-gold Chronometer Royal reference 6694 from 1971, while the steel case with number 382 366 originated from an automatic with date reference 6562, which was produced in 1962. What was going on in the mind of the person who made this mysterious manipulation? Was he after the solid gold in the case and oscillating weight?

Despite all the limitations, this *mariage* was entrusted to the hands of longtime Vacheron watchmaker Martin Becker, and an overhaul of the movement and case was commissioned. Because the aforementioned golden oscillating weight is no longer available, one was taken from a rotor of an AP automatic caliber P 2499 on hand in the spare-parts warehouse and screwed onto the rotor bearer plate of the Vacheron caliber. Since in the fifties, the sixties, and even still in the seventies both AP and V&C purchased reserved raw movements (*ébauches*) from Jaeger-LeCoultre, the comparable calibers were identical in many respects. Only two obvious differences should be highlighted. Some of the Audemars Piguet movements were regulated in only four positions, while the V&C movements were always regulated in five positions. The AP rotor bearer was decorated with Geneva stripes, while the V&C had a sunburst finish. The overhaul of the *mariage* cost $360. Besides the golden oscillating weight, some screws and small parts also had to be replaced but were included in the previously mentioned price.

As with all the other watches, a balance sheet must be drawn up. Was the $2,855 paid for the purchase price and overhaul of a *mariage* at all justified and worthwhile? A professional dealer of noble old watches could possibly answer this question in the negative, while a collector, who perhaps might like to understate the case, would probably have his justified joy in it. The dial and movement from a Chronometer Royal, safely encased in a modest-looking but nevertheless noble steel case—wouldn't this perhaps be the secret wish of many a sovereign watch lover? After all, we all are familiar with the slogan "Outside steel, inside gold," which was the motto of Audemars Piguet's Royal Oak sports watch launched in 1972.

A gold Chronometer Royal Ref. 6694 could not be found at auctions or watch fairs in recent years. A steel automatic/date Ref. 6562 has been offered several times by a reputable vendor at the Munich Watch Fair without finding a buyer at an asking price of $5,967. Finally, to give our readers an idea of what the 6694 Chronometer Royal in yellow gold and 6562 automatic in stainless steel looks like in real life, we have included photos of all these watches.

8.10. Rolex Oyster Day-Date Ref. 1803 in Yellow Gold

Opinions about this watch are always divided. Some are proud of their dream watch and obviously like to flaunt it. Others cannot afford this timepiece and condemn it on every suitable and unsuitable occasion. For those in the third group, the modest connoisseur, only a Day-Date with a white-gold or platinum case comes into question. This is the only way they see their extraordinary taste confirmed. In preparation for this article, we looked for a suitable Day-Date and finally found a favorable opportunity on the internet. Here the seller, a collector from Berlin, not only advertised a newly overhauled watch but, in particular, emphasized the top condition of this item. The bid of $3,810 was accepted, and a few days later the watch was sold. Visually this Day-Date was in very good condition. The

Illustration 493
Vacheron & Constantin *mariage* on its arrival from the USA. The dial and caliber K 1072/1 movement came from a Chronometer Royal Ref. 6694 made in 1971, the steel case from an Automatic with Date Ref. 6694 from the year 1962.

Illustration 494
For comparison, a Vacheron & Constantin Ref. 6694 Chronometer Royal in original condition

Illustration 495
The caliber K 1072/1 from Vacheron & Constantin without the gold oscillating weight, and below it the complete rotor of an Audemars Piguet caliber P 2499. Jaeger-LeCoultre supplied the raw movements (*ébauches*) for both watches. The guilloché gold weights were identical for both customers, but not the movements. Those of AP were initially regulated in four positions only, those of V&C in five positions. The rotor carrier plate of the Audemars Piguet was decorated with Geneva stripes, that of Vacheron & Constantin with a sunburst finish.

Illustration 496
View of the caliber K 1072/1 movement number 630,032. Not only is this movement installed in the wrong case, but it also lacks the 750 yellow-gold oscillating weight.

Illustration 497
The Vacheron & Constantin *mariage* after its overhaul. The rotor again has its gold oscillating weight, and damaged screws and other components have been replaced. Along with the previously described Patek caliber 27-460 M, this Vacheron K 1072/1 automatic is one of the most beautiful Swiss-made watch movements. Both types of movement have a Gyromax balance wheel, even though the one used by Patek has eight *masselotte* weights and that by Vacheron ten.

Illustration 498
Vacheron & Constantin Ref. 6562 in stainless steel, and our *mariage* in a case like that of the model shown on the left

blue dial in combination with the yellow-gold hands and the yellow-gold indices gave the watch a special look. Furthermore, the invoice from a Berlin Rolex concessionaire for a complete overhaul was reassuring. A total of $380 had been charged for this. After the usual inspection, the rate test on the Witschi Micrometre Professional followed. While the daily rate (MOY) of + 004 sec. was in a good range, the average rate in six positions (DEL) of 014 sec. was not satisfactory. A consultation with the Berlin concessionaire was not very helpful and resulted in the involvement of Rolex Cologne. After the watch was sent in, it was quickly discovered in Cologne that the Berlin concessionaire had overhauled the movement himself and sent only the case to Rolex for refurbishment. Now the caliber 1556 also underwent a professional overhaul and immediately achieved its former chronometer values. After some persuasion, the Berlin Rolex concessionaire took over the costs charged by Rolex Cologne and made sure that this purchase could be concluded with a happy ending. For this reason, neither the name of the seller nor that of the Rolex concessionaire will be mentioned.

If you take this deal into account and add the fact that the corresponding Oyster gold bracelet was delivered for another $867, you can be satisfied with this new acquisition. A total of $4,678 for a well-preserved and recently overhauled Rolex Oyster Day-Date with gold and leather strap, papers, and box can be called a satisfying investment. Even if this classic watch with caliber 1556 movement does not have a date quickset, with a dark-blue business suit and yellow-gold cuff links it always helps to make a perfect appearance.

Illustration 499
Not only the dream of many collectors, but often the watch desired by many successful managers and businessmen: the Rolex Oyster Day-Date. On the German market this model was offered exclusively with a gold bracelet. In Austria it was also sold with an alligator strap because of that country's luxury tax. This blue version is one of the many dial variations. It certainly goes well with a dark-blue business suit.

Illustration 500
The caliber 1556 automatic movement of our Day-Date. Precision, robustness, and reliability were included in this technically superior movement's specifications. Here we would like to point out a simple but technically relevant detail, because it has never been properly highlighted in the relevant literature. The gears (ratchet wheels) of the automatic device module are made of a special aluminum alloy, and the surface is anodized. This anodized layer is as hard as glass and very wear resistant, even without lubrication. The movement's gears and pinions, on the other hand, are made of brass and steel, case-hardened and ground. Thus, the gear train of the automatic device module does not focus on precision, but on wear resistance and high emergency-running properties. The gears and pinions of the movement, on the other hand, require high pitch accuracy, precise module formation, and proper lubrication. Thus for efficient, wear-free winding, on the one hand, and high rate accuracy, on the other, the company has reached deep into its technological bag of tricks.

Illustration 501
Reverse side of the Rolex Day-Date with screw-down case back, which can be opened only with a special tool. One can recognize the export version by its gold hallmarking: the Helvetia alone is stamped on the upper left lug, 18K on the right one.

8.11. IWC Automatic Ref. 3205 in Stainless Steel

Even though this watch looks very modest at first glance, it is in fact a small technical marvel. First of all, the elaborately designed case with a measured diameter of 1.35 inches (34.3 mm, factory specification 1.31 inches or 33.5 mm) and height of just 0.287 inches (7.30 mm) is indicative of this. The screw-down case back with fine thread and a special key lug with eighteen edges makes such a thin design possible. A version of this reference 3205 with a yellow-gold case can be found in chapter 2 ("IWC") of this book on page 86.

The manufacturer of this watchcase, Firma C. R. Spillmann SA, Chêne-Bourg, supplied similar case designs to Vacheron & Constantin. An example of this can be found in chapter 7, "Vacheron & Constantin," reference 7512 on page 283. The model presented was offered on the internet by a German used-watch dealer and purchased for $1,735. Since the case and movement were in good condition, no use was made of the granted right of return. Because it had probably been idle for a long time, the IWC caliber 3254 required a professional overhaul and was submitted in person to IWC in Schaffhausen for this purpose. An accompanying letter with precise instructions, including a request to check the rate values of this watch with the Witschi measuring instrument after the overhaul and to enclose the printouts, was submitted as always.

After several months of waiting, I collected this watch in person at the IWC boutique in Schaffhausen and paid the steep overhaul fee of $1,788. However, I did not receive the measurement certificate I had requested in writing. The watchmaker who had overhauled this watch was immediately mobilized.

Presumably it had been forgotten that this watch had been delivered with a measurement certificate to compare the rate values after the overhaul with the previously measured values.

Since this catch-up process probably took more than an hour for organizational reasons (the IWC manufacturing plant and the service department are now two legally and organizationally separate companies), the requested measurement certificate could be delivered only after a long lunch break. Since the chronometer-capable caliber 3254 (based on the Jaeger-LeCoultre 900) exhibited worse rate values after the expensive overhaul than before, I complained about this state of affairs, and the watch was returned to the service department for rectification. The author's companion, Dr. G.T., suspected that the service department had not read the accompanying letter and the author's instructions. After another three weeks, IWC sent a notification of completion and confirmation that the requested measurement records of the rate values would be enclosed. Due to the travel and visit bans imposed by the coronavirus pandemic, this IWC reference 3205 was picked up by a confidant of the author and held in Switzerland until further notice. At the time of going to press, the results of the reexamination of this watch were not yet available. However, given my previous favorable experience with IWC, it can be assumed that the small mishap had been promptly rectified.

8.12. Patek Philippe Ref. 2551J-200

In chapter 5, "Patek Philippe," I wrote about two copies of this beautiful automatic from Geneva. In this noble collector's item is the automatic caliber 12‴-600 AT, which was the first of its kind, introduced in 1953. Exactly 7,100 examples of this movement were produced between 1953 and 1960. This caliber was used in the following models: reference 2526, 2540, 2551, 2552, 2583, 2584, 2585, 3403, and 3415, the latter only in the first series. The movement of our watch, with the number 764,770,

Illustration 502
Even this stainless-steel version of the IWC Automatic Ref. 3205 has a very elegant appearance. With a measured case diameter of 1.35 inches (34.3 mm; factory spec. 1.31 inches or 33.5 mm) and an overall height of just 0.287 inches (7.3 mm), it has a special screw-down case back and is waterproof to a depth of 98 feet (30 m).

was probably made in 1957. The model shown was purchased via eBay from a private individual for $12,476, and because of its worn condition it was handed over to the overhaul department of Patek Philippe in Geneva by way of the Munich branch. The written order also included refurbishment of the dial. While the overhaul was carried out at the 1228 Plan-les Ouates site, the dial was refurbished at Cadrans Flückiger SA, the company's own dial factory in St.-Imier, Canton of Bern. After six months this reference 2551 arrived back in Munich freshly overhauled. Exactly $3,153 had to be paid for the complete job. This watch passed the usual inspection with flying colors; this was true for

Illustration 503
IWC reference 3205 with the screw-down case back removed. The IWC caliber 3254 is based on the Jaeger-LeCoultre 900 automatic movement. It has thirty-four jewels, a diameter of 1.04 inches (26.5 mm), and a height of 0.171 inches (4.35 mm). The balance frequency is 28,800 vibrations per hour (movement no. 2,215,490, case no. 2,219,639). On the left is a detail enlargement of the case number.

Illustration 504
The case of the IWC Ref. 3205 has a special screw-down case back with fine thread. This is opened by means of a star-shaped, eighteen-sided bottom that requires a special tool. This case was provided by C. R. Spillmann SA from Chêne-Bourg.

the movement performance as well as for the quality of the dial. Collectors of valuable watches from the house of Patek Philippe are also advised to put all desired procedures down on paper when ordering an overhaul, to avoid possible misunderstandings. Also, some informative pictures of the case, dial, and movement should be taken before the planned overhaul, especially since the lenses installed in new mobile devices enable one to take sharp photos.
Even though the overhaul of the Patek Philippe Ref. 2551, also known as Disco Volante (Flying Saucer), cost almost $3,254, this amount may be considered a good investment. On the internet these models are offered between $20,000 and $30,000, but it is not guaranteed that all of them are in the same mint condition as our Patek Philippe shown.

8.13. Rolex Cosmograph Daytona Ref. 116.520 in Stainless Steel

The most sought-after cult watch of all, the Rolex Cosmograph Daytona with the reference no. 116.520, deserves to be included in this chapter. For this sought-after model in stainless steel, watch dealers are now quoting delivery times of about six years, so you should make a note of it right away. While the list price for this watch was still $12,259 in 2019, an even-higher price of $13,290 had to be paid for it as of January 1, 2020, a price increase of around 8.4%! The current selling prices of used watches on eBay range from around $20,600 to $26,000. International offers (e.g., at Chrono24.com) range from around $24,000 to $38,000.

Illustration 505
Close-up of the 12‴-600 AT caliber's fine-adjustment mechanism under roughly 5× magnification. Fine adjustment of the gear is accomplished via micrometer screw and regulator arm. The counterpressure is provided by a round-bent spring element. A second way to fine-tune the gear is provided by the Gyromax balance with eight mass balance weights. The system is thus doubled, as they say in the vernacular. A Gyromax balance wheel alone was used in the subsequent automatic movements from Patek Philippe. Between the balance wheel and balance cock, one can still see the Breguet hairspring. The balance cock itself is positioned on the plate by two invisible straight pins and is secured by a cylinder head screw. The small gap between the plate and the balance cock is used by the watchmaker, or his screwdriver, to lift off the balance cock.

The Daytona presented was purchased from a dealer about ten years ago but had rarely been worn. Nevertheless, for reasons of functional reliability, this timepiece was entrusted to the Rolex branch in Cologne and a complete overhaul was commissioned. At the same time, a written request was submitted that this watch should correspond optically and technically to a new specimen. This wish was also fulfilled in full, and the amount of exactly $1,208 was charged.

Illustration 506
The 12‴-600 AT, Patek Philippe's first automatic movement, with all the subtleties that such an excellent caliber can contain. A solid-gold guilloché rotor, Gyromax balance wheel and Breguet hairspring, changer for bidirectional winding, and much more ensure the very best quality.

Illustration 507
The Patek Philippe Ref. 2551, also called the Disco Volante or Flying Saucer, after an overhaul by the manufacturer in Geneva. The dial has been refurbished at Cadrans Flückiger SA, Patek Philippe's own dial factory, so that it can be called like new. Despite being almost sixty years old, the movement achieves chronometer values again after this overhaul. Only the case was left as it was, because refinishing and polishing always involve material erosion and thus the loss of gold (movement no. 764,770, case no. 698,306).

Illustration 508
The Rolex Daytona as it appeared after passing final inspection in Cologne following its overhaul. The center links of the Oyster strap are masked with gold protective foil. The refinished lugs are covered with a red-striped protective foil. The bezel is covered with a plastic ring, just like a new watch on delivery. This special care is probably given only to watches that are overhauled in-house by Rolex.

So that the interested reader can also take a look behind the scenes of an overhaul, some details of Rolex Cologne are also revealed. The author had the opportunity to visit the workshop and production area of Rolex Cologne several times during the time of the former technical director and authorized signatory Georg Ressel and to talk to the experts there. In all, this technical area resembled a factory where everything was available for the production of a complete watch. All the operations and production steps were carried out in the same way as in the Biel and Geneva production facilities.

Though he has no desire to sidestep Rolex concessionaires and their workshops in Germany, the author would always prefer to send Rolex watches directly to Cologne for an overhaul. This is because concessionaires also pass on most of their overhaul work to Cologne.

One of our pictures shows the condition of Daytona and the markings it receives after a successful overhaul of its movement, case, and bracelet. Both the refinished sides of the case middle and the strap receive special protective foils after

Illustration 509
Close-up of the Rolex Chronograph caliber 4130. In the center is the column wheel, which controls all chronograph functions.

an overhaul has been performed. A plastic ring is placed over the bezel to protect it.

After having had numerous wristwatches overhauled by Rolex Cologne, the author has had no complaints despite comprehensive initial and final inspections.

However, the author has also experienced other watchmakers and case specialists who achieved the same standard of quality in keeping with exact specifications; for example, watchmaker Martin Becker from Straubenhardt, who unfortunately retired, and case specialist Alfred Hermann in Neuenburg, who is still active.

Illustration 510
For more than twenty years, the Rolex Cosmograph Daytona Ref. 116.520 has been at the top of the popularity scale for watch enthusiasts. With its Rolex-developed chronograph caliber 4130 movement with automatic winding and seventy-hour power reserve, it has—from an engineer's point of view as well—technology so advanced that it can hardly be topped (case no. 452X3523; the movement number could not be seen because the watch was not opened).

Data on the Overhauled Watches Presented								
Brand	**Model**	**Reference**	**Caliber**	**Movement No.**	**Case No.**	**Case Type**	**Purchase Price (EUR)** [1]	**Overhaul Costs (EUR)** [1]
Audemars Piguet	Automatic	5112 BA	K 2070	74 754	17 756	750 GG	ca. 3,250	ca. 392
IWC	Yacht Club	R 811 AD	C. 8541 B	1 959 737	1 929 857	Steel	ca. 775	ca. 608
Jaeger-LeCoultre	Memovox Automatic	E 855	825	1 936 863	1 131 186	Steel	ca. 710	1,298.62
Patek Philippe	Automatic Date	3445	27-460 M	1 118 030	328 637	750 GG	3,710	1,347.34 [2]
Vacheron & Constantin	(Chronometer Royale)	--	K 1071/1	630 082	382 366	Steel	ca. 2,300	332
Rolex	Oyster Day-Date	1803/8	1556	DD 279043	2 092 352	750 GG	3,512.02	685 DM
IWC	Automatic	3205	3254	2,215,490	2,219,639	Steel	1,600	1,627.50 CHF [3]
PPC	Automatic	2551	12-600 AT	764,770	698,306	18 K YG	11,500	2,906.81
Rolex	Chronograph	116520	4130	[4]	452X3523	Steel	11,000	1,113.99
[1] Approximate values are conversions from USD to EUR [2] 2 Amount without gold buckle, strap and box from PPC [3] Overhaul costs at IWC employee rate [4] Watch was not opened								

Illustration 511
Potpourri of the dials of valuable Swiss wristwatches (*from top left, clockwise*: Audemars Piguet Royal Oak, IWC Ref. R 817 AD, Jaeger-LeCoultre Geomatic Chronomètre, and Vacheron & Constantin Ref. 6782 Q)

9. New Automatic Movements from Switzerland

New Automatic Movements from Switzerland in the Mirror of Earlier Automatic Calibers

9.1. How Does a New Self-Winding Movement Come About?

First, let's take a look back. As late as the beginning of the 1970s, when nearly 100,000 people were employed in the Swiss watch industry, almost every major watch manufacturer had self-developed automatic movements. In addition to the upper class, which at that time included OMEGA and Rolex, more than thirty other self-sufficient manufacturers bustled about in the flourishing Swiss watch business. But as early as the mid-1970s, with the advent of quartz technology, this situation changed fundamentally and meant the inevitable end for most companies! This also put an end to the colorful variety of self-winding movements produced at that time.

After this almost brutal but technologically unavoidable market adjustment, only a handful of Swiss watchmakers were left with their own automatic calibers. In addition to market leader Rolex, these included Patek Philippe, Jaeger-LeCoultre, and Zenith. Chopard, Girard-Perregaux, and others joined them later. After the aforementioned collapse, almost all other makers of Swiss wristwatches made use of the automatic calibers made by large-scale manufacturer ETA or, in the higher price range, the raw movements from Frederic Piguet. Today, both are part of Nicolas Hayek`s Swatch Group and have almost a monopoly position.

Today, when a watch manufacturer decides to introduce a new automatic caliber, it either wants to add another, arguably even-better model to the movements already in production or to make itself independent of its existing raw-movement suppliers. The former is true for the new JLC caliber 975; the latter is true for the new Audemars Piguet movement.

Of course, it is easier for an experienced movement manufacturer to develop a new automatic caliber. After all, they can draw on the extensive experience they have gained with previous movements. A newcomer, on the other hand, does not have the aforementioned advantages and must first create the infrastructure for the development, design, and production of the new self-winding movement. This includes recruiting highly qualified employees, providing suitable premises, procuring an extensive machine park, and setting up the ateliers required for watch assembly. Both Vacheron Constantin and Audemars Piguet had to make these large investments to be able to launch a new automatic movement from scratch.

These diverse processes are subject to a high level of entrepreneurial risk, because all the measures listed are prerequisites for, but not a guarantee of, a successful new movement. That's why decisions on such projects are usually made more quickly and less bureaucratically in owner-managed companies than in group-controlled companies, where even the CEO is an employee embedded in the corporate hierarchy, albeit with a great deal of responsibility.

9.2. What Steps Precede This Undertaking?

First of all—as with any successful marketing effort—the following three questions must be asked: Where do I come from? Where am I today? Where do I want to go? With these considerations in mind, the first step is a needs analysis ("Why do I need a new automatic caliber?"), then a market analysis ("What kind of automatic caliber can meet current and future demands?" "Which design would put me ahead of the competition?"), and finally a feasibility study. The latter should provide the decision-making bodies with a statement on technology, costs, and deadlines and also prove the chances of success. When it comes to decision-making by business people, who often put their faith in computers, it is always an advantage to have a high-profile technician standing by. Because even an extensive, colorful, and three-dimensional CAD representation requires careful examination, because even a drawing, no matter how impressive it is, does not necessarily make a good design.

Information about the latest movements from the competition is also required to complete this task. The wheel was invented more than six thousand years ago; nevertheless, some technicians are still trying to reinvent it.

If one finally knows the exact state of the art and starts with one's own design drafts, one should know that in technical mechanics, almost all fields are already covered. The innovations that can still be imagined today can be expected only from especially creative engineers, but they also rely on the application of new machining technologies and materials, and further reduction of manufacturing tolerances. Software and CAD systems, such as Pro/Engineer, permit complicated designs that also include strength calculations and allow functional verification. Simulation, which is possible here, is also of great importance.

With the integration of the company's own movements from earlier production, the level of the most-important competitors, and the—it's hoped—new ideas, the intended new product is transformed into a feasibility study. New ideas or design principles

Illustration 512
Lineup of the new watches (*left to right*): Audemars Piguet Royal Oak with caliber 3120, Vacheron & Constantin with caliber 2475, Jaeger-LeCoultre Hometime with caliber 975, and IWC Ingenieur with caliber 80110

are executed and subjected to laboratory tests as components or subassemblies, and their functional reliability and practical everyday suitability are verified. This also includes confirmation once again of statements about technology, deadlines, and costs. On this basis, management can make the decision on and approve the new product. This approach significantly reduces risks and costs for the company.

9.3. How Does Such a Process Work?

Caesar already demonstrated to us in *De Bello Gallico* (*The Gallic War*) that well-thought-out planning and comprehensive preparation are the most-important prerequisites for success. This rule has survived to this day and also applies to the launching of a new watch with a new automatic movement. The internal process for the successful production of a new movement can be roughly described with the following steps:

- On the basis of a performance specification, new technical concepts on the basis of a set of specifications, taking into account previous in-house movements, those of the market-leading competitors (rivals), and own new ideas (innovations)
- Analytical evaluation of one's own concepts in comparison with competitors with regard to own goals
- Management statement on technology, costs, and deadlines
- Management decision with approval of technique, budget, and schedule
- Designation of the person responsible for the project and the project team
- Development, design, prototype construction and testing, various application tests, detail improvements, and additional modifications
- Approval of the technical concept with definition of the prototypes, and the quantities and variants to be produced, as well as preparations for production, including securing new or additional production processes with the involvement of suppliers

Illustration 513
Lineup of the old watches (*left to right*): Audemars Piguet Automatic with caliber K 2072, IWC Automatik with caliber C. 8541 B, Vacheron & Constantin with caliber 1072/1, and Jaeger-LeCoultre Geomatic with caliber K 883 S

- Construction and testing of the prototypes
- Evaluation of the tests and completion of the design drawings
- Official approval of the technology and quantity production
- Start of quantity production and market introduction

Depending on its scope, such a process can take more than five years. The respective period depends on the requirements (complexity of the plant), the development effort, the co-use of components from plants already in production, existing know-how, the engineering capacity, and the available budget. The costs for such a project can be around \$5.5 to \$8.8 million or even more, depending on the stage of expansion. It should be noted that the reuse and optimization of existing and proven plant components can significantly reduce development costs and risk. On the other hand, there is the obligation to come up with new ideas and innovations to be able to prove the benefits of the watch, the lead over one's competitors, and one's own competence.

Even the most advanced mechanical and thermal endurance testing, with the aim, as in the automotive industry, of being among the best in breakdown statistics, cannot replace innovative creativity. Because if this were the case, many drivers would—just as an example—prefer the cars from the temporary world market leader from Japan and not the two noble brands from Stuttgart-Untertürkheim or Zuffenhausen.

9.4. Efficiency and Level of Innovation

The most-promising and most-economical solutions are achieved with a project team that has a high level of in-house know-how, creative engineers, and precise knowledge of its own previous successful calibers, as well as the current movements of its competitors. In addition, knowledge from other technical sectors is required, which can be of importance for the constructive design, the choice of materials, and the manufacturing technology.

A manager with a commitment to success is required to lead this team: someone who has received their education at a technical university and who convinces through impressive qualifications the ability to act as a personal role model, with levelheaded foresight, knowledge from other industries, such as mechanical and lightweight engineering (aerospace), and consistent action. With their sovereign leadership ability, they provide the necessary internal freedom that the team needs for creative and innovative work, and stand for the success that the company needs. Working in close cooperation and communication with the production, service, and marketing departments, this manager occupies a key position in the company. Like few journalists, the author himself has adequate in-house expertise in the areas of development, design, the use of CAD, and production.

9.5. The New Automatic Movements

To put it in a nutshell, choosing a watch is often like choosing a good wine: it is always an expression of personal taste. One person loves a full-bodied Burgundy, another swears by a noble Bordeaux, and a third is happy with a simple Trollinger. When it comes to the small micromechanical marvels that reliably and unpretentiously do their job in wristwatches every day, the focus is more on the technology. And it is precisely this part of a watch that we want to take a closer look at, using four examples and making a concrete comparison with earlier movements. The most-important criteria will be design, suitability for daily use, ease of service, movement performance, and innovative new features. The well-known Rolex 3135 automatic caliber will serve as a standard, because in the author's opinion it is one of the best large-production movements available on the market. The most-important functional elements of this movement, as well as those of the four new ones, are listed in table 1 and allow an interesting comparison to be made, even by the uninitiated reader. The fact that one also encounters different technical opinions among the three watch brands that belong to the Richemont Group lends a special charm to this examination.

To further enhance this chapter, earlier self-winding movements from the four watch brands are presented to the reader in detail. This comparison clarifies the most-important design differences between yesterday and today and also allows a statement to be made as to whether or why the new automatic movements turned out better or not. These facts will certainly be of interest to our younger readers, who may not have been in this world some forty years ago or who were still plagued by elementary school. Now, let's have a look at what the new automatic movements look like and how they can be seen in the mirror of earlier calibers. The many photographs that accompany this chapter not only prove the author's claim to have held each of these watches in his hands, but also provide the reader with deep insights into the various designs.

9.6. Audemars Piguet, Caliber 3120

The first self-winding movements came into use in 1953. Here, the detour via designs with rotation-limited oscillating masses, popularly known as "hammer automatics," was not chosen. Even the first automatic calibers had 360° rotating rotors. And that was not all. Guilloché oscillating weights made of 18-karat gold, a rotor carrier plate with Geneva stripes, a screw balance wheel with Breguet hairspring, fine adjustment via micrometer screw and swan-neck spring, bidirectional winding, and the finest workmanship for plate, bridges, gears, and pinions were among the attributes of the caliber P 2498. It was delivered as a reserved raw movement by Jaeger-LeCoultre from Le Sentier, only 3 kilometers away. While the caliber P 2498 had a small second hand, the P 2499, manufactured in parallel, had a center second hand. From 1959, the reworked automatic movements K 2070 (small seconds at six o'clock) and K 2071 (center seconds) were introduced, followed by the caliber K 2072, with date indication at three o'clock from 1961 onward.

The year 1967 saw the appearance of the K 2120 caliber, which was the thinnest automatic movement in the world at that time, with an overall height of 0.096 inches (2.45 mm). The version with date display (K 2121) followed in 1970 (see also chapter 1). While until then all calibers were supplied by Valjoux or as reserved raw movements by Jaeger-LeCoultre, there was also an interlude in the 1980s when calibers were acquired from raw-movement manufacturer Frederic Piguet (caliber K 2131). In all, the Jaeger-LeCoultre movements of the 888 and later 889 series played the leading role.

The entire partnership situation between Audemars Piguet and Jaeger-LeCoultre changed abruptly when the former raw-material supplier of VDO/Mannesmann was sold to the Richemont Group. Audemars Piguet's 40% share in Jaeger-LeCoultre was also sold for a considerable sum. The most important supplier of raw materials was thus removed from Audemars Piguet's direct grasp, but the company's coffers were filled to the brim for a new start of its own.

Initially, rights and drawings were acquired from Jaeger-LeCoultre for the automatic caliber K 2120/2121. This was followed by development of its first own manually wound caliber 3090. The next step was as clear as day: a new self-winding movement of the very highest quality. The experts in Le Brassus got down to work quickly and had only three important specifications to consider. The installation diameter of the Jaeger-LeCoultre caliber 889 of 1.02 inches (26 mm) (total 1.04 inches or 26.6 mm, corresponding to 11½") as well as its overall height of 0.169 inches (4.3 mm) had to be maintained, but everything else was to be better. Only in overall height did the specifications make a few concessions, although a little more would have brought unmistakable advantages. For the Audemars Piguet technicians involved in the project, this was a demanding

Illustration 514
One could describe this timepiece as "elegant and of simple beauty." This automatic dates to the 1960s and was delivered on April 4, 1964. Such a watch is an extremely rare find. The gold case's small refinements are apparent only on closer inspection (Ref. 5209, case no. 29 251, movement no. 86 332).

Illustration 516
A familiar face that has not lost its expressiveness even after many years: the latest version of the Audemars Piguet Royal Oak, reference no. F 40842, here with the new automatic caliber 3120

Illustration 515
Audemars Piguet caliber K 2072, which was launched in 1961. A rotor with four ruby roller bearings and guilloché oscillating weight in 750 gold, a screw balance wheel with Breguet hairspring, fine adjustment via micrometer screw with swan-neck spring, and other attributes give this movement a top ranking. It came as a reserved raw movement (*ébauche*) from Jaeger-LeCoultre in nearby Le Sentier.

Illustration 517
The finest of everything: symmetrically formed bridge with Gyromax balance wheel with eight *masselotte* weights to adjust the rate, radial slotted holes to adjust the hairspring stud, and a solid 920 gold rotor. Pins for positioning the bridges are nowhere to be seen, since flanged bushings with internal threads (called "screw feet" in watchmaking parlance) were used.

challenge. The disadvantages of a low overall height will be discussed later. From a purely technical point of view, the result of this work, which took around five years, is impressive. It included a double gold rotor supported by ball bearings, an ingeniously simple and maintenance-free rocking bar changer (which had already proven itself in the caliber 2120/2121), a Gyromax balance wheel housed in a bridge, and much more.

The design of this caliber includes some interesting features. The automatic bridge is not superimposed but is integrated into the movement. On the plate, one finds the mainspring barrel bridge on the left, the gear train bridge with rotor bearing in the center, and the automatic bridge on the right. The balance bridge is nicely symmetrical and gives this micromechanical wonder movement a special elegance. This is underlined by the beveled edges of plate and bridges, the Côtes de Genève finish, and the look of the golden rotor, which is supported by ball bearings. The new fine-adjustment mechanism deserves special attention. As in the caliber 2120/2121, a Gyromax balance wheel (a former patent of Patek Philippe's) again provided an accurate rate. Deceleration is regulated by an adjustable hairspring stud, which in turn is held in two cylinder head screws in radial oblong holes—an elegant solution. Among other innovations, the design of the stop-seconds mechanism and the quickset date mechanism deserves mention. The latter allows the switching of two days with one crown rotation.

Bridges and plates are largely connected by means of flanged bushings with internal threads, known as "screw feet" in watchmaking parlance. The plates and bridges are usually made of brass alloy, but the flange bushings are made of steel. Compared to earlier designs, the latter have a double function: they are a dowel pin "foot" and a screw connection in one. Anyone who has studied mechanical engineering at ETH Zurich or another technical university will be familiar with this design principle from lectures on machine elements. It has been used for decades in "lightweight construction." Lightweight construction means structures made of aluminum or other low-density materials, which in turn are held together by flanged bushings and bolts made of steel for strength reasons. The second aspect of this technique is serviceability. If such a flange bushing is damaged, it can be easily replaced and the bolted joint renewed without reworking. It is a mystery to the author why this technique, which is common in the aerospace industry, was introduced to watchmaking so late. Now a word about the problem of the low case height. This was also one of the reasons why the automatic bridge had to be integrated into the movement of this new AP caliber. A separate automatic bridge would have required more height but would have allowed more space in the actual movement. In contrast to all other new automatic movements, our AP runs at just 21,600 A/h. Since the power reserve is inversely proportional to the number of semioscillations, our AP would have a power reserve of forty-five hours instead of sixty hours at 28,800 semioscillations. To achieve higher values, a much-larger barrel would have been necessary. And this in turn would have required more construction volume, which was not available in the current design specification. The test watch handed over for this article, a Royal Oak 0128 in stainless steel (reference no. F 40842), immediately felt at home on the author's left wrist, since a Royal Oak in steel and gold with the number 911 is one of his everyday watches. The results of the standard initial test, which includes functionality, workmanship, and performance, were positive. We were quickly convinced that we were looking at a very high-quality product in terms of technology and design. The practical test was also convincing, albeit with a small interlude. On the third day the watch unexpectedly stopped while I was mowing the lawn. Our timepiece immediately ended up back in the hands of a renowned watchmaker from AP, who analyzed the reason for the failure and replaced the critical component. And so this unintentional endurance test also helped eliminate what was probably the last weak point of this new automatic movement.

But how does it compare with an earlier AP caliber? For this we selected a beautiful automatic from the 1960s. The rarity with the reference 5209 (case number 29 251, movement number 86 332, delivered on February 26, 1964) impresses with its dial and elaborate case. But only after opening the screw-down case back can one see which beautiful and elaborate self-winding movement was previously used by AP. The caliber K 2072 was unveiled in 1961 and delivered as a reserved raw movement (*ébauche*) from Jaeger-LeCoultre in nearby Le Sentier. The diameter of this caliber is 1.14 inches (29 mm); the height is 0.248 inches (6.3 mm). The automatic bridge represents a separate unit and is screwed to the base movement. Obvious attributes of this movement are the *guilloché* oscillating weight in 750 yellow gold, rotor bearing by means of four ruby rollers, and a rotor bearer plate with Geneva stripe finish. The caliber, which is adjusted in four positions, has a screw balance wheel with Breguet hairspring and a fine-adjustment mechanism with micrometer screw and gooseneck spring. Even the escapement wheel shaft was given shock protection with a cap jewel. Further technical data can be found in table 1.

This beautiful collectible was purchased at an auction at Christie's and underwent an expert overhaul. This can be seen not only from its impressive external appearance but also from the movement values, which almost reached chronometer values again. Comparing the two designs, which are more than forty years apart, one cannot miss the technical progress displayed by the caliber 3120. Especially the successful arrangement of the three bridges (mainspring barrel, gear train, and self-winding bridge) with the harmonious symmetrical balance bridge gives a very aesthetic overall impression. A new movement from one cast that can worthily continue the tradition of the house of Audemars Piguet. Compliments!

9.7. IWC International Watch Co., Caliber 80110

At the end of the 1940s, many Swiss watch manufacturers were concerned with the development and introduction of automatically winding wristwatch movements. This usually involved winding mechanisms with a limited oscillating weight, known as hammer automatics. Albert Pellaton, who took over the technical management of IWC in 1944, was in charge of this development at IWC. Although the first design drafts were still based on the hammer automatics mentioned, after several years of development a patent application was filed in 1950 for a particularly successful design with a rotor that wound on both sides. In 1951 the first automatic caliber C.85 was unveiled and was used in various automatics. The best known of these timepieces was probably the IWC Ingenieur, which used soft-iron protection to shield against magnetic fields and was water resistant. In general the Ingenieur was designed for the tough everyday work of technicians and was also highly valued by doctors. However, not every representative of those professions would have needed an Ingenieur or in fact had the salary necessary to purchase it. For example, the monthly salary of a locomotive engineer or a power station electrician would hardly have enabled them to purchase an Ingenieur. Even for the simple stainless-steel version with leather strap and steel pin buckle, Ref. R 666 AD, the price was $500 in 1955. The Ingenieur product line, the classic wristwatch with a cage made of ferromagnetic soft iron to protect against magnetic fields, was first unveiled in 1954. This principle was also used in the Ingenieur's predecessor, the Pilot's Watch Mark 11. The trendsetting caliber 85, with which the first Ingenieur was equipped in the 852/8521 version, was consistently further developed and continued under the caliber designations 853 and 854 (version with date 8531 and 8541). For the comparison with the new IWC automatic caliber, we have chosen an IWC Ref. R 817 AD, which is more than thirty-five years old. This watch uses a caliber C. 8541 B movement, the last version before production ceased in 1976.

In the early 1980s, IWC was acquired by VDO/Mannesmann and placed under new management. Günter Blümlein, who initially started out as a consultant and subsequently took over the management of IWC, immediately ceased in-house production of movements and ordered the scrapping of all punching and cutting tools. This meant the end of independent caliber production at IWC. Henceforth only purchased calibers (i.e., those from other manufacturers) were to be used. ETA in Grenchen was chosen as the movement supplier and subsequently supplied IWC with the automatic caliber 2892 and the chronograph caliber 7750 in various versions. Later raw movements from Jaeger-LeCoultre were added, with the automatic movement 900/901 and the subsequent model 889 being used. One IWC Ingenieur model was equipped with the latter, although IWC was not always entirely happy with this very sensitive caliber.

But now to our new IWC caliber 80110. When the author first laid eyes on the new IWC Ingenieur, it evoked memories of the IWC Ingenieur Ref. 1832 in stainless steel, also called the Jumbo. This model, whose exterior was created by designer Gerald Genta, was the last Ingenieur with an IWC manufacture caliber C. 8541 B. This visual impression was followed with excitement by the question of the likewise new-manufacture caliber 80110. Its spring-loaded automatic bridge and Pellaton winding system immediately identified it as an IWC product. How good that old virtues had been retained here. But a look at the balance cock and the regulator caused the first irritations. No Breguet hairspring, no balance wheel with *masselotte* weights, no eccentric fine adjustment of rate and click? It required a detailed study of the plate, bridges, and other components to confirm our suspicions. A detailed comparison with the ETA caliber Valjoux 7750 finally brought clarity. The chronograph, as well as the automatic device module, had been taken from this widely used chronograph automatic, which was almost thirty years old in its origins, and the basic movement had been reworked and modified to adapt the IWC Pellaton winding system. But did that explain everything? We wanted to know exactly and not only took the new IWC Ingenieur and an unnamed chronograph with the caliber ETA Valjoux 7750 to hand but also checked and compared the exploded views, the parts lists, and the assembly instructions for both calibers. In the following we list not all but, for our readers, probably the most-important differences between the two movements.

Winding, Hands, and Date Mechanism

As far as can be seen from the drawings, only the intermediate handwheel was modified by IWC. Only a few ETA components are used for the date mechanism (e.g., the driver cannon pinion, change gear, cannon pinion, hour wheel); the rest of the date mechanism comes from IWC.

Main Plate and Interposer Plate with Associated Components

Here are some relevant modifications that are not immediately apparent at first glance. To accommodate IWC's own barrel, it is housed in bushings made of copper beryllium, thus increasing smoothness and wear resistance. For the rear screw connection of the new escape wheel bridge, the bore pattern was expanded and an additional flanged bushing with an internal thread was installed. The most-obvious modifications are to be found in the interposer plate. The upper escapement wheel bearing is no longer integrated here but is taken over by a newly designed escapement wheel bridge. This is again arranged in a recess and precisely positioned with the main plate by two flanged bushings (screw "feet"). With this design the escapement wheel can now be removed easily and quickly. On the ETA caliber 7750, the complete intermediate plate with winding bridge and chronograph module must be disassembled for this purpose. In addition to changing the outer contour of the intermediate plate, another important modification required the attachment of the complete IWC Pellaton winding system. This included four additional flanged bushings—one of them in the modified crown wheel core—as well as a redesigned fourth wheel bridge.

Balance Cock with Balance Wheel and Fine Adjustment

Another novelty is the newly designed balance cock and the fine-adjustment mechanism with IWC balance wheel. Fine adjustment is no longer performed by a two-piece regulator arm (ETA) but by the Triovis system, which consists of a toothed rim segment and a fine-thread screw. The three-spoke balance wheel is excited by a flat hairspring and is shock-protected by Incabloc bearings. An adjustable hairspring stud allows the adjustment of the symmetry of the oscillation process (drop).

Illustration 518
An IWC Automatik Ref. R 817 AD with three-part case and screw-down case back made of 750 yellow gold made in 1970 (movement no. 1 976 636, case no. 1 788 768)

Illustration 520
The new Ingenieur from IWC literally makes a powerful impression. The 0.574-inch-high (14.6 mm) case houses the automatic caliber 80 110, which is 0.284 inches (7.23 mm) high.

Illustration 519
The IWC caliber C. 8541 B in detail. In the left foreground are the three-spoke balance wheel with three weight screws and Breguet hairspring. On the balance cock is the regulator arm with eccentric screw adjustment. Next to it is the eccentric screw for adjusting the drop. The cocks for the balance and escapement wheel are simply screwed together. Not visible are the two dowel pins ("feet" in watchmaking parlance) that position these two components exactly. The pins are not visible because they are inserted from below into blind holes.

Illustration 521
After opening the massive screw-down case back, one can admire the new caliber 80 110. Here one can see four of the newly designed components for this movement: the anchor wheel bridge, balance cock, fourth wheel bridge, and rotor.

Automatic Winding

The author makes no secret of the fact that he still considers IWC's Pellaton winding system, now more than fifty years old, to be one of the most efficient and innovative technical solutions he has come across. The basic kinematic processes combined with the inventive detailed mechanical solutions are among the highlights in the world of self-winding watches. This ingenious system was adopted for the new Ingenieur caliber, albeit with some modifications. For the rotor bearing, Albert Pellaton once chose the bearing principle, which we know from Machine Elements lectures: "The big bearing carries, the small one leads!" In principle, all the front wheels of our automobiles are also designed in this way. A purely cylindrical axle with two bearing blocks of the same diameter is, of course, much cheaper to manufacture but is considered a "mortal sin" in mechanical engineering. The automatic bridge, which can still be found as "shock absorber spring for automatics" in the caliber C. 8541 B parts list, was recalculated and modified in such a way that almost the same spring constant should be available for the rotor over the entire rotation range of 360°. The author's request to see this calculation was unfortunately denied. According to IWC information, the rotor bearing has a spring travel of ±0.025 mm in the plane of rotation. Axial shocks are also absorbed by this bearing, and damage to the intermediate plate is prevented by means of rubber buffers embedded in the lower side of the rotor. The question as to who carried out this calculation and whether it was determined using the finite-element method unfortunately remained unanswered. The rest of the Pellaton winding mechanism's design was adopted almost unchanged. On the automatic wheel, the wolf teeth module was halved. The two pawls received widened teeth. Of course, doubling the number of teeth theoretically improves efficiency. The fact that smaller teeth wear more quickly must also be taken into account. The widened teeth provide greater safety in the event of axial impacts but are many times more expensive to manufacture. In the context of a new project, the author procured more than thirty automatics with calibers C.8531, 854, 8541, etc. and had them overhauled at IWC. In none of these watches, which had at least thirty years of service behind them, had a latch slipped or the automatic wheel been damaged. Nevertheless, the current movement variant is better protected against impacts and other stresses.

In IWC's magazine *Watch* 1 (April 2005), the following can be read about the properties of the newly designed rotor: "The rotor itself is also of 'inherently sprung' design. It has been provided with specially shaped cutouts to allow it to twist in response to vertical shocks and impacts and to absorb shock-like kinetic energy without stressing the bearings." The author was unable to comprehend this statement. According to IWC, the rotor carrier plate is made of brass and has a material thickness of 0.022 inches (0.56 mm). Even if only one of the two segments had been calculated in its effective length as a "beam clamped on one side with external point load," there would not have been any noticeable deflection, even with very high forces. Brass is a nonferrous metal made of copper and zinc and does not have the high modulus of elasticity of spring steel (modulus of elasticity of brass 9,000 kg/mm², of spring steel 22,000 kg/mm²).

According to IWC, vibration tests and homologation were successfully carried out, and vertical impacts and blows did not result in any damage to the movement.

All in all, the new IWC caliber 80110 can be said to have a rock-solid and very robust design. It might be understandable that there are compromises in the structure and bearing of the rotor as well as in the construction of the regulating system.

Illustration 522
The shock absorber spring, later called the automatic bridge, of the caliber 8541 B shown on the left, magnified approximately 9×. The two different bearing diameters can be seen on the journal.

Illustration 523
The shock absorber spring or automatic bridge of the IWC caliber 85110, with a journal that has one continuous bearing diameter

Comparison of the Most-Important Characteristics of the Automatic Device Module, Oscillation System, and Fine Adjustment (Selection), Table 1					
	AP 3120 (new)	AP K 2072 (old)	IWC 80110 (new)	IWC C. 8541 B (old)	Rolex 3135
Movement Diameter1 (inches/mm)	1.04/26.60 (11¾''')	1.14(1.16)/29(29.5)	1.18/30	1.10/28	1.10(1.12)/28.1(28.5)
Movement Height	0.167/4.26	0.248/6.3	0.284/7.23	0.232/5.9	0.236/6
Rotor Type	110° segment, completely in 920 gold	180° segment with oscillating weight in 750 gold	180° segment with heavy metal oscillating weight	180° segment with heavy metal oscillating weight	180° segment with heavy metal oscillating weight, rotor barer plate with spring effect
Rotor Housing	ball bearings	4 ruby roller bearings	2 rubies	2 rubies with different diameters	2 rubies with different diameters
Rotor Function	bidirectional	bidirectional	bidirectional	bidirectional	bidirectional
Changer	rocking-bar changer	rocking-bar changer	Pellaton winding	Pellaton winding	ratchet wheel changer
Balance Wheel Type	two-spoke with flat hairspring	two-spoke with Breguet hairspring	three-spoke with flat hairspring	three-spoke with Breguet hairspring	four-spoke with Breguet hairspring
Balance Wheel Fine	Gyromax	micrometer screw with gooseneck spring Adjustment	Triovis regulator	3 masselotte weights plus fine adjustment by means of eccentric screw	4 weight screws, Microstella system
Balance Wheel Housing	bridge	balance cock	balance cock	balance cock	bridge
Number of Vibrations (A/h)	21,600	18,000	28,800	19,800	28,800
Power Reserve (hrs.)	60	40	44	40	50
Rate Deviation (DEL sec./day)[2]	8	9	5	005[3]	3
D/h Ratio	6.24	4.6	4.75	4.75	4.68
Jewels	40	29	28	23	31
Introduced in	2005	1962	2005	1968	1988

1. Installation diameter / overall diameter 2. In six positions (average value from three measurements) 3. After overhaul by Martin Becker

Now a question about the production of plates and bridges. If the movement manufacturer ETA in Grenchen produces the Chronographene caliber 7750 in large quantities, how can a small series of a modified version for IWC be produced economically? Quite simply! Plates and bridges are conventionally stamped in the standard version and then finished on various machines in several further operations. In the latter processing steps, IWC's precise design data are then successively incorporated. This process is known in technical jargon as CAM (computer-aided manufacturing). The stainless-steel Ingenieur Reference IW 3227 test watch provided by IWC was very bulky and, with its voluminous size, barely fit under the left shirt cuff, but it went well with the author's white-gold cuff links. Such weighty watches are probably better suited for rolled-up sleeves, which further emphasize the dynamics of this Ingenieur.

The newcomer from Schaffhausen passed the usual initial test and shone with astonishingly good rate results. Even with several measurements, a position difference (DEL) of 005 to 006 sec./day was consistently achieved. Thus, the Ingenieur clearly beat the other three test watches and also underlined the special solidity of this new automatic movement. But how does it compare to the former IWC caliber C. 8541 B? In previous publications the author has consistently emphasized that the IWC caliber C. 8541 is one of the best automatic movements ever made in Switzerland. Certainly an innovative wideness would have allowed even-better products. Even then, the Pellaton winding system was one of the most robust technical solutions. The fine-adjustment mechanism with Breguet hairspring, balance wheel with three *masselotte* weights, and separate adjustment of rate and decay via eccentric screws were considered state of the art even then. It is true that the caliber C. 8541 B sometimes showed critical position deviations in one or two of the six positions, but these were due to the design and the manufacturing tolerances of the time. If the basic design had been imaginatively reworked to incorporate today's manufacturing methods and tolerances, IWC would certainly have scored a "direct hit" that would have restored the Schaffhausen manufacture to a leading position in the world of Swiss watches. Already under Günter Blümlein the course was set in a different direction in the mid-1990s, which ultimately led to today's caliber 80110.

The story of the gold IWC men's wristwatch used for comparison is also interesting, since it allows us only a glimpse of the C 8541 B manufacture caliber after opening the screw-down case back. This rare collector's item was posted on the internet in England and went to its new owner for around $933. A real bargain! Its identity was proven by its movement number

1 976 636 and case number no. 1 788 768. The hallmarks and other IWC trademarks were also present inside the screw-down case back. The beautiful piece had been sold by the IWC London branch on September 22, 1971. With the reference designation R 817 AD, this timepiece was unveiled in 1968 but had been listed in IWC's price list for only a full two years. The price of this R 817 AD in 750 yellow gold and leather strap was 1,040 CHF.

The overhaul commissioned from IWC was booked at $477, to which was added 16% EU import sales tax, customs duty, and freight costs. The automatic, which came back almost as new, passed the usual arrival test and the visual inspection with flying colors. Only the mediocre gear behavior, which was hardly better than before the overhaul, dampened my enthusiasm. A position difference (DEL) of 015 sec./day seemed to the author to be too high. To save me another trip through customs, this timepiece went to trusted watchmaker Martin Becker. In close cooperation with IWC's service department in Schaffhausen, our reference R 817 AD was disassembled and analyzed again. The reassembly with a completely new fine-adjustment mechanism as well as the corresponding adjustment solved the problem and helped the more-than-thirty-five-year-old collector's item to achieve sovereign chronometer values. The position deviation in the six positions came to a value of 005 sec./day, the drop to 0.1. The author was particularly proud of this result, since it was on the same level as the new Ingenieur with the caliber 80110 movement.

To conclude this review, most readers will want to know the results of a comparison of old and new. If we look at the Pellaton winding system with shock-absorbing automatic spring bridge highlighted in the advertising, these are not exciting innovations, but familiar designs that have been improved only in part. In the opinion of the author, some details can be regarded as missed opportunities, such as the design of the rotor arm plate or the rotor axle. Also the fine-tuning of the balance with the Triovis system may not necessarily be seen as an improvement but, according to IWC, a relief for watchmakers. The same applies to the balance rim and hairspring. The three regulation weights of the Gyromax system formerly used were not adopted, nor was the proven Breguet hairspring. Here the constructive focus on simplicity has probably displaced some old virtues but not brought any real innovations. In the end, the question arises of whether our caliber 80110 can be considered a really new movement and, above all, a new manufacture movement. It is certain that the entire German trade press has given this new caliber high praise. But no editor had noticed that this new movement was only partly new and was based on the ETA-*ébauche* 7750.

9.8. Jaeger-LeCoultre, Caliber 975

Noble timepieces from Jaeger-LeCoultre are among the best-known watch brands in Germany today. The REVERSO model, which appears in many different forms, can be considered a "cult watch." In the 1950s and '60s this brand was not yet so widespread in most European countries. What might have been the reason for this? To answer this question, we must first consider all the facts, which reveal the Jaeger-Le-Coultre production figures in the 1950s and '60s. Between 1944 and 1965, exactly 83,000 examples of the manually wound caliber 478, which was also used in military watches, were produced. Of the caliber 497, the heart of the Futurmatic, a popular and advanced automatic watch with a power reserve display and a crown positioned on the case back, the manufacture produced exactly 52,000 movements between 1951 and 1957. For customs reasons, around 80% of these calibers were shipped to the USA unadjusted. Even 115,000 examples of the manually wound caliber 814 with an alarm feature were produced between 1953 and 1964, and again, 60% of this production went to the US market. In addition to these movements, which were produced in large numbers, Jaeger-LeCoultre also produced raw movements (*ébauches*) for other brands; for example, for Audemars Piguet in Le Brassus, for Vacheron & Constantin in Geneva, and at times even for Patek Philippe. In the first two cases these were reserved raw movements, with both Audemars Piguet and Vacheron & Constantin helping to develop them. For Patek Philippe, the *ébauches* from Jaeger-LeCoultre were only a stopgap until the production of the new and again-self-developed automatic calibers had started. It should also be mentioned that until 1968, Vacheron & Constantin and Jaeger-LeCoultre were related through family ownership. Inheritance disputes and disagreements between the owners at the time led to the separation of the previously smooth-functioning corporate alliance.

Jaeger-LeCoultre was not spared the difficult times caused by the "quartz watch revolution." Only through targeted acquisitions of new business areas, such as medical technology, was the tradition-rich company able to survive this difficult period.

The course for a new manufacture caliber was set in 1998, during Günter Blümlein's lifetime. Even then it was clear that the well-established 889 (subsequently 889/1 and 889/2) automatic movement could not meet the challenges of the future on its own. With a diameter of 1.02 inches (26.00) mm and a height of 0.127 inches (3.25 mm), this automatic movement met the requirements of flat-built men's wristwatches. Among engineers, this low construction height in relation to the outer diameter is regarded as a technical constraint based only on management specifications, but not on the laws of technical mechanics. How good it is that these disadvantages have been recognized and that the path of virtue already proven earlier has been chosen again. Even at first glance the new caliber 975 reveals the constructive division into base plate, bridges, and automatic device module.

The author wanted to see this for himself and visited Jaeger-LeCoultre in Le Sentier, Switzerland. Here he was able to talk to the designers, visit the development laboratories, and also take a closer look at production. Even a look into the eyes of the hardworking employees was not to be missed. Almost nothing remained hidden from the curious visitor. The depth of production realized at Jaeger-LeCoultre is quite remarkable and can be compared to that of Rolex. Whether it is the still-traditional plate production, the welding of tiny movement components by laser, or the mechanized setting of the new Nivarox balance

Illustration 524
This rare automatic came from New Zealand and amazed everyone during the initial inspection. Its rate deviation in the six positions was only 005 sec./day. This almost-forty-year-old chronometer was thus much more accurate than our new Hometime.

Illustration 526
The new Hometime by Jaeger-LeCoultre, a self-winding watch with the new caliber 975 movement. Under the hour hand is another hand that appears only when a second time zone is set.

Illustration 525
View of the automatic caliber K 883 S of our Geomatic from Jaeger-LeCoultre. No spectacular finishing, but a rock-solid design with remarkable accuracy. As was common in the past, the automatic bridge is screwed onto the base movement.

Illustration 527
The new caliber 975 incorporates not only modern design principles but old virtues as well. The automatic bridge, for example, is once again mounted as a separate assembly. The asymmetric balance bridge has an adjustable hairspring stud clamped in a similar way as on the Rolex caliber 3135. The balance wheel, which is excited by a flat hairspring, has four spokes and four weight screws. If they would make the opposing adjusting screws in two different colors or provide them with screwheads of different lengths, they could reduce the risk of confusion when adjusting the rate.

with four adjusting screws—these all were solutions of the finest order. Asked about the four most important criteria of the new caliber, Jaeger-LeCoultre made the following statements:

- The most precise caliber on the market
- Unbeatable in terms of reliability (i.e., no components can be adjusted by impact or shock)
- Increase in service intervals from the previous three to five years to six to ten years
- All watch functions adjustable via the crown (very low-wear solution)

The basic philosophy of the 970/975 family not only includes supplementing the previous self-winding caliber 889/2 but also represents the basis for further developments and complications. The decision in favor of one-directional rotor winding is based on several comprehensive tests. Two comparable movements, one with one-directional and the other bidirectional winding, were put on a watch winder, and their winding performance was measured. On this winder, bidirectional winding proved superior to one-directional. In tests conducted on the left wrist of various employees, however, one-directional winding was again superior to bidirectional. Jaeger-LeCoultre experts also let slip that this one-directional winding would soon be used in the well-known 889/2 automatic caliber as well. When asked provocatively why Jaeger-LeCoultre would describe the ceramic ball bearings in the rotor bearing as a novelty, when ceramic ball bearings had long been used in gardening tools, Jaeger-LeCoultre replied that previously it had not been possible to manufacture the required tiny ceramic balls, in this case made of zirconium oxide.

All in all, the new caliber 970/975 is a clean and professionally designed automatic movement. Its large diameter and the available movement height ensure sufficient construction volume to make the plate and bridges stable and to guarantee the necessary strength. There is sufficient space for the large mainspring barrel and the following gear train. For gears and pinions there is enough height to allow sufficiently large bearing distances. After removing the automatic bridge, the entire gear train, as well as other components, becomes accessible. The balance wheel and balance bridge are also well sized and easily accessible. The drop is easily adjustable by turning the hairspring stud. The gear is adjusted by means of four weight screws, each positioned opposite the other. If these are turned inward vis-à-vis each other, the moment of oscillation (GD^2) is reduced. If, on the other hand, they are turned outward, the moment of oscillation increases. Since the moment of oscillation is inversely proportional to the rate, the watch then either advances or retards. We already know this adjustment of the rate from the four microadjustment screws, which ensure an exact rate in the balance of the Rolex caliber 3135. On Rolex watches these screws come in two different sizes and are adjusted from the inside. On Jaeger-LeCoultre timepieces, the four adjusting screws can be adjusted from the outside. Incidentally, this is an important prerequisite for being able to automate such a step later on.

The rotor is of the same design as that of the caliber 889.2, except that the ball bearing is of a different design and larger in outer diameter. The rotor carrier plate is basically identical and again acts as a spring, only the flywheel weight is double stepped and has a greater mass. What is not immediately obvious—but still has to be mentioned—are the spatial reserves of this design. There is a lot of unused space under the rotor, as well as in the outer area of the plate. Surely a good starting point to enable this caliber still many additional functions. Let's be surprised!

But how does this compare to an earlier Jaeger-LeCoultre caliber? To answer this question, we chose a steel-cased Geomatic Ref. E 339/1, a very rare and much-sought-after chronometer. Inside this timepiece is the caliber K 883 S with the movement number 2 019 073. This rarity was tracked down at a dealer in New Zealand and changed hands for $1,500. During the usual arrival test, the gear performance was also checked. What were the results? Sovereign chronometer values! The deviation in the six positions was measured repeatedly and remained at (DEL) 005 sec./day! Only the maximum amplitude of 275 signaled deficits in the lubrication of the now-more-than-forty-year-old watch. The externally not very impressive automatic K 883 S movement was developed from the basic caliber 880, which was launched in 1959. With an outer diameter of only 1.02 inches (26.00 mm) but a height of 0.241 inches (6.14 mm), it is a rock-solid movement whose virtues are not so much its beauty, but the precision with which it runs.

The basic design of this movement includes plate, gear train, and mainspring barrel bridge. The double-spoke balance wheel is guided by a balance cock and has a Nivarox hairspring. The hairspring stud is movable, but a fine adjustment for the regulator arm was omitted for unknown reasons. The self-winding bridge is screwed to the base movement. The rotor winds in both directions, which is transmitted via a rocking-bar mechanism. All in all, a very solid design.

If one looks at the movements side by side, the almost forty years that separate them become more than clear. The fundamentally different design principles prove the modernity of the new caliber 975 with no ifs, ands, or buts. Our Jaeger-LeCoultre Hometime Ref. 1628420 with caliber 975 and movement no. 2 321 526 test watch did not shine at the beginning with the best rates. It lost about seven seconds per day, and the deviations in six positions (DEL) with 009 sec./day did not arouse enthusiasm. During the previously described visit to Le Sentier, they were able to correct the daily rate but not the deviations in six positions.

Finally, the author wants to emphasize that the new caliber 970/975 is a great success. In all relevant disciplines the experts in Le Sentier have consistently taken modern design principles into account and also come up with new ideas. The only thing that still needs to be brought up to scratch is the movement performance. That Jaeger-LeCoultre in Le Sentier has also mastered this discipline is proven not only by the old Geomatic presented, but also by another of the author's everyday watches. The steel Master with perpetual calendar, Ref. 140.8.80 with the number 553, serves day after day with a precision that is consistently 001 to 002 sec./day in the six positions. From Grand Prix racing, we know cases where the high-performance engines lack nothing except for the exact adjustment. Is this perhaps also the case with our new caliber 975?

9.9. Vacheron & Constantin, Caliber 2475

Watches from Vacheron & Constantin have always been among the best made in Switzerland. Especially in the 1960s, the models of this long-established Geneva manufacturer were in no way inferior to those of Patek Philippe and were on the same level in terms of design, technology, and diversity. The annual production figures of these two Geneva watch manufacturers were also of the same order of magnitude. For its 250th anniversary in 2005, this renowned Geneva watch brand gave itself the most beautiful gift: its own movement production and its own automatic movement. But what all had to be done beforehand to make this wish come true? Before answering this question, let's first take a look at the past. The hour of the first self-winding wristwatches struck at Vacheron & Constantin as early as 1949, with the raw movement (*ébauche*) supplied by Jaeger-LeCoultre in Le Sentier. It was largely identical to the Jaeger-LeCoultre caliber 476 but was designated by Vacheron & Constantin as movement 477 (modified 477/1). Still equipped with a limited oscillating weight (popularly called a "hammer automatic"), it wound in one direction only. It was first used with a gold men's wristwatch with the reference number 4871, which was unveiled at the end of 1949. This was followed in 1955 by the first automatic models with a 360° rotating rotor. In the following years Jaeger-LeCoultre remained a supplier of reserved raw movements. With the change of ownership in 1988 and the subsequent restructuring of the company, the circle of movement suppliers included first Frederic Piguet and later Girard-Perregaux. Neither in technology nor in quality could these two manufacturers reach the level of Jaeger-LeCoultre, but they were considerably cheaper.

When designing the new in-house automatic movement, its diameter and height were prescribed, since it had to replace three movements that had been used so far. With a diameter of 1.03 inches (26.2 mm) and a movement height of 0.141 inches (3.6 mm), the dogma of low construction height was not abandoned, and this again forced the designers to accept compromises that clearly stood in the way of a real new beginning. Thus, the basic components of the movement consist of a rather flat plate, mainspring barrel bridge, gear train bridge, and balance cock. The self-winding mechanism was integrated into the two bridges and is one-directional. A changer could hardly have been accommodated in the limited space available. The double-spoke balance wheel uses a flat hairspring, while the fine adjustment is done by a Triovis system. However, Vacheron & Constantin made clear that this Triovis system is only a temporary solution until Nivarox is able to deliver the screw balance wheel, which we already know from the Jaeger-LeCoultre caliber 975. On the dial side, a separate plate allows for the arrangement of the various dial indications, including the power reserve. The overall design of the new caliber 2475 was kept conventional and unmistakably bears the signature of the responsible designer. In his previous professional life, he had also served the companies Nivarox, Frederic Piguet, and Jaeger-LeCoultre. Unfortunately, he was not familiar with Vacheron & Constantin's earlier automatic calibers; otherwise he might have adopted some of their excellent technical solutions in terms of design and style. We are thinking in particular of the caliber K 1072/1, which will be mentioned again.

Illustration 528
We consider this automatic Ref. 6782 Q from Vacheron & Constantin to be among the most-beautiful and most-noble watches of the 1960s. A silver-plated dial with dauphine hands and hour indexes in 750 yellow gold characterize the face of this rare collector's item. In addition, there is the rotating bezel and the special crown, which were reserved only for this model.

Illustration 529
For comparison, the earlier Vacheron & Constantin caliber K 1072/1 from 1968. It was purchased as a reserved raw movement from Jaeger-LeCoultre. A rotor with four ruby roller bearings, guilloché oscillating weight in 750 gold, a Gyromax balance wheel with Breguet hairspring, and other details helped this caliber to a level matched only by movements from the house of Patek Philippe.

Illustration 530
Truly a noble appearance! Here Vacheron Constantin presents us with one of the first models with the new, in-house automatic caliber 2475. The date, day of the week, and power reserve display give this gem a high value. Even the Geneva Seal, in the past modestly stamped only on the plate and bridge, is proudly emblazoned on the dial between four and five o'clock.

Illustration 531
This shows us many details of the new Vacheron & Constantin caliber 2475. In the foreground is the double-pinned balance cock with double-spoke balance wheel and flat hairspring. This movement version still has a Triovis regulator. As soon as Nivarox can deliver it, a four-spoke balance wheel with four weight screws like that of the new Jaeger-LeCoultre caliber 975 will be used in its place.

Comparison of the Most-Important Characteristics of the Automatic Device Module, Oscillating System, and Fine Adjustment (Selection), Table 2					
	JLC 975 (new)	JLC 883 S (old)	VC 2475 (new)	V&C K 1072/1 (old)	Rolex 3135
Movement Diameter (inches/mm)[1]	1.18/30	1.02/26	1.0(1.03)/25.6(26.2)	1.14(1.16)/29(29.5)	1.10(1.12)/28.1(28.5)
Movement Height	0.224/5.7	0.241/6.14	0.133(0.220)/3.4(5.6)	0.248/6.3	0.236/6
Rotor Type	180° segment with oscillating weight in 920 gold	180° segment with heavy metal	180° segment entirely in 920 gold	180° segment with oscillating weight oscillating weight in 750 gold	180° segment with heavy metal oscillating weight, rotor bearer plate with spring effect
Rotor Housing	ceramic ball bearings	2 rubies	ball bearings	4 ruby roller bearings	2 rubies with different diameters
Rotor Function	one-directional	bidirectional	one-directional	bidirectional	bidirectional
Changer	–	rocking-bar changer	–	rocking-bar changer	ratchet wheel changer
Automatic Bridge	superimposed	superimposed	integral	superimposed	superimposed
Balance Wheel Type	two-spoke with flat hairspring	three-spoke with flat hairspring	two-spoke with flat hairspring	two-spoke with Breguet hairspring	four-spoke with Breguet hairspring
Balance Fine Adjustment	4 weight screws	regulator arm with no fine adjustment	Triovis	Gyromax	4 weight screws, Microstella system
Balance Wheel Housing	bridge	balance cock	balance cock	balance cock	bridge
Power reserve (hrs.)	50	40	44	40	50
Number of Vibrations (A/h)	28,800	19,800	28,800	18,000	28,800
Rate Deviation (DEL sec./day)2)	9	5	0083	10	3
Jewels	29	23	27	29	31
Year Introduced	2004	1967	2005	1968	1988

1. Installation diameter / overall diameter
2. In six positions (average from three measurements)
3. After installation of new fine-adjustment mechanism

In all, the new Vacheron Constantin caliber 2475 represents a solid design that also pleases with its particularly successful finishing. However, if Vacheron Constantin wants to regain the leading role it had in the 1950s and 1960s, the new caliber 2475 should play only a transitional role until a truly innovative and promising automatic movement is available.

The test watch, which was handed over in the still-new factory in Le Sentier, impressed by its fitting-case diameter as well as the indications for day of the week, date, and power reserve. Even the Geneva Seal, previously present only on the plate and bridge and therefore invisible from the outside, was proudly emblazoned on the dial. The now-fashionable glass case back was also not missing and allowed a view of the beautifully finished automatic movement. During the initial test, all relevant criteria were on the positive side; only the rate deviation in six positions was not convincing. It was between 012 and 014 sec./day and thus not within our expectations. What was the reason for this? In consultation with Vacheron Constantin in Geneva, the noble piece was taken to watchmaker Martin Becker, who himself had provided service for the German branch of Vacheron Constantin for more than sixteen years. There our test watch was disassembled and analyzed and a completely new fine-adjustment mechanism was installed. This work was followed by a lively exchange of ideas between Vacheron Constantin and Martin Becker regarding possible modifications to this new movement. What seemed even more important to us were the improvements achieved. After the return of the watch, it was tested again. The rate in six positions was now only 008 sec./day, which is more than one-third better than on delivery.

But how does this compare to an earlier automatic movement from Vacheron & Constantin? For this purpose we selected a very rare piece of jewelry from 1968. The reference number 6782 Q (movement no. 598 456, case no. 433 445) contains the caliber K 1072/1, which is also a reserved raw movement from Jaeger-LeCoultre. It is largely identical to the Audemars Piguet caliber K 2072 described, except that the rotor carrier plate has a sunburst finish instead of Geneva stripes. Finally, our Vacheron & Constantin uses a Gyromax balance wheel and can do without fine adjustment via micrometer screw and gooseneck spring. This rare collector's item was acquired more than ten years ago by watch dealer Manfred Josten from Mönchengladbach; the price at that time was $4,411. This watch is noteworthy for several reasons. First, it is the only Vacheron & Constantin model that has a rotating bezel. Second, it has the original crown from the 1968 production year, which for unknown reasons is not listed in the available Vacheron & Constantin parts list.

Comparing this almost antique caliber K 1072/1 (/1 stands for Gyromax balance) with the new caliber 2475, the newcomer is in no way inferior to the old one from a purely aesthetic point of view. Only on closer inspection does one notice the deficits that result from flat-design dogma. Neither the Gyromax balance wheel with ten *masselotte* weights nor the Breguet hairspring was used. The material thickness of the gears and the length of the ruby bearings also had to be reduced. These all are measures that inevitably cause reductions in resistance to wear.

Now a word about the connection of the plate and bridges. The K 1072/1 caliber has a conventional design. The respective connections consist of two dowel pins ("feet") and one or two cylinder head screws. On the balance cock of our K 1072/1, the two dowel pins protrude from the plate but the upper fitting holes are designed as blind holes, so that our dowel pins are effective but not visible. To ensure that the clamp sits neatly with only one screw, it has a centrally located clearance milling on the underside. This also allows the clamp to be lifted off easily. This connection method was also used for the new caliber 2475, although with visible dowel pins ("feet"). Apparently the design with flanged bushings with internal threads, as we know it from the other three new calibers, was deliberately omitted here. Summing up all pros and cons of the two calibers, one can't help but ask if the tradition of Vacheron & Constantin, on the one hand, and today's technical progress, on the other, could not have been combined more effectively with the new caliber 2475? But what is not yet can still be.

9.10. Rolex, Caliber 3135

First, a word of clarification. The author does not have a side job as a PR consultant at Rolex, nor does he allow himself to be paid for airline tickets to Geneva or have watches serviced free of charge. The fact that our reference caliber 3135 comes off so well in the evaluation is solely due to the technical facts. The Rolex automatic caliber 3135 has been described by the author elsewhere several times. Nevertheless, the basic design of this movement should be summarized. The bridges for the mainspring barrel, the gear train, and the balance are on the sturdily designed plate. The self-winding bridge is a separate unit attached to the base movement. The oscillating weight of the bidirectional rotor is made of heavy metal and is secured by a thin plate of rhodium-plated brass. This has offset apertures and thus provides the desired spring action in the axial direction. The rotor axis is guided by two ruby ball bearings of different diameters, and the vertical clearance is adjusted by a clamping wedge. Basically this design principle corresponds to the rotor bearing that Albert Pellaton previously chose for his IWC caliber 85.

The robust changer consists of two ratchet wheels made of a special aluminum alloy and anodized in red. For this function it is not the pitch accuracy that is required, but exclusively the hardness of the tooth flanks. In fact, the anodized layer is as hard as glass and allows long-term operation without lubrication. The gear train of the basic movement has special properties.

Illustration 532
Encased in this new Rolex Turn-O-Graph model in steel and gold is the caliber 3135 movement.

The bearings of gears and pinions are designed in such a way that there is almost the same bearing friction both in vertical and horizontal positions. The balance bridge and the balance itself require special mention. Not only is the balance symmetrically arranged along the longitudinal axis of the watchcase, giving this caliber a special aesthetic appeal, but its design is also considered to be particularly well thought out. The drop can be adjusted using the adjustable hairspring stud, the vertical play of the four-leg balance by two special knurled screws on the bridge bearing. The rate of time itself is adjusted by the four

Illustration 533
The caliber 3135's balance wheel with Breguet hairspring and four Microstella timing screws under approximately 7× magnification

Illustration 534
The four-spoke balance wheel with Microstella timing screws, Breguet hairspring, and adjustable hairspring stud. On the bridge on the right, one can see the small plate that locks the hairspring stud. On the right and left sides of the bridge, you can see the knurled nuts that are used to adjust the vertical play of the balance staff. The rotor bearer is made of thin rhodium-plated brass, and its staggered apertures provide the desired absorption of axial shocks.

Microstella screws, a mechanical principle that corresponds both to the Jaeger-LeCoultre set screw system and the Audemars Piguet Gyromax system. The sum of all virtues is complemented by the Breguet hairspring, which has been tried and tested for decades and can "breathe freely" thanks to the ample space available. Taken together, all the aforementioned details bring about the great robustness and high precision that are achieved almost only by watches from the house of Rolex.

9.11. Conclusion and Outlook

A Comparison with Surprising Results

While reading these lines, some readers will probably secretly ask themselves why all watch manufacturers do not build self-winding movements like those of Rolex. One answer is likely to be justified by the behavior of engineers: "Three engineers, three different opinions." The fact that there are fundamental differences of opinion in the watch industry, which are actually not always understandable, may be explained here with two examples. The first relates to the escapement wheel bearing.

Illustration 535
Disassembled and inverted, the automatic device module of the caliber 3135 with conversion gearing and rotor. The colored gears are made of special aluminum and are hard-anodized. The small pinion transmits the torque from the automatic device module to the winding mechanism.

The person responsible for this at IWC was of the opinion that in this case, a bearing using a combi-chuck was purely cosmetic. Audemars Piguet's opinion was the direct opposite and was immediately confirmed by Rolex. In the case of a simple anchor wheel bearing using a ruby, strong impacts would lead to a reduction or complete failure of the lubricating oil. With the combi-chuck, sufficient oil and lubrication would be guaranteed even after unforeseen stresses. The Rolex technician in charge also stressed the importance of this system's vertical play, adjustment, and oil retention. His final comment was "Rolex never does anything for free!" The second case concerns the rotor bearing. Here, Jaeger-LeCoultre came to the opinion that a roller bearing (ball bearing) with ceramic balls is the best long-term solution. The person responsible for this at IWC was of the opinion that ball bearings expire after a manageable period of time and that ruby bearings are thus far superior to this design principle.

With this statement, the question arises of why Rolex now has the rotor supported by a ball bearing in its caliber 4130, which is encased in the Daytona. The answer from Rolex was spontaneous: one can achieve a lower height with a ball bearing!

But how does the final evaluation of the four new automatic movements turn out? The two main tables contain all relevant design data and the measured rate deviations. If one takes a closer look and compares the details of the movements, one will reach a reliable conclusion. Of course, the entire analysis is only a snapshot. It will take four to five years for the new automatic movements from Switzerland to prove their capabilities. There is no doubt that the automatic movements from Audemars Piguet and Jaeger-LeCoultre have sufficient potential for further development. The fact that both IWC and Vacheron & Constantin have proceeded hesitantly and have not fully exploited their potential can already be seen from the comparison tables. The positive thing about all technical solutions is the chance to make what already exists even better. That's why it is time for those who have been approached—and this in both senses of the word—to get to work now!

Illustration 536
The Rolex caliber 4130 movement installed in the Daytona. *Rolex photo*

Illustration 537
The test watches (*top left clockwise to bottom left*): Patek Philippe Annual Calendar Chronograph, Jaeger-LeCoultre Master Compressor Chronograph, IWC Da Vinci Chronograph, and the watch used as a reference, a Rolex Daytona Cosmograph

10. Current Chronographs from Switzerland

IWC, Jaeger-LeCoultre, and Patek Philippe Must Be Measured against Rolex

10.1 Introduction

The renaissance of mechanical wristwatches began more than thirty years ago. In terms of new chronographs with manufacture calibers and self-winding, however, watch lovers have had to wait until the new millennium. We took a look at the current models from IWC, Jaeger-LeCoultre, and Patek Philippe; took them apart; and tested them. We then compared them to the product made by market leader Rolex, the Daytona Cosmograph.

In addition to the constructive (i.e., watchmaking) design of the movements, the test list also included rate accuracy, utility value, wearing comfort, and external impact.

The involvement of six expert watch enthusiasts as test subjects led to some remarkable conclusions.

A chronograph not only is considered an adornment of every man's wrist but has also always been understood as a special watchmaking achievement. Among experts, chronographs are therefore often and rightly referred to as "complication watches."

The fact that the design and manufacture of a chronograph movement is indeed a special achievement is demonstrated by the fact that, until the launch of the new automatic chronograph movements presented, the three Swiss noble manufacturers IWC, Jaeger-LeCoultre, and Patek Philippe offered "time recorders" only with purchased raw movements. Even if the latter were reworked in an artful and technically very high quality, they always remained "foreign objects" in the manufacture caliber family.

At IWC, the Da Vinci already had a firm place on the chronograph market since 1985, especially since, in addition to the automatic caliber Valjoux 7750 from ETA, it also had a module for a perpetual calendar developed in-house. Despite a great willingness to innovate, Jaeger-LeCoultre did not have a self-developed chronograph movement on its urgency list for years. The friends of time recorders of this brand used to have to be satisfied with a watch with the quartz caliber 630, which, by the way, was also used by IWC (as the caliber 601).

Patek Philippe has a long tradition of making chronographs, but these watches always used manually wound Lemania or Valjoux movements. To their credit, these *ébauches*, as raw movements are called in French, were so modified and embellished at Patek Philippe that their original origins were barely recognizable.

The simplest solution for designing a new chronograph with automatic winding is considered to be the use of a module that can be bought on the market that is placed under the dial on an existing automatic movement. Such combinations are easy to recognize (e.g., by the very low position of the date indication or the offset of chronograph pushers and winding crown), which are, after all, on two different levels.

Our three candidates opted for a different path, and each developed a completely new chronograph movement. Such a step can take a period of around five years and involve an investment on the order of $5,600,000. Despite this major investment of time and money, a great success (i.e., a particularly successful design) cannot be expected in every case. This also depends to a large extent on the know-how, size, and level of the development team, as well as the technical resources available to the respective manufacture. Added to this is the current patent situation, which can inherently restrict the technical options available to newcomers. Certainly the development engineers at IWC, Jaeger-LeCoultre, and Patek also had to cope with this situation.

Illustration 538
Because the Daytona in 18K white gold provided by Rolex Cologne was not by itself a reliable basis of evaluation, two further Daytonas were obtained from specialist dealers. These provided a broader basis for comparison.

10.2. Three Current Chronographs—with Rolex as the Reference

The IWC, Jaeger-LeCoultre, and Patek Philippe brands are undoubtedly among the best watchmakers in Switzerland. In this regard, Patek Philippe can confidently be granted a special role, that of the "leading position in the upper league," so to speak. The self-winding chronograph with annual calendar presented by Patek Philippe can also almost be called exotic. The cost of this masterpiece, which is produced only in small numbers with a platinum case, is more than $54,000. In addition to the small number produced, delivery wait times can be one, two, or three years.

Table 1: The Test Watches

Brand	Model	Reference Number	Caliber	Case Version	Movement Diameter (inches/mm)	Movement Height (inches/mm)	Number of Parts	List Price (Euros) (in 2008)
IWCF	Da Vinci Chronograph	3764	89360	stainless steel	1.19/30.4	0.293/7.46	movement 269 calendar module 20	10,600
Jaeger-LeCoultre	Master Compressor Chronograph	148.8.25	751	stainless steel	1.00/25.6 installation 1.03 overall	0.221/5.63	movement 251 calendar module 21	7,700
Patek Philippe	Annual Calendar Chronograph	5960P	CH 28-250 IRM QA 24H	950 platinum	1.18/30 base movement 1.29/33 with date plate	movement 0.204/5.2 calendar module 0.097/2.48 total height 0.302/7.68	movement 302 calendar module 154	51,370
Rolex	Daytona Cosmograph	116519	4130	750 white gold	1.20/30.5	0.255/6.5	201	15,640

10.3. Initial Inspection and Practical Test

Upon receipt, all the test watches were subjected to a functional test that measured their accuracy. They were measured on the Witschi Chronoscope S1 with Mikromat S while fully wound, and four measurements were taken, from which average values for daily rate, daily rate deviation, decay, and amplitude were then determined. We have summarized these measured values in table 2 and also placed the respective maximum and minimum values in brackets.

While measurements in the laboratory are always performed under the same conditions for all watches, a practical test can be quite different. For this reason we chose not four but six experts to perform the wear test for us. It is common knowledge among engineers that a measurement is not a measurement: because we had chosen the Rolex Daytona as a reference, on the one hand, but, on the other hand, we had only one watch available, we purchased two more Rolex Daytona watches anonymously: one from a German and one from an Italian dealer. Thus, a statistically secure basis could be created to prevent possible under- or overvaluations of the Rolex brand.

In addition to the watch, a form was prepared for each of the six test subjects, on which were entered their respective measurement results. Thus, the daily rate was to be determined and entered in the morning and evening over a period of fourteen days. The daily morning and evening times were also recorded. The respective daily rate was determined via clock radio or telephone time. The practical values determined by the test persons are summarized in tables 3a and 3b. However, our watch testers were interested not only in the accuracy of the watches. They also commented on the evaluation criteria "general impression, practical value, accuracy, wearing comfort, and outward appearance." These individual statements and the test results are summarized on page 342.

The extensive wearer tests and the rate values determined in these tests differed significantly in some cases from the values measured in the laboratory. For example, the "crown up" test position may be of significance only to busy bus and streetcar drivers. The "dial down" position will also be relevant only for those contemporaries who often go to fortune-tellers and have their future prophesied from their palm lines.

After the fourteen-day wear test, the IWC advanced about +29 seconds, the Jaeger-LeCoultre about +24 seconds, and the Patek Philippe about +22 seconds. The Rolex completed the test with ±0 seconds. The two additional Daytona models held their own, advancing a full three and seven seconds after two weeks. The bottom line is that the three Daytona watches from Rolex showed the smallest differences between practical values and laboratory measurements.

10.4. The Chronograph Movements

The following is a quotation from the lectures by Prof. Dr.-Ing. Klaus Ehrlenspiel, professor emeritus of the Chair of Design in Mechanical Engineering (now Chair of Product Development) at the Technical University of Munich: "Implementing a complex design task in such a way that the result is functional reliability, minimal technical scope, favorable manufacturing costs, low maintenance, and an attractive design requires special innovative talent, a high level of expertise, team spirit, and leadership responsibility."

This quote, which is more than twenty years old, is still valid today for the development of new movements. In addition to pure design experience, a development team also requires knowledge of materials science, parts production, assembly, and logistics. Equally important is knowledge of the current state of the art and the products of competitors.

First of all, the specifications for a new self-winding chronograph movement must include the most-important criteria, such as functional content, accuracy, power reserve, additional functions, and external dimensions. Once the technical characteristics have been finalized, costs and deadlines still have to be determined and secured. To obtain a reliable answer to all these questions, we not only consulted the designs and exploded drawings of the four movements under consideration but also had the latter disassembled and reassembled by the manufacturers.

Table 2: Measurements Each in Six Positions[1]

Maker	Daily Rate sec./t (x)	sec./day (x)	Maximum Daily Deviation (sec./day)	Average Amplitude in Degrees Drop 4)
IWC	+3.7 (3.5/4.1)	+2.9 (2.4/3.6)	285 (271/291)	0.2 (0.1/0.4)
Jaeger-LeCoultre	+3.9 (2.9/4.5)	+12.7 (11.5/13.9)	260 (247/270)	0 (0/0.1)
Patek Philippe	+0.1 (0.1)	+5.7 (5.0/6.6)	261 (255/264)	0.1 (0/0.3)
Rolex (WG) 2)	+2 (1.9/2.1)	+3.7 (3.3/3.9)	269 (252/276)	0.3 (0.2/0.9)
Rolex (SZ) 3)	+1.3 (1.1/1.5)	+4.0 (3.3/4.8)	270 (267/273)	0.3 (0.2/0.5)
Rolex (WZ) 3)	+0.6 (0.3/1.1)	+6.2 (5.4/7.1)	259 (252/263)	0.5 (0.2/0.9)

1. Average from four measurements, minimum and maximum values in brackets (min./max.)
2. Test watch provided by Rolex used as reference 3. Anonymously procured test watches (SZ = black dial, WZ = white dial)
4. Oscillation symmetry

We have summarized supplementary technical values in table 2. We also found important the different overhaul times, as well as the differences between normal operation and operation with the chronograph switched on.

10.5. The Caliber CH 28-520 IRM QA 24H from Patek Philippe

The designers at Patek Philippe also broke new ground when they set about developing the self-winding chronograph. If you subtract the module for the entire annual calendar, they were left with a diameter of 1.18 inches (30 mm) and a movement height of 0.204 inches (5.2 mm)—that is, a volume of 0.223 in.^{3} (3.67 cm^3) with which to realize the task at hand. Also in this technical solution, the chronograph module is a fully integral part of the movement, and only the automatic bridge next to the mainspring barrel bridge is attached. The core of the mechanism is the vertical clutch, which is directly on the stopwatch axis. The chronograph functions are also controlled here by a switching wheel.

The technical and aesthetic level of Patek Philippe movements is beyond discussion. Timepieces with this brand name are produced in manageable numbers. Each watch is a valuable, incomparable, unique piece and should be appreciated and worn as such. The annual calendar chronograph from Patek Philippe is available only with a platinum case. Annual production numbers are likely to be only in a three-digit range. As to delivery time, we had been told by the concessionaire two to three years. Anyone who is ready not only to accept such a lengthy waiting period for an annual calendar chronograph from the house of

Illustration 540
The four-spoke balance wheel is held in place by a balance cock and has four mass balance weights for rate adjustment. The cylinder head screw in the oblong hole adjusts the hairspring stud.

Illustration 539
The caliber CH 28-520 IRM QA 24 H movement. It has forty jewels and runs at 28,800 vibrations per hour. This watch's platinum case has an elegant screw-down case back with sapphire crystal. On the bottom right, one can see the three auxiliary pushers for setting the annual calendar.

Illustration 541
Patek Philippe Annual Calendar Chronograph Ref. 5960 P in platinum. With the development of the first chronograph caliber with automatic winding, the company also broke new ground in the design of the display: instead of being shown on separate subdials, chrono minutes and hours are displayed on a stacked register at six o'clock.

Illustration 542
Watchcases with style are always considered a matter of course at Patek Philippe. The case middle and bezel have a diamond finish.

Patek Philippe, but also to put a whopping $54,000 on the counter, must be a very convinced watch enthusiast—and have the necessary small change.

In all, you can calmly call this time recorder an exotic—an absolute top product of the Swiss watch industry. The watch even shines with high utility value and very good rate performance.

10.6. The Caliber C. 89360 from IWC

Following on from the 80110 automatic caliber installed in the IWC Ingenieur, IWC introduced a new chronograph movement with automatic winding in the Da Vinci line in 2007. Unlike the aforementioned three-hand caliber 80110, the chronograph movement described here, caliber 89360, is a genuine new development. Only the movement's external dimensions, with a diameter of 1.19 inches (30.4 mm) and a height of 0.293 inches (7.46 mm), are comparable to those of the ETA/Valjoux caliber 7750, which was the principal movement used in previous IWC chronographs.

When a new movement has to be developed, the responsible designer first asks themselves how much space is available for the implementation of the technical requirements. The respective volume is calculated from the diameter and height. In the case of the IWC caliber 89360 described, a generous 0.330 in.^{3} (5.41 cm^3) was available. Jaeger-LeCoultre allowed its developers

Illustration 543
The IWC chronograph caliber 89360 has a diameter of 1.19 inches (30.40 mm) and a height of 0.293 inches (7.46 mm).

just 0.176 in.^{3} (2.89 cm^3), little more than half the IWC specification. In the Patek Philippe movement the available space is 0.223 in.^{3} (3.67 cm^3), and in the Rolex Daytona 0.289 in.^{3} (4.75 cm^3), still 68% and 88% of the volume of IWC's caliber, respectively.

Stefan Ihnen set the tone for the technology of the IWC chronograph movement. The design of the new Da Vinci line was the responsibility of Guy Bove, creative director at IWC. The specifications for the new chronograph called only for the creation of a completely new movement with automatic winding, integrating all functional areas in a space-saving and functionally reliable manner.

In the specifications for the 89360 caliber, management specified the following, among other things: unconventional design, including the integrated arrangement of automatic winding, chronograph module and movement, balance regulated by four screws (as in the caliber 5000), mainspring barrel for sixty-eight-hour power reserve, direct-acting pushers with spring-loaded zeroing lever, zeroing with flyback function, directly driven minutes and seconds, double counter mechanism, chronograph function control via switching wheel, new changer for automatic winding, and much more. For production and cost reasons, the developers were also able to adopt some components of the caliber 5000.

In creating this new caliber, the designers benefited from its ample volume. Thus, the basic static components of this movement, such as the main plate, gear train bridge, and mainspring barrel bridge, are amply dimensioned and thus very stable. The automatic spring bridge was redesigned and made somewhat stiffer. It was thus possible to omit the rubber buffers, which were still used on the caliber 80110.

Illustration 544
IWC Da Vinci Chronograph Ref. 3764 in stainless steel with leather strap. With this new creation, IWC convincingly underlines its position as a watch manufacturer. The unique selling points of this mighty watch are the double-counter mechanism, the automatic winding mechanism in a new design, and the elaborately designed case.

Since the old Pellaton winding system is quite tall, the new double pawl winding system is used. These two superimposed double pawls, which are housed in a double crank mechanism, act together on the automatic wheel but are offset axially and radially. According to IWC, this refined system has significantly higher efficiency and also allows for a lighter rotor than the old Pellaton drive. From a technical-mechanics point of view, even though it takes up much more space than the new solution, the latter should be more robust and durable.

Illustration 545
Caliber 89360 fine-adjustment mechanism with a balance wheel with four weight screws for rate adjustment. Compared to the Patek Philippe fine-adjustment mechanism, it lacks a bit of elegance. The escape wheel shaft is held in place only by an artificial ruby and not by cover plates.

Illustration 546

Illustration 547

Illustration 548

Illustration 549

Illustration 550

Illustration 551

Figures 546 to 551, IWC Caliber C. 89360

Fig. 546: Overall view of the IWC chronograph caliber C. 89360 movement after removal from the watchcase—the top is dominated by the powerfully executed rotor. Its apertures reveal part of the automatic bridge. **Fig. 547**: The movement with the rotor removed. In the right foreground is the automatic bridge. The ratchet wheel, the four pawls, and the ratchet wheel bridge of the new double-ratchet winding mechanism can also be seen. On the left, one can see the complete automatic bridge, and at the top the first components of the integrated chronograph mechanism (zeroing lever min./hr., zeroing lever sec., etc.). Further explanations can be found in the caption with fig. 552. **Fig. 548**: View of the mainspring barrel bridge after removal of the automatic bridge. On the left is the balance cock with balance wheel. Two of the four regulating screws are still visible. A successful but elaborate detail: the winding wheel in the form of a gear rim, which is supported and guided by three rubies. **Fig. 549**: After the mainspring barrel bridge is removed, this view shows the mainspring barrel on the left, the chrono center wheel in the middle, and the shift or column wheel half hidden on the lower right. **Fig. 550**: In the next step, one can now see the complete switching wheel (see also fig. 547). **Fig. 551**: After removing the gear train bridge, we see the main plate with the remaining components. At twelve o'clock, one can see the zeroing slider and the oscillating lever. At two o'clock, one can see the start-stop slider and, below it, the shift wheel. The elongated part is the oscillating lever, which is not in functional position and ends at the minute counter wheel. The three large gears, viewed from the center toward ten o'clock, are the minute wheel, third wheel, and fourth wheel. On the left side, above the balance, one can see the escape wheel and one of the two escapement pallets.

Illustration 552

Automatic bridge with the flanged connecting rod gear for the double pawl drive. Because the pawls of this drive cannot follow the vertical movement of the rotor axis (the Z direction), the automatic bridge was subsequently locked by means of a cylinder head screw. This compensated for a design error. If the kinematics of the Pellaton drive had been studied intensively beforehand, this error could have been avoided.

Illustration 553

Switch wheel for the chronograph function with six grooves. At the top left is the mainspring barrel. Of interest to a designer are the flanged bushings with internal threads. These components combine the function of dowel pins and connecting screws. The technique was first used in aircraft construction.

In IWC's caliber, the chronograph section for the minute and hour counters also runs permanently and is set into operation directly via the vertical clutch. The seconds counter is started by an oscillating drive.

All in all, the new IWC caliber 89360 is a conventional, solid, and well-dimensioned design that can replace the already dated ETA/Valjoux 7750 caliber with a clear conscience. Innovative flights of fancy were hardly to be expected, since IWC had already stopped developing its own movements a quarter of a century ago and has concentrated on developing additional functions and partial improvements in recent years.

10.7. The Caliber 751 from Jaeger-LeCoultre

This self-winding chronograph movement was created by a three-man development team that included designer Jean-Claude Meylan, designer Magali Métrailler, and watchmaker Richard Rapin. In addition to the basic specifications for a classic chronograph with three counters and ratchet wheel control, the targets included high precision, a vertical clutch, diameter of 11½ *lignes* (1 inch, or 25.6 mm), date display, high frequency, three-day power reserve, and ceramic ball bearings for the rotor. To realize a chronograph movement with automatic winding as well as a date display with a diameter of just 1 inch (25.6 mm) and a height of 0.221 inches (5.63 mm) can be regarded as a very big and difficult challenge.

However, no one was told that this new chronograph movement was based on the caliber 1185 from Frederic Piguet (movement diameter 1 inch [25.6 mm], overall height 0.218 inches [5.55 mm]).

It was indeed not easy to accommodate the individual subassemblies, given the aforementioned spatial conditions. The rotor and automatic bridge, chronograph module, movement with two barrels, and date switching had to be combined in such a way that a spatially optimized arrangement combined with safely functioning processes was guaranteed. Automatic winding takes place in both directions of rotation, using not the changer already known from the caliber 889, but a pawl changer comparable to IWC's. The chronograph module is housed mostly above the movement. All of its gears are constantly engaged, right up to the vertical clutch. Again, the chronograph functions are controlled by a switching wheel. Overall, this chronograph movement is neatly and thoughtfully designed, even if the given dimensions involved many compromises in the constructive design as well as in the dimensioning of the individual parts. It is perhaps also due to the miniaturization of this caliber, which was taken to the extreme, that the rate performances did not turn out as one is normally accustomed to with watches from Jaeger-LeCoultre.

Illustration 554
The Jaeger-LeCoultre in-house thousand-hour test signet in the form of a gold-colored medallion adorns the case back, which is held in place by four screws.

Illustration 555
Jaeger-LeCoultre Master Compressor Chronograph Ref. 146.8.25 with stainless-steel bracelet. A Munich-based industrial designer described the appearance of this new timepiece from Le Sentier as elegant and successful. The name Compressor refers to the sealing mechanism: by turning the compression key in front of the crown and the pusher, a special seal is compressed or relieved. This technique and the complex case design are also responsible for the watch's water resistance of up to 328 feet (100 m).

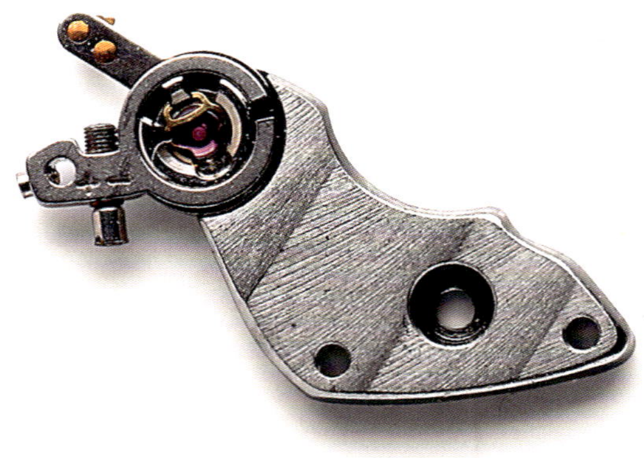

Illustration 556
Balance cock of the chronograph caliber 1185 by Frederic Piguet with Triovis regulator and Incabloc shock protection. This chronograph caliber was also used as *ébauche* by OMEGA, Breguet, Blancpain, and others. For Jaeger-LeCoultre, this movement was redesigned and fitted with two barrels. However, the balance and balance cock were retained in all versions. The space available would not have been sufficient for the installation of a complete JLC regulator (see also fig. 546).

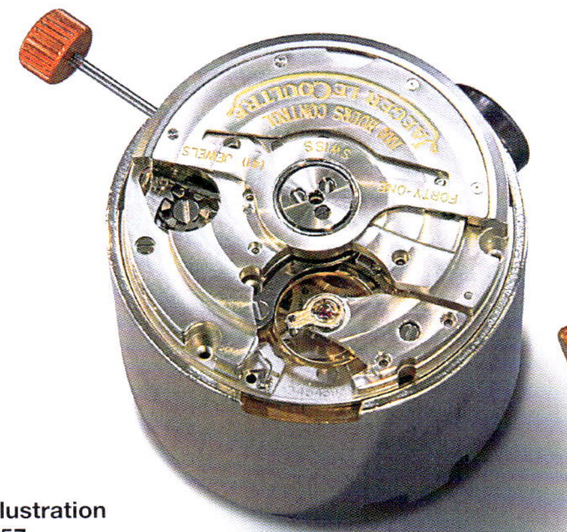

Illustration 557

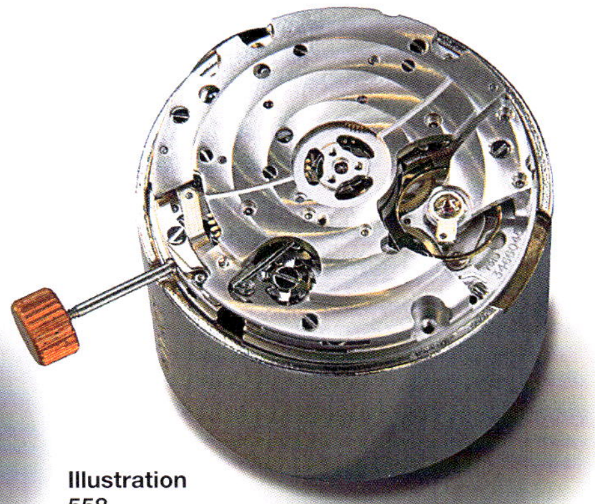

Illustration 558

Illustration 559

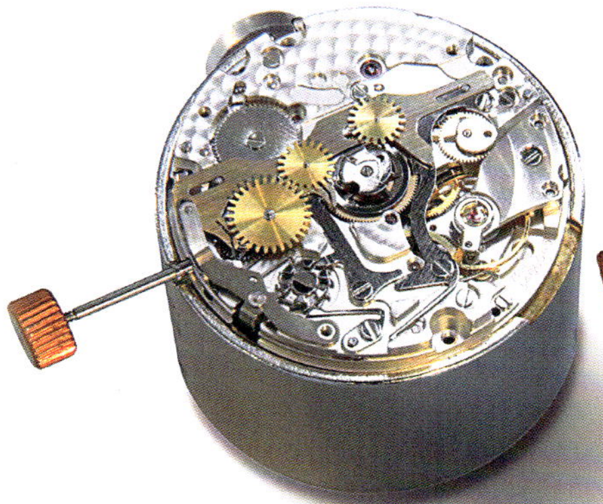

Illustration 560

Illustration 561

Illustration 562

Fig. 557 to 562, Jaeger-LeCoultre Caliber 751
Fig. 557: Overall view of the uncased Jaeger-LeCoultre Caliber 751 chronograph movement. In this picture the rotor (unfortunately without the gold oscillating weight), the automatic bridge, the chrono bridge below, and the balance cock with balance wheel are dominant. On the left, in the opening of the chrono bridge, one can still see the switching wheel. **Fig. 558:** Here is the caliber with rotor removed. In the upper part one can see the automatic bridge. Start-stop pushers are now positioned at nine o'clock, reset-pushers at seven o'clock. **Fig. 559:** After the chronograph bridge is removed, the components of the chronograph mechanism are now visible. The heart lever is arranged diagonally, with the clutch rocker (*right*) and clutch lever (*left*) below it. **Fig. 560:** In the center of this image is the center chronograph wheel, a rather complicated component. For repairs, it must not be disassembled but only replaced. **Fig. 561:** As an intermediate stop, the gear train bridge can be seen prior to removal. **Fig. 562:** After the gear train and mainspring barrel bridge have been removed, only the oscillating system, gear train, and two mainspring barrels remain on the main plate.

Illustration 563
Balance cock of the caliber 751. Fine adjustment of the escapement is handled by the Triovis regulator. Compared to the picture on the left (illus. 556), from the top of the balance cock one can see that the movement manufacturer Frederic Piguet provided the inspiration for this caliber. Blancpain, Breguet, and OMEGA use similar raw movements.

Illustration 564
Fine-adjustment mechanism of the caliber 889/1 with flat hairspring and Triovis regulator. The difference compared to the Frederic Piguet balance cock is clearly visible. Here the escape wheel shaft is mounted in cover plates.

10.8. The Reference: The Caliber 4130 from Rolex

Illustration 566
Rolex Daytona Chronograph Ref. 116.519 in 18K white gold with alligator strap and folding clasp. This model exudes a reserved elegance, to which the harmony between the black dial and the leather strap of the same color also contribute. The matte sheen of the case tells the connoisseur that he is dealing with a noble precious metal and not alloyed steel.

Reference means a benchmark for a measured value; for an object with defined properties, the term means a comparative value. When Rolex launched a new Daytona with its own caliber 4130 chronograph movement in 2000, demand and curiosity were high. Especially in those circles that regarded the previous model with the modified caliber from Zenith as nothing more than a compromise, the joy over Rolex's own caliber had almost euphoric proportions. "Where Rolex is on it, Rolex must be in it" was the slogan at the time.

A diameter of 1.18 inches (30 mm) and a height of 0.255 inches (6.5 mm) were specified for the design of the new caliber 4130. This meant that the available volume was 0.289 in.^{3} (4.75 cm^3), about 12% less than the IWC caliber 89360. From the outset, the designers at Rolex were aware of the rules that had to be followed to realize this new chronograph caliber: a simple, clear design; manufacturability in high quantities; consistent ease of maintenance; seventy-two-hour power reserve; robustness; and chronometer accuracy. All these goals were obviously achieved.

Many functions are already integrated in the caliber 4130's main plate. The gear train bridge and chronograph bridge are mounted on top of it. This is followed by the automatic bridge

Illustration 565
The Rolex caliber 4130 is installed in the Daytona Cosmograph chronograph. For the first time, the rotor has ball bearings. The balance wheel with bridge and Microstella regulating weights and the blued Breguet hairspring with the name Parachrom are also striking.

with the almost ingeniously simple ratchet wheel changer. The balance and mainspring barrel bridges can be removed even when the movement is powered on, so that a mainspring change takes only half an hour. A trained watchmaker can manage a complete overhaul of the caliber 4130 in two and a half hours.

The balance wheel runs at 4 Hertz and has a Breguet hairspring. It is regulated by four microadjustment screws. The chronograph module is fully in the power train and is switched on via a vertical clutch.

With the clean technical solution, on the one hand, and the small number of 201 individual parts, on the other, an outstanding result was achieved with this caliber. It can be considered not only as a reference for other chronograph movements, but also quite generally as a model.

Prof. Ehrlenspiel would have been delighted at the sight of this movement.

Illustration 567
The massive case back protects and covers the 4130 chronograph movement, rendering it "bombproof."

Illustration 568

Illustration 569

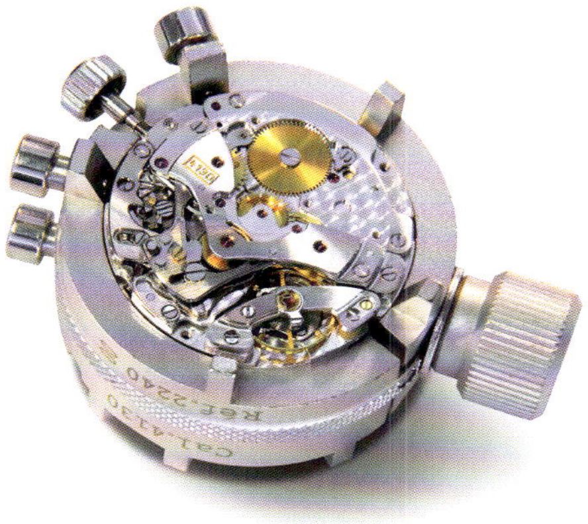

Illustration 570

Illustration 571

Illustration 572

Illustration 573

Fig. 568 to 573, Rolex Caliber 4130

Fig. 568: The Rolex caliber 4130 movement in the Rolex Ref. 2240 movement holder. The rotor and balance bridge form the initial impression of this successful chronograph movement. Unlike with Rolex's other automatic movement 3135, the rotor's axle is not supported by two rubies but guided by a ball bearing. **Fig. 569:** After removing the rotor, one can see this movement's clean design. On the upper right is the automatic bridge, recognizable by the red-and-yellow anodized aluminum gears. The bridge for the chronograph next to it is also engraved with the reference number 4130 of this caliber. **Fig. 570:** In this picture the automatic bridge has been removed. Below one finds the balance bridge with balance wheel. This movement can be adjusted precisely thanks to its Breguet hairspring and the four Microstella timing screws. **Fig. 571:** After removal of the chronograph bridge, the almost V-shaped hammer can be seen on the left. On its right end you can see the hour wheel, and at the left end the minute counter wheel. Between them can be found the elaborate vertical clutch, whose gear wheel engages the chronograph seconds wheel. **Fig. 572:** Removal of the chronometer components reveals the Y-shaped gear train bridge. On the left, one can see the column wheel, and at the top the crown and ratchet wheel. **Fig. 573:** Finally, the main plate with oscillating system, gear train, and mainspring barrel. Next to it are the minute, third, and second wheels.

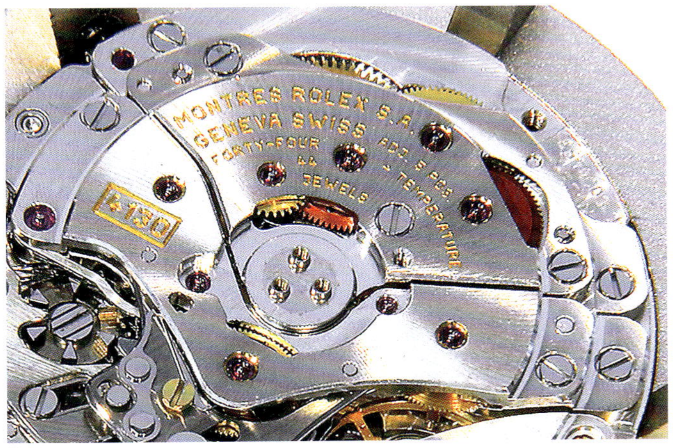

Illustration 574
Caliber 4130 with the rotor removed. At the top is the automatic bridge with the company inscription. The red-and-gold hard-anodized gears are clearly visible. To the left is the chronograph bridge, with the caliber 4130 engraving. On the far left is the column wheel for the chronograph mechanism.

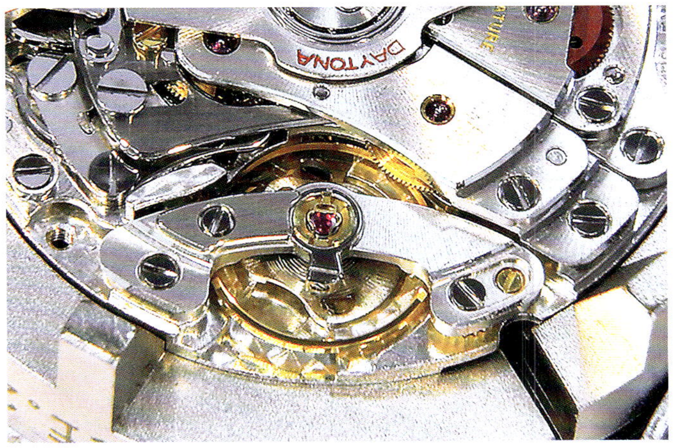

Illustration 575
Balance bridge with balance wheel, Breguet hairspring, and Microstella timing screws. Unlike on the caliber 3135, the locking plate for the hairspring stud is not on the right side of the bridge but on the left. The vertical play of the balance staff is adjusted only on the right side of the balance bridge.

Illustration 576
We would not have been able to carry out such an extensive practical test without the high level of commitment shown by our knowledgeable watch enthusiasts. This group photo was taken after completion of the practical test (*left to right*): Ingolf Menius, Daimler-Benz Stuttgart; Dipl.-Ing. Harald Dörenbach, manager of a medium-sized company; Birte Zweiniger, service manager of Patek Philippe in Munich; Prof. Dr.-Ing. Joachim Heinzl, formerly of the Technical University of Munich; and Peter Braun, editor in chief of *Armbanduhren* ("Wristwatches").

10.9 How the Testing Was Done

After the arrival of the test watches, which were provided by the four watch brands, the first step was an initial test. The daily rate, daily rate deviation, drop, and amplitude of the four test pieces were determined on the time grapher (Chronoscope S1 with Mikromat S).

The "daily rate" is the deviation that occurs from normal time within twenty-four hours. If a watch is thus two seconds ahead per day, the daily rate is +2 seconds. The "maximum daily rate deviation" (delta or D) is the greatest difference between the individual values determined in six positions. For example, if the greatest positive value in one position is +4 and the greatest negative value in another position is −1, the maximum rate deviation is five seconds.

The term "drop" or "symmetry" refers to the uniform drop of the escape wheel tooth from the ruby pallets of the lever at the input and output sides of the lever. Ideally, the drop should be equal in all positions, but unequal up to a maximum of 0.7 milliseconds. Amplitude or "amplitude of oscillation" means the largest angle of amplitude of a balance wheel to both sides. This angle should be around 270°. If the oscillation of the balance wheel is higher and goes toward 360°, it is called "bouncing" of the balance.

When judging the precision of a watch, the largest daily rate deviation (delta) is of primary importance. The smaller this value is, the more independent the oscillation and drive system of a movement is of its position. The daily rate itself is often overrated by some testers. If a watch advances or retards evenly, the daily rate can easily be corrected by the balance wheel's regulation system. The daily rate alone cannot give a direct indication of the design quality of a movement.

In our search for suitable test subjects, we wanted to recruit not only knowledgeable but also particularly interested watch enthusiasts for this task.

Prof. Dr.-Ing. Dr. h.c. Joachim Heinzl, professor emeritus of the Chair of Fine Instruments and Microtechnology at the Technical University of Munich, and later president of the Bavarian Research Foundation, took over the testing of the IWC Da Vinci Chronograph in stainless steel.

It should be mentioned that Prof. Heinzl has almost never worn a wristwatch since 1961. With the Da Vinci on his arm, he therefore felt like a newly banded bird. He was enthusiastic about the watch's accuracy but not about its wearing comfort, because operating the buckle was not easy for him. Overall he found the watch too big, too thick, too angular, and ultimately too extroverted. He would also have had to buy new shirts with wider cuffs to go with this watch.

Harald Dörenbach, a graduate engineer and managing director of a medium-sized company, took the Master Compressor Chronograph from Jaeger-LeCoultre under his wing for two weeks.

Dörenbach found the wearing comfort very good and the accuracy good, but the external appearance a bit too clunky and flashy. As an everyday watch, Dörenbach wears a Master-Date from Jaeger-LeCoultre in stainless steel with a stainless-steel bracelet. He is very satisfied with this watch, and its performance is unbeatable.

Birte Zweiniger, master watchmaker and head of service at Patek Philippe Germany, took the test of the rare platinum Annual Calendar Chronograph in her hands.

Ms. Zweiniger nobly refrained from criticizing the watch, which received consistently positive reviews among the testers. All of them rated its utility value very highly, and its operation was described as very simple. Accuracy was rated as very good on the basis of the protocols, and the wearing comfort was praised by all.

Table 3: Additional Technical Information

Maker	Caliber	Vibrations per Hour (A/h)	Number of Jewels	Power Reserve (hrs.)	Overhaul Time (hrs.)	t with Chrono sec./day switched on[2)]
IWC	89360[3)]	28,800	40	68	6	1.4
Jaeger-LeCoultre	751[3)]	28,800	41	72	7.5	0.6
Patek Philippe	CH 28-520 IRM QA 24H	28,800	40	55	[1)]	[1)]
Rolex	4130	28,800	44	72	2.5	0

1. No information 2. Average from four measurements 3. With flyback function

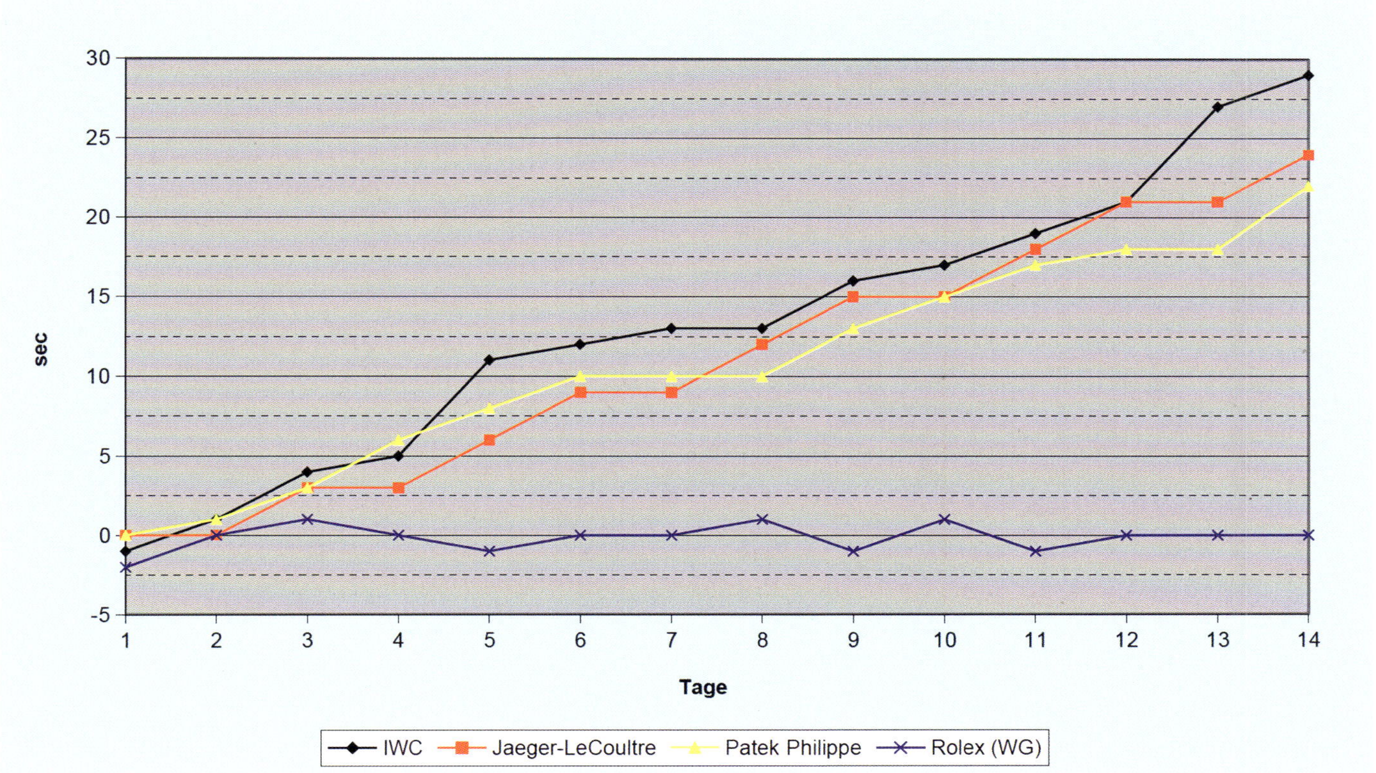

This graph illustrates the daily rates of the four different test watches. These time recorders were worn on the left wrist of the test subjects during the day and spent the night in the "dial-up" position on their respective bedside tables. The exact time was taken daily in the morning and evening as a basis, and the respective deviations were recorded.

Peter Braun, chief editor of *Armbanduhren* magazine, did not miss the opportunity to take a closer look at the precise movement of the white-gold Rolex Daytona Cosmograph.

The pampered journalist found the chronograph to be of pleasantly restrained dimensions, fitting snugly on his left wrist. Only the safety bar on the folding clasp occasionally opened. The rate performance was described as excellent. The external appearance was quite positive and did not possess the otherwise usual "Rolex factor."

Peter Reiser, a graduate economist and management consultant from Cologne, and Ingolf Menius, Mercedes-Benz Sales in Stuttgart, were responsible for checking the two Rolex Daytona watches that were included for a statistically reliable statement. Reiser's enthusiasm for the Daytona in steel and yellow gold with black dial could not be missed. Only top marks were given to the general impression and the utility value. Accuracy was rated as excellent. Wearing comfort and external appearance also received high praise. Menius rated the general impression and utility value of the Daytona in steel and yellow gold with a white dial as "good," while accuracy and wearing comfort received the top rating of "very good"! The external impression was also positive—customers and business partners conveyed a feeling of admiration and respect, but by no means of envy.

All test subjects dutifully and reliably noted the respective rate of their watches in the morning and evening, wrote this down on the test sheet provided, and also always placed their timepieces in the "dial up" position on their nightstand at night.

Although only the timepieces from Rolex were designated as certified chronometers, the two test watches from IWC and Patek Philippe were also confidently within the tolerance range of officially approved precision watches. Only the Master Compressor from Jaeger-LeCoultre exceeded the strict specifications of the Swiss testing offices in the test criterion "maximum daily rate deviation."

Depending on personal taste, technical evaluation, and available budget, each reader can now choose "their" watch.

Illustration 577
Patek Philippe Nautilus Ref. 3700/1 in stainless steel, probably the most sought-after sports watch at the moment

11. The Noble Sports Watches

The Audemars Piguet Royal Oak, Patek Philippe Nautilus, and Vacheron Constantin 222

11.1. Introduction

The sensational price development that the Nautilus by Patek Philippe has recently undergone prompted president Thierry Stern to discontinue the stainless-steel version in spring 2021. The first Royal Oak model in stainless steel from Audemars Piguet now fetches prices that are ten to fifteen times higher than the original selling prices. Even the 222 model from Vacheron Constantin, which played more the Cinderella role among these three brands, is now experiencing unexpected demand.

These watches achieved sensational results at the Phillips auction in Geneva held on May 8–9, 2021. A Patek Philippe Nautilus of the original model Ref. 3700/1 in stainless steel from 1981 was sold for $126,199 (movement no. 1,308,247, case no. 538,817). A yellow-gold version from 1982, Ref. 3700/11, sold for a sensational $252,395 (movement no. 1,310,255, case no. 559,031). Finally, there was a Ref. 5711/1A in stainless steel that was offered and sold to a new owner for $154,241 (movement no. 3,647,386, case no. 4,945,299).

The three Royal Oaks from Audemars Piguet that were offered at this auction also sold for considerable prices. The one example of the reference 5402ST from the year 1973 was knocked down at $77,120 (movement no. 127,082, case no. 67,840, and additionally A 840). The second model, presumably from the year 1975, sold for $91,142 (movement no. 47,117, case no. 99,015). The third Royal Oak in stainless steel from 1978 sold for the scarcely imaginable price of $189,296 (movement no. 159,322, case no. 99,954, and B 1954).

There was a special surprise at this auction provided by the call of a Vacheron Constantin 222 Ref. 44018 in stainless steel with dark-blue dial. With a starting price of $22,257, there was an unexpected bidding war, and in the end the highest bidder was willing to pay $112,175 for this model (movement no. 685,208, case no. 522,120)!

For this reason, our chapter is highly topical.

Sports watches often played an important role in the range of products offered by Swiss watch manufacturers. This was true not only for the well-known chronograph manufacturers, such as Breitling, Heuer, Zenith, Universal Geneve, and others, but also for such companies as IWC, OMEGA, Longines, Ebel, and many more. Only a few Swiss noble brands, which will be covered, have so far been closed to this trend. Their market segment was formed by the noble, timeless, and conservative timepieces, which in their simplicity were hard to surpass. This situation changed abruptly in 1972, with the appearance of the striking Royal Oak sports watch from Audemars Piguet. "Steel on the outside and gold inside" was one of the well-known advertising slogans of the time. AP's most important success story had begun! Four years later, Patek Philippe launched the Nautilus. The third in a series of new luxury sports watches, the 222 from Vacheron & Constantin, was unveiled in autumn 1977. But what did the representatives of this new generation of watches have in common apart from their sporty appearance? Well, the answer may surprise some readers. First, both Audemars Piguet timepieces were designed by the same designer, Gerald Genta, and second, the same automatic movement was present in all three watches. Contrary to the widespread opinion that Vacheron Constantin's model 222 was also designed by Gerald Genta, the case and strap of this watch were in fact created by German designer Jörg Hysek, who was just twenty-four at the time.

Gerald Genta (†), who later had his own watch brand, was already considered a gifted designer of beautiful watches. One of his first successes was the famous Polerouter pilot's watch by Universal Genève. However, the design of the Royal Oak may be considered his first great success. Gerald Genta subsequently approached Patek Philippe and received from them the unplanned commission for the design of the Nautilus.

The fact that the "driving engine" of all three sports watches was also identical had several reasons, as we will see. In any case, the right decision was made, especially since this extremely flat but well-designed automatic caliber experienced an unexpected renaissance thirty years after its launch and is now part of Audemars Piguet's production program.

Illustration 578
Noblesse in stainless steel—a group photo of the three sports watches presented: the Nautilus Ref. 3700/1 from 1978, the Royal Oak Ref. 5402 SC from 1982, and the 222 Ref. 44018 from 1981. The respective public prices were $4,700, $4,038, and $5,462.

11.2. Sports Watches—Trend of the Time?

Let us begin by looking back at the overall economic situation that characterized the Swiss watch industry at the beginning of the 1970s. The unstoppable triumph of quartz technology and the immense quantities of cheap goods coming from the Far East had created market and sales situations that the traditional Swiss watch industry was unable to cope with. In particular, the cheap watches from the Far East, often overloaded with all kinds of bells and whistles due to the use of microelectronics, were very popular with consumers and often left no chance for the elaborate Swiss micromechanics! Numerous Swiss watch manufacturers could not offer effective alternatives to the changed market and had to close their doors—and those of their suppliers.

It was precisely during this difficult period that Audemars Piguet, a fine watchmaker based in the Jura village of Le Brassus, took the plunge and in 1972 launched a sports watch that was a quantum leap in terms of design and technology and clearly stood out from all its competitors. It was the birth of a success story that continues to this day!

The octagonal case, also known as the "porthole" look; the eight typically placed, recessed, and nonrotating white gold screws; and the integrated strap gave this timepiece a recognition value among watch enthusiasts that we perhaps know only from watches made by Rolex.

The first internal prototype was made of white gold, since it cost less to make than a steel one!

The next sports watch from a noble house was the famous Nautilus from Patek Philippe, which was unveiled in 1976. Gerald Genta decided on an oval-shaped case design for this model with the lugs for the screw-down bezel on the right and left. Here, too, the integrated strap went well with the case. This successful design has endured to the present, and the Nautilus is still one of Patek Philippe's most successful models, even if other sporty model lines have since been added to the program.

The conclusion of this trilogy was the model 222, unveiled by Vacheron Constantin in 1977. Here, however, it was Berlin-born designer Jörg Hysek, then just twenty-four years old, who selected the main stylistic elements. In this case a round case was again chosen, whereby the integrated bracelet was somewhat based on that of the watch's two predecessors. The design orientation of the case was also rather different. While the two predecessor models had a one-piece case, the 222 has an elaborately designed three-piece case. But here, too, the metal strap connector was integrated into the case middle, just like on the Nautilus. Why this sports watch line—in contrast to the two aforementioned competitor models—failed to achieve lasting success was probably due to the more playful design of its bezel and also the public taste of the time. The stainless-steel version remained in production only until 1981, while the steel-and-gold (mixte) version presented, as well as the model in gold, were made only until 1985. Perhaps the reason for the short life of this creation is the well-known comparison between high-quality watches and the female gender: whether beautiful watches or pretty ladies, it always remains a question of personal taste.

11.3. The Movements—Commonality No. 1

Following the trend of the time toward ever "lower" movements, Jaeger-LeCoultre in Le Sentier developed a completely new caliber in the mid-1960s. It was given the caliber designation 920 but was never installed in the company's own watches and was initially supplied as a raw movement only to Audemars Piguet in Le Brassus and former sister company Vacheron Constantin in Geneva.

In 1967, Vacheron & Constantin launched this caliber under the designation K 1120 as the world's flattest automatic movement, with a total height of only 0.096 inches (2.45 mm). The following version with date indication at three o'clock was given the

Illustration 579

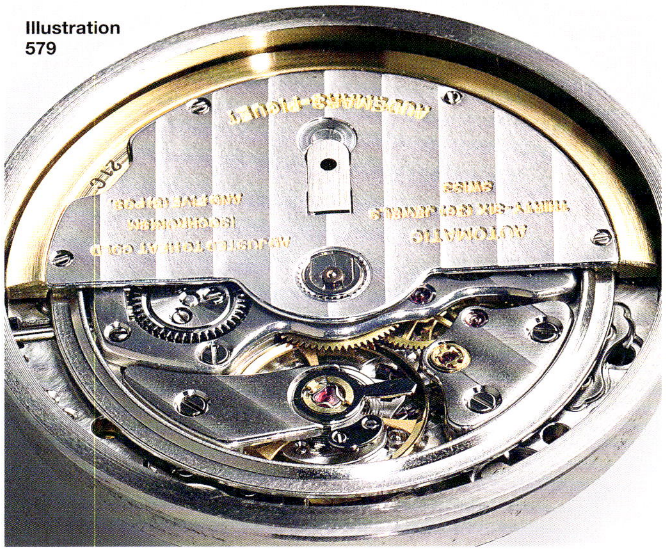

Illustration 580

Illustration 581

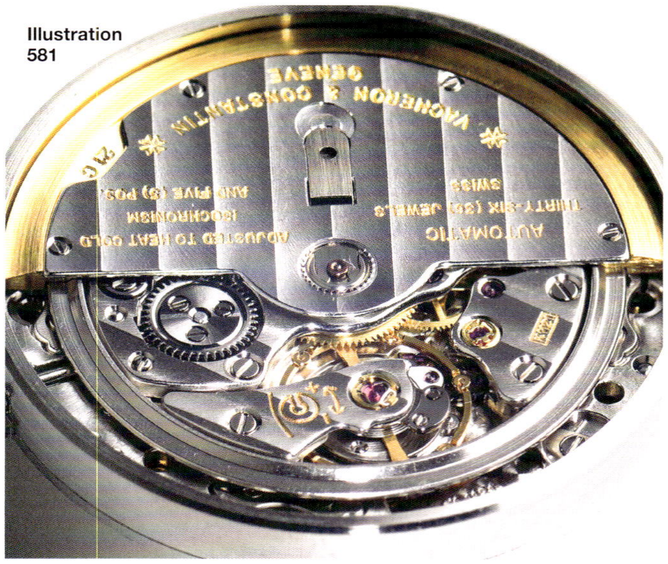

caliber designation K 1121 and had an overall height of just 0.120 inches (3.05 mm). At the same time, AP launched this movement as the caliber 2120 with no date display, but with a perpetual calendar in a classic leather strap version.

Somewhat later, Patek Philippe also received this reserved raw movement, which required some modifications, and the caliber designation was 28-255 or 28-255 C (C indicating date).

We wanted to know why this caliber was used in so many different ways.

Due to the lack of its own raw-movement production, Audemars Piguet had a tradition of purchasing them from Jaeger-LeCoultre in the neighboring town of Le Sentier, only 4 miles (7 km) away.

To follow the trend of the time for watches with the flattest possible construction, the calibers K 2071 (hours, minutes, and center seconds) and K 2072 (as before, plus date display at three o'clock), which had been used up to then, were simply too large, with total heights of 0.216 and 0.248 inches (5.5 and 6.3 mm), respectively.

What could be more obvious than to create several new watch collections using the new, ultraflat automatic movements 2120 and 2121? It was almost inevitable that the Royal Oak would also be powered by this new caliber.

The reason why renowned watchmaker Patek Philippe in Geneva also relied on this caliber on a large scale is often overlooked there today. At the beginning of the 1970s, the proven automatic movements 27-460 and 27-460 M (Monodate) were discontinued. They were to be replaced by the newly developed caliber 350 (1-350, an improved version) and later by the calibers 310 SC and 335 SC, which were still in the planning stage. Since the caliber 350 was designed as a "backwinder," it did not provide the required water resistance and did not have a date display, and therefore it was ruled out for the Nautilus. The critical period of more than five years during which Patek Philippe had no reliable automatic caliber of its own was bridged by the use of this proven, ultraslim caliber. From Patek Philippe's movement directories it can be seen that during that difficult period, about 20,000 examples of the 28-255 and 28-255 C (Calendrier, or calendar), for a total of 40,000 movements, had been delivered to the noble Geneva watchmaker! Only the

Illustration 579
View of the AP caliber K 2121 housed in the movement retaining ring. The beryllium ring, rotor arm, and oscillating weight are held together by four screws. The beryllium ring runs on four vertically mounted ruby roller bearings. The rotor is centrally guided by the inner bore of the pinion riveted to the rotor, whose axle consists of a well-dimensioned beryllium hollow flange. In the foreground, the three-spoke balance wheel with adjustable hairspring stud and regulator arm without fine adjustment.

Illustration 580
The variant used by Patek Philippe with the caliber designation 28-255 and 28-255 C (C = *calendrier*, or calendar) has significant differences. The surfaces of the bridges and rotor are decorated with the sunburst finish. The rotor arm has two recesses and is not a semicircular segment. Also, the "hat" on the rotor arm with stamped 21 C was omitted and the fine-gold hallmark 21K was placed on the right cut surface. Due to the modified rotor design and higher swing weight, the Patek Philippe movement is 0.0003 inches (0.1 mm) higher than the others and thus has a total height of 0.124 inches (3.15 mm).

Illustration 581
Compared to the AP version, the caliber K 1121 from Vacheron & Constantin has a Gyromax balance wheel, which does not require a separate fine-adjustment mechanism. The symbol on the balance cock indicates a mass balance weight with adjustment indications for faster (+) and slower (–) rate. The old company name, VACHERON & CONSTANTIN, is still engraved on the rotor.

successor to the first Nautilus with Ref. 3800/1 was again fitted with a self-winding movement, the caliber 335 SC.

Like Audemars Piguet, Vacheron & Constantin did not have its own automatic caliber at that time, instead relying on its tried-and-tested relationship with movement supplier Jaeger-LeCoultre in Le Sentier. When the 222 was launched in 1977, Vacheron & Constantin could already look back on ten years of experience with the K 1120 and 1121 calibers and could confidently use the latter in the new 222 sports watch.

Although at first glance the three movement versions look almost identical, there are still differences to be described. But first, let's take a look at the design of the automatic movement. The entire design, including all relevant components, is strictly designed for low overall height. The mainspring barrel is cantilevered in the plate.

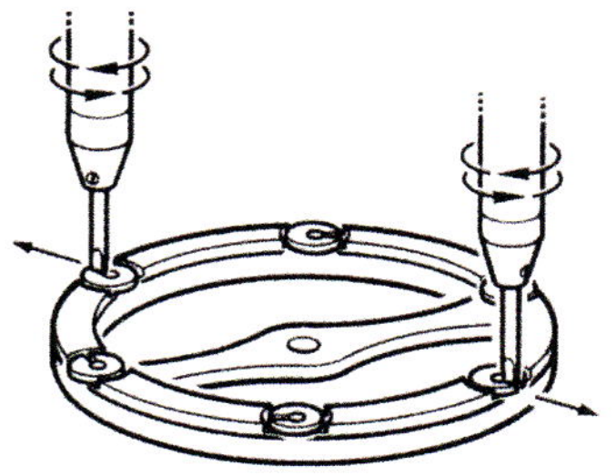

Slots facing outward = balance oscillates faster

Illustration 1

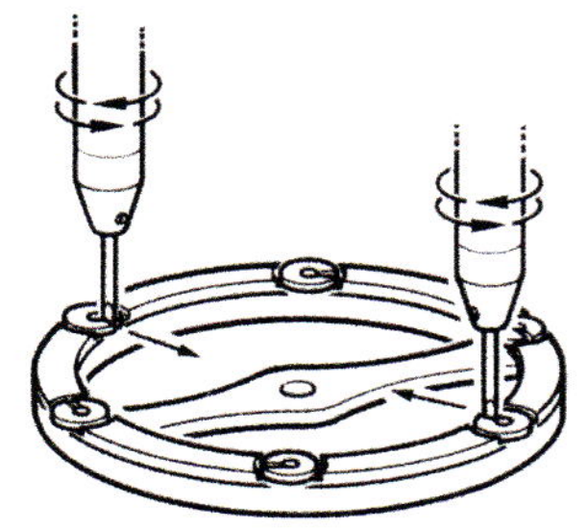

Slots facing inward = balance oscillates slower

Illustration 2

Illustration 582
With the Gyromax balance, the rate of the watch can be adjusted by means of six mass balance weights. If you turn these weights—two opposite each other—so that the slots face outward, the moment of oscillation (GD^2) is reduced and the balance oscillates faster. If you turn the slots inward, the moment of oscillation increases and the balance oscillates more slowly. In contrast to adjustment by means of a regulator arm, the entire symmetry of the hairspring and escapement lever is preserved here.

The changer is also integrated under the mainspring barrel bridge. The gear train bridge and balance cock complete the "static" components. A two-spoke Gyromax balance wheel includes six mass balance weights, Nivarox hairspring, Kif-Ultraflex shock protection, and adjustable hairspring stud, which are standard equipment. At Audemars Piguet and Vacheron Constantin, movements with a conventional balance and pushers were also used at times. A second indication had to be omitted due to lack of space.

The design of the very flat rotor was completely atypical. The rotor arm is guided by an L-shaped beryllium ring, which is in turn screwed to the 21-karat gold oscillating weight and the rotor arm. Four equally spaced ruby roller bearings mounted vertically in the plate support the rotor ring described. The centric bearing of the complete rotor is provided by the inner bore of the drive (gear wheel) riveted to the rotor arm. Above this drive, the locking latch (latch for the oscillating mass) engages the hollow flange and is held axially by a shoulder. The geometric design of the latch creates a preload in the end position that fixes it securely. With this innovative design, the case movements can be transmitted directly to the rotor as winding energy without any major frictional losses.

Now to the differences that distinguish these three automatic movements. In terms of technology and finishing, the Audemars Piguet calibers (K 2120 and 2121) and those of Vacheron & Constantin (K 1120 and 1121) are identical, except for balance variations and inscriptions. As one can see from the Royal Oak shown, Audemars Piguet sometimes used a three-spoke balance wheel with regulator arm instead of the two-spoke Gyromax balance. Due to a lack of space, the regulator arm had no fine position. This variant was used by AP for technical reasons. The extremely flat design often led to contact between the Gyromax weights and the hairspring on the balance cock, or on the winding rotor. After the balance cock was redesigned, AP returned to the proven Gyromax balance wheel. However, as you can see in the list of fittings as well as in the instruction manual, a later conversion to the proven Gyromax balance is also possible. The design and function of the movement-holding ring are the same on AP and V&C, while on Patek's Nautilus this ring is integrated into the case.

The Patek caliber 28-255 and 28-255 C also differ from the previously described designs. This is clearly visible in the surface finish. While on AP and V&C the bridges and cocks have a Geneva stripe finish, on the Patek movement the surfaces of these components are decorated with a sunburst finish. Also noteworthy are the variations in the rotor. The rotor carrier has two distinct recesses and is not segmented as a semicircle as on those of the AP and V&C movements. The 21-karat oscillating weight is radially attached and screwed without interruption. The "cap" with the stamped "21C," which is present on AP and V&C, was omitted. Instead, the gold hallmark "21K" is found on the vertical surface of the oscillating weight at Patek.

Patek must have realized that the gold "cap" on the oscillating weight of the AP and V&C not only caused a lot of additional manufacturing effort but also visually affected the rotor symmetry.

All versions of the Patek caliber 28-255 or 28-255 C were equipped with a Gyromax balance wheel—a Patek patent. This also applies to the use of Kif Elastor shock protection. However, the escapement wheel bearing was different. Here, Patek had chosen a combi-chuck with a simple cap jewel, while the other two manufacturers had used a combi-chuck with a mainspring (Duobil). Another difference, which was often asked about, is to be found in the design height. With the same movement diameters of 1.10 inches (28 mm), the construction heights of AP and V&C are given as 0.096 and 0.120 inches (2.45 and 3.05 mm), respectively, but for Patek they are given as 0.10 and

Illustration 583

Illustration 584

Illustration 585

Illustration 586
Exploded view of the case structure of the Royal Oak; *from bottom to top*: fixing screws with single-shell case, movement retaining ring, gasket in elaborate flange shape, crystal with ring and gasket, and bezel with hexagonal screws

Illustration 587
This exploded view of the Nautilus clearly shows the entire case structure. The model shown is from the reference 3800, guaranteed waterproof to 393 feet (120 m).

Left column

Illustration 583
The Royal Oak Ref. 5402 SC. A timepiece that confidently suggests masculinity and dynamism like no other. The design created by Gerald Genta helped Audemars Piguet achieve overwhelming success. Each of the eight hexagonal screws is mounted in a "hexagonal bed," so that the screw slots serve only as a stylistic element with no actual function.

Illustration 584
In 2021 the Nautilus celebrated its forty-fifth birthday. Striking and impressive is this jumbo Ref. 3700/1, which experienced a renaissance at the 2004 Basel Watch Fair (Basel World). The new Nautilus Ref. 3711/1 in white gold has a diameter of 1.67 inches (42.5 mm) and a new black dial.

Illustration 585
Why neither the case design with screw-in bezel nor the dial of the V&C (VC) Ref. 44018 was well received by the public was probably due to the taste of the time. After all, this is a rock-solid sports watch from the oldest watch factory in the world.

Illustration 588
This close-up of the reference 3700/1 Nautilus conveys the shapes and surfaces of its somewhat idiosyncratic design. The Nautilus had not been envisaged in Patek Philippe's product planning but was offered by watch designer Gerald Genta. He had previously created the famous Royal Oak sports watch at Audemars Piguet.

0.124 inches (2.55 and 3.15 mm; the latter for the date version). If you study the design drawings and the precise designs of these calibers, you come to the conclusion that for PPC, this additional 0.003 inches (0.1 mm) in height results from the changed rotor design. The reasons for this deviation are described in detail in chapter 5, "Patek Philippe" (caliber 28-255 and 28-255 C on page 149).

11.4. The Cases—Two Created by Designer Gerald Genta, One by the Young German Designer Jörg Hysek

Certainly no one at Audemars Piguet could have imagined that patent application no. 559 929 of December 6, 1971, would herald an unprecedented success story. The Royal Oak was born!

Apart from the Jumbo format, the two designers Gerald Genta and Jörg Hysek created three distinctly different watches. Gerald Genta, then forty-six years old, was well known in Switzerland; however, the 222 was the first movement designed by Jörg Hysek, who was then just twenty-four. Which design deserves the highest rating shall remain open here. The decisive factor is always the success with the public. And here the Royal Oak has won first place! But it was later overtaken by the Nautilus!

The Royal Oak and Nautilus share the basic principle of the single-shell case, but the solution for attaching the bezels is completely different. As can be seen in the exploded views, the case of the Royal Oak is held together by eight screws. Here, each of the upper hexagonal white-gold screws is inserted into a hexagonal bed in the bezel, and "sleeve screws" are attached to them from below. According to Christoph Guhl, Audemars Piguet in Le Brassus, it has happened that particularly skilled watchmakers have tried to unscrew the Royal Oak from above.

The two-piece winding stem is designed as a "rip crown" and is inserted after the case is screwed down. The lower part of the case, the half shell, was machined from a solid metal block and is especially sturdy. Also integrated there are the fittings for the metal watch strap, which is attached on both sides with two white-gold screws.

Patek decided to use a completely different design. Although a single-shell case is also used, the bezel is mechanically designed as a bridge. After the movement is mounted, the bezel is placed on the case and connected on both sides with two set screws. The truncated cone attached to the tip of these set screws finds its negative form in the case and thus fixes the position of the bezel. In conjunction with a specially shaped gasket, water resistance of 393 feet (120 m) is guaranteed. From an engineer's point of view, this design may not be regarded as entirely uncritical. The very elaborately crafted metal bracelet is screwed into the case via a center link.

The links are connected with pins riveted and polished on the outside. Only on the folding clasp, some of the links are screwed to adjust the respective required length of the strap.

A completely different case construction was chosen for Vacheron & Constantin. Here we have to deal with a three-piece case. In this case a cartridge-shaped movement mount is slid into the middle part of the case from below and screwed to the bezel, which is designed as a ring nut. This also requires a special ring key, without which the watch cannot be opened. On the 222 presented, a rash watchmaker actually tried to open the case at the bottom with a cup knife. The design of this case may be understood as a low-cost solution because of its reduced manufacturing costs.

As with the other two watches, the 222 has a breakaway crown. Like the Nautilus, the metal band is attached to the case middle section by a center link and screwed into place. A special note is necessary for the steel-and-gold-band version. In contrast to the Nautilus, the center links were not made of solid gold but—probably for reasons of economy—gold plates were used instead. In the beginning these gold plates were only soldered

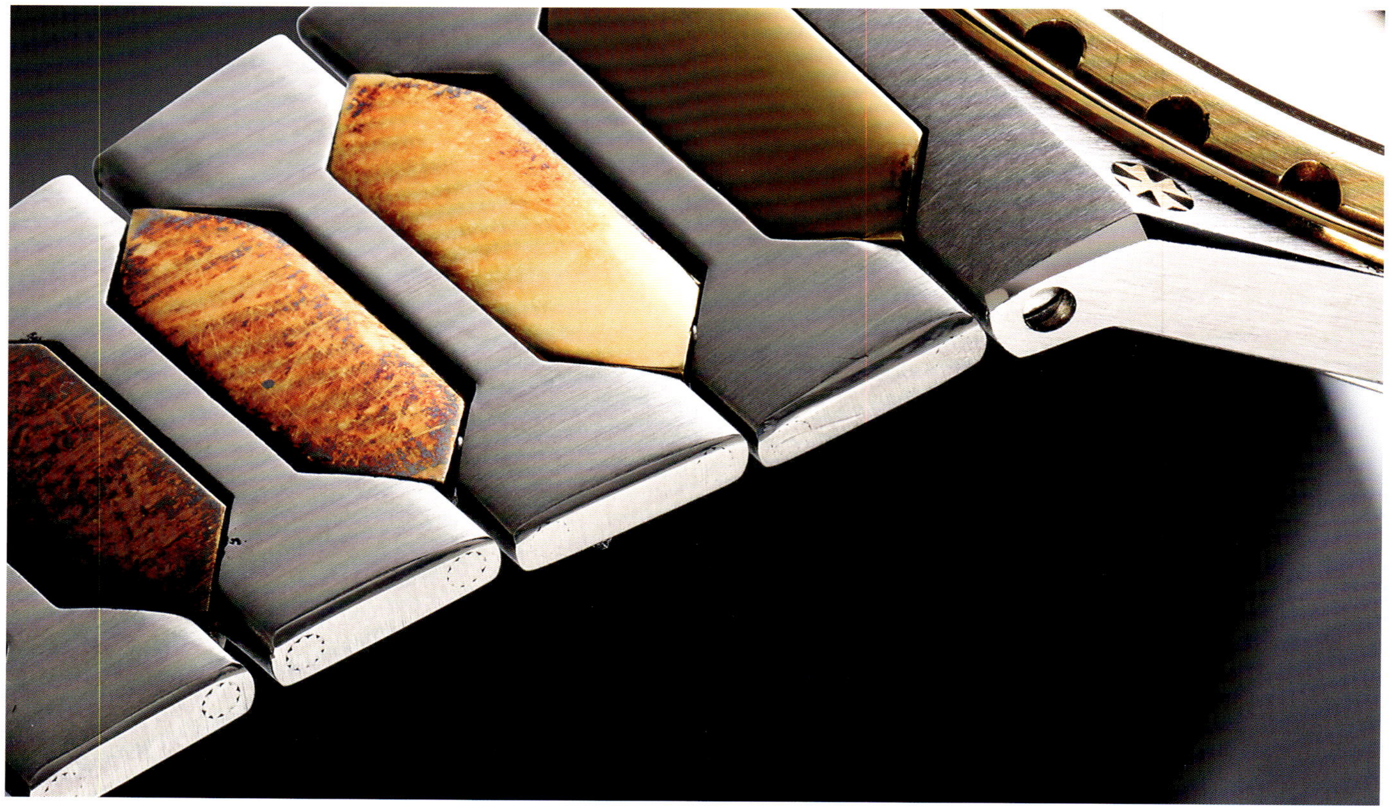

Illustration 589
Detail enlargement of the lug on the case middle part of the model 222 by Vacheron Constantin. The last intermediate piece of the metal bracelet is screwed to the case middle via a set screw. You can even see the screw slot. At the top you can still see a small part of the ring nut, as well as the Maltese cross, Vacheron Constantin's emblem, embedded in the case. The intermediate pieces of the band are made of steel and have only one gold segment (i.e., gold platelets) attached. These gold plates are made of 14K yellow gold. Their surface is already slightly oxidized. These gold segments or intermediate pieces could also have been made solid. Whether one can also save manufacturing costs with the technical solution seen here remains to be answered.

on and later fell off quite often. Because of this bad experience, the gold plates were pinned several times in the later version, which is also the case on the 222 presented.

For the steel-and-gold-case versions, Audemars Piguet was the first manufacture to use a special 18-karat yellow-gold alloy for the intermediate links and the bezel. For wear reasons, competitors at the time used only 14-karat yellow gold.

11.5. Closing Remarks

The result is all that counts! Applying this clear statement to our three different timepieces, the V&C 222 had the least success with the public. Launched in 1977, the steel version of this model remained in production only until 1981. Production of the steel-and-gold version, as well as the all-gold model, ended in 1985.

While the Nautilus was initially unable to match the Royal Oak's numbers, it has always held its own over the years. The Jumbo Ref. 3700/1 was followed by the somewhat smaller Ref. 3800/1. Once again equipped with the in-house automatic movement caliber 335 with date display and center second hand, it is one of the best-known "perennial favorites" from the house of Patek Philippe SA in Geneva. The current model Ref. 3900/1 has the caliber 330 SC encased.

At the beginning of 2021, by the order of president Thierry Stern, production of the steel model was halted.

Audemars Piguet's Royal Oak probably doesn't need much commentary. The "original model" launched in 1972 is still in production today. In recent years, new model variants have been added, so that today the "Royal Oak design" includes everything from simple timepieces and dual-time zone watches to chronographs and the Grande Complication, allowing it to answer almost every watch wish. Of course, time did not stand still in the choice of movements either. Besides the caliber 2121 and 2126 (2125), which are produced today by Audemars Piguet in Le Locle, the new manufacture automatic caliber 3120, which was introduced in 2004, is also installed in the watches.

If you exclude the quality of the automatic calibers used in all three watches and take only wearing comfort, suitability for daily use, and readability as evaluation criteria, all of them should get almost the same grade. In terms of manufacturing technology, the 222 is likely to be the least expensive version, since its case design requires only relatively simple turning, milling, and grinding work. The design of the other two brands is much more complex, since their single-shell cases require quite complicated manufacturing steps.

For the benefit of our readers, we have looked into the question of how accurate a watch that is already getting on in years can be. For this reason the German representatives of Audemars Piguet in Wiesbaden and Patek Philippe in Munich were contacted and an overhaul was commissioned. The 222 was entrusted to watchmaker Martin Becker, who had worked for Vacheron & Constantin Germany for decades. We then waited anxiously for the results. After the watch was returned, we tested the rate performances with the Witschi Wicometre Professional and Micromat P testing devices. These are the measuring devices used by manufacturers for the final testing of their watches before they are delivered.

Before we go into the results, it is necessary to briefly discuss the target orientation of this ultraslim caliber. The customer's specifications state that this automatic movement was not intended as a chronometer caliber. The latter would also have required a seconds display, which had to be dispensed with for lack of space. All three timepieces achieved almost comparable values, although the Nautilus was the best regulated. It reached a rate of +002 sec./day (MOY) and showed a total difference of 10 sec./day (DEL) at six positions. For details, see the table in illustration 591.

We know from marketing experts that price is often not the primary consideration when selling luxury watches. As we have also seen, our three sports watches shared not only the same movement and designer, but to a large extent also their selling price.

So what caused the success or failure in these models? Probably only the successful design? Readers can make up their own minds here.

Finally, we wanted to know how these three sports watch models are traded at auctions and watch exchanges. We discovered that the Jumbos from AP are rarely offered. The respective prices ranged mostly between $55,600 and $66,700. The search for the large Nautilus was similar, with prices now averaging more than $111,285.

Although rare, the variants of the reference 44108 or 222 proved to be hard to sell. At the Munich Watch Exchange in 2005, an all-gold model was available for only $6,677. However, this has changed drastically. Steel models now go for $33,300 to $44,500, and gold models for $22,250 to $33,385. The *mixte* presented comes from Sotheby's and went under the hammer, including buyer's premium, for the equivalent of $2,170. In spring 2022, at the Dr. Crott Auction House, a similar model was called and sold for no less than $35,600 plus buyer's premium.

Sometimes it seems to be the same with valuable watches as with expensive automobiles. It is not an eight-cylinder hidden under the hood that decides, but the brand's name recognition, the current fashion, and the chic design.

Illustration 590
Production Start-Up
The production series of the three models presented here began with these numbers.

Producer	Model	Reference	Caliber	Movement No.	Case No.	Production Year	Public Price (steel version in Swiss Francs)
Audemars Piguet	Royal Oak 5402 ST	2121	127 006	67 001	Series No. 1	1972	3,200
Patek Philippe	Nautilus	3700/1	8-255 C	(1)	(1)	1976	4,200
Vacheron Constantin 222	44018/411	K 1121	521 982	(1)	1977	4,375	4.375

1. Could not be traced by the manufacturer

Illustration 591
Rate Performances of the Three Sports Watches after Overhaul

	PPC	AP	V&C	Comments
DEL	010	013	013	maximum rate deviation in six positions (sec./day)
MOY	+002	+014	+012	average daily rate in six positions (sec./day)
BEAT ERROR	0.2	0.3	0.4	Fall (or symmetry: deviation from the uniform fall of the escape wheel tooth from the pallet stones on the entry and exit sides of the lever).
AMPLITUDE	263	267	265	Amplitude of the balance wheel in degrees of arc

12. The Secret of the Dials—the Face of a Watch

12.1. The Meaning and Value of a Watch Dial

At a well-known auction house in Geneva in 2017, a special Patek Philippe with perpetual calendar from the 1950s was offered at 500,000 to one million CHF and was knocked down at 2.911 million CHF. Despite the considerable sum, no one had noticed that the dial of this watch was not an original. Neither had this dial been made of gold plate—as was customary at the time—nor had it been manufactured by the supplier at the time, Cadrans Stern Frères SA.

In addition, there were craftsmanship errors in the execution and printing of this dial. This case is described in more detail on page 395.

That this is not an isolated case is shown by the second example. At the same Geneva auction house in 2016, a rare Patek Philippe with chronograph and perpetual calendar was offered with a call price of 600,000–900,000 CHF and was knocked down for CHF 965,000. Although the accompanying catalog shows the dial from the back in addition to this watch, no one had noticed that this dial was not an original, nor was it made of gold plate by the dial supplier Cadrans Stern Frères SA. More details about this example can be found on page 399.

The fact that even at Patek Philippe the question of dial authenticity cannot always be answered exhaustively is described by using the example of an antique gold wristwatch with the reference number 3448.

Even in the case of valuable wristwatches that are now part of the Patek Philippe museum inventory, there have been surprises. In the case of some watches that were purchased at auctions a few years ago to complete the museum inventory, it was discovered only later that the dials mounted on these watches were not originals.

This chapter was written because nowhere in the literature has the subject of dials been treated in the depth offered here. Now collectors and experts have a solid and reliable knowledge base, since all data are precisely documented.

The dial of a watch is often like the human face: it conveys the first, lasting impression. This impression is also immediately imprinted on the dial, even if one has not yet read the manufacturer's name. Dials have to meet aesthetic criteria above all, so visual effects are a priority for all brands.

The dials of Patek Philippe watches often shine because of their simplicity, and it is only on closer inspection that the high quality and nobility of these noble timepieces become apparent. This is especially true of the exquisite enamel dials used in the classic references 2526 and in the successor model 3428 in the 1950s and 1960s. Even the extremely rare reference 2577 was delivered with an enamel dial.

Dials that have many complications to display, like those on chronographs and perpetual-calendar watches, can also be impressive. The emphasis here is not on simplicity, but on showing how the dial serves a variety of purposes.

12.2. Swiss Watch Dial Makers

In Switzerland, the making of watch dials has always been a domain of the middle classes. Today, around twenty independent, mostly family-run dial manufacturers still operate in Switzerland. However, the large watchmaking groups have taken precautions and, in their quest for independence from the market, have bought up previously independent dial factories. For example, the Richemont Group acquired Stern Création SA in Satigny in 2000. The dial manufacturer Beyeler & Cie. in Geneva was taken over by Rolex in 2001. The dial factory Flückiger & Sohn SA, founded in 1869 by Zelim Jacot (ZJ), was acquired by Thierry Stern, the current president of Patek Philippe, in November 2004. Today this company operates under the name Cadrans Fluckiger SA. In October 2006 the Swatch Group acquired dial manufacturer MOM Le Prélet in Les Geneveys-sur-Coffrane in the canton of Neuchâtel (Netherlands).

Despite the above acquisitions, most of the manufacturers promised their existing customers that they would continue to

Illustration 592
The reference 2526 also looks very good with a black enamel dial. According to a Patek Philippe employee, exactly 580 pieces of the reference 2526 were made: 480 in yellow gold, fifty in rose gold, thirty in white gold, and twenty in platinum. Ten each of the yellow- and rose-gold versions are said to have received black enamel dials. The watch shown is one of them.

Illustration 594
The IWC reference R 817 AD looks good with the blue dial. This timepiece was offered by an Israeli dealer at the Watch Exchange in Munich. It owes its brand-new appearance to a complete overhaul.

Illustration 593
A caliber 3135 movement is encased in this new Rolex Turn-O-Graph model in steel and gold. The red on the second hand, the labeling, and the date display are reminiscent of earlier models from the 1950s.

Illustration 595
The only automatic by Vacheron & Constantin that has a case with rotating bezel is the Ref. 6782Q. This model was available from 1967 to 1973.

supply dials in the same form as before. Due to the many watch brands that have been added, as well as the numerous new models, the aforementioned medium-sized companies also have enough to do. This is especially true for unusual special requests, as well as special series produced by the watch industry. These include the following companies, but this list makes no claims to completeness: ArteCAD SA, Tramelan; Cadrans Design SA, La Chaux-de-Fonds; Donzé Cadrans SA, Le Locle; Erma SA, Boécourt; Eurocadres Sàrl, Vevey; Fabbrica Quadranti SA, Novazzano; Fehr & Cie. SA, La Chaux-de-Fonds; Fraporlux Swiss SA, Pruntrut; Gemmes-Tech SA, Chavannes-près-Renens; Les Cadraniers de Genève SA, Geneva; Liss SA, Grenchen; Pieterlen, Maître Frères SA, Glovelier; Métallique SA, Biel; Montremo SA, La Chaux-de-Fonds; Natéber SA, La Chaux-de-Fonds; Pierre Galli, La Chaux-de-Fonds; RvK Guillochage SA, Chézard; Siamcad SA, La Chaux-de-Fonds; Jean Singer & Cie. SA, La Chaux-de-Fonds; Taratec SA, La Chaux-de-Fonds; Vaucher Manufacture SA, Fleurier; Vicro, Pruntrut; and Wintraders SA, Chêne-Bourg.

Some of the companies listed are also prepared to manufacture special and one-off products. In this case the respective prices can range from $660 to $2,200 and more.

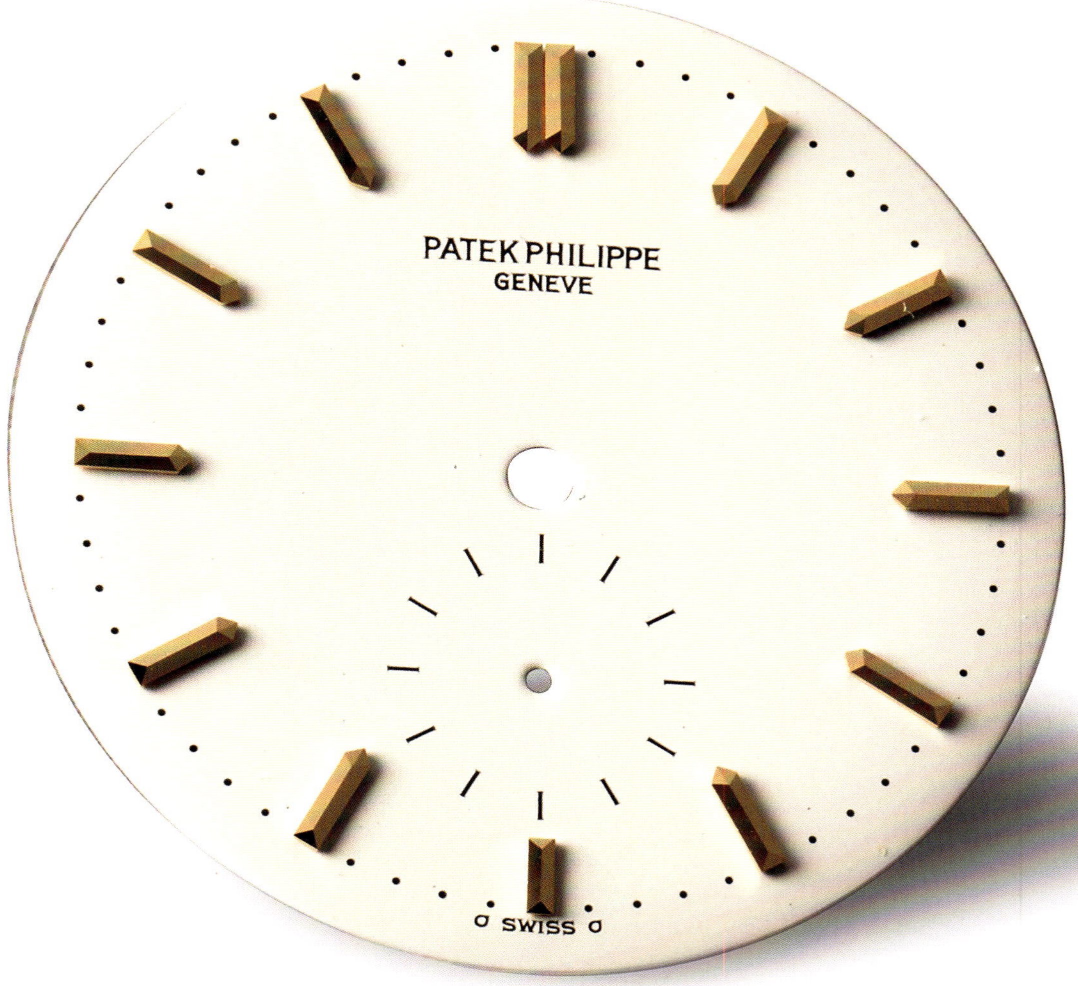

Illustration 596
A mint original replacement dial with a white lacquer finish for the Patek Philippe references 2526 and 3428 is shown. Here, the Ó SWISS Ó label documents the gold fineness only for the thirteen indices. The plate itself is made of a nonferrous alloy. The Ó SWISS Ó required by APRIOR regulations also confirms that the dial was made after 1973.

12.3. The Making of a Dial—Design and Construction

The making of a dial is preceded by concept and design. The design of this most obvious part of a watch is of particular importance. Two special requirements have to be met at the same time: the dial must be an effective eye-catcher, and it must ensure immediate brand recognition. Thus, the appearance of a dial has a special significance.

Once the concept of the dial has been decided upon, design work can begin. This includes selecting the materials, the method of attachment to the movement, the number and shape of the openings on the dial surface, the attachment of the indices, and the sequence of the printing process.

The finished design drawing and the associated parts list then contain all the specifications needed to manufacture a dial.

Once all of the concept and design work has been checked again and approved, the dial supplier receives these documents and, after the order has been placed, sets to work producing a prototype. Once the new dial is delivered, it is installed in a prototype case, but without a movement. This is often referred to as a dummy. If the dial in the dummy meets with the approval of all the decision makers, including the managing director or CEO, production approval can be given. At this stage the design may also be completed. The materials, manufacturing processes, dial attachment method, and many other specifications must be determined. Should the enclosed bill of materials also specify the assembly sequence, the individual components are numbered accordingly.

Illustration 597
The reverse of this replacement dial shows that it was made of nonferrous metal by Stern Création SA. The particularly neat workmanship on the indices "feet" is proof of professional manufacturing. In the center of the picture is one of the two feet for attaching the dial to the movement.

Illustration 598
The two window cutouts for the day of the week and month of a Patek Philippe reference 3448 with perpetual calendar. In the center is the hole for the hand tube and shaft.

12.4. Materials for the Making of Watch Dials

A simple dial usually consists of a round piece or base support, also called a plate. It also includes the indices (*appliqués*) for displaying the hours and minutes, dots for the seconds, and the intended paint colors. The plate (base support) is usually made of metal. If the dial is to protect the movement from magnetic fields, soft iron is used. Since soft iron, a pure iron alloy, is very susceptible to corrosion, either a protective lacquer is applied or the dial is silver-plated.

Usually most dials are made of nonferrous metal (i.e., brass, a copper alloy, silver, or even sheet steel). In special cases, such as some watches from Patek Philippe, dials made of gold plate were also used in the past. Depending on the watch model (reference), a sunburst finish was applied to the front of the dial or it was left natural. On some references the gold dial was coated with clear varnish or even electroplated with silver.

Gold dials were always stamped. Either the Swiss precious metal stamp or, later, just 750 was stamped on the dial (see fig. 611 on p. 367 and fig. 646 on p. 380).

From August 1973 until 2007 at the latest, the APRIOR label "Ó SWISS Ó" was also printed on the dial below six o'clock (see fig. 605 on p. 362). This seal of quality was intended to prove to the viewer that the dial was made of 18-karat gold. Furthermore, the APRIOR mark was also used when only the indices, and not the plate, were made of 18K gold (illus. 596 on p. 356).

12.5. Manufacturing Processes and Production Sequence

Once a plate has been selected for the dial, it is delivered in the form of a sheet metal roll (known as semifinished product in technical jargon). The sheet metal strip is unrolled and punched out with the help of a punching tool—an automatic machine. Depending on the design of the dial, not only its shape but also the holes for the various hands, the date window, and the moon phase window cutout are punched out. Such a punching device is called a cutting tool in technical jargon. Not only are the edges of this dial blank then deburred, but in special cases the date windows are also chamfered (illus. 598). Furthermore, the two dial surfaces are gently sanded to produce a flawless surface. The through-holes for the indices and the blind holes for the minute track are then drilled in accordance with the design specifications. Next, the "feet" for positioning the dial on the movement are soldered onto the back of the dial. In the past few

Illustration 599
Reverse of the dial of a Patek Philippe annual calendar Ref. 5146 G. No hallmarks are visible. The two "feet" for fixing the dial are still present, but the two holes at three and nine o'clock are now used for this purpose. Nonferrous metal was used as the base material.

Illustration 600
The same dial from the front. The baton indices and the Arabic numerals for the 3, 9, and 12 are not yet attached. The two holes for attaching the dial are clearly visible at three and nine o'clock. Contrary to earlier practice, the proof of origin label under the date window reads SWISS MADE.

years, some watch manufacturers have abandoned this mounting method and instead screw the dials directly onto the movement. For this purpose the necessary holes are drilled in the dial, usually at three and nine o'clock (illus. 599 and 600).

In the case of fine watches, after all this work is completed, the dial is galvanically silvered to provide optimum protection against corrosion. This is also true if the dial is to be painted.

Various methods can be used to paint a watch dial. The simplest method is spray painting, which is usually used to apply a base coat. Depending on the quality requirements and budget constraints, the base coat can be sanded down again to achieve a particularly high surface quality. The next step is the fine coating, which can consist of one or more layers.

As in the case of automobiles, the paint can also be applied by dip coating. In this electrochemical coating process, the dial is dipped or passed through an electrically conductive, water-dissolved dip coating. DC voltage is applied between the (conductive) dial and a counter electrode. The paint, in the form of a water-soluble binder, adheres to the dial surface and forms a thick film of paint there. The desired film thickness is achieved via the reaction time.

The next production step is printing for the minute scale, second scale, company signature, and, below six o'clock, the SWISS proof-of-origin label. In the case of chronographs, the printing of the subdials, the tachometer scale, and other information is often done in multiple colors. All these work steps are carried out using the pad-printing process. In this printing technique, a sharp-edged printing form with a milled print image, also known as a stamp, is filled with ink, precisely marked off, and then transferred 1:1 to the dial by pad. With the resilient pad it is even possible to print over bumps or depressions without errors.

Finally, the dial is given a coat of clear varnish (zapon varnish) to protect it from oxidation. Only then are the indices, numbers, and symbols, with their tiny feet, inserted into the provided holes and riveted from behind. After another visual and dimensional inspection, the quality assurance department gives the go-ahead for assembly.

Patek Philippe and other watch brands know of another technique that is used on many of the watches shown in this book. Instead of printing the minute scale, second scale, and the company signature, these are burned in with enamel. For this purpose the graduation marks for the minutes and seconds, as well as the company signature, are first milled in, filled with enamel powder, and then fired. This results in transparent lettering and markings that are clearly raised above the dial level. Even the SWISS label below six o'clock is occasionally created using this process. Only after this elaborate preparatory work is the dial painted several times, and the indices and numerals are applied and riveted in place. Since the varnish also has to follow acceleration due to gravity, it flows off the raised enamel lettering during this process and remains only on the flat dial surface. On page 166 (illus. 267), we see a section of the dial of a reference 2508 whose lettering is enameled and the surface lacquered.

12.6. The Enamel Dial

In addition to the methods described above, A. Lange & Söhne, Patek Philippe, and other watch brands still have the royal class—the enamel dial. More than a hundred years ago, it adorned almost every pocket watch.

One of the dream watches of every collector is probably the masterpiece that heralded the age of self-winding at Patek Philippe in 1953. The reference 2526 remained in production from 1953 to 1959 and was sold with yellow-gold, rose-gold, white-gold, and platinum cases (illus. 601, *following page*). With few exceptions, most of the watches left the production facilities in Geneva with the cream-colored enamel dial. Probably only twenty timepieces received black enamel dials, as a long-deceased employee revealed to the author many years ago. But why is there so much secrecy about these enamel dials? The author looked into this matter, speaking with a retired former employee of Cadrans Stern Frères SA, which then supplied dials to Patek Philippe, and visiting several Swiss companies that still offer dials with enamel coating. In the process I discovered some methods by which these enamel dials were made in the 1950s. As for the pure enameling of the dials, all three processes were done the same way. First, the two through-holes for the hands were made on the dial blank, then the two dowel pins (feet) were soldered on, the edges were deburred, and the dial side was finely sanded. Then the twenty-six holes for the thirteen indices (eleven single indices, two for the 12) were drilled with a diameter of 0.007 inches (0.2 mm) by using a drilling jig. In the next step the recesses for the minute scale, the strokes for the second scale, and the lettering were milled in and filled with colored (gold) enamel powder.

After firing the minute scale, second scale, company signature, and SWISS proof-of-origin label, the surface was covered with glue and evenly sprinkled with enamel powder. In the kiln, the adhesive burned off and the enamel melted and spread evenly over the dial. This process was repeated several times until the desired quality was achieved. Because of the often-uneven stress ratios in the enamel layers, it could happen that one of the layers cracked during cooling and the dial became scrap.

Once the enamel dial had gone through the entire working process without damage, the holes for the indices were reworked with a diamond drill and brought back to the correct diameter. The feet of the previously mentioned indices (*appliqués*) were then placed in the holes and riveted from behind (illus. 602).

For the complex production of an enamel dial, there were two other methods at Cadrans Stern Frères SA. In one the dial was completed without drilling the indices, and only then were all twenty-six holes drilled, using a drilling jig and diamond drill. The rest of the work was done as described.

In the third production method, the twenty-six holes were not drilled at all, and after the enamel layer was finished, all indices were positioned with the help of a template and glued onto the dial (illus. 601, p. 360). Even today you can often find examples of the reference 2526 that have a dial mounted in this way. Ignorant of the various processes, many an expert assumed that these dials were fakes.

If you have ever held an enamel dial in your hands, you probably noticed that the reverse is also enameled. This usually rust-brown coating is called counter enamel (see fig. 621, p. 372). This second enamel coating on the back is applied to prevent the dial from warping or bending during the firing process (tension compensation).

Finally, it should be mentioned for the layman that only straight or flat dials can be enameled. On externally rounded (i.e., curved) dials, the enamel would run during firing and not be positioned where it is needed.

Enamel dials were also used by Patek Philippe for the references 2577 and 3428. The model Ref. 5078 P, a minute repeater with a caliber R 27 PS automatic movement and platinum case, has *EMAILLE* (enamel) printed on the dial above six o'clock. The Roman numerals, the minute scale, and the seconds scale are enameled, while the transparent indices, as we know them from the reference 2526, were omitted.

Some watches made by A. Lange & Söhne of Glashütte also have enamel dials. These elaborate dials come from a small subcontractor in Upper Bavaria who wishes to remain anonymous.

Another way to produce enamel dials is to use enamel paste. Here the enamel powder is not applied to the fine adhesive layer, but enamel paste is used. For this purpose the enamel pigments are ground into a fine powder and mixed with water or oil. This paste is then very carefully applied to the prepared dial plate by using the finest of brushes. The dial is then fired and, depending on the work specification, another enamel layer is applied.

Illustration 601
Patek Philippe Ref. 2526 with enamel dial. In the through-hole for the second hand, one can see the typical constriction caused by the enamel that flowed off during production.

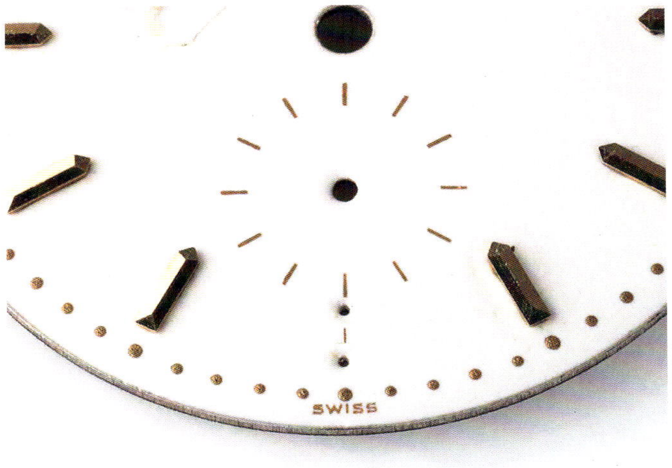

Illustration 602
On this enamel dial, the index for the 6 has yet to be inserted. The holes for the two index feet are just 0.015 inches (0.4 mm) in diameter. Here, too, the constrictions are visible in all the holes created by the enamel running off during firing.

Illustration 603
On this enamel dial, the indices have been glued on, using a stencil. The index for the 9 was carefully detached here by the dial maker for illustrative purposes but glued back on after this picture was taken.

12.7. The Guilloché Dial

The raw dial is clamped in a guilloché machine, and, depending on the specification, a series of circles and straight lines are applied to the dial surface. In illustration 604 we see a guilloché machine from 1884. Nowadays, the graphic image of a guilloché dial can also be created with a special press and associated stamp. In this process the negative image of the guilloché dial is engraved in a stamp that uses great pressure to impress on the dial the desired contours. Such presses require a closing force up to 300 tons.

The punch or the tool for the reproduction of a guilloché dial is made of hardened tool steel and can be used for the production of thousands of dials. All the manufacturing methods and procedures noted are described in detail in the captions accompanying the photographs of the dials.

Illustration 604
This guilloché machine, which dates to 1884, is still used today to produce these artistic dials. Behind the machine is longtime CEO of Vacheron Constantin Juan-Carlos Torres, and on the right the author.

12.8. SWISS and SWISS MADE Dial Labels

Almost all Swiss watches had the word SWISS printed on the dial below six o'clock. Depending on the design of the respective dial with numbers or indices, this imprint was placed symmetrically (i.e., in the center), so that the "I" of SWISS was always positioned on the dial's cross-axis.

If the dial had luminous dots or numerals, another "T," for tritium, was printed on the left and right.

The later change from the SWISS proof-of-origin label to SWISS MADE was handled differently by the watch manufacturers listed in this book.

While Patek Philippe largely stayed with SWISS until the turn of the millennium, Rolex switched to the SWISS MADE label as early as the 1960s. Most other watch brands, such as Patek Philippe, stayed with the SWISS label for a long time and, in an illogical sequence, switched to SWISS MADE only around the turn of the millennium. The best way to understand the two different labels is to look at the watches shown and described.

12.9. Association pour la promotion industrielle de l'or (APRIOR), a High-Quality Feature No Longer Known Today

This topic should be of particular interest to the reader because it cannot be found either in the literature or on Google. APRIOR was a noncommercial company that had the task of marking items or parts thereof that were made of gold with a special, protected seal. This seal consisted of the Greek letter "sigma" (Ó), which was either printed on articles made of gold or stamped as a hallmark. In French, "Le sigle Sigma" (the Sigma logo) was also called "Goutte" (the drop).

This society was joined not only by almost all the well-known Swiss watchmakers, but also by jewelry producers. Here is an excerpt from the statutes: "A l'origine il s'agissait d'une campagne pour la promotion du cadrans suisse avec heures en or" ("Originally it was a campaign for the promotion of Swiss dials with gold hours").

APRIOR was registered in the commercial register on August 17, 1973, and had its headquarters at 2300 La Chaux-de-Fonds, Avenue Leopold-Aubert 67. For a wristwatch, the sigma (Ó) had a special meaning. If the dial plate or the associated indices (or both) were made of 18K/0.750 gold, the imprint below six o'clock was "Ó SWISS Ó." If, for example, a Patek Philippe reference 3445 wristwatch, unveiled in 1961 and produced until 1972, has a dial with the imprint Ó SWISS Ó, it can be identified without a doubt as a later replacement dial (illus. 605). This dial was probably installed during an overhaul at the manufacturer's factory or in the workshop of a concessionaire, because the old dial was possibly damaged or no longer met the owner's requirements. This indication is even more interesting for very valuable (i.e., expensive) collector's watches. If, for example, a Patek Philippe reference 3448 was made in 1977 and does not have the imprint Ó SWISS Ó on the dial, but only SWISS, it is likely that a dial that does not correspond to the production year would have been installed later. Here one must determine whether a replacement dial could have been installed—one that was produced before 1972 and taken from remaining stocks. This dial would have had only the SWISS label below six o'clock. For the complications, which were produced only in small quantities by Patek Philippe at that time, the corresponding dials were ordered and delivered only in small series.

Illustration 605
This blue replacement dial for the Patek Philippe references 3445, 3514, 3541, and 3558 is made of 18K/0.750 gold plate. The indices are also made of gold. It was supplied by dial manufacturer Fritz Flückiger & Sohn SA in St. Imier. The Ó SWISS Ó label corresponds to APRIOR regulations and indicates that the dial was made after 1973.

12.10. Eric Clapton's Ref. 2499/100 Chronograph as an Example of the APRIOR Label

Since the meaning of APRIOR still is known to only a few, we offer a rather spectacular example here.

At Christie's auction house in Geneva, a Patek Philippe Ref. 2499/100 came up for bid on November 12, 2012, under lot no. 151. This perpetual-calendar chronograph had been fitted in a platinum case. With movement no. 869,308 and case no. 2,817,876, this watch was produced in 1987. Below the minute scale at six o'clock, which is of course hidden by the circular date display, there is the Ó SWISS Ó label (illus. 606, *right*). The fact that this valuable chronograph came from the estate of the British musician Eric Clapton appeared in large letters in the auction catalog. This watch went to its new owner with a hammer price equivalent to $3,655,757.

The reference 3970 chronograph shown on page 215 (illus. 328) was previously part of the author's collection. This watch was introduced in 1985 as a successor of the reference 2499/100 shown. The case is also made of platinum, the only difference being that the case diameter was reduced from 1.45 to 1.141 inches (37 to 36 mm). With movement no. 3.054,717 and case no. 4,025,538, this beautiful timepiece was produced in 1997. Worth knowing should be the differences on the dial. Although on the dial of this watch the indices are made of 18K/0.750 white gold, the usual Ó SWISS Ó label below six o'clock is

Illustration 606
This Patek Philippe perpetual calendar chronograph, Ref. 2499/100, sold at a Christie's Geneva auction on 11/12/2012 for $3,655,757. This watch was made in 1987 and has Ó SWISS Ó printed below six o'clock in accordance with APRIOR regulations. The case is made of platinum, but the indices are only white gold. This watch belonged to British musician Eric Clapton. *Christie's Geneva photo*

Illustration 607
The dial of the Patek Philippe Ref. 3970P. It followed the Ref. 2499/100 after production of that model ended in 1985. The associated watch was manufactured in 1997. By then, the APRIOR regulations had already been disregarded, although they were still valid for many years. The indices were made of 18K white gold.

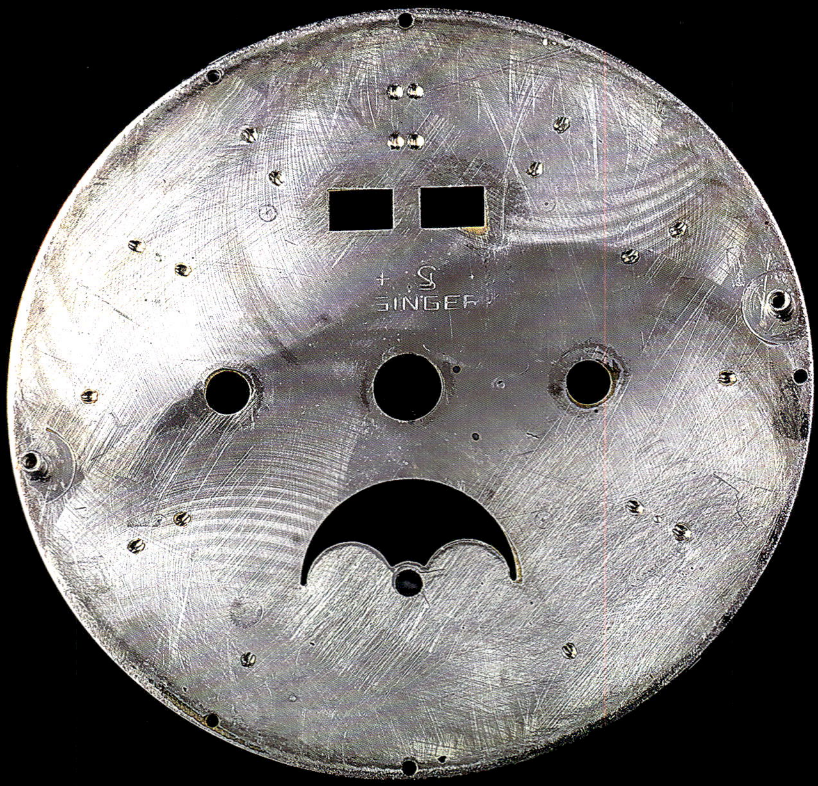

Illustration 608
This view of the reverse of the Ref. 3970P's dial reveals that it was produced by Jean Singer and the material used. Despite the watch's retail price of more than 100,000 DM, nonferrous metal was used to make the dial.

Illustration 609a
This shot of a Patek Philippe Ref. 3448, an automatic with perpetual calendar in 750 yellow gold, should be highly interesting. With movement no. 1,119,009 and case no. 313,779, this beautiful piece was made in 1964 and sold on August 27 of the same year. At the very bottom of the dial is stamped APRIOR label "Ó SWISS Ó." Since the APRIOR regulation was only introduced as of August 1973, when this watch was already nine years old, the evidence suggests that a replacement dial was installed on this watch at a later date. At first glance, one could assume that this replacement dial was made by Cadrans Stern Frères SA. Unfortunately, however, some small errors have crept in, which contradict this assumption. The printing for day of the week and month as well as for Patek Philippe is not executed according to type (i.e., not as concise as on the original). The "Ó SWISS Ó" label also does not correspond to the original design.

(see Ill. 609 B, *below*).

Illustration 609b
In the lower detail enlargement, the APRIOR label "Ó SWISS Ó" is printed but not positioned exactly as on an original Cadrans Stern Frères SA dial (*above image*). There, the two "Ó" characters are set exactly under the minute markers for "29" and "31." This enlargement shows a Ref. 3448 in 750 white gold with movement no. 1,119,379 and case no. 331,540 from 1974. The dial is original, but the varnish around the outer edge has been damaged due to the watch having been disassembled several times.

missing here. The minute scale is made with quite-long graduation marks. The SWISS proof of origin is inserted in a recessed area of the minute scale. In contrast to Eric Clapton's chronograph, which is about ten years older, despite having white-gold indices, the label specified in the APRIOR regulations was omitted here (illus. 607, *top left*).

A look at the reverse of this dial reveals that it is made of nonferrous metal and was supplied by dial maker Jean Singer (illus. 608, *lower left*). Beneath the anticorrosion finish, in places one can still make out the base plate made of nonferrous metal (brass).

During the era of Roland Tille (†), once managing director of the dial factory Cadrans Stern Frères SA in Geneva, all complicated watches with gold cases were always equipped with dials made of gold plate. This topic will be discussed in more detail later.

12.11. Conclusion concerning the APRIOR Label

With the launch of this noncommercial company on August 17, 1973, as a marketing tool to highlight articles and components made of gold, the dial label APRIOR was also consistently applied to Patek Philippe watches.

It was not until the demise of Cadrans Stern Frères SA—a dial supplier that had been active for decades and had close ties to the company—in January 1981 that other dial manufacturers had to step in very quickly. As can be seen in the following explanations, probably not all dial manufacturers active after 1981 were aware of APRIOR's rules, observed them, and consistently adhered to them.

The following list should be understood only as a small insight but clarifies without doubt the practice in use at Patek Philippe at that time.

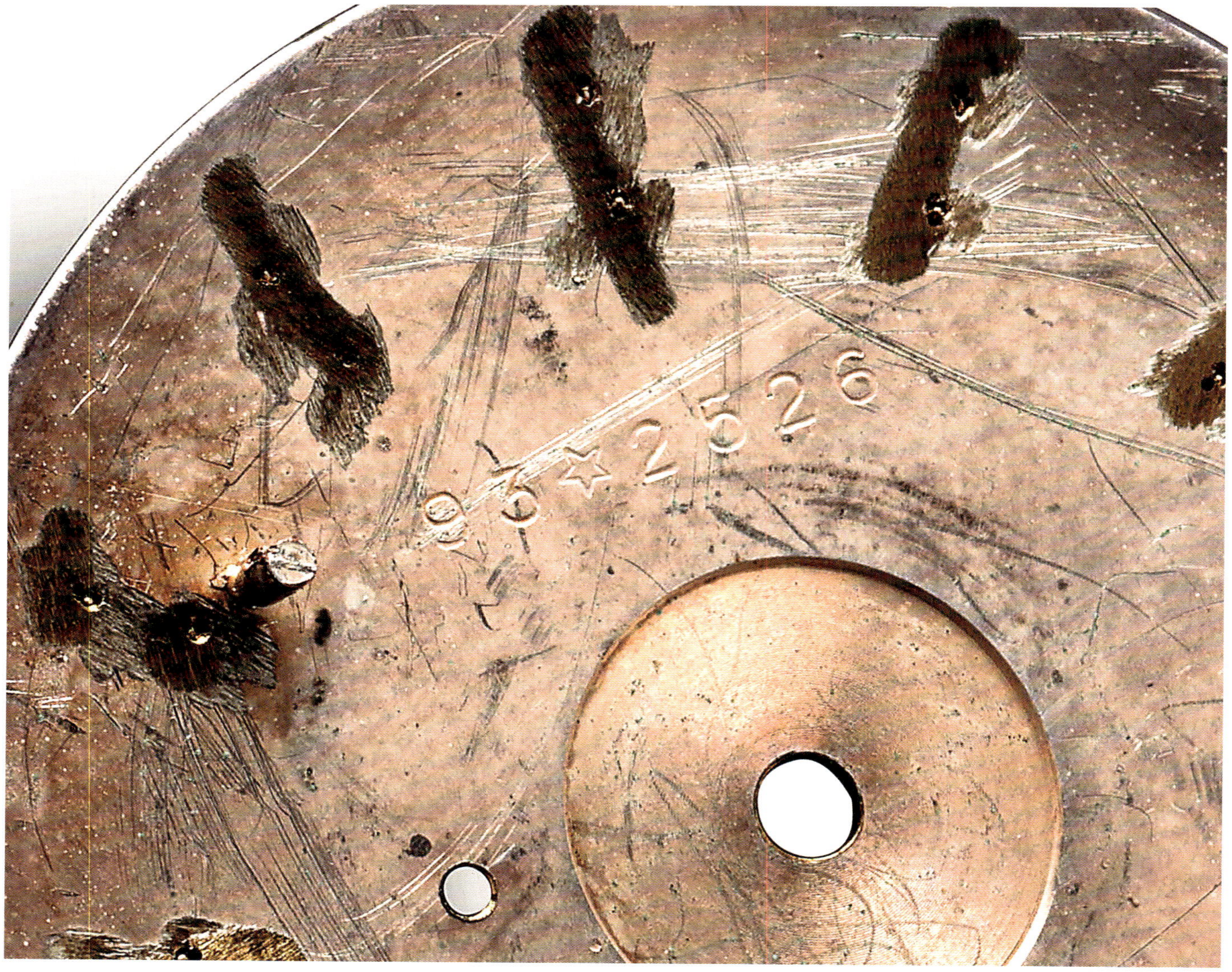

Illustration 610
The marking 93 ☆ 2526 reveals that this is a replacement dial for a Patek Philippe Ref. 2526 made by Cadrans Stern Frères SA. In this case, the dial was once again refurbished and the indices "feet" were sanded down very carelessly.

The reference 5050 with movement no. 1,957,350, an automatic with perpetual calendar, was made in 1995 and bears the Ó SWISS Ó label below six o'clock (see p. 296 in the book *Patek Philippe Genève* by Martin Huber and Alan Banbury, 2nd ed., 1998).

A Ref. 3448, an automatic with perpetual calendar with movement no. 1,119,064 made in 1967, appears on page 365. The Ó SWISS Ó imprint is visible under the month-date circle, although this watch was made in 1967 and APRIOR was not introduced until 1973.

It is certain that a new (i.e., a replacement) dial was mounted here at a later date.

The Patek Philippe (see fig. 609) Ref. 3448 shown in this chapter has the movement no. 1,119,500 and was produced in 1977. The fact that the dial of this watch bears the Ó SWISS Ó imprint, which was common at that time but is not an original, has a special and also spectacular reason. This highly exciting story will be told a little later in this chapter. Once again, it shows how important it is that all the watches presented and described have been in the author's possession. On June 26, 2007, the APRIOR association was deleted from the trade register. Thus, the previously used Ó SWISS Ó label officially ceased to exist as of this date.

Many watch manufacturers had already abandoned this quality feature some years earlier. Further details are explained in the chapter on dial analysis.

12.12. Suppliers of Dials to the House of Patek Philippe

1. Cadrans Stern Frères SA, Geneva

Until its liquidation in January 1981, the company Cadrans Stern Frères SA in Geneva was owned by the brothers Charles († October 11, 1976) and Marcel Stern.

Illustration 611
On this gold dial with the number-letter combination 93G3514 C and the Swiss gold hallmark, the star was probably forgotten. Nevertheless, due to the 93 it can be identified as a dial made by Cadrans Stern Frères SA for Patek Philippe. The official Swiss gold fineness hallmark is stamped 18K/0.750.

This honorable company became known in 1932, when Cadrans Stern Frères SA, then represented by brothers Jean and Charles Henri Stern, took over the economically troubled watch manufacturer Patek Philippe in Geneva.

As part of a later division of the Stern family inheritance, Cadrans Stern Frères SA and Patek Philippe SA were subsequently run separately.

The dials from Cadrans Stern Frères SA had a special signature. This consisted of the respective customer number, an engraved star, and the corresponding reference number. The examples of the watches shown are the dials with the signatures 93 2526 and 93 G 3514 C (see figs. 610 and 611).

According to then managing director Roland Tille (†), dials made of 18K/0.750 gold plate were always used for watches with gold cases and indices. The striking of the Swiss gold hallmark was required by law.

Only at a later stage was the reference number on the dial sometimes replaced by an order number. The number 93 was always the customer number for Patek Philippe.

The decline of the once-wealthy dial manufacturer Cadrans Stern Frères SA began when Charles Stern became more involved with financial investments than with his dials. In addition, he trusted Bulgarian-born financial juggler Eli Pinkas (see also "Pinkas Affair," *Spiegel* 36 [1980]) and because of him lost almost his entire fortune. After Charles Stern took his own life on October 11, 1976, his younger brother Marcel Stern had to continue running the company, unfortunately with little sales experience and even less fortune. The end followed in January 1981, when Cadrans Stern Frères SA was forced to file for insolvency.

2. Stern Creation SA, Rue de la Bergère, 1242 Satigny

The demise of Cadrans Stern Frères SA meant that many of the established watch manufacturers had lost their main dial maker, so quick action was therefore required. On April 14, 1981, with the financial support of several watch manufacturers (Patek Philippe, 300,000 CHF; Rolex, 200,000 CHF; Corum, 100,000 CHF; and many others), the new company Stern Création SA was established. About 130 employees from the insolvent Cadrans Stern Frères SA were hired at that time. The share capital of the new company Stern Création SA was then 1,500,000 CHF. Its new address was 1242 Satigny, Rue de la Bergère 25.

The signature of Stern Création SA was struck vertically as follows:

S
☆
C

Customer number 93 was assigned to dials for Patek Philippe. The order number was then stamped to the right or left of it with a separation of about 0.20 inches (5 mm; fig. 612). Gold dials were produced only in special cases. As a rule, nonferrous metal was used, even if a reference's original dial had been made in gold. These nonferrous metal dials were silver- or rhodium-plated before further processing to prevent corrosion. Examples are shown. Due to the different signatures dials from the manufacturers, Cadrans Stern Frères SA and the new company Stern Création SA can be easily distinguished.

3. Jean Singer & Cie. SA, Rue des Crétets 32, 2300 La Chauxde-Fonds

Dials from this house can be recognized by its maker's mark. This usually consists of a capital *S* with a superimposed small *J* (illus. 613). The SINGER + SINGER + SINGER + identifier was also used. The dial of a yellow-gold Patek Philippe Nautilus can be seen on page 380. The Jean Singer maker's mark can be seen here as well.

Illustration 612
Markings on a new nonferrous metal dial for a Patek Philippe Ref. 5009. On the left is the hallmark of Stern Création SA, on the right the customer number 93, and next to it probably the order number 468 A. The indices feet have been professionally ground.

4. Fritz Flückiger & Sohn SA, Rue de la Chef 12, 2610 St. Imier

This dial factory was founded by Zélime Jacot in 1860 and later taken over by his brother-in-law and employee Fritz Flückiger. The dials are recognizable by the maker's mark JZ for Zélime Jacot (illus. 615). On November 12, 2004, the company Fritz Flückiger & Sohn SA was bought by Patek Philippe and renamed Cadrans Flückiger SA. The dials produced today no longer bear any sort of signature and can thus be assigned to the respective reference only by the face of the dial and its appearance.

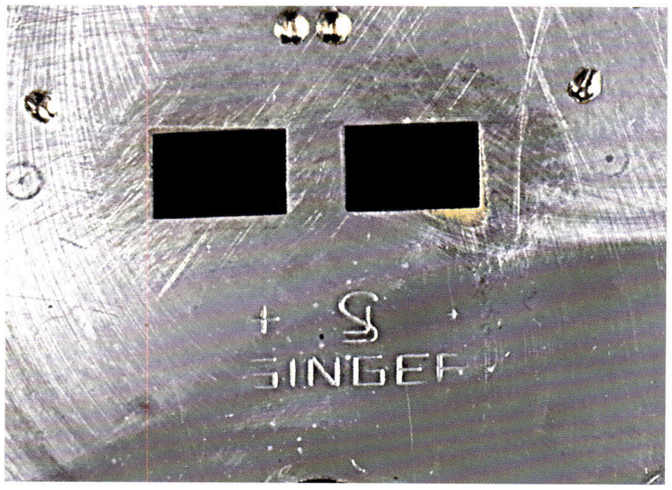

Illustration 613
The reverse of the dial of a Patek Philippe Ref. 3970 P. The signature JS SINGER reveals that the supplier was Jean Singer & Cie. SA from La-Chaux-de-Fonds. Although the case of this perpetual calendar chronograph is made of platinum and more than $110,400 was paid for the watch, the dial is made only of nonferrous metal.

5. Cadrans Beyeler & Cie. SA

This company is at Rue de deux Ponds 2, 1205 Geneva, and was founded on 09/10/1960. Beyeler was a dial supplier for various companies for decades and had Rolex in Geneva as its biggest customer.

On 07/2/2010 this company was bought by Rolex and incorporated into the group.

Beyeler has been deleted from the commercial register.

6. Cadrans Fluckiger SA (rebranding after takeover by Patek Philippe)

Even before the takeover by Patek Philippe, this company's dials were not only punched but sometimes also turned "pot-shaped" from a single piece and fixed in position by "notches" or milled grooves.

Illustration 614
The reverse of the dial of a Patek Philippe Ref. 5056 automatic with annual calendar. There are no identification markings whatsoever to be seen. The holes drilled at three and nine o'clock are new. These allow the dial to be screwed directly on to the movement (also see fig. 599, p. 358).

Illustration 615
The reverse of a dial from Fritz Flückiger & Sohn SA made for IWC. A 1 is visible above the hole for the central hand, and below it the identification marking 454373 ZJ. ZJ stands for Zélime Jacot, founder of the company.

As mentioned, the newer dials from Cadrans Flückiger SA (illus. 614) do not have any signatures. On these new dials, there are additional holes at three and nine o'clock for screwing the dial to the case. Thus, the dials can now be screwed from the top in an exact position. However, the fasteners are hidden by the bezel and are therefore not visible to the observer.

7. Change of Ownership of the Stern Group in 2000

In June 2000, Stern Création SA in Satigny, now known as the Stern Group, together with its subsidiaries Stern Manufacture and Stern Appliques in La-Chaux-de-Fonds, which were owned by Orior Holding SA, was taken over by the Richemont Group.

This finally gave Richemont its own dial manufacturer, but the company promised to continue supplying its existing customers. These included Patek Philippe, the company's main customer for many years.

This change of ownership, especially to its largest competitor, also led, among other things, to Thierry Stern of Patek Philippe buying longtime dial supplier Fritz Flückiger & Sohn in fall 2004. This strategic move served as a wise step for the future of Patek Philippe, allowing it to maintain its previous independence.

12.13. Watch Dials: Definitions

Original Dial

This term is used when referring to the dial that was used in the manufacture and assembly of the watch. It is thus contemporaneous with the making of the watch.

Original Replacement or Spare Dial

This term is used when the original dial is replaced in the course of a repair or overhaul. It does not always originate from the same period in the watch in question was made, but can also result from a later order if the reference was issued in several series.

Usually, watch manufacturers always order slightly more dials than are needed for a series. The remaining dials are then stocked as original replacement dials after a reference is discontinued.

Replacement Dials Based on the Original Template

These dials were procured at a later date (i.e., after the production of a particular reference had ceased). They are considered replacement dials for future needs in case of, for instance, later overhauls.

Here it may happen that the watch manufacturer does not order these dials from the original supplier, but from another dial producer. Such cases occur when a dial manufacturer has ceased production or the watch manufacturer wanted or had to switch to another dial manufacturer. This is what happened in the case of the Patek Philippe Ref. 3448, movement no. 1,119,500, case no. 332,858. The original dial, which was fitted when the watch was manufactured, was replaced at an unknown point in time by a third-party dial (i.e., a dial that was not appropriate for the type). On the occasion of an overhaul at Patek Philippe in Geneva, this third-party purchase was replaced by a dial manufactured later by another dial supplier. In this case the original dial supplier was Cadrans Stern Frères SA in Geneva, which was liquidated in January 1981 and thus could no longer provide a replacement.

The supplier of the third dial now on the watch on the occasion of the overhaul at Patek Philippe was the successor company Stern Création SA in Satigny. According to Roland Tille this dial, made of gold, could have been included in its raw form in the bankruptcy estate of Cadrans Stern Frères SA and later acquired by Stern Création SA.

The different signatures of Cadrans Stern Frères SA and Stern Creations SA make it easy to distinguish the two makers of the two dials without any doubt.

Externally Purchased Replacement Dials

Dials of this type are those that do not have a maker's signature and therefore do not have the markings defined in the explanations.

The mentioned Ref. 3448 was purchased with a dial that was neither made of gold plate nor had signatures as described. Such a dial is called a foreign purchase. It can also be called a fake.

The Dial Analysis, a Novelty in Watch Literature

In the previous remarks it was emphasized several times that for an objective presentation, the author must have the watches one wants to describe. This high demand is even clearer if one wants to write about their dials. One should also be able to hold these in one's hand. Before that, these "watch faces" must first be removed by the concessionaire or at the watchmaker's factory. This labor-intensive process requires the disassembly of the movement from the case, the lifting off of the hands, and finally the removal of the dial itself. As a rule, lifting off the hands does not pose a problem for the skilled watchmaker, but there is always a risk that the hands will bend when they are lifted off. It can be much more difficult when removing the dial of an older wristwatch. These are usually attached with their two "dial feet" inserted in two holes in the movement and screwed there with studs, also called ISO grub screws. These studs have a screw slot and are screwed in and out with a small screwdriver.

Dealers of antique watches are well aware of the most frequently asked questions that can be expected from potential buyers. Here, the interest in the authenticity of the dial is often at the top of the list. If the dial is undoubtedly still in its original condition, an experienced dealer can usually state this convincingly, but without providing evidence. However, they could make a binding statement only if they had been informed about the contents of this chapter. Thus, by using the following explanations, both the dealer and the buyer will benefit. The former can come up with professionally substantiated statements, and the latter can easily comprehend them.

12.14. Results of the Examination of Patek Philippe Dials

On Patek Philippe watches, as of August 1973 dials that were made of gold plate with numerals printed on them or gold indices bore the APRIOR marking Ó SWISS Ó. After 1981, dials that were made only of nonferrous metal but had gold indexes were also marked with this Ó SWISS Ó label. In the following documentation, very specific examples are provided as to how the marking of the respective dials was executed.

After the insolvency of Cadrans Stern Frères SA in January 1981 and the loss of this source of supply, other dial manufacturers, such as Jean Singer & Cie. SA in La-Chaux-de-Fonds, Fritz Flückiger & Sohn SA, Rue de la Chef 12, St. Imier, and others, received orders from Patek Philippe for the production and supply of dials. As also stated in our documentation, the instructions given by Patek Philippe to these dial manufacturers were not always completely passed on or were not always fully complied with or executed by them. It seems that there was no one left at Patek Philippe who either had the necessary expertise

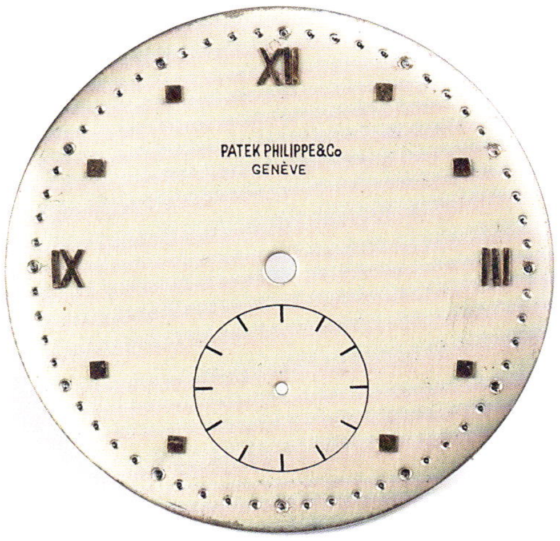

Illustration 616
Dial from a Ref. 570 manually wound watch first made in 1938. The dial has been refurbished, but not correctly. The second scale should have a double circle with second markers. Also, the SWISS label is missing below six o'clock. This dial would negatively affect the selling price of an otherwise highly sought-after Ref. 570.

Illustration 617
The reverse of the dial from the Ref. 570. 93 was the Cadrans Stern Frères SA customer number for Patek Philippe, the number 10130 perhaps being the production or order number. The work on the indices feet was only moderately well done. One can just make out the letters PPC scratched into the surface.

or who monitored these procedures closely. We hear from dealers of antique watches that the appearance and the authenticity of the dial are often the primary reasons for the purchase decision. This is a good reason to take a closer look at this topic.

In this chapter, all relevant criteria on the subject of dials are examined and explained so that the potential watch buyer can be fully informed when considering a purchase.

Dial of the Reference 570

With a diameter of 1.25 inches (32.0 mm), our first dial is from the reference 570, a manually wound watch that was unveiled in 1938. Even at first glance, one can see that this dial has been refurbished. First of all, the seconds scale, the circle for the small seconds at six o'clock, is not professionally done. Among other things, the inner circle is missing here, and only the twelve graduation marks were printed (illus. 616). The SWISS proof of origin is also missing below six o'clock. The number 93 10130 is stamped on the reverse; 93 was the customer number for Patek Philippe at Cadrans Stern Frères SA. The second number could not be identified—perhaps it was the serial production number. The most important indication of a refurbished dial is the reworked indices feet. When the indices were put back in place, they were only cursorily ground over, a clear proof for a rather moderate refurbishment (illus. 617).

Dial of the Reference 1450

The rectangular dial with eleven diamonds comes from the Reference 1450 from the year 1954. Its dimensions are 1.04 × 0.76 inches (26.5 × 19.5 mm). This dial was also not typically refurbished. The second scale's graduation marks should be arranged circularly, and the SWISS label would be correct, not SWISS MADE (illus. 618). The reverse bears no signature; only the traces left by the reinsertion of the diamonds can be recognized immediately (illus. 619).

Dial of the Reference 2526 in Three Different Versions

In this case a valuable Patek Philippe Ref. 2526 unintentionally landed dial side down on rocky ground (illus. 622). In illustration 620 we can see a Ref. 2526 with a beautiful and undamaged dial. A look at the back of this dial (illus. 621) with diameter 1.17 inches (29.8 mm) proves that this is an original dial with the number 93 ✫ 165; 93 was the customer number and 165 the consecutive production number. The SWISS label is also correct. A replacement dial—probably from the 1970s—which was also refurbished is shown in illustration 626.

Below six o'clock we find the APRIOR marking Ó SWISS Ó, proving the minimum age of this dial. On the back we find the number 93 ✫ 2526, thus customer number, the maker's mark for Cadrans Stern Frères SA, and the reference 2526.

The feet of the indices, which were reinserted after the refurbishment, are only moderately worn (illus. 629).

An original replacement dial for a Ref. 2526 in new condition and still in its original packaging was discovered and purchased by the author in 2012 from an antique-clock dealer in Denver, Colorado. The dial is painted creamy white and has gold indices. Below six o'clock is the Ó SWISS Ó label (illus. 623). The reverse confirms that it is a newly made dial (illus. 624 and 625). There we find the stamp 93 0053 A as well as the maker's mark for Stern Création (i.e., S ✫ C). This dial was therefore made by Stern Création SA in Satigny. The "feet" of the indices are neatly finished and prove the new condition of this dial.

Illustration 618
A particularly noble version of the reference 1450 with a white-gold or platinum case and diamond dial. However, the version shown was refurbished and has two weak points: the second scale should be circular, and the proof of origin should read SWISS.

Illustration 619
Here is the reverse of the diamond dial shown above. Such watch dials were usually made by special companies specializing in the setting of stones. Since no identification marking is present, it is impossible to say who the maker was. It certainly wasn't Cadran Stern Frères.

Illustration 620
Simply beautiful—that could be a comment upon first seeing the dial of a Patek Philippe Ref. 2526. It is not for nothing that this reference is one of the most sought-after wristwatches ever.

Illustration 621
This photo shows us the reverse of this enamel dial; 93 is the customer number for Patek Philippe, 165 is the running production number. Also interesting is the brown counter enamel (*contre-emaille*), which equalizes the stress on both sides of the nonferrous metal sheet (copper alloy) during production.

As on the dial of the Ref. 3940 (illus. 647, p. 380), the holes for the indices "feet" are slightly countersunk. The "feet" were caulked after insertion in these holes, an elegant solution. The purchase price was $1,500, later assessed by German customs at $1,414.26. The final invoice presented by the logistics service provider FedEx, including $37 customs duties, $268.71 import sales tax, and $13.73 administrative fees, came to a total of $319.44. Including the PayPal payment of $1,338.46, exactly $1,657.90 had to be raised for this dial. For such a valuable special version, one would probably have to pay a bit more today.

Illustration 622
This valuable Ref. 2526 automatic was probably accidentally dropped on stony ground. The impact energy ($m/2 = v^2$) of the movement behind the dial caused it to break in several places.

Illustration 624
Here is a close-up of illustration 625. Both the identification marking 93 0053 A and the Stern Creation ☆ hallmark on the right are clearly visible.

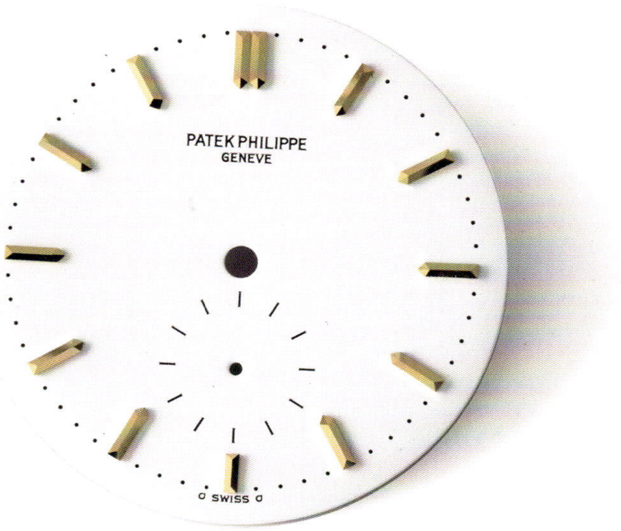

Illustration 623
This dial for the reference 2526 looks to be in absolutely new condition, but unfortunately it has a high-gloss finish instead of the previous enamel coating. In keeping with the production period and the gold indices, the Ó SWISS Ó label is found below six o'clock as per regulations.

Illustration 625
On the reverse of this like-new dial are the identification marking 93 0053 A and, *below and to the right*, the S ☆ C hallmark for Stern Creation.

Dial of the Reference 3417 Amagnétique

Illustration 627 on page 374 shows an original dial from the second series of the reference 3417, which was launched in 1959. On the dials of the second series, the imprint Amagnétique was omitted and the seconds were simplified. Illustration 290 on page 186 shows a watch from the first series. The SWISS imprint is correct. The diameter is 1.16 inches (29.5 mm). The reverse side of the dial, made of soft iron of the permalium type, still has the sticker with the words "Scientifiquement amagnétique" applied after production (illus. 630, p. 374). The feet of the white gold indices are neatly polished.

Dial of the Reference 3445

An original replacement dial made by dial maker Fritz Flückiger & Sohn for the reference 3445, unveiled in 1961, can be seen in illustration 628 on page 374. The dial is blue lacquered and has white-gold indices and the Ó SWISS Ó label below six o'clock. On the back there is the gold fineness hallmark 18K/0.750 and the numbers 2 303355. The 2 was Flückiger's customer number for Patek Philippe; the second six-digit number was the consecutive production or order number. The feet of the indices are also neatly polished (illus. 631, p. 374). The dial's diameter is 1.16 inches (29.5 mm).

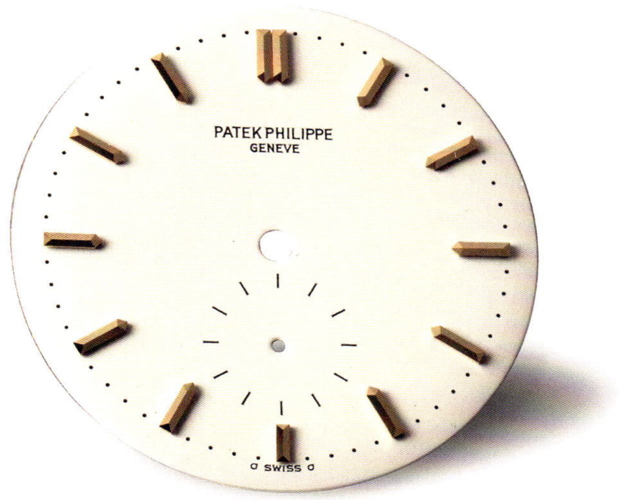

Illustration 626
This shows us a visually convincing replacement dial for the reference 2526 without the typical enamel finish. The Ó SWISS Ó label specified by APRIOR regulations undoubtedly confirms that this dial was not made until after 1973.

Illustration 629
The reverse of the replacement dial for the reference 2526 reveals that it has been refurbished once before. The indices feet have been less than masterfully ground. The identification marking 93 ☆ 2526 is interesting, telling us that the dial originally came from Cadrans Stern Frères SA.

Illustration 627
Dial for a Ref. 3417 from the second production series, a watch with manual winding and protection against electromagnetic fields. This unpretentious watch with white-gold indices is made from a special soft iron called permalium.

Illustration 630
The reverse of this dial for the reference 3417 made from a special soft iron. This dial has not yet been installed and still bears a *scientifiquement amagnétique* (scientifically nonmagnetic) sticker. It was not made by Cadrans Stern Frères and has no hallmarks.

Illustration 628
An original replacement dial made after 1973 for the reference 3445, which was unveiled in 1961, with the Ó SWISS Ó label present below six o'clock as per APRIOR regulations. The dial, indices, and date window frame are made of yellow gold.

Illustration 631
The maker's identification mark, 2 303355, and the federal precious-metal certificate can be found on the reverse. The Cadran Stern Frères ☆ hallmark is not present here, because the 2 is the Fritz Flückiger & Sohn SA customer number for Patek Philippe.

Illustration 632
This dial for a Ref. 3445 is an absolute rarity. The Ó SWISS Ó label below six o'clock indicates that the dial was not made by Fritz Flückiger & Sohn SA until after 1973.

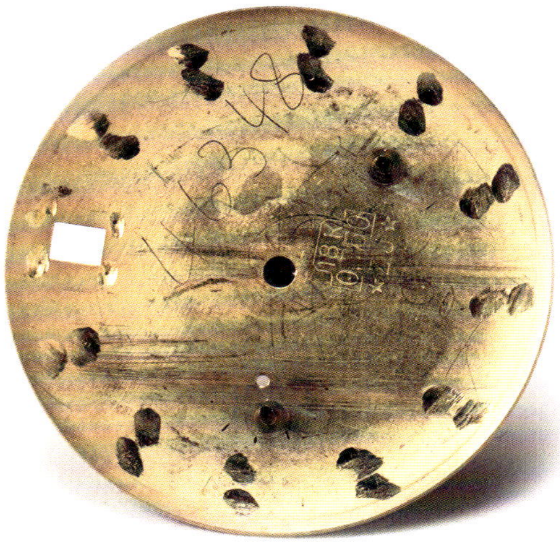

Illustration 633
Stamped on the reverse of this special watch dial is the official Swiss gold fineness hallmark, 18K/0,750, together with the identification marking ☆ ZJ ☆. ZJ stands for Zélime Jacot, founder of the company. This dial may be worth several thousand dollars.

A Special Dial on a Reference 3445

This dial, which belongs to the reference 3445 in illustration 632, may be regarded as a minor sensation. It too is a replacement dial made of gold by Fritz Flückiger & Sohn and has as a special feature an engraved Calatrava cross below twelve o'clock. Below six o'clock is the Ó SWISS Ó label. The dial's diameter is 1.16 inches (29.5 mm). On the back there is only a gold fineness hallmark 18K/0.750, with the marking *ZJ* for Zélime Jacot integrated at the bottom of this stamp (illus. 633).

Illustration 634
This detail shows the Patek Philippe Calatrava cross-milled below twelve o'clock.

Illustration 635
On this dial of the reference 3514, both the dial and the indices are 18K yellow gold. This dial is identical to that of the model 3514 shown on page 199. The SWISS label is present below six o'clock, an indication that it was made prior to August 1973.

Illustration 637
Instead of indices, the second dial from a Ref. 3514 shown has only printed Roman numerals. Several of them had been damaged and unprofessionally repaired. The SWISS label was also damaged. The second scale is similar to that on the dial in the photo on the left.

Illustration 636
Stamped on the reverse of this dial are the noble metal hallmark 18K/0,750 and the identification marking 93 Y 3514 A. The indices feet have been neatly sanded.

Illustration 638
The reverse of the dial of this Ref. 3514, which is made of nonferrous metal, has the identification marking 93 Y 3514 A but no gold fineness hallmark. This is clear proof that until the beginning of the 1970s, only dials made of gold also had indices made of gold.

Two Examples of Dials for the Reference 3514, One in Gold and One in Nonferrous Metal

The next two dials, both from reference 3514s, illustrate the differences between gold and nonferrous-metal versions. The first dial (illus. 635) is made of yellow gold, as are its indices, but bears only the SWISS proof-of-origin label. Accordingly, this piece was made prior to August 1973. On the back, next to the gold fineness hallmark 18K/0.750, are the numbers 93 G 3514 Y; 93 is the customer number for Patek Philippe, 3514 the reference number. No explanation was found for the letters *G* and *Y* (illus. 636). The dial's diameter is 1.15 inches (29.4 mm). This dial came from the Australian used-watch dealer Harrington & Co. and cost about $720, including the hands. In addition there were import duties amounting to $149.43, so the total cost was $871.31. For a gold dial in mint condition, including hands, this is a reasonable price, especially since such spare parts are no longer available from Patek Philippe in Geneva.

The second reference 3514 dial is made of nonferrous metal and has Roman numerals printed on it. Some of them were damaged and have been improperly repaired. The SWISS proof-of-origin label, which is also damaged, can still be faintly seen (illus. 637). The marking 93Y3514A is stamped on the reverse. Here, too, the meaning of the letters *Y* and *A* could not be determined (illus. 638). It is important to note that dials with printed numbers were not usually made of gold.

This dial was incorrectly mounted on a Patek Philippe Ref. 3445 (see fig. 640, p. 378) and was replaced by a correct dial during an overhaul performed at Patek Philippe.

Illustration 639
The official Swiss gold fineness hallmark 18K/0.750 and the maker's mark ZJ are stamped on the reverse of this special watch dial (illus. 635). ZJ stands for Zélime Jacot, founder of Fritz Flückiger & Sohn SA. The significance of the numbers 46348 scratched into the dial could not be determined. Perhaps they are the last digits of the associated movement number.

Illustration 640
At the bottom right, the incorrect dial, while at the top left is the reference 3445 with a suitable replacement dial.

Illustration 641
For many years the golden Ellypse has been one of the most popular Patek Philippe watches. Here is a dial from the reference 3589. The Ó SWISS Ó label is present below six o'clock and confirms, despite the printed Roman numerals, that the dial is made of gold.

Illustration 642
Lasered on the reverse of this dial is the identification marking 18K 0750 ☆ ZJ ☆, and beneath it the numbers 2 101544. The 2 is the customer number for Patek Philippe; ZJ again stands for Zélime Jacot, meaning the watch dial maker Fritz Flückiger & Sohn SA from Saint-Imier.

Dial of the Reference 3589

Another notable dial comparison comes from analyzing the two dials of the references 3589 and 3788, both of the gold Ellypse type from the 1970s and 1980s. The first-mentioned reference 3589 has a gold dial covered with clear varnish. This in turn has printed Roman numerals and the Ó SWISS Ó label below six o'clock (illus. 641). On the back, there is the engraved identification marking 18K 0750 *ZJ* 2 101 544: ZJ stands for Zélime Jacot, the founder of the Fritz Flückiger company, 2 is the customer number for Patek Philippe, and 101544 is probably the running production number (illus. 642).

Illustration 643
The dial of the gold Ref. 3788 Ellypse with the Ó SWISS Ó label and gold indices is also made of gold.

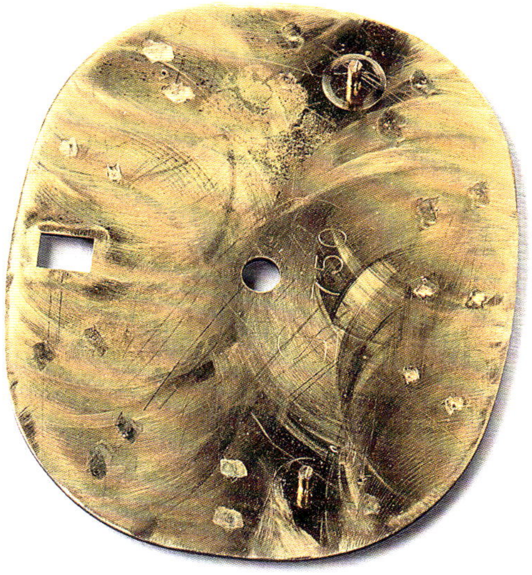

Illustration 644
Lasered on the reverse of this dial are the hallmarks JS for Jean Singer and 750 for 18K yellow gold. Here 750 is used instead of the earlier gold fineness hallmark 18K/0,750.

Dial of the Reference 3788

The dial of the reference 3788 is blue gold and has eleven gold indices. Below six o'clock is the Ó SWISS Ó label (illus. 643). The automatic caliber 28-255 C with date ticks is inside the watch belonging to this dial. Only the number 750 and JS (for Jean Singer) are engraved on the back (illus. 644). The feet of the indices are neatly ground. The two dials measure 0.95 × 1.15 inches (24.2 × 29.3 mm; 3589) and 0.92 × 1.12 inches (23.5 × 28.7 mm; 3788). These references remained in the program until the 1980s and 1990s, well after the end of former dial supplier Cadrans Stern Frères SA in January 1981.

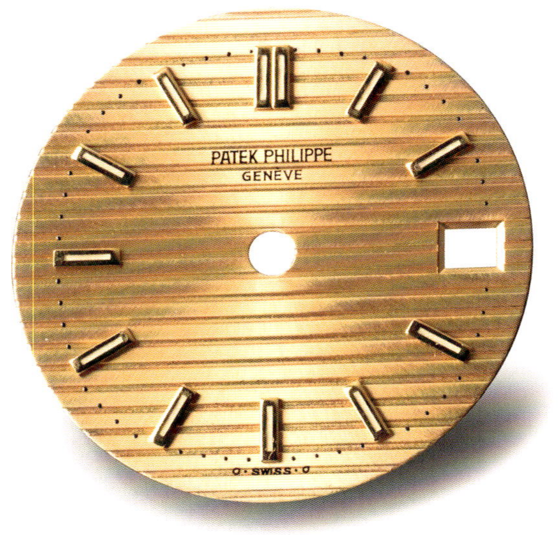

Illustration 645
The Nautilus model is one of the most popular Patek Philippe watches at the present time. This dial is solid yellow gold and has luminous indices. Below six o'clock is the Ó SWISS Ó label.

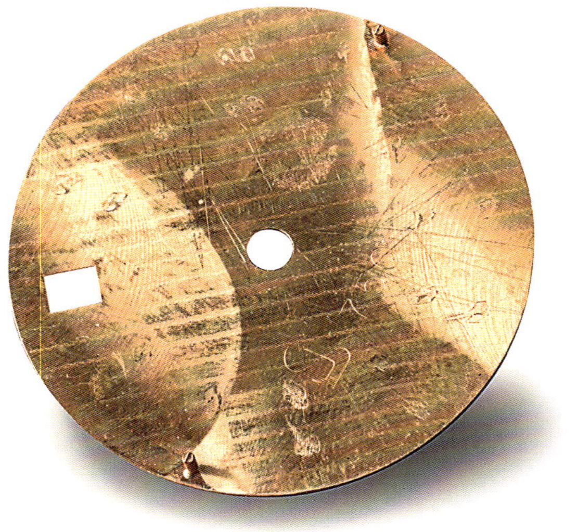

Illustration 646
Engraved on the reverse of this very thick dial, measured at 0.019 inches (0.5 mm), are the hallmarks JS for Jean Singer and 750 for 18K gold. In this case, the solid-gold dial goes well with the massive-looking gold Nautilus men's wristwatch.

Illustration 647
The dial of the reference 3940 J with perpetual calendar, unveiled in 1985, also debuted with minor changes in the current model 5140. This dial is from a watch made in 1990 and has the Ó SWISS Ó label below six o'clock.

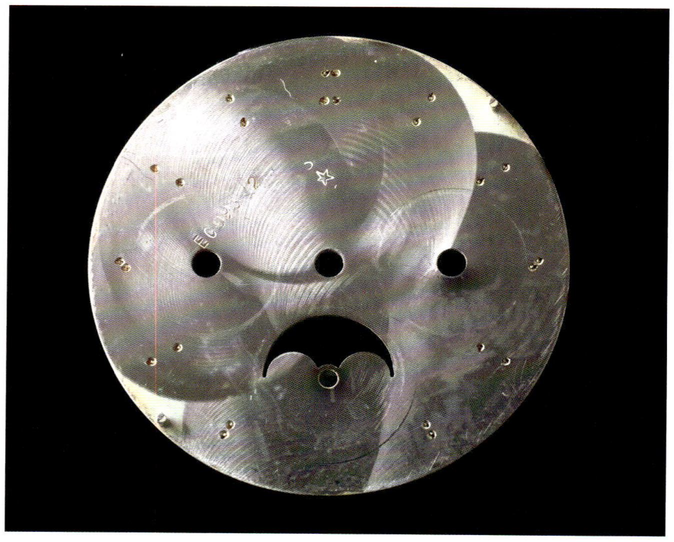

Illustration 648
The identification marking ECO 93 2 and the hallmark S ☆ C for dial maker Stern Création are present on the reverse of the dial. The 2 indicates that the dial markings are in English.

Dial of the Reference 3800/1

One of the most sought-after watches today is the sporty Nautilus. Here we present a gold dial from the reference 3800/1. With a diameter of 1.05 inches (26.8 mm), the quite massive gold dial also features gold indices with luminous material, as well as the Ó SWISS Ó label at six o'clock (illus. 645). Engraved on the back are the gold fineness hallmark 750 and the identification marking JS for Jean Singer (illus. 646). This is probably a mint-condition replacement dial.

Dials of the References 3940 J and 3970 P

Two of the author's watches were also used for this study. The first is a reference 3940 J, a Calatrava with perpetual calendar in yellow gold. Produced in 1990, this watch (movement no. 772,398, case no. 2,884,663) has a dial made of nonferrous metal but gold indices. The Ó SWISS Ó label is clearly emblazoned at six o'clock (illus. 647). On the back, there is the ECO 93 2 marking and the signature S ☆ C for Stern Création (illus. 648).

The chronograph reference 3970 P with perpetual calendar was the author's second watch to be included in this study. This time recorder (the translation derived from Greek for a chronograph) was produced in 1997. Although the case is made of 950 platinum, the dial itself is made only of nonferrous metal. The indices are

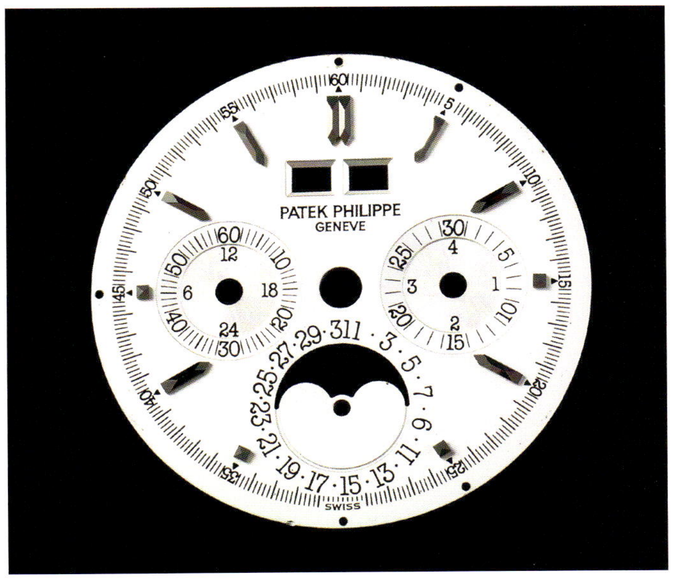

Illustration 649
The dial of the reference 3970 P chronograph with perpetual calendar has the SWISS proof of origin below six o'clock, printed in a recess in the minute scale. Although the associated watch was produced in 1977, the Ó SWISS Ó label as per APRIOR regulations is missing here.

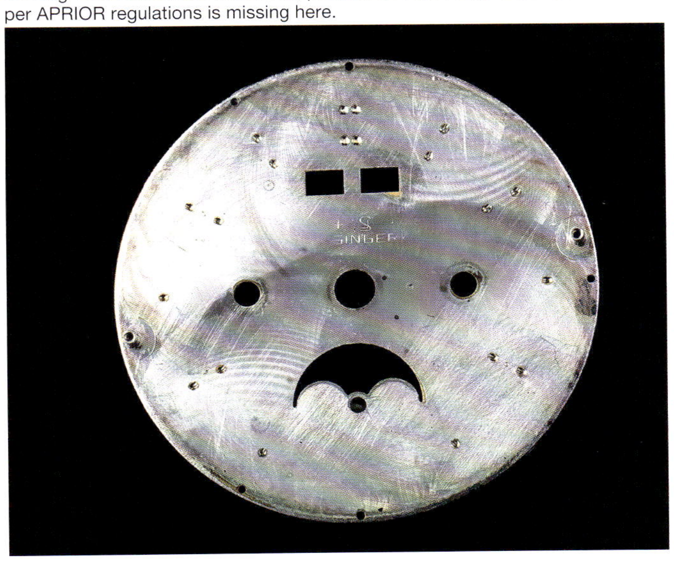

Illustration 650
The JS plus SINGER hallmark on the reverse of the dial, which is made of nonferrous metal, identifies the producer as Jean Singer. One could actually have expected a gold dial for a watch that cost more than 100,000 DM in 1997. But what eager-to-buy watch lover would think about this seemingly unimportant detail?

made of 750 white gold, and the SWISS label must suffice as proof of origin and is in a recess in the minute scale (illus. 649). On the back of this dial are the signatures JS for Jean Singer, with a plus sign placed in front of it, and the lettering SINGER. The APRIOR regulations were probably forgotten here or were no longer prescribed. Also noticeable are the small holes at twelve, one, five, six, and nine o'clock, although the usual "feet" for fixing the dial are still present (illus. 650). They probably wanted to do two things at once when attaching the dial: covering both the older and the newer attachment methods.

Illustration 651
One can describe the creamy-white dial of the reference 5009 Gondolo with its yellow-gold Arabic numerals as a simple beauty. The Ó SWISS Ó label below six o'clock refers to these gold numbers.

Illustration 652
Stamped on the reverse of this nonferrous-metal dial is the identification marking 93 468 A and, on the left, arranged vertically, the Stern Création hallmark.

Dial of the Reference 5009 Gondolo

The next dial is from the Gondolo model. It is new and has dimensions of 0.74 × 1.04 inches (19 × 26.5 mm). The dial, with twelve yellow-gold Arabic numerals, is made of colored metal. At six o'clock is the Ó SWISS Ó label as a compulsory exercise for the golden numerals according to APRIOR regulations (illus. 651). This reference was launched in the early 1990s. On the back, one finds the number group 93 468 A as well as the signature S ☆ C for Stern Création; 93 is the customer number for Patek Philippe, 468 either the running production number or perhaps the order number for the reference 5009, and A probably for the version with yellow-gold Arabic numerals. The feet of the Arabic numerals are neatly ground (illus. 652). However, the last references are only assumptions.

Illustration 653
Its cream-colored finish and the gold Breguet numerals give this dial of the reference 5040 a very noble appearance. On the right subdial, the inner numerals 1 to 4 indicate the leap year cycle. The Ó SWISS Ó label is below six o'clock.

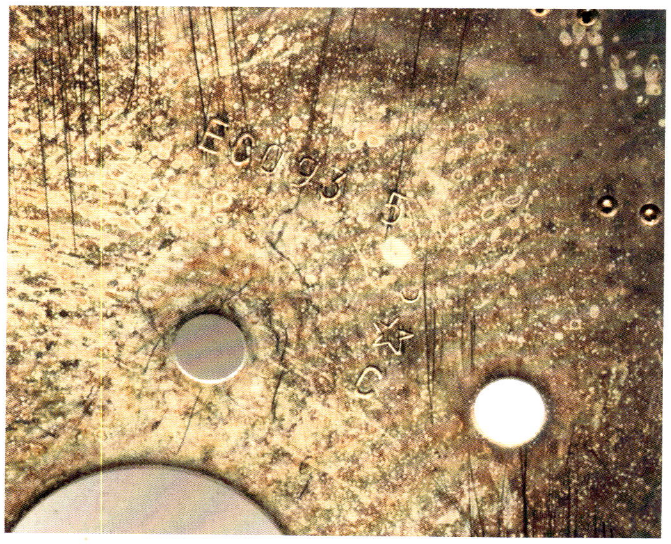

Illustration 654
On the reverse of this dial, at the bottom right, is the Stern Création SA hallmark, and above and left of it the identification marking ECO 93 5.

Dial of the Reference 5040

The new dial of the reference 5040 with perpetual calendar is from a watch from the production year 1999. The Breguet numerals are made of 18K yellow gold. The lettering is in English. Printed below the minute scale is the Ó SWISS Ó label (illus. 653). Stamped on the reverse are the code ECO 93 5 and the Stern Création maker's mark S ✩ C. For a new dial like this one, one would have to pay about $1,600 to $2,200 during an overhaul (illus. 654). Comparing this 1999 dial from Stern Création to the 1997 dial from Jean Singer used in the Ref. 3970 (illus. 649), the former and younger dial still has the APRIOR label, while the latter, two years older, no longer has it. One would have to ask oneself who was responsible for this process at Patek Philippe at that time. Or were these specifications no longer known there?

Illustration 655
The cream-colored dial of the reference 5056 immediately reveals a reference 5056 watch with annual calendar. The gold Roman numerals give the dial a noble look. The date display window is below six o'clock. Despite the gold numerals, only the SWISS proof-of-origin label is present below the date window.

Illustration 656
There are no identification markings on the reverse of the dial. The feet for positioning the dial on the movement are also gone. Holes drilled at three and nine o'clock are now used to attach the dial to the movement with screws. This dial was certainly made by Cadrans Flückiger SA.

Dial of the Reference 5056

The dial for the reference 5056 with luminous Roman numerals is new but has a small error at the seventh minute marker. Below the date window, the SWISS label is quite small. At the beginning of the new century, Patek Philippe also discontinued the APRIOR specifications. There are two small holes at three and nine o'clock, which are used to fix the dial to the movement (illus. 655). The dial's diameter is 1.18 inches (30.0 mm). There are no longer signatures on the reverse. Neither the manufacturer nor manufacturing details can be identified there. Also, the two feet for fixing the dial to the movement are missing. According to the author's research, the dial was made by now-in-house dial manufacturer Cadrans Fluckiger SA (illus. 656).

Illustration 657
This reference 5059 dial has a retrograde date display where the hand jumps back to the one at the change of the month. The window at nine o'clock indicates the day of the week, the one at three o'clock the month. Below twelve o'clock, the Roman numerals I to IV indicate the leap year.

Illustration 658
The back of this dial shows that it was turned pot shaped, probably milled from solid metal. The five milled recesses are for the corrector pushers, as well as the positioning of the dial on the movement. The identification marking 93 955 and the Stern Création SA hallmark are clearly visible.

Illustration 659
Patek Philippe Ref. 3448, from the second production series (1973 to 1982). On the one hand, it has a sapphire crystal; on the other hand, the Ó SWISS Ó label below six o'clock proves that, as per APRIOR regulations, the dial and the indices are made of gold.

Illustration 660
This close-up of the dial reveals flaked paint between the 1 and 2. The two o'clock index has also been damaged. Even though the author has been involved with Patek Philippe watches for more than thirty-five years, he has never experienced damage to the finish like this.

The reference 5056 with annual calendar was launched in 1998 and remained in the Patek Philippe program for almost ten years. Even in 1998, the APRIOR specifications would still have been valid but must have been disregarded.

This mint dial was purchased for the equivalent of $765.62 from American used-watch dealer Alexander Blasik of Erlanger, Kentucky. This price also includes $16.55 customs duty and $119.60 import sales tax.

Dial of the Reference 5059

This dial is from current production and is from the reference 5059 with perpetual calendar. The dial is painted matte white and has Roman numerals for the time and Arabic numerals for the retrograde date indication. At six o'clock, one can just make out the SWISS proof-of-origin label (illus. 657). A look at the reverse reveals some special features of this dial. It is pot shaped and turned from a round, nonferrous metal blank. The outer diameter is 1.16 inches (29.6 mm); the pot depth is about 0.086 inches (2.2 mm). Striking are the five cutouts, each 0.086 inches (2.2 mm) wide, which are used to position the dial on the movement. Stamped on the back are the numbers 93 955 and the signature S ☆ C of Stern Création SA. This fact is somewhat surprising, because Stern Création SA was bought by the Richemont group in 2000 but probably continued supplying dials to Patek Philippe (illus. 658). The name Stern Création SA was finally deleted from the commercial register on August 5, 2005.

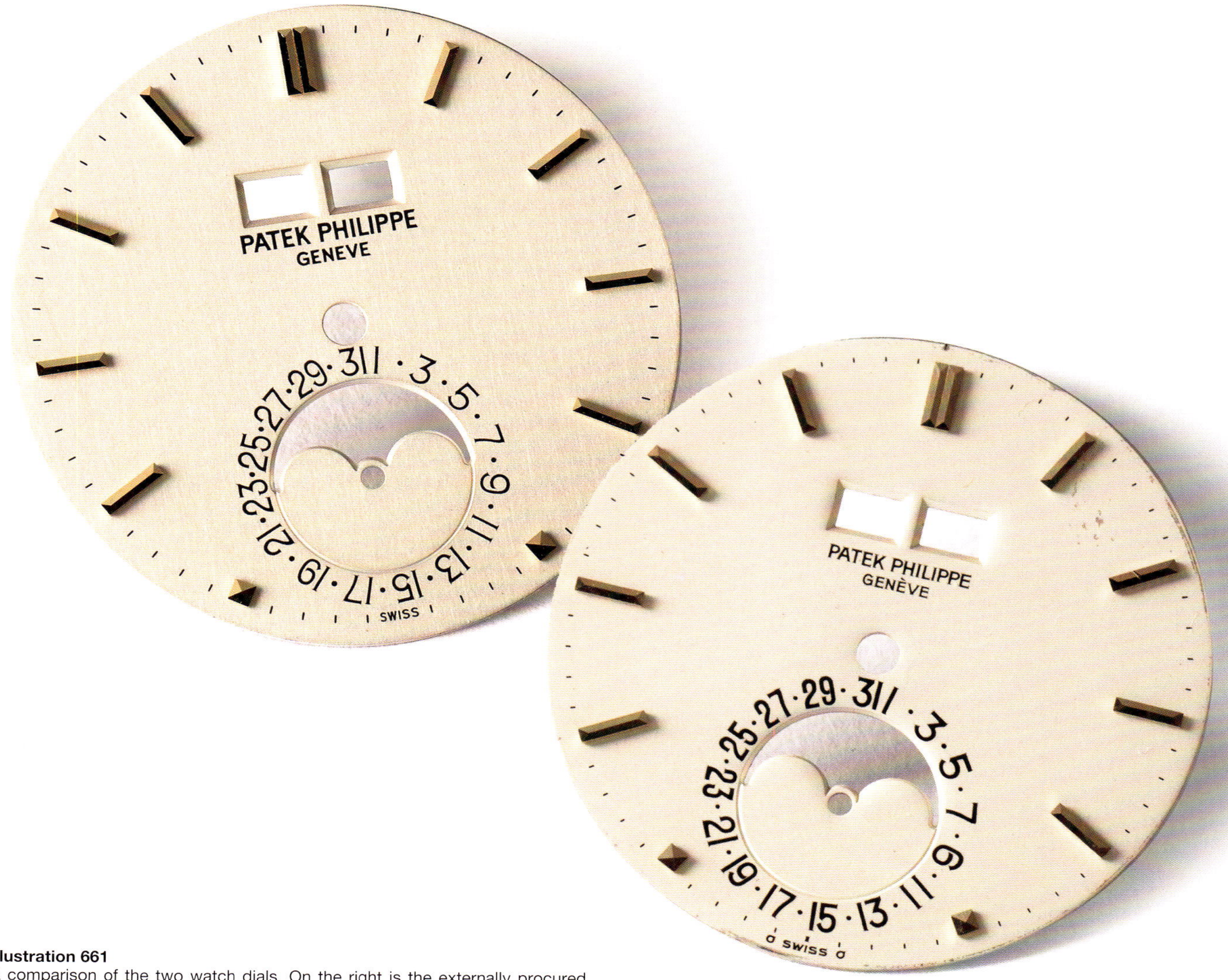

Illustration 661
A comparison of the two watch dials. On the right is the externally procured article—the dial that was on the watch when purchased—while on the left is the new dial provided by Patek Philippe. Note the differences in the print intensity of the days of the month, in the design of the windows for the day-of-the-week and month displays, and in the company signature. On the original replacement dial on the left, the Ó SWISS Ó label should actually have been present below six o'clock, since the corresponding watch was manufactured in 1977.

Dial of the Reference 3448 with Perpetual Calendar, First a Big Question Mark, Then a Revealing Solution to a Puzzle

Probably the most exciting story about the dial of a Patek Philippe watch was experienced by the author himself. The whole thing started quite harmlessly. A German auction house had offered a Patek Philippe Ref. 3448 in yellow gold, a very sought-after model with perpetual calendar, which was produced over a period of about twenty years with only 586 copies made (illus. 659). After this valuable piece did not find a buyer at this auction, the author went to the after sale at the auction house to take a look at said watch. During his examination of this watch, it was noticed that the varnish on the dial was flaked between the indices 1 and 2. Furthermore, index 2 was damaged, a notch having been made on its upper side by application of force (illus. 660).

In addition to the description in the auction catalog, the auctioneer mentioned that this watch was from its original owner and that it had undergone an overhaul at the Wempe headquarters in Geneva in May 2001. A quote from Wempe dated May 21, 2001, and the corresponding invoice for $1,383 would be submitted later.

After both parties had agreed on an acceptable purchase price, the author took possession of this watch. The following week, an overhaul in Geneva was ordered via Patek Philippe in Munich, which also included an exchange of the dial for an original from the production year of this watch (1977). The quote for this overhaul was issued on August 11, 2014, and amounted to exactly $4,136.27. The stated delivery time was twelve months. The order was confirmed immediately; only the shortening of the delivery time was requested. This topic will be dealt with in chapter 5, "Patek Philippe."

Illustration 662
The backs of the two dials. The one on the left, provided by Patek Philippe, is made of yellow gold and is stamped with the hallmark of Stern Création SA and the identification marking 93 1023. The externally procured dial (= fake) on the right is made of nonferrous metal and has none of the usual manufacturer's information.

The author's request that he receive both the old dial, which was on this watch when purchased, and the new dial, which was now to be mounted, was granted promptly. When the two dials arrived, they were analyzed in detail and documented in Peter Milde's photo studio (illus. 661 and 662).

I was in for a big surprise! The dial that had been mounted on the movement of this watch at a probably unknown time was actually a foreign product and thus a fake (illus. 663 and 664). Had this fact not been noticed during the overhaul in Geneva in 2001, or had it been overlooked out of ignorance? A comprehensive comparison with the new dial, which was also delivered with the watch, is made in this chapter, and all relevant differences, as well as other features, are documented (illus. 663 to 666, p. 386). The evidence reproduced here with detailed photographs is certainly unique in watch literature and a novelty for watch enthusiasts and experts alike. Helpful in this analysis were once again the statements made by Monsieur Roland Tille (†) in Geneva. When Monsieur Tille was shown both dials for comparison, his statements were unambiguous. The dial that was originally mounted on this reference 3448 did not come from the former Cadrans Stern Frères SA dial factory. Here are his unambiguous quotes: "The dials we used to produce for Patek Philippe were always made of gold plate for complicated watches and bore the well-known hallmarks." In addition, he noted the following technical errors in the claimed third-party dial: "First, the strokes of the minute track are made too short. Second, the print used for the numbers of the date calendar around the moon phase is too bold. Third, the bevels of the windows for the days of the week and months are not clean and not quite true to size in height and width. Fourth, the feet of the indices have been ground completely uncleanly."

Illustration 663
The externally procured (fake) dial has only one marking, JJ 86, which is scratched into the back. Researching the background of this case might be a job for a criminal investigator. Serious dial manufacturers do not sell dials with no proof of manufacture.

Illustration 664
The workmanship of the indices feet, especially those of the twelve o'clock index, is also poor and deserves to be called a botch-up. Anyone hoping to sell such a dial as an original from Patek Philippe simply has no idea.

Illustration 665
The replacement dial provided by Patek Philippe has clear proof of manufacture. First there is the Stern Création hallmark, then the Patek Philippe customer number 93 with the number 1023, probably the order or production number. The indices feet are also neatly ground.

Illustration 666
The federal fineness hallmark for 18-karat gold, 18K/750, issued by the Federal Department of Finance in Bern, Department of Precious Metals Control, also confirms beyond doubt that the dial is made of yellow gold. The newer gold hallmark 18K/750 has replaced the older 18K/0.750. Today people often settle for just 750.

Monsieur Tille also had some interesting remarks to make about the second dial. This dial would have belonged as a blank to the stock of his old company Cadrans Stern Frères SA and would have been transferred to the newly founded company Stern Création SA in Satigny in the course of the takeover of machines and material in April 1981. As a rule, replacement dials were produced only in small numbers for spare-parts stock or were commissioned as individual pieces.

Since only the SWISS proof-of-origin label is printed on this dial, and this print was set in the area of the minute scale, this dial could have been finished at a time when the APRIOR specifications were no longer used. Lack of knowledge on the part of the employees at Stern Création SA, as well as at Patek Philippe, could also be a reason why, despite the production year of 1977, the labeling as per APRIOR regulations was no longer observed. Despite this overwhelming evidence, Patek Philippe's service director at the time remained rock solid in his statement that the dial originally mounted on the watch was an original. In 2017 this employee left Patek Philippe quite unexpectedly—an extraordinary event for this company.

Only the watchmaker who had the watch in his hands at the time could say where the faulty dial, the so-called foreign product, came from and when it was mounted on the movement of the reference 3448 shown. And which owner of such a valuable watch would get the idea that the watchmaker, who was in charge of overhauling his noble piece, would install the wrong dial? Especially since he could not check which dial was actually mounted on his watch when he received his overhauled timepiece. Real trust is required here, especially with such a valuable collector's item, which today is traded in the range of $165,000 to $220,000. In case of doubt, it is worth paying to have the respective dial disassembled and checked by the manufacturer. The estimated cost of such a job is about $760 to $1,100. Finally, here is another quote from Roland Tille:

"It must be noted, however, that Patek Philippe was not as well organized in the earlier years as it is today. Also, at that time, the company was spread over five different locations in Geneva. If a replacement dial was needed in a hurry and we—i.e., Cadrans Stern Frères—could not deliver one at such short notice, another manufacturer was sometimes asked to provide it. This was also the case when the dial had been damaged during repair work and this self-inflicted damage had to be repaired quickly. It's easy to imagine that this alternative dial supplier was cheaper than us but didn't always achieve the quality that was our standard."

12.15. Jaeger-LeCoultre Watch Dials

Dial of a Jaeger-LeCoultre Three-Hand Watch

This probably new dial in silvered sunburst finish has gold indices (illus. 667). The SWISS MADE label is placed at six o'clock. There is no hallmark on the back; only the hand-scratched number 569 can be seen (illus. 668). The bore for the hand tubes and seconds shaft is 0.07 inches (1.8 mm) in diameter, and the outer diameter is 1.20 inches (30.5 mm). Thus, the supplier of this beautiful piece cannot be identified.

Dial of a Jaeger-LeCoultre Three-Hand Watch with Date Window

This sunburst dial (illus. 669) with clear signs of wear is gold colored. The indices are not applied, but faceted. Here the dial is clamped onto a dividing head, and the indices are inserted into the indentations with the help of a drawing steel with a diamond tip (i.e., pulled through). At six o'clock there is again the SWISS MADE label. On the back of this dial (illus. 670), which is made of nonferrous metal and deep-drawn, the signature JS SINGER can be found, although it is very faintly visible. Furthermore, there are manually scratched markings and the numbers 591 and 11. This dial with an outer diameter of 1.10 inches (28.0 mm) must have come from a self-winding movement from the 1960s.

Illustration 667
This lovely piece could probably be called a new watch dial. With its gold indices and SWISS MADE label, it is from a three-hand watch.

Illustration 668
The reverse of the above dial will win no beauty contests. Excessive fingerprints and the lack of identification markings leave us in the dark as to the origins of this dial.

Illustration 669
This dial with date window is typical of watches in demand in the late 1960s. The indices are faceted rather than superimposed. This type of production was considered inexpensive, since the hour indices only had to be milled into the dial plate by using an indexing head.

Illustration 671
This dial of the Odysseus model with perpetual calendar looks quite elaborate. The SWISS MADE label is below six o'clock.

Illustration 670
There are numerous identification markings scratched into the reverse of this dial, and their meaning is unknown. The hallmark JS SINGER is faintly visible.

Illustration 672
There are no identification markings present on the reverse of this dial.

Jaeger-LeCoultre—Dials for the Odysseus Perpetual with Perpetual Calendar (Ref. 166.7.80) and the Grand Reveil Perpetual with Perpetual Calendar (Ref. 180.1.99), a Watch with Alarm Function

With these two models from the 1990s, Jaeger-LeCoultre wanted to enter the top class of wristwatches. The first model, an automatic with a perpetual calendar, can be considered very successful in its design, especially since the entry-level price for this watch was very reasonable compared to other brands (illus. 671). There is no hallmark on the back of the dial, which is made of nonferrous metal (illus. 672). The SWISS MADE label shines at six o'clock. The inner bore for the various transits has a diameter of 0.22 inches (5.8 mm); the outer diameter is 1.10 inches (28.0 mm). The feet of the golden indices are very neatly caulked.

The dial of the Grand Reveil, a watch with perpetual calendar and alarm function (illus. 673), is even more noble. Here, too, the indices are made of yellow gold, but they are provided with luminous material so that the user can recognize the set alarm time even at night. Here the dial is again made of nonferrous metal, and there is no hallmarking (illus. 674). In this case, too, the feet of the indices are very well riveted by hand. The diameter of this dial is 1.22 inches (31.0 mm).

Dial of the Jaeger-LeCoultre Master Perpetual Calendar Ref. 140.8.80

This attractive watch has an exceptionally beautiful dial. It was turned from light metal and is of the highest quality. The dial appears quite noble in its black color, a special glossy paint, and

Illustration 673
The dial of the Grand Reveil is quite noble looking. This watch, with alarm function and perpetual calendar, was quite popular in the 1990s. The dial is made of nonferrous metal, but the indices are yellow gold.

Illustration 675
The dial of the Jaeger-LeCoultre perpetual calendar in black is very stylish. The noble, glossy finish emphasizes its distinguished appearance.

Illustration 674
The reverse of this dial provides no information about who supplied it. The quality of the index feet installation is worthy of mention.

Illustration 676
The reverse of this dial bears no clues to its origin. It is turned from light metal. The installation of the indices feet is particularly well done.

the subdials and the moon phase display go together very well (illus. 675). The back reveals that this is a pot-shaped rotating part that has a total depth of 0.031 inches (0.8 mm) (illus. 676). The feet of the indices are perfectly caulked in the turned recesses. There are no hallmarks of any kind. The origin of this beautiful dial thus remains obscure—the inner bore of this dial is 0.70 inches (1.8 mm), the outer diameter 1.20 inches (30.5 mm).

12.16. IWC Watch Dials

Dial of the IWC Yacht Club Ref. R 811 AD, Early Version

The first versions of this remarkable sports watch from Schaffhausen had only the IWC signature on the dial at twelve o'clock. The dial shown can be considered as new and cost about $220, including hands. It is made of nonferrous metal and has gold indices (illus. 677). On the back you can find the identification 454373 ZJ. ZJ stands for Zélime Jacot; thus it is a product of Fritz Flückiger (illus. 678). At six o'clock is the SWISS proof-of-origin label. The diameter of this dial is 1.16 inches (29.7 mm).

Dial of the IWC Yacht Club Ref. R 811 AD, Later Version

This used dial with applied IWC logo comes from Beyeler Geneva. The T SWISS T label at six o'clock tells us that the indices are provided with luminous paint (tritium). The dial is made of nonferrous metal and is silvered and ray-cut (illus. 679). The diameter of this dial is only 1.11 inches (28.4 mm). However, the distance between the two feet (from center to center) is 0.98 inches (25 mm) in both cases (illus. 680).

Illustration 677
The IWC Yacht Club was one of the most successful sports watches from the Schaffenhausen watchmaker from the late 1960s until the early 1970s. This almost brand-new dial has gold indices and pips. Here the IWC company logo is still printed.

Illustration 679
The second dial of an IWC Yacht Club has an applied IWC logo but no pips. The outer diameter is also 0.059 inches (1.5 mm) less than the version shown on the left.

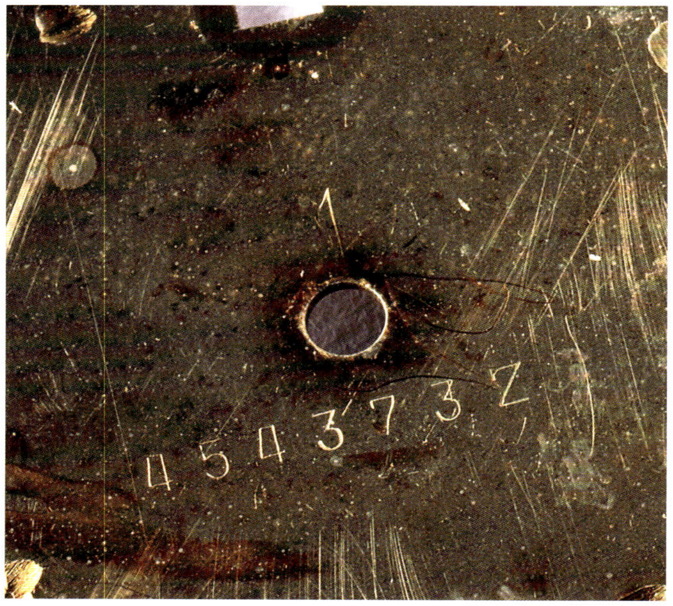

Illustration 678
The reverse of the dial has an identification marking. There is a 1 stamped above the hand shaft hole, and below it the number-letter sequence 454373 ZJ. ZJ stands for Zélime Jacot, which identifies the dial as a Fritz Flückiger & Sohn SA product.

Illustration 680
On the reverse of this dial, the only permanent marking is the stamped Beyeler Génève signature, with no other identification. Thus, this dial comes from Beyeler in Geneva.

Dial of the IWC Portugieser with Perpetual Calendar, Ref. IW 5022

With a diameter of 1.47 inches (37.5 mm), this dial tells us that the case of the corresponding wristwatch has a diameter of 1.65 inches (42 mm). The gold Arabic numerals and the corresponding indices give the watch a pleasing appearance (illus. 681). There are no hallmarks on the back of this dial (illus. 682). The feet of the numerals and indices are cleanly ground. This dial is made of nonferrous metal and probably comes from Stern Création SA, the current dial factory of the Richemont group.

12.17. Vacheron & Constantin Watch Dials, Now Vacheron Constantin

Dial of a Vacheron & Constantin Three-Hand Watch with Small Seconds

This stylish dial has a diameter of 1.18 inches (30.2 mm) and has gold indices (illus. 683). In the six o'clock position there is the Ó SWISS MADE Ó label. The letters MU are engraved on the back of the dial, which is made of nonferrous metal. The feet of the indices are neatly polished (illus. 684). The reference of this watch could no longer be determined; the watch itself is likely a model from the 1990s.

Illustration 681
This dial is from the large model IWC Portugieser Perpetual Calendar. Here the moon phase is below twelve o'clock. The power reserve lasts up to seven days.

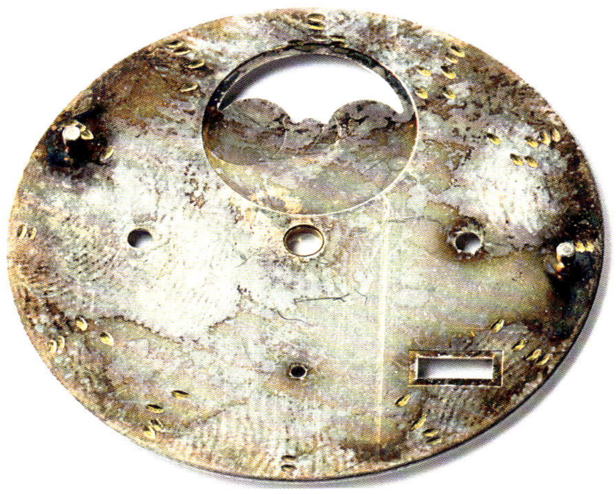

Illustration 682
The reverse of this impressive dial contains no information of any kind. Only a metalworking expert would notice that the moon phase circle is also turned out in the back as it is in the front, to avoid material stresses.

Dial of a Vacheron & Constantin Three-Hand Watch

This exemplary refurbished dial by Vacheron & Constantin would have been almost complete, except for the omission of the SWISS label at six o'clock. The gold indices are neatly set (illus. 685). On the back are the numbers 116 476; however, they do not make it possible to immediately identify the manufacturer (illus. 686).

Illustration 683
This simple but stylish dial from a Vacheron & Constantin three-dial watch has gold indices underlined by the Ó SWISS Ó label. The small seconds is proof of a manually wound movement.

Illustration 684
The reverse of this piece reveals that it is a new dial. We were unable to identify the origin of the engraving MU. Thus it was impossible to identify the maker of this dial.

Dial of a Vacheron & Constantin Automatic with Date Display

This like-new dial with rose-gold indices is from a watch with a K 1072 or K 1072/1 automatic movement. At six o'clock is the SWISS proof-of-origin label (illus. 687). Stamped on the reverse of this dial made of nonferrous metal are the numbers 116 421 1 (illus. 688). The number 116 could have been Vacheron & Constantin's customer number with its then supplier of watch dials. According to the seller, F.Q. Watches of Guayaquil, Ecuador, this dial is that of a Ref. 6394. The total purchase price with associated costs was about $659. This is perhaps a high price for a nonferrous metal dial, but acceptable given its rarity.

Illustration 685
Here we see an almost like-new dial from the reference 7394, an automatic without date. Almost everything is correct on this dial with gold indices. Only the SWISS proof-of-origin label below six o'clock was forgotten.

Illustration 687
This is what a brand-new dial looks like. The sunburst finish has a special charisma, matched very well by the date window frame and the indices in rose gold. The SWISS label below six o'clock indicates that this dial was produced before 1973.

Illustration 686
The reverse of this dial, which is not quite as well finished, has only the number 116 476. No manufacturer can be identified from this group of numbers. The number 116 was probably this supplier's customer number for Vacheron & Constantin.

Illustration 688
On the reverse of this dial made of nonferrous metal, one can see only heavily ground indices feet. The stamped identification number reads 116 421 1. The number 116 was probably one of this supplier's customer numbers.

12.18. Rolex Watch Dials

Dial of a Rolex Datejust with Diamonds

Probably no other brand's dials are imitated as often as those of Rolex. The as-new dial presented was probably already installed once. You can see this by the slight marks on the edge of the dial (illus. 689). This beautiful piece has a diameter of 1.10 inches (28.0 mm). There are no hallmarks on the reverse, but the dial is very well finished (illus. 690).

Dial of a Rolex Submariner Ref. 16610 in Original Packaging

This dial was delivered with a replacement bezel, gasket, and hands. The small tin box and the described contents can be seen in illustration 691. Initial corrosion damage can be seen around the indices of this typical Submariner (illus. 692). The back of this dial is neatly finished. It bears the number 121 104 1 (illus. 693).

This Submariner dial came from the internet retailer Wristwatchseller in Dallas, Texas, and cost $78.85, including freight. The seller had not indicated that the dial around the indices was already corroded. Because of the importance of this dial and also because of the documentation in this book, the right of return was waived. The diameter of this dial is exactly 1.08 inches (27.5 mm).

Illustration 689
A dial with audience appeal from a Dayjust still looks impressive, even though it has been used before. You can see this in the small traces of wear.

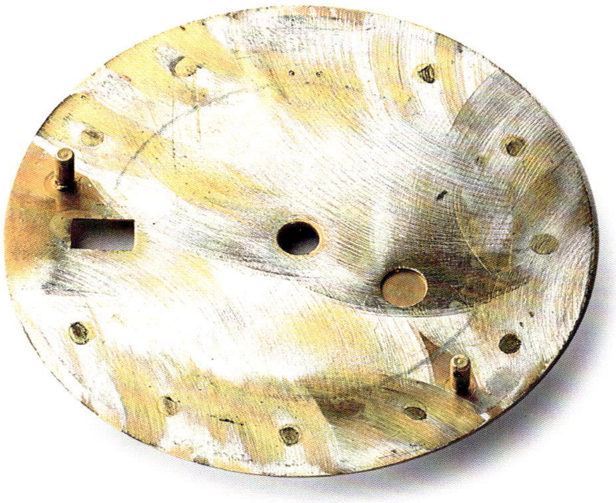

Illustration 690
The cleanly finished reverse is praiseworthy but contains no clues to the identity of the manufacturer.

Illustration 691
The Submariner is one of Rolex's most popular models. In the past, replacement dials were superbly packed in a small metal box. Here one can see the replacement bezel, gasket, and well-packaged hands. The dial itself is still safely packed inside the metal box.

Illustration 692
This impressive dial is already showing the first signs of corrosion damage, which is not supposed to happen to Rolex dials. After more than thirty years, it is difficult to assign blame. Perhaps this dial was simply improperly stored.

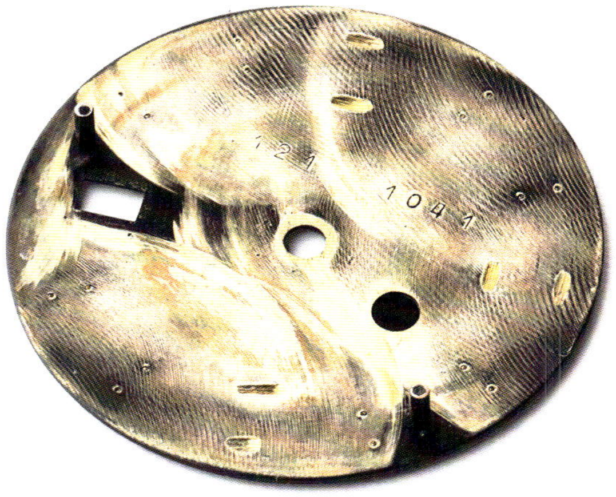

Illustration 693
Here is the cleanly finished reverse. The stamped identification marking 121 104 does not provide any usable information.

Rolex Day-Date Dial in New Condition

To be absolutely sure that this dial is an original, it was sent to Rolex Cologne for verification. Its authenticity was confirmed, especially since the engraving B 320 is present. The applied Roman numerals are in 18K/0.750 yellow gold and look very transparent (illus. 694). The diameter of this dial is 1.12 inches (28.7 mm). The reverse side is cleanly polished (illus. 695). This dial was purchased on the internet from a seller in the USA for $250 and cost approximately $288 with freight and all duties; a good price for a mint original Day-Date dial.

Illustration 694
A dial for a Day-Date in new condition. The Roman numerals are yellow gold. The SWISS label is present below six o'clock. Rolex confirmed the authenticity of this dial.

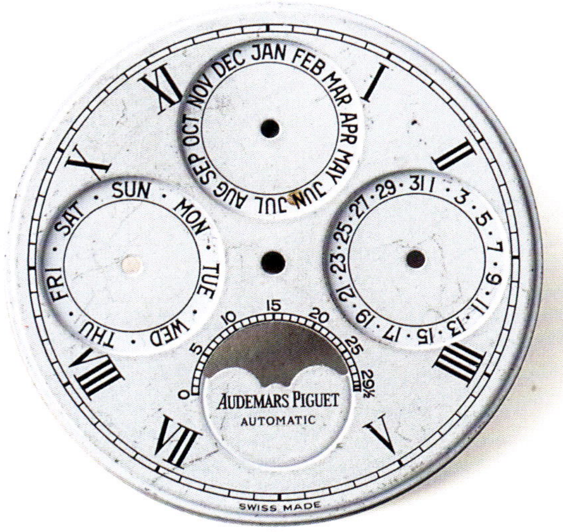

Illustration 696
This dial from an Audemars Piguet Perpetual Calendar exudes reserved elegance. The SWISS MADE label is below six o'clock.

Illustration 695
The reverse of the Day-Date dial is cleanly finished, the number feet expertly ground. The identification marking B 320 identifies the manufacturer Beyeler in Geneva.

Illustration 697
The reverse of this dial made of nonferrous metal looks quite simple. The hallmark S ☆ C with the identification marking 6 0497 identifies the dial maker Stern Création from Satigny.

12.19. Watch Dials from Audemars Piguet

The final exhibit in our dial show is a dial for an Audemars Piguet with perpetual calendar. The creamy-white lacquered dial has a noble look, with Roman numerals for the hours. The subdials are set lower. At six o'clock we find the small SWISS MADE label (illus. 696). On the back of the dial, which is made of nonferrous metal, we find the mark 6 0497 A and the hallmark S STAR C for Stern Création SA. The 6 represents AP's customer number, 0497 probably the order number, and A for the respective version (illus. 697 and 698).

Still in its original packaging, the dial was purchased from Australian used watch dealer Harrington & Co. in Woolloongabba for about $527 including transport and import duties.

Summary

The reader will certainly be able to understand why Patek Philippe dials are in the foreground of this study. At auctions, watches from this manufacturer fetch the highest prices. And in this game, the dial often plays the most important role. Here's a quote from Stefan Muser of the Dr. Crott auction house: "The more beautiful the dial, the higher the price!" For this chapter the author not only examined well more than one hundred dials of the watch brands discussed, but also evaluated the offers of the leading auction houses over many years.

12.20. Three Imposing Examples of the Special Significance of Dials

A Replacement Dial for a Patek Philippe Ref. 2499/100 for $32,550

At an auction house in Geneva, a replacement dial for a Patek Philippe reference 2499/100 with perpetual calendar came up for bid on 11/05/2015. The dial, offered under lot number 308a, was knocked down after a short bidding battle, including buyer's premium for a full $32,550 (illus. 699). The marking on the back is 93 ☆ 255 HJ; 93 is the customer number for Patek Philippe, the ☆ the maker's mark of Cadrans Stern Frères SA, 255 probably identifies the current production or order number, and HJ is the code for *heures jaune* (yellow-gold indices). This sheet was probably produced a few months before Cadrans Stern Frères SA went bankrupt. At that time, the era of gold dials had also long since come to an end (illus. 699 and 700).

Illustration 698
Close-up of the hallmark on our Audemars Piguet dial. The 6 is Stern Création's customer number for this brand (see fig. 697).

Patek Philippe Ref. 2497, a Watch for $3,256,148, but without the Original Gold Dial from Cadrans Stern Frères SA

A Patek Philippe reference 2497 with perpetual calendar was offered at an auction house in Geneva on 11/9/2015. According to the catalog description, this watch had been owned by Emperor Haile Selassie of Ethiopia. The dial of this watch, which is shown in an enlargement in the catalog, is obviously a foreign dial (illus. 701). On the one hand, the two indices in dot form, which should have been present for the 5 and the 7, were missing. Second, the numbers for the thirty-one days around the moon phase, from 9 to 21, were not printed "turned." Furthermore, the graduation marks for the minute scale were much too long, and the SWISS label was forgotten. Also, atypically, the reverse of the nonhallmarked dial was painted black (illus. 704). The auction catalog had emphasized that the dial was made of solid gold. This cannot be true, because the legally required Swiss gold fineness hallmark 18K/0.750 is not present. The manually engraved movement number 888,058 may serve as a further indication. The dial supplier at that time, Cadrans Stern Frères SA, produced the series dials according to Patek Philippe's specifications and always had the customer number 93, the order

Illustration 699
One can see how expensive original replacement dials can also be from this example. The dial, which fits a Patek Philippe reference 2499/100, was sold at an auction house in Geneva on May 11, 2015, including buyer's premium, for a total of 30,000 Swiss francs. The Ó SWISS Ó label proves that the indices are made of 18K yellow gold. The dial itself is made of nonferrous metal only. *Christies Geneva photo*

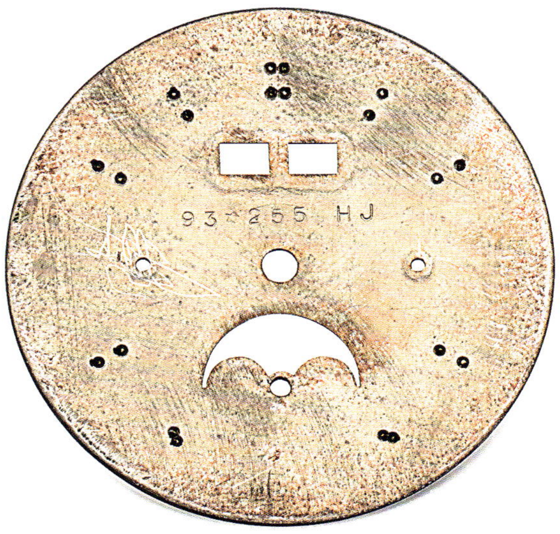

Illustration 700
The identification marking 93 ☆ 255 HJ is stamped on the reverse of this dial. The number 93 is Cadrans Stern Frères SA's customer number for Patek Philippe, 255 could be the order number, and HJ stands for *heures jaune* (gold indices). *Christies Geneva photo*

Illustration 702
Patek Philippe reference 2497 with an original dial. It is made of gold and was manufactured by Cadrans Stern Frères SA. On this dial, you can clearly see the differences from the black-painted replica seen on the left. On the latter you can also see the luminous material that was applied to the numerals. Neither the final inspection by Cadran Stern Frères SA nor the quality expert at Patek Philippe would have accepted such a botch-up. *RobertMaronWatches photo*

Illustration 701
Patek Philippe reference 2497 with perpetual calendar in yellow gold once owned by Emperor Haile Selassie of Ethiopia, but with a nonoriginal dial. *Christies Geneva photo*

Illustration 703 (*left*)
This dial from a Patek Philippe reference 2497 with perpetual calendar was recognized as a fake. Here the two round indices for the 5 and 7 are missing, as is the SWISS label below six o'clock. The printed numbers for the thirty-one days around the moon phase are not turned from 9 to 21, and the graduation marks for the minute scale are too long. Finally, the treatment of the two windows for the day of the week and month is untidy and almost clumsy. *Christies Geneva photo*

Illustration 704 (*right*)
The reverse of this dial has only the movement number 888,058 manually engraved, as seen on other replacement dials. The description of this dial stated that "the dial is made out of solid gold," but there is no Swiss gold hallmark. Finally, the black finish on the reverse stands out; a completely uncommon practice in dial manufacturing. The Cadrans Stern Frères hallmark with the customer number 93, the star, and the reference or order number are also missing here. *Christies Geneva photo*

or serial number, and the company's own star hallmark engraved. A manually engraved movement number never existed because the dial supplier could not know on which movement the commissioned dials would be set. This manual marking was used only for replacement dials that were not made by Cadrans Stern Frères SA, but by another supplier, and therefore were not originals.

The movement number of the caliber 27SC used in this watch was 888,058, the case number 679,792. Two other reference 2497 watches from the same production period 1954–55 were used for comparison. The first one had the movement number 888,055 and the case number 680,004, the second one the movement number 888,090 and the case number 683,860. Both still had original gold dials and the known markings.

Thus, the reasons given by the auction house for the discrepancies in this dial are simply not tenable. Emperor Haile Selassie, who also called himself Negus, can unfortunately contribute nothing to the evidence himself. He died at age eighty-three on August 27, 1975.

First offered on September 11, 2015, this watch was withdrawn from the auction for unknown reasons. However, it was offered again in the auction on May 15, 2017. Here again the same text description was printed, only the rear view of the dial was omitted. This time the watch, which was listed at $560,000 to $1,118,375, was knocked down. To its new owner it was worth $3,255,589! Presumably this buyer was not bothered by the inaccurate description of this watch's dial.

Illustration 705
Patek Philippe reference 2499 with perpetual calendar and chronograph from 1956. The indices for the 5 and 7 are missing from the dial. On very close inspection, one can also see that the innermost circle of the tachymeter scale to the left of the 25 touches the first 1 of the day-of-the-month indication 11. Furthermore, the 3 in the number 130 at the bottom of the scale has an unusual thickening in the upper area. These errors were found on several reference 2499 watches and suggest the same dial supplier and stamp. Despite the incorrect replacement dial, which no one noticed, this watch sold for 965,000 CHF. *Christie's Geneva photo*

Illustration 706
A look at the dial of the watch shown on the left, photographed from behind, provides the final proof that it is not an original dial. Not only the customer number 93, which Patek Philippe received from the supplier Cadrans Stern Frères SA, is missing, but also the serial or order number, as well as the ☆, the company hallmark of Cadrans Stern Frères SA. Finally, the original dial was made of 18K gold plate and bears the federal gold fineness hallmark. The counterfeit mounted here is made of nonferrous metal only. A manually engraved movement number, such as 868557 here for the caliber 13‴ Chrono Q, never existed at Cadrans Stern Frères SA, because they didn't even know what the number of the movement to be encased was when the dial was delivered. Such a thing existed only with a third-party supplier of dials. They had to mark these individual pieces for their customer with the respective movement number to ensure that the watch could be properly reassembled at Patek Philippe. *Christie's Geneva photo*

Patek Philippe Ref. 2499 with Incorrect Replacement Dial, Auction Price $1,079,231

At the same auction house, a Patek Philippe Ref. 2499 with chronograph and perpetual calendar was offered on May 16, 2016, under lot no. 174 (Ill. 705). This valuable piece supposedly came from a significant private collection. The lot estimate was $671,000 to $1,006,500 and was knocked down at $1,079,231.

Two quotes may be reproduced from the catalog description of this watch. First, it says, "case, dial, and movement signed." Second, it reads, "Over the course of its 60 years of existence the dial has undergone an ever so gentle cleaning and has accumulated a most charismatic, uniform ivory-colored patina since the hard enamel signature and scales are as raised as the connoisseur would expect." The movement number of the 13‴ Chrono Q caliber is 868,557, and the case number is 696,499; 1956 is listed as the year of production. According to Christies, this reference 2499 was produced from 1950 to 1985 and was issued in four series totaling 349 pieces. The gem listed here is from the second series, which was put into production with round chronograph pushers instead of the previous flat ones starting in 1956.

Even a first glance at the dial reveals to the connoisseur that something is missing from this watch: namely, the square and pyramid-shaped indices for the numbers 5 and 7. Several examples of the second series of this reference 2499 were used for comparison, all of which have these indices on the dial. The watch with movement number 868,747 and case number 2,611,725 was examined closely. This watch had the two indices mentioned. The watch was sold on November 16, 2002, also in Geneva, for $717,437.

The final proof that a second dial was mounted on this watch can be found on the reverse of the dial, which is shown in the catalog. Instead of the usual Cadrans Stern Frères SA gold plate, this dial was made of nonferrous metal. Missing once again were the Patek Philippe customer number 93, the serial or order number, and a ☆, the company hallmark.

Also, the movement number 868,557, which is engraved on the dial, never existed at Cadrans Stern Frères SA, but only at a third-party supplier of dials. The latter had to mark these individual pieces for his customer with the respective movement number to ensure that they could be assigned to the watch when it was reassembled.

Presumably the lucky buyer of this watch is unlikely to learn about the special role of his watch's dial. Also, the two-year period in which one can still complain about hidden defects or faults of a watch bought at auction expired for this watch on May 16, 2018.

Patek Philippe Ref. 2499 with Original Dial, Auction Price $394,786

That this auction was also not without delicacy should be mentioned in conclusion. Under lot no. 116 in the same catalog was another Patek Philippe Ref. 2499, this time from the third series, with a call price of $279,500–$503,268. The movement number was 869,489, the case number 2,700,441. Even a first practiced look tells the expert that it is almost certainly an original dial. The fitting of the month and weekday windows was done with the usual Cadrans Stern Frères SA precision (0.279 × 0.102 inches [7.1 × 2.6 mm] outer dimensions), and the square indices for the 5 and 7 were applied true to type. A beautiful watch!

With a hammer price of $394,786, only about a third of the price of the previously described version without the original dial, the buyer must have been completely satisfied.

Illustration 707
The author found this stylish enamel dial in the familiar cream-colored finish on the internet.

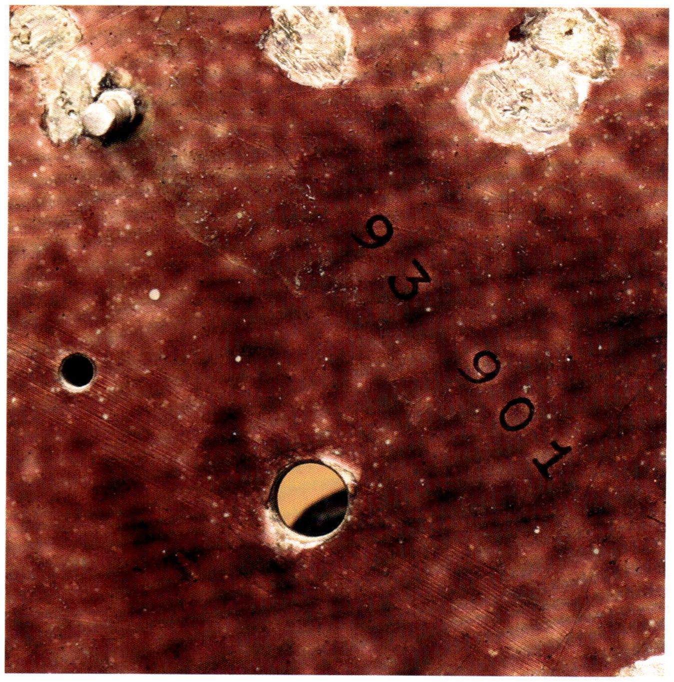

Illustration 708
The reverse of this dial turned out to be a great sensation. Since these enamel dials were always numbered consecutively during production, it proves that more than 901 pieces were made. Such a number has never been mentioned in the literature before.

The Enamel Dial with Breguet Numerals

Almost rarer than copies of this rare collector's watch with a black enamel dial are the versions with an enamel dial and the particularly memorable Breguet numerals.

One such watch offered at an auction in Geneva as late as 2021 was withdrawn because the asking price of $134,200 was not met at that time.

In the same year, California used-watch dealer Robert Maron Watches of Thousand Oaks, California, sold a reference 2526 with Breguet numerals, formerly owned by artist Andy Warhol, for a six-figure sum. Why this watch was equipped with very unusual and therefore probably not-so-suitable dash hands could not be explained.

Date	Auction House	Movement No.	Case No.	Sale Price (Swiss Francs / USD)
April 9, 1989	Antiquorum Geneva	762,975	694,742	Not known[1]
June 23, 2005	Antiquorum New York	764,367	681,534	23,575 USD
Dec. 15, 2004	Christies' Geneva	760,903	682,039	83,650/95,637[2]
Nov. 17, 2008	Christies' Geneva	760,903	682,039	87,000/99,467[2]
Mar. 23, 2005	Antiquorum New York	764,367	681,534	23,575 USD
May 10, 2010	Christies' Geneva	760,179	678,315	72,500/82,889
Apr. 7, 2011	Sotheby's Hongkong	764,367	681,534	30,000 / 250,000 HKD

1. Sale price not known; sold in a postauction sale
2. As one can see from the table, the same watch was sold at Christie's and four years later was auctioned again, this time at a slightly higher price.

Illustration 709
This enamel dial with Breguet numerals is an absolute rarity. This noble piece was delivered by Patek Philippe to Swedish concessionaire Bo Berggren Urhandel AB in Stockholm as a spare part, probably more than sixty years ago. The dial later turned up in the estate of a former watchmaker of this concessionaire and was put aside by his heirs. At an unknown date, it was subsequently rediscovered by the heirs and offered on the internet. The reverse of this enamel dial reveals something not previously known to us. Presumably there were other suppliers besides Cadrans Stern Frères SA who supplied Patek Philippe with special versions of enamel dials. The rediscovery of this dial represents a minor sensation that should not be withheld from the reader.

Illustration 710
This comparison of two models of the reference 2526 with different enamel dials illustrates the imposing difference between the standard dial with baton indexes and the one with Breguet numerals. The price difference between the two models today can be a factor of three to four.

The following table illustrates how rare this version of the Patek Philippe Reference 2526 with Breguet numerals was even in years past:

On November 6, 2022, another dial for a Patek Philippe Ref. 2526 was offered by Antiquorum Geneva, but with yellow-gold indices. The hammer price was $18,173. Its production number was 93 337, and, in contrast to the previous example, it has evenly spaced digits.

In December 2022 a cream-colored enamel watch dial for a Patek Philippe Reference 2526 was offered on eBay. This dial had serious cracks on the enamel surface between indices 10 and 11 and 11 and 12, making it unusable. Nevertheless, it was sold for $210. This dial was marked 93 288 and was probably made in the mid-1950s.

All the enamel dials available to the author so far have had yellow-gold indices and a three-digit production number.

For many years the author has maintained search programs on the internet to track down specific and sought-after items. In fall 2022 an enamel dial for the Patek Philippe Ref. 2526 unexpectedly appeared on the internet. The flawless dial has yellow-gold indices (illus. 707) and bears—a real sensation among connoisseurs and experts—the production number 93 901 (illus. 708). This astonishing discovery supports the author's assumption that Patek Philippe produced far more wristwatches with enamel dials than previously known. To the mentioned 580 pieces of the reference 2526, there might be added about three hundred dials for the subsequent reference 3428, bringing the total number to 880 pieces. If you add an estimated twenty to thirty spare dials, which were handed out as repair reserve at that time, you get a total of 910. With the current example, at least positive proof for the production of 901 dials can be provided pictorially!

Regarding the numbered enamel dials, it should be noted that most Reference 2526 watches with white-gold or platinum cases were not equipped with enamel dials but with white-gold dials. The guesswork about the total number of Reference 2526 watches may continue, but only above the number 901.

12.21. The Results of This Analysis

Of the twenty-eight Patek Philippe brand dials analyzed in this chapter, only two specimens require further explanation. The diamond dial of the Ref. 1450 (Ill. 618) was not made by Cadrans Stern Frères SA, but by another manufacturer. The setting of stones, especially of valuable diamonds, was entrusted to specialized companies. In addition, there is the erroneous SWISS MADE label, which is out of place on this watch and was used only around the turn of the millennium.

The second reference is to the enamel dials with Breguet numerals. If only a few pieces were required or special requests had to be fulfilled, as may have been the case for the dial in Ill. 709, Patek Philippe was permitted to order enamel dials from other suppliers as well. At that time, the parent and later sister company Cadrans Stern Frères SA was so busy that special or spare-part requests could be fulfilled only with longer delivery times.

Illustration 711
Enamel dials are extremely sensitive to impact and cannot withstand compressive forces. This dial, which exhibits considerable damage, came from a Ref. 2526 that was probably accidentally dropped on a stone floor. Even if the case and movement can still be repaired after such an accident, the dial must always be replaced by a new one. For such cases, one can obtain a replacement dial only from Patek Philippe, which is no longer enameled but only lacquered (see fig. 623, p. 373).

Illustration 712
The reverse of the damaged enamel dial shown on the left exhibits no signs of the damage on the dial face. Important to the watch enthusiast is the stamped marking with the numbers 93 321, which indicates it was produced in 1955.

12.22. Summary

Patek Philippe

In general, the following conclusion can be drawn for Patek Philippe dials:

Except for the two examples mentioned, which should be considered exceptions, all Patek Philippe dials had one of the markings described here on the reverse until the end of the 1990s. Dials without these markings were probably supplied directly or as replacements by other dial manufacturers.

Especially for valuable watches, which have a current value of more than $44,700 or even higher, one should have the dial checked at a suitable opportunity. Probably the best opportunity is by means of a commissioned overhaul. If the overhaul of the watch takes place at a concessionaire authorized for this purpose, one can take a personal look at the reverse of the dial in question. If the overhaul takes place at a national branch or in Geneva, one can request a photo of the back of the dial by means of image transmission via email. Today's image creation and transmission techniques have no limits.

Illustration 713
The reverse of the enamel dial with Breguet numerals shown in illustration 709 also reveals some previously unknown details. As can be read in the text on page 403, Patek Philippe also had the option in the 1950s and '60s of purchasing enamel dials from other dial manufacturers in exceptional cases. The special art of producing enamel dials for watches was almost a thing of the past by that time, but the know-how had survived from the days of pocket watches. In this case, the manufacturer used a lighter color mixture as the counter enamel rather than the rust-brown customary for Cadrans Stern products. It is unlikely that the meaning of the 4N marking can be determined today.

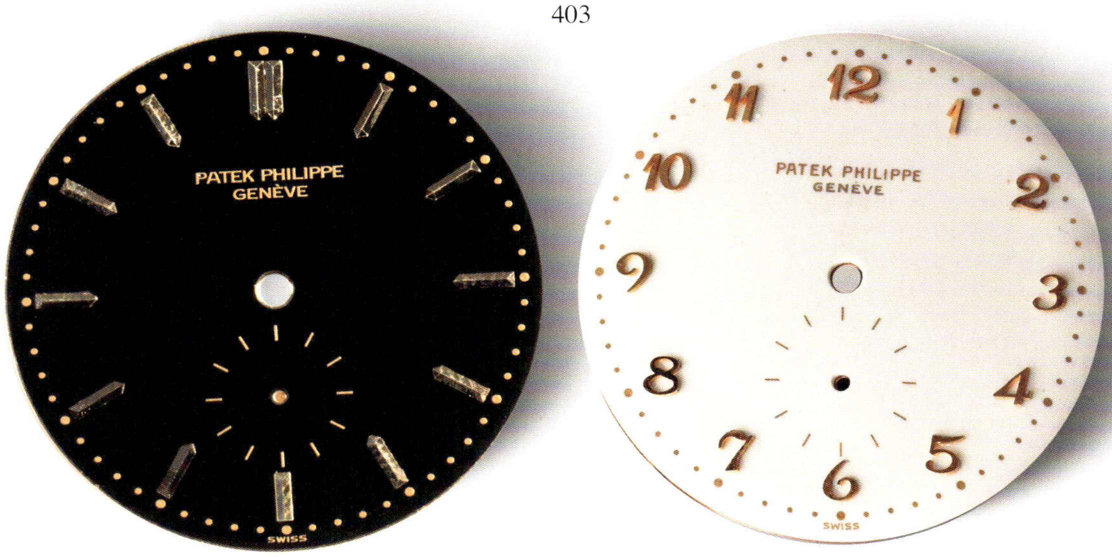

Illustration 714
The two enamel dials of this Patek Philippe reference 2526 represent a small fortune. Both the black enamel rarity and the one with Breguet numerals would add a special shine to any stylish wristwatch. As already described on page 401, a single enamel dial with rose-gold indices was auctioned for no less than $55,800 on May 7, 2022. The reader may now imagine at what price these two dials would change hands.

Illustration 715
This shows us how valuable dials were packaged and transported in the period from the 1950s to the 1980s. Both the upper and lower parts of this small plastic box were provided with a foil cushion to provide the best possible protection for both sides of the dial during transport.

Illustration 716
This shows us today's method of packaging valuable watch dials. The dial is secured both in the central hole for the hand axis and on the outer edge. This means that even the most-valuable dials can be transported safely.

If one has any doubts about the authenticity of a dial on a particularly valuable Patek Philippe watch, one can have the dial removed and examined by a licensed concessionaire or the Patek Philippe branch in question. The author followed this procedure with his references 3940 J and 3970 P. The cost for this should be around $780–$1,100, because on complicated watches the hands of the subdials can be damaged when dials are lifted off and may have to be replaced.

Exercise Caution

On the basis of the many watches that the author has personally dealt with and examined, the risk of purchasing a watch with an incorrect dial should generally be low. Caution should probably be exercised with older models, such as references 130, 533, 1436, 1463, 1518, 1526, 1563, 1579, 2438, 2497, 2499, 3448, and 3450. This is because watches that are already at or approaching retirement age must be treated with caution. If, on the one hand,

Illustration 717
The author found this stylish enamel dial in the familiar cream-colored finish on the internet.

Patek Philippe recommends an overhaul after every five to six years of operation, and, on the other hand, the watch in question is already around fifty years old, it should theoretically have undergone eight to ten overhauls. If the watch in question remained in possession of the original owner even after so many overhauls, a low risk can be assumed for the new buyer. However, if this watch has gone through several changes of ownership and has been spruced up by the new owner (e.g., an auction house) to achieve a higher selling price, caution is advisable for the prospective buyer. Using the case of his reference 3448 as an example, the author has described in detail what can happen to a new owner quite unexpectedly.

Jaeger-LeCoultre
There is no way to identify with certainty the origins of Jaeger-LeCoultre dials. As was learned from a Sentier employee from earlier times, the respective dials were ordered from the manufacturer in batches, but without a special hallmark. Exceptions were practiced only for special designs. A regular system is therefore not recognizable and was probably not practiced.

IWC
No system has been found for the marking of dials by IWC (International Watch & Co.) either. Only watches with a gold case and the De Luxe label on the dial had gold dials.

Vacheron & Constantin
The dials of watches from the house of Vacheron & Constantin were hallmarked with numbers until the early 1970s and are thus easily identifiable. The dials of the later versions, especially those that were installed in watches after the company was rebranded Vacheron Constantin, are usually no longer identifiable.

Rolex
Because of their diversity, the dials of Rolex timepieces rarely provide proof of origin. Today the group's own Beyeler dial factory supplies all the dials required without any special marking.

Audemars Piguet
Although just one Audemars Piguet (AP) dial is described here, all the other dials from this maker that were examined bore a proof of origin until the 1990s. Dials made of gold were used only in exceptional cases. The main supplier of dials used to be Cadrans Stern Frères SA, and later Stern Création SA.

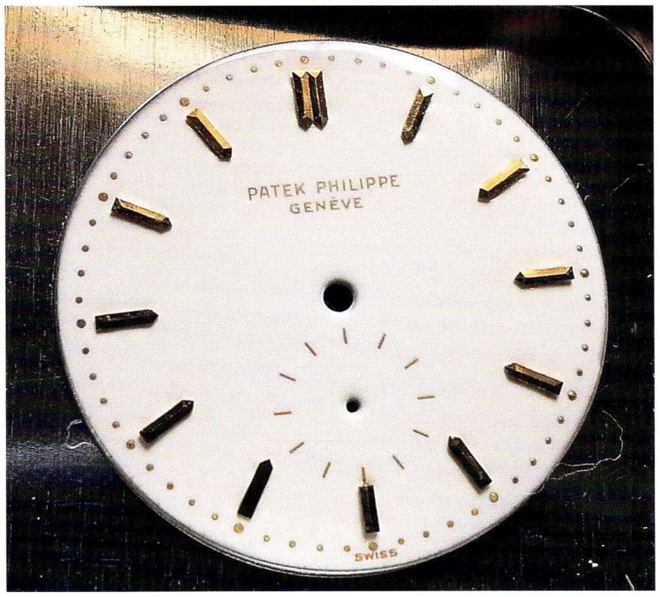

Illustration 718
The reverse of this dial turned out to be a great sensation. Since these enamel dials were always numbered consecutively during production, it proves that more than 901 pieces were made. Such a number has never been mentioned in the literature before.

Illustration 719
This enamel dial with Breguet numerals is an absolute rarity. This noble piece was delivered by Patek Philippe to Swedish concessionaire Bo Berggren Urhandel AB in Stockholm as a spare part, probably more than sixty years ago. The dial later turned up in the estate of a former watchmaker of this concessionaire and was put aside by his heirs. At an unknown date, it was subsequently rediscovered by the heirs and offered on the internet. The reverse of this enamel dial reveals something not previously known to us. Presumably there were other suppliers besides Cadrans Stern Frères SA who supplied Patek Philippe with special versions of enamel dials. The rediscovery of this dial represents a minor sensation that should not be withheld from the reader.

Conclusion

In conclusion, it can be stated that only at Patek Philippe is a reliable identification of dials possible, but this statement is valid only until the turn of the millennium, 2000.

The fact that today, at their own dial factory Cadrans Fluckiger SA, they take a different approach and no longer mark the dials of current models is probably a business decision.

For the reader, an important note must be emphasized in closing. If the dial of one's own watch has been damaged, the owner can either have it refurbished or remade by a dial manufacturer of their choice. In this case the exact imprint of the respective watch brand is also permitted. But only for this dial and only for this specific case is this practice considered legally compliant. If, for example, other dials of this brand were to be manufactured and sold, this would not be permitted under current law. Here the trademark rights of the watch manufacturer in question would be violated. One more reason to always check with the manufacturer as to what options are available if a dial is damaged and needs to be replaced.

Foreword

With the beginning of the Renaissance, portable, spring-powered watches became a luxury that no man of standing could do without. These first mechanical movements, decorated with engravings and ornaments, became not only a symbol of one's status, but an emblem of early engineering. To this day technical mechanics is the royal road for creation of a smoothly functioning timepiece: precision, outstanding knowledge of the subject and the will for technical progress are the hallmarks of the wristwatches presented in this book.
In order to literally "move with the times" and pave the way for the next generation of visionaries, we at the Technical University of Munich (TUM) focus on interdisciplinarity, talent development and excellence in research and teaching. Our success proves us right: TUM is a university of excellence, ranks among the best universities in Europe, and is connected worldwide through strong alliances with scientific institutions and companies. At the same time we always align our actions with the values, needs and expectations of society. This guiding principle permeates research, innovation and teaching and teaching "Made at TUM."

President Thomas F. Hofmann

Illus. 720

13. Physical Principles for the Accuracy of Wristwatches

13.1. Introduction

In the fifties and sixties it was considered an unwritten law that the higher the purchase price of a wristwatch, the more accurate it was. Following the triumph of the quartz watch and the radio-controlled watch, this rule no longer applies. Nevertheless, mechanical movements still exert a fascination that few people can resist. From a mechanical point of view, a wristwatch that one has to reset only once every fourteen days because it has a rate deviation of one minute represents a masterpiece of precision that has never been equaled by any other mechanical device in the hands of a customer.

A rate deviation of the aforementioned magnitude means a rate that is too fast or too slow by about five seconds per day. The accuracy problem of a watch lies in the fact that it keeps running (i.e., it is not reset to zero) after the time has been set, as is the case with a stopwatch, and thus the "time measurement error" keeps adding up. In addition, rate deviation can be checked at any time without any problems.

One day has 86,400 seconds, so the deviation of five seconds per day corresponds to 0.006%. The odometer in a premium car, on the other hand, has an accuracy of +3%. More is also not useful because of tire wear. So if the odometer shows 62 miles (100 km), in reality the car has driven only 60.32 miles (97.087 km). Only you don't notice that. A wristwatch with this level of accuracy would have a daily rate deviation of 2.592 sec./day (i.e., forty-five minutes a day).

Another example is the engine that powers a high-performance automobile. Vehicle engines with a rated power of 300 hp, manufactured according to the most-accurate methods, can have differences in performance of up to 7%. In practice this difference corresponds to the power of a small car engine (e.g., in the Trabant). If the engines had the accuracy requirements of the wristwatch mentioned, the mean error would be 12 watts, which is the power of a model railroad locomotive.

Each component of a mechanical watch must therefore be manufactured with extraordinary precision to achieve an accuracy of 0.006%. In the following we will explain the problems that a watchmaker has to deal with and the means by which a centuries-old mechanism can achieve such perfection.

13.2. The Balance Spring

Technically the watch movement is a speed-controlled, intermittently operating spring movement. The mainspring provides the driving energy to operate the numerous gears for the hands and the mechanical devices (e.g., for changing the date or indicating the day of the week). The energy stored in it determines the "power reserve" (i.e., how long the watch can be operated without being wound up again).

The heart of the wristwatch, which determines the speed of rotation of the hands, is the balance or fine-adjustment mechanism. It forms a torsional oscillator consisting of a flat spiral spring and a small flywheel, the balance ring. This arrangement should produce a harmonic oscillation, which, either at the end amplitudes or at zero crossing by means of a so-called escapement, moves the movement forward by a small defined amount each time.

A harmonic oscillation is excellently suited as a "clock generator" for the watch. It fulfills a very important condition. The oscillation time T (i.e., the time that elapses until the balance has rotated back and forth once) is independent of the oscillation amplitude. No matter whether the balance turns 270° or only 90°, the time for this always remains constant. This is called

Illustration 721
Balance with four spokes, four Microstella regulating screws and Breguet hairspring with curved outer end.
Photo by Rolex

Illustration 722
Fine adjustment mechanism of the Rolex caliber 3035 automatic movement with balance, Breguet hairspring and Microstella screws. The Breguet hairspring shown on the left is installed here.

isochronism. It is decisive for the accuracy of the watch, because the oscillation time T must have the accuracy of 0.006% mentioned at the beginning. And it must do so at least 345,600 times a day.

So you would "only" have to generate a perfect harmonic oscillation, then you would actually be theoretically finished. But only theoretically.
The all-important period of oscillation of a harmonic oscillation amounts to

$$T = 2\pi\sqrt{\frac{J}{M}}$$

But since the escapement releases the gear twice during a complete oscillation, the cycle time is halved to

$$T = \pi\sqrt{\frac{J}{M}}$$

M describes the elasticity of the hairspring, J the moment of inertia of the oscillating mass (i.e., the balance wheel).

For harmonic oscillation to occur, the following basic condition must be met: the restoring force of the balance spring must be exactly proportional to the angle of deflection. If this is not the case, then isochronism is not guaranteed and the period of oscillation is not correct. The accuracy of M must therefore be at least 0.76%. At the same time, the spiral spring bent from the spring wire weighs only about 2.5 mg. That is, four hundred spiral springs together weigh 1 gram. The elasticity of the spiral spring is calculated to

$$M = \frac{E \times h \times s^3}{12 \times L}$$

E: Coefficient of elasticity of the spring material	
h: width of the spring	typically 0.0035 to 0078 inches (0.09 mm to 0.2 mm)
s: thickness of the spring	typically 0.0007 to 0.0021 inches (0.018 mm to 0.055 mm)
L: length of the spring	typically 8.25 inches (210 mm)

The thickness s of the spring is particularly critical, since it goes into the elasticity M with the third power, meaning that minimal changes trigger a disproportionate effect. If the wire is too thick, the clock runs too fast; if it is too thin, the clock runs slow. If the wire exhibits thickness variations of 0.0003 inches (0.001 mm), isochronism is no longer guaranteed and the spiral spring is worthless.

But that is not all. For the barrel to be truly isochronous, the hairspring must work completely symmetrically like an inflating circle during its interplay; it must "breathe." Since it is driven by the balance wheel in the center but is attached to the immovable hairspring stud on the outside, the force vectors cause the center of gravity of the hairspring to shift; it not only inflates, but it actually swings back and forth as a whole. As a result, the restoring force of the hairspring is not fully used to drive the balance wheel, as required, but is also partially used to shift the center of gravity of the hairspring, which worsens the rate accuracy because the isochronism is lost.

A partial remedy is the use of a so-called Breguet hairspring. In this case the end of the hairspring that is normally attached to the hairspring block is bent over or bent up so that its end point can give way, and thus the shift in the center of gravity can be almost completely avoided. Making a Bréguet hairspring requires more effort and, if it is to be made manually, requires a great deal of experience and craftsmanship. Since the bent end of the hairspring can be moved inward only for reasons of space, it must be passed over the hairspring, resulting in a greater overall height of the movement.

The required accuracy in the production of the hairspring depends on the quality of the machine tools used, which can usually be controlled well enough. The elimination of temperature dependence, on the other hand, is difficult. The first thing that comes to mind is the dimensional change of the components due to temperature. If the length of the spiral changes, the movement runs slower at higher temperatures, but faster at lower temperatures. Since at the same time the thickness *s* increases with the third power and the dimensional change is isotropic in all directions, the influence of the change in thickness predominates, and the clock would run faster overall with an increase in temperature.

Far more difficult to control and of greater importance is the temperature dependence of the hairspring material. The modulus of elasticity of metallic materials changes with temperature, but this is almost imperceptible in the daily use of metallic objects at room temperature. These minimal changes play a decisive role in the precision required by a watch movement. If, for example, the balance spring were made of bronze or brass, a temperature change of only 1 degree Celsius would mean a movement rate deviation of ten seconds per day due to the change in the elasticity modulus of the material.

It is therefore not surprising that the material properties of the balance wire are of major importance. The material composition is therefore also one of the best-kept secrets of the watch industry. The most commonly used alloy is called Nivarox (nonvariable, nonoxidizing) and consists of iron, nickel, chromium, beryllium, and other elements in a secret mixing ratio. When using a Nivarox hairspring, the rate deviation is only 0.5 second per degree Celsius and day.

Nivarox hairsprings are made by Swiss company Nivarox-FAR, which equips about 95% of all Swiss watches with its hairsprings. The reason for this is obvious. The minimum amount of Nivarox to be taken from a metal smelter is 771 pounds (350 kg). That is enough for 280 million balance springs. The material price per hairspring would be $0.22. The manufacturing costs and the investments account for the lion's share of the selling price. Sophisticated industrial manufacturing methods and appropriate machine tools enable Nivarox-FAR to offer a balance spring at a single-digit dollar amount. For a watch manufacturer who builds about 10,000 units in-house, production is not worthwhile.

To the chagrin of traditionalists, more and more new high-tech materials have been used in *haute horlogerie* in recent years, ones that are more commonly used in the electronics or aerospace industries. One of these is silicon, a material well known above all from semiconductor technology. Silicon is specifically light, extremely hard, nonferromagnetic, and largely wear-free and slides particularly easily due to an exceptionally smooth surface, which is why very little lubricant is required. At first glance it is an ideal material for the manufacture of balance springs. The manufacturing process no longer has anything in common with classic balance spring production, which sometimes causes skepticism among older experts. Instead of the complex drawing process of the Nivarox wire, with subsequent winding, the etching process commonly used in semiconductor production is employed. The basis is an ultrathin silicon wafer from which the spiral is etched. This opens up completely new design possibilities, since the hairspring with add-on parts can be made from a single piece (illus. 713).

However, silicon has a serious disadvantage: its coefficient of elasticity is excessively temperature dependent. This means that a hairspring made of pure silicon is not suitable for a precision watch. Extensive materials research, reminiscent of the development of metal alloys in the previous century, has nonetheless made it possible to exploit the advantages of silicon technology. One process involves lightly oxidizing the surface

Illustration 723
Silicon hairspring ready for installation.

of the silicon hairspring. This then produces silicon oxide, known as quartz, which has an extremely low coefficient of expansion. In effect the balance spring has been "glazed," combining the advantages of silicon with the coefficient of expansion of quartz.

The further the manufacture of the components moves away from classic watchmaking, and materials and processes from the semiconductor industry are used, one has to ask what this will ultimately lead to. Perhaps straight to the quartz watch?

Illustration 724
Silicon wafers. *Rolex photo*

13.3. The Oscillating Weight, the Balance Wheel

In addition to a return spring, a torsional vibration system also requires an oscillating weight so that the interplay between spring tension energy and kinetic energy can take place. The oscillating weight has a moment of inertia calculated as

$$J = \frac{1}{2} \times m \times \left(r_1^2 + r_2^2\right)$$

r1 inner radius and r2 outer radius of the balance wheel, m = total mass of the balance wheel

It can be seen that the moment of inertia increases quadratically the farther the rotating mass is from the center.

During manufacture, particular attention must be paid to the careful balancing of the balance wheel. If this is not the case, pendulum forces will arise in the vertical position due to gravity, depending on whether the unbalance moves up or down. In addition, there is the correct choice of oscillation frequency. If the watch is moved back and forth several times while being worn—for example, if the wearer stirs a pot—then this external movement can add to the movement of the balance if both are in approximately the same frequency range. Therefore one tries to make the balance oscillate at a high frequency if possible, so as not to come into the range of such disturbances under any circumstances. However, this increases energy consumption and reduces the power reserve. The compromise is about five to eight vibrations per second.

Similar to the hairspring, the balance wheel is also subject to the influence of temperature fluctuations. The increase in temperature causes an expansion of the mass rim. As a result, the radius of the balance wheel increases, the mass moves farther outward, the moment of inertia increases due to the quadratic dependence on the radius, and the watch slows. Unfortunately, this is superimposed on the temperature sensitivity of the balance spring, which also produces a retardation at higher temperatures due to "material softening." With a brass balance and a steel spring, both effects add up to a deviation of about twelve seconds per degree and day.

The so-called compensation balance was to provide a remedy. The slotted balance ring consists of two different materials (illus. 715) so that a bimetal effect is created. When heated, they move the free ends inward, shifting part of the mass to a smaller radius and thus reducing the moment of inertia. This is to compensate for the decreasing spring force of the hairspring as the temperature increases. Adjustment is made by moving the weight screws to a range of stronger or weaker deformation. How the balance is to be maintained in the process is one of those well-kept secrets of the watchmakers' guild. A look at the dependencies also shows that this can work only to a limited extent.

The formula for the period of oscillation T includes both values for M and J under the root. For the period of oscillation to remain constant, the numerator and denominator must change to the same extent:

$$T = \pi\sqrt{\frac{J}{M}} = \pi\sqrt{\frac{12 \times L \times m \times r^2}{E \times s^3 \times h}} \quad \rightarrow \quad T \sim \frac{r}{\sqrt{E}}.$$

m = mass of the balance ring, r = mean radius of the balance ring, L = lengths of the hairspring, s = thickness of the hairspring, h = width of the hairspring, E = Young's modulus of the hairspring

But since the moment of inertia changes with the square of the radius, there is a linear dependence of the oscillation time on

Illustration 725
Compensation balance with split balance rim and 14 screws.

the radius, in this case a decrease, while the elastic modulus of the spring decreases as a root function. A straight line and a concave line cannot be made to coincide; at best, they intersect at two points (i.e., at only a maximum of two fixed temperatures does the compensating unbalance function). Watchmakers call this mathematically trivial fact the "secondary error of the hairspring."

Today's wristwatches no longer use compensating balances. The Nivarox hairspring in combination with a special material for the balance ring allows monometallic items to be manufactured. By careful selection of the components, called classification, the partners concerned are combined in such a way that the manufacturing tolerances largely balance each other out, thus optimizing rate accuracy. The rate deviation is then theoretically around 0.5 seconds per degree Celsius and day.

13.4. Lubrication of the Watch

An ideal harmonic oscillation is technically impossible to represent. Because of the constantly occurring frictional forces, every oscillation without energy supply comes to a standstill after a more or less short time. Instead of a harmonic oscillation described by the formula

$$A(t) = A_0 \times sin\left(\frac{2\pi \times t}{T}\right)$$

, a damped harmonic oscillation takes place, which corresponds to the formula

$$A_d(t) = A_0 \times e^{-\delta \times t} \times sin\left(\frac{2\pi \times t}{T}\right)$$

The damping element $e^{-\delta \times t}$ with the decay coefficient δ leads to a continuous decrease of the amplitude *A(t)*; the oscillation amplitude thus becomes a little smaller with each osillation. This loss is returned to the balance wheel by the escapement at each zero crossing. If sufficient energy is available, the oscillation amplitude thus remains practically constant. The energy must be provided by the watch's spring mechanism, whose tensioning force decreases significantly toward the end of the power reserve. As a result, the small impulse that the balance receives at the zero crossing is no longer sufficient, and the amplitude inevitably becomes smaller and smaller. If the spring movement is not wound up in time, the watch eventually stops.

It is because of the resulting change in oscillation amplitude that isochronism (i.e., the independence of the oscillation time from the amplitude) is so enormously important. If this were not the case, the rate accuracy would be extremely dependent on the residual tension of the mainspring. Fortunately, the isochronism exhibited by a harmonic oscillation also occurs with a damped oscillation.

As a result, a new difficulty is added. If one correctly calculates the period of oscillation T_d of a damped oscillation, one finds that it depends on the decay coefficient δ.

$$T_d = \pi \times \sqrt{\frac{J}{M}} \times \frac{1}{\sqrt{1 - \left(\frac{\delta}{\omega_0}\right)^2}} = \frac{T_0}{\sqrt{1 - \vartheta^2}}$$

T_0 is the period of oscillation of the undamped harmonic oscillation.

The period of oscillation T_d of the damped oscillation becomes larger (i.e., the watch runs down).

$\vartheta = \frac{\delta}{\omega_0}$ is the damping factor,

is the frequency of the undamped oscillation.

Because of the necessary lubrication, the degree of damping is not constant in time, and there is no satisfactory technical solution to this problem. The bearings of the balance and the escapement must be lubricated to ensure wear-free running of the balance wheel. The viscosity of the lubricant determines the level of damping and thus directly the rate deviation. The lubricant should therefore not exhibit any noticeable change in viscosity as a function of temperature. As the lubricant ages, gumming may occur due to oxidation caused by atmospheric oxygen, or volatile components may migrate. Both of these slow down the watch cycle.

The ideal lubricant must therefore be resistant to aging and at the same time have a viscosity that is independent of temperature. Unfortunately, this lubricant has not yet been developed. One reason for this could be that a medium-sized watch manufactory needs only a few kilograms of it per year, and thus development for the chemical industry is not the first priority.

To make matters worse, the operation of the watch produces extremely fine dust due to wear, which is in fact "sand in the gears" for the extremely fine bearings. The main cause of the abrasion is the automatic winding mechanism, which uses various levers and gears to create the tension for the mainspring and transmits relatively high forces in the process. The watch industry is countering this problem by increasing the use of ceramic materials, which, on the one hand, require less lubrication and, on the other, are said to produce only minimal abrasion.

To ensure that the precision of 0.006% mentioned is permanently maintained, the watch must therefore be sent to the manufacturer's factory for revision at least every four to five years, where it must be carefully cleaned, re-oiled, and fine-tuned again. This involves a not-inconsiderable cost of ownership.

13.5. Dependence on Air Pressure

Another frictional force that can influence the running of the watch is the air resistance of the balance wheel during rotation. Here the first thing that comes to mind is the dependence on air pressure. This dependence can be virtually eliminated by hermetically sealing the case, which is required for water pressure tightness anyway. The daily changes in air pressure, even when traveling from the plains to the mountains, are always much smaller than one atmosphere, which corresponds to a depth of 33 feet (10 m) of water.

On the other hand, it is worth taking a look inside the watch. The following formula generally applies to air resistance:

$$K_L = \frac{1}{2} \times c_x \times F \times \rho \times v^2$$

F = frontal area, cx = drag coefficient,
ρ = density of air, v = velocity

Since frontal area and drag coefficient are not variable, we summarize the formula as follows:

$$K_L = C_1 \; x \; \rho \; x \; v^2$$

For the velocity or speed of a harmonic oscillation, $A(t) = Ao \times sin(\omega t)$

$v(t) = A0 \times \cos(\omega t) \times \omega$ applies, and thus for the air resistance of the balance wheel

$$K_L = C_1 \text{ x } \rho \text{ x } A_0^2 \text{ x } \omega^2 \text{ x } (\cos(\omega t))^2$$

The air resistance is therefore dependent on the square of the amplitude and the oscillation frequency. This is critical when the amplitude changes due to decreasing mainspring tension. Things become complex: a smaller amplitude immediately results in less air resistance. This would increase the frequency, since the frequency depends on the damping. The higher frequency immediately results in higher air resistance, which in turn slows down the rate of the watch. The two mutual effects thus keep the rate deviation within limits. However, there is an additional difficulty. The isochronism of a damped oscillation occurs only if the damping is proportional to the speed, in this case the velocity. In the case of air resistance, damping that depends quadratically on the velocity is superimposed on the proportional damping. This disturbs the isochronism in principle.

From a physical point of view, a different solution suggests itself: the balance ring consists of a torus, which is connected to the axle not by spokes, but by a thin surface. As a result, no air is moved during the rotation, and thus the quadratic dependence on the speed is omitted. A laminar flow remains whose resistance depends only on the viscosity of the air and its density and is proportional to the velocity. Therefore no disturbance of the isochronism occurs, and all is well with the world again.

Also, the energy requirement of an oscillation, which would result in

$$E = \frac{4}{3} \times A_0^3 \times \omega^3 \times \rho \times C_1$$

, would be considerably reduced, which would benefit the power reserve.

13.6. Fine Adjustment of the Complicated Watch

More than 90% of all Swiss movements are equipped with an escapement group (i.e., balance spring, balance, escapement, and escapement wheel) made by Nivarox-FAR. Depending on the desired quality, standard *exécution normalisée* balances are supplied at approximately $14 apiece, and chronometer *exécution chronomètre* balances at approximately $27 apiece. Why the watch manufacturers charge the end customer up to $1,340 for a spare part for an escapement group will probably remain a manufacturing secret of the Swiss watch industry.

The escapement groups supplied by Nivarox-FAR all have a reasonably uniform rate deviation of −120 sec./day. They are subjected to fine adjustment at the watchmaker's after completion. Since most movements have the same escapement group, it is not surprising that, as an elaborate wear test has proven, the rate differences between the individual specimens from a manufacturer are just as large as those between the manufacturers themselves.

Fine adjustment of a wristwatch is an extremely time-consuming step in the manufacture of a precision watch. The reason lies in the position dependence of the gear. Depending on the position of the watch (there are usually six), the bearings are subjected to different loads, the bearing friction is not the same in all planes, and this results in a degree of damping that depends on position. Thus the watch changes its rate depending on its position. The art now lies in finding a practical compromise for the majority of watch wearers, in which the differences in rate are compensated as far as possible in the daily use of the watch.

Since the beginning of the nineteenth century, there has been a partial remedy for this problem. The *tourbillon* was developed to at least avoid position dependence in one plane (illus. 716). The escapement group (i.e., balance wheel, balance spring, lever, and lever escapement) is built into a cage that can rotate in the plane of the hands and is driven by the fourth wheel. With the second hand, the cage rotates around the hand axis once a minute. This avoids the position dependency in the hand plane, since the cage cannot assume a preferred position. Originally the *tourbillon* was developed for pocket watches, because this type of watch is always worn vertically with the crown pointing upward, so the cage always rotates only around the horizontal axis. Therefore, for a wristwatch that does not permanently assume a preferred position when worn, the *tourbillon* is more of a technical and usually visual refinement. *Tourbillons* are technically very complex to manufacture and difficult to maintain, whereby the gain in accuracy is disproportionate to the effort.

However, the fine-adjustment possibilities are quite limited.

The harmonic oscillation has an oscillation time

$$T = \pi \times \sqrt{\frac{J}{M}},$$

; therefore, one can change only the values of J and M to influence the oscillation time. The elasticity of the spring is given by

$$M = \frac{E \times h \times s^3}{12 \times L}$$

Only the length L can be changed; the finished spring's other dimensions can no longer be adjusted. The change in length is made with the aid of a so-called retainer (illus. 717). Here the rear end of the spring is simply clamped between two movable pins, and the effective length L is changed. This changes the elasticity reciprocally, and the period of oscillation changes proportionally to the root $T \sim \sqrt{L}$.

The other possibility is to change the moment of inertia J. The moment of inertia is calculated physically as

$$J = \int r^2 \times dm$$

For a punctiform mass m with radius r from the point of rotation, we get

$J = m \times r^2$ and for a balance ring $J_u = \frac{1}{2} \times m\,(r_1^2 + r_2^2)$

At first glance, the seemingly obvious solution would be to change the total mass m of the balance oscillator (e.g., by attaching or removing additional small weights), similar to balancing the wheels of an automobile. The disadvantage is that this cannot be done continuously, but only in the selected mass units. For watches in the upper price range, a different, more elegant approach is taken. The total mass is left constant and the distance of small partial masses from the center is changed.

In practice, so-called mass balance weights are used for this purpose (illus. 718). Small semicircular, horizontally rotatable masses are arranged on the balance rim. One can turn the semicircle so that it points outward; the moment of inertia, which is an integration of all masses, would thereby become larger because the small additional masses rotate on a larger radius. The clockwork would go slower. The advantage now is that one can adjust all positions between the masses turning outward and turning completely inward and thus change the moment of inertia J continuously.

13.7. Results of the Wear Test

Modern manufacturing methods have also found their way into the watch industry in recent decades. With the aid of state-of-the-art, numerically controlled machine tools, it has been possible, according to the general manager of a well-known Swiss watch manufacturer, to achieve a high-precision manufacturing accuracy that can lead to a dramatic reduction in rate deviation. Added to this is the processing of synthetic materials, which should ensure this rate precision over a longer period of time due to less wear. Careful quality controls with optical measuring instruments, some of which are new to the watchmaking guild, can detect even minimal tolerances and offer an enormous advantage when "classifying." If a watch manufacturer were to actually exploit these possibilities, these advantages should also be apparent to the customer in the form of a lower rate deviation.

This test was carried out with an IWC Portugieser Ref. IW500114 made in 2018 and an IWC Ref. R 817 AD made in 1969.

The IWC caliber from 1969 was considered one of the most precise Swiss-made movements of its day. The comparative test was conducted over a period of sixteen days.

Illustration 726
IWC Portugieser with Tourbillion

Illustration 727
Balance cock with regulator arm, micrometer screw, and gooseneck spring. On the right is the KIF shock protection system. From a caliber K 1071 made by Vacheron & Constantin in the 1960s.

Measured rate deviation after sixteen days (1,382,400 sec.):

1st IWC Portugieser (year of manufacture 2018), deviation 108 sec. (i.e., 6.8 sec./day)
2nd IWC Ref. R 817 AD (year of manufacture 1969), deviation −1 sec. (i.e., −0.09 sec./day)

These are the relevant data for the wearer, because they show how often the watch must be set. If the deviation of one minute can just be tolerated, the IWC Portugieser must be set every seven days (i.e., once a week). The 1969 IWC showed virtually no rate deviation, surpassing even that of a quartz watch. Progress?

Another quality feature that is of secondary importance to the customer is the stability of the rate. The movement does not always run smoothly, since this is influenced by the environmental conditions described in the previous chapters. In fact, these fluctuations partially cancel each other out, as intended.

The results were calculated by statistical methods, using the industry standard of 95% probability.

Rate deviation IWC Portugieser:
Total deviation over running time:
5.4 sec./day ≤ 7.1 sec./day ≤ 8.8 sec./day
Deviation when worn:
3.5 sec./day ≤ 5.4 sec./day ≤ 7.3 sec./day
Deviation while lying down:
5.7 sec./day ≤ 8.2 sec./day ≤10.6 sec./day.
(to understand the statement: the rate deviation is between 5.4 sec./day and 8.8 sec./day with a probability of 95%, for example)

13.8. Position Dependence

Dial up: 6.2 sec./day ≤ 8.2 sec./day ≤10.2 sec./day
Dial down: Average approx. 11 sec./day*)
Crown down: average approx. 16.8 sec./day*)
Crown forward: average approx. 7.1 sec./day*)
*) insufficient data

Rate deviation IWC year of manufacture 1969:
Deviation total over running time
−1.9 sec./day ≤ −0.07 sec./day ≤ 1.7 sec./day

Power Reserve:
IWC Portugieser Ref. IW500114: Running time without winding six days
No influence on rate deviation
IWC Ref. R 817 AD: stopped after thirty-six hours
Winding efficiency: IWC Portugieser gains 1.5 days of power reserve per day (i.e., it is always wound with seven days of power reserve when worn daily)

13.9. Quality Impression

IIWC from 1969: no negative findings.
There were complaints about the IWC Portugieser's date change: no quick date change (i.e., the date change begins at 23:30 and in some cases is not completely finished by 10:00 the following day). The new date is not centered in the window. The date display moves to the final position only when the watch is tapped.

Presumably a small ridge or an uneven sliding track somewhere stands in the way of a flawless date change. Irritating for a wristwatch in the $15,300 range. Where is the high-precision quality assurance?

Illustration 728
Close-up of a balance wheel with *masselotte* weights

Illustration 729
IWC Automatic Ref. R 817 AD in stainless steel with automatic caliber C. 8541 B movement, a very rare and valuable collector's watch

The bracelet of the watch combines the disadvantages of the folding clasp with those of the tang buckle in a very elegant way. The folding clasp is disadvantageous when the wearer has relatively large hands and a narrow wrist. Putting on the watch is then very laborious. The adjustment of the inside width is done by an additional buckle, whose disadvantage was immediately noticed the first time the watch was worn. Since the folding clasp on the new watch is relatively stiff, the buckle came loose when the watch was put down due to the force applied, and the bracelet opened abruptly along its entire length. The brand-new watch came within a hair's breadth of falling to the ground.

Due to its enormous height, this watch constantly slips out of the cuff when worn and does not slide back on its own. It is therefore eminently suitable for social distancing, provided the other person knows something about wristwatches.

13.10. Summary

Isochronism is the key to the accuracy of a wristwatch. Not only due to manufacturing tolerances, but above all due to external influences, an absolutely perfect isochronism is impossible to achieve. Nevertheless, through enormous efforts both from the manufacturing side and from the material selection with centuries of experience, watchmakers have succeeded in producing a mechanical "marvel" with a precision of 0.006%. Although there are now quartz watches that are considered disposable that have a power-of-ten higher accuracy, this does not diminish the achievement of watchmakers. In particular, a high-priced watch with all the complications and finest engravings is the ideal product of a late industrial society. The material requirements are minimal, so there are no raw-material problems and resulting environmental pollution. On the other hand, the handicraft effort, which cannot be represented by machines, is enormous and secures the jobs of many skilled workers without the need for large amounts of energy and without problematic waste. The buyer receives a precision instrument that, if the quality of workmanship is right, will give them pleasure for many years and represents an antithesis to our throwaway society. In this sense we wish the classic watch industry continued success in the future. Here is our final picture of high-tech from Schaffhausen.

Illustration 730
IWC Automatic Ref. IW500114 with date display and seven-day power reserve; one of the test watches

Illustration 731
This IWC Ingenieur with perpetual calendar and sixty-eight-hour power reserve is a testament to the art of watchmaking. The only reminders of the earlier successful Pellaton drive are the not-very-innovative double pawls and the shiny gold automatic wheel.

rexroth
A Bosch Company

FMS
MONTAGETECHNIK

Certified Excellence
rexroth
A Bosch Company

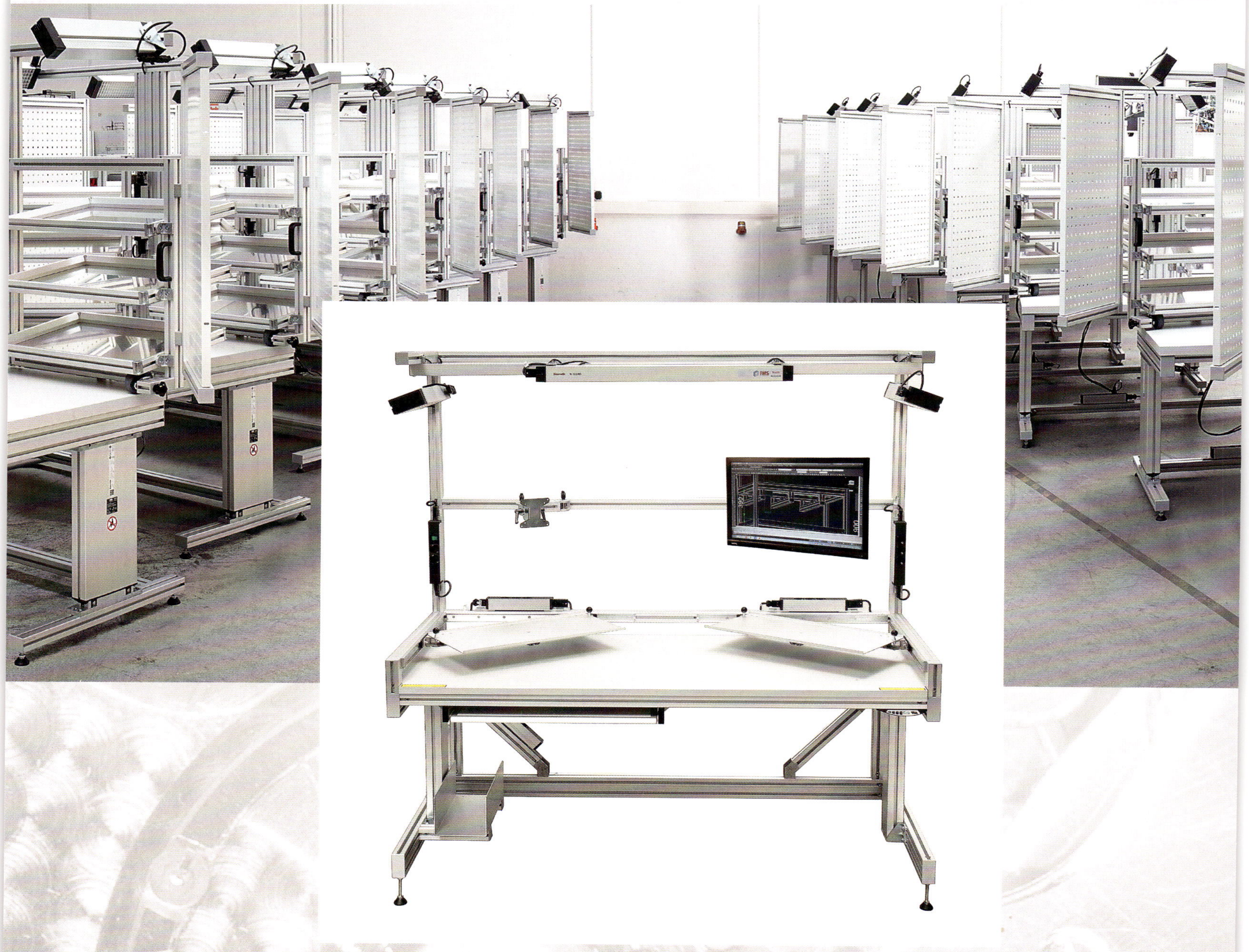

Illustration 732
The assembly of mechanical wristwatches is always considered a technically highly complex process and has always been carried out by hand (i.e., manual assembly). Not all the relevant criteria for this demanding activity are always satisfactorily considered and implemented. Even if the watchmakers' tables at some manufacturers are made of high-quality mahogany wood, and visitors to the watchmaking facilities are very impressed by this equipment, it is still far from certain that all the important prerequisites for an overall economically successful production have been met. There are often various gaps in the areas of ergonomics, assembly process control, and logistics. There is also often a lack of promising and innovative optimization ideas. FMS Montagetechnik GmbH deals with the development of such new techniques and has acquired more than sixteen patents in recent years. As a sales partner for assembly technology products, FMS is the Certified Excellence Partner of Bosch Rexroth AG for the southern Bavaria region. On the top right and left are height-adjustable manual workstations equipped with the highly ergonomic EASY TOUCH system. In the center is a manual workstation with EASY TOUCH system, also height adjustable. Very precise measuring and control tasks are performed at this workstation. For the digitization of assembly—also in the watchmaking industry—the EASY TOUCH system represents a significant future component because, in contrast to manual workstation systems offered on the market, it can be fully digitized due to its function and geometry (www.fms-montagetechnik.de).

14. Technical Notes—from the Engineer's Point of View

Buying a wristwatch is an emotional matter for most enthusiasts. Dial, case shape, and image play the main role in the decision-making process. Even if one can often see a beautifully decorated movement through the glass back of expensive watches and praise it favorably, most of the technical content remains hidden from the viewer.

For this reason, it will be well worth knowing for a friend of beautiful and valuable wristwatches if these timepieces are also once considered from a technical point of view.

14.1. Accuracy—a Supreme Discipline of the Wristwatch

As already mentioned several times, a watch movement is a machine, albeit a rather small one. The following observations and investigations should also be seen from this perspective.

First a look back:
In the 1960s, OMEGA was still considered the undisputed market leader in the watch industry, even ahead of Rolex.

Movements were still designed on the drawing board, very clear and transparent. Blanks, bridges, cocks, and other components were usually mass-produced with punching tools—also called cutting tools—with high precision. The production of these cutting tools in particular required especially skilled toolmakers who were paid far more than the specialists employed in watchmaking.

As proof of this successful working method, 100,000 automatic movements, 10,000 of them caliber 551s and 90,000 caliber 561s, were produced in one go in 1966, and all passed the chronometer test with distinction without any gaps. This unique record of serial accuracy, issued on February 28, 1966, and signed by Max Wyss, president of the Association of Swiss Institutes for Official Chronometer Testing, attests to the high level of design, precision parts manufacturing, and near-perfect watch assembly at the time.

Even if today, in the watch industry, design is carried out exclusively using CAD (computer-aided design), this means a modern, helpful, and technically highly precise method of working, but by no means a creative process characterized by innovation! Particularly good ideas cannot be forced even on well-trained engineers and are still in short supply today in a world determined by technology.

The author likes to wear, as an everyday watch, an OMEGA Seamaster Chronometer Ref. 168 022 in stainless steel. It contains the chronometer caliber 564. With a movement diameter of 1.09 inches (27.90 mm), a height of 0.198 inches (5.05 mm), 19,800 A/h, and a power reserve of about fifty hours, this caliber can keep up with the accuracy of new watches, even those from Rolex.

This precision watch, which dates from 1967, was overhauled at OMEGA in Biel and runs in six positions with a delta of 5.0 sec./day (i.e., just as well as a Rolex GMT-Master II that was worn in parallel and was about three years old). Despite the fifty-year age difference, both watches run equally well.

14.2. What Do Design and Manufacturing Technology Contribute to Rate Accuracy?

This topic is treated scientifically and comprehensively in chapter 13, "Physical Principles for the Accuracy of Wristwatches," by Dr. rer. nat. Gerhard R. Thoma.

The following hints and considerations are without direct connections or correlations, but more or less products of coincidence, which the author encountered during his numerous visits to Swiss watch manufacturers.

First, let's shed some light on the prerequisites for high-precision manufacturing. When manufacturing components, we start with machining (i.e., milling, turning, and grinding). Milling is the production of surfaces, recesses, and curves by using numerous types of milling tools. Turning is used to create shafts, wheels, flanges, and many other items that need to be made into rounded shapes. Grinding is the creation of high-precision surfaces by means of a grinding wheel.

In addition, there are electroerosive machining processes, also called EDM (electrical discharge machining). This process is used when particularly complicated shapes are required that cannot be produced by conventional machining. In addition, there is the advantage that, for example, the hardness of the material does not play a role in wire and spark erosion. The use of lasers also plays an important role. The laser beam can be used for cutting, welding, and material deposition.

We should also mention chipless forming, such as the deep-drawing of grooves on the screw-down case back of a watchcase. We are already familiar with this technique used on the OMEGA Seamaster Chronometer from the 1960s, or in the current Club Sport model from NOMOS in Glashütte. We will return to this watch later.

The deep-drawing process was also used in the manufacture of watchcases when the price was particularly low. An example of this is the Model 5000 from Jaeger-LeCoultre, presented on page 117.

In today's watchmaking, especially the manufacture of movements, machining centers have found their way everywhere. These versatile machines can turn, mill, drill, and grind fully automatically. They can work twenty-four hours a day, seven days a week, without interruption. For this permanently running production, only the supply of material and the removal of finished parts need to be considered. The tool change also takes place automatically.

In the meantime, components can be manufactured with an accuracy of one µ; for example, at machining centers of Kern Microtechnik in Eschenlohe, Upper Bavaria. Rolex is one of this company's good customers.

Stamping and cutting tools still play a significant role in large-scale production because they are able to manufacture components in large quantities inexpensively and precisely.

Thus, with today's manufacturing quality and the associated repeatability, it can certainly no longer be down to the manufacturing process to produce good or even outstanding automatic movements.

The special contents and properties that help an automatic movement achieve a top position are therefore determined solely by development and design, and thus ultimately by the engineers responsible. Here we refer to chapter 10, "Current Chronographs from Switzerland," on page 331. There it is proven that the chronograph caliber 4130 from Rolex has the fewest components but scores best in the disciplines of accuracy, serviceability, and maintenance.

14.3. The New Rolex Sea Dweller Ref. 126600 with the New Caliber 3235 and the NOMOS Club Sport Neomatik 42 Datum Ref. 781 with Manufacture Caliber DUW 6101

A Somewhat Unusual Comparison

It should be quite a challenge to find a watch brand that could actually take on the market leader. If you take a look at the annual watch catalogs published by German trade magazines and see how many brands are now on the market, it is hard to find a suitable model for this comparison. In this search, the author looked at several watch manufacturers that had attracted positive attention in the sector of new movement development. The new DUW 6101 automatic movement from NOMOS won the race in the evaluation criteria of level of innovation, accuracy, and smart constructive ideas, even though the rules of technical mechanics had been strained with a movement height of only 0.141 inches (3.60 mm) and a movement diameter of 1.38 inches (35.20 mm; 35.20/3.60 = 9.78).

When determining the rate values, the Rolex Sea Dweller and the NOMOS Club Sport were tested several times on the Witschi measuring device, and comparable results were achieved

Illustration 733
Rolex Sea Dweller Ref. 126600, a much-sought-after model

Illustration 734
NOMOS Sport Club Ref. 781. An impressive sports watch from Glashütte in Saxony with amazingly good rate values.

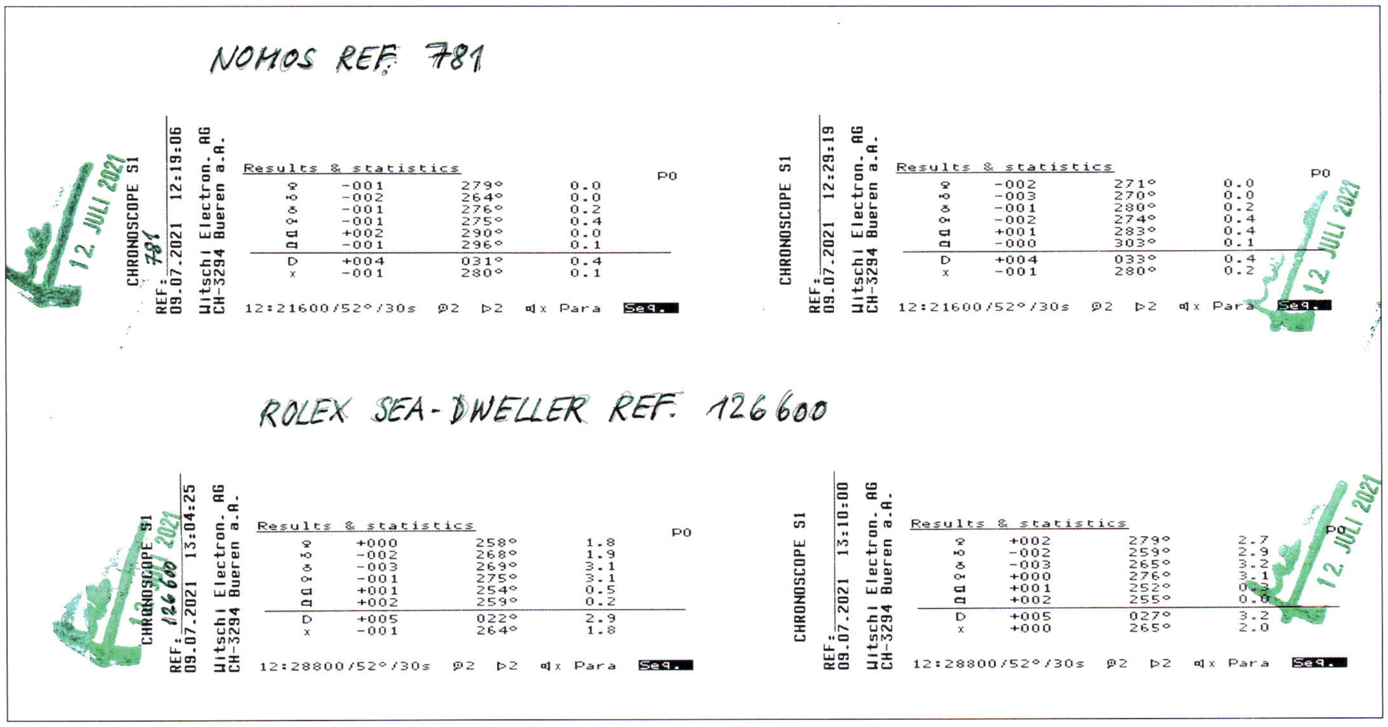

Illustration 735
Comparison of the rate performances on the Witschi measuring instrument between the Rolex Sea Dweller Ref. 126600 and the NOMOS Sport Club Ref. 781. As one can see, the watches run with similar precision. Both automatic movements were developed and designed by engineers.

in each case. The only question that remains open is whether the NOMOS caliber DUW 6101 would still achieve these excellent rates in five years. With movements from Rolex, there's probably no doubt about that.

Illustration 736
The reverse of the NOMOS Sport Club. Running beneath the screw-down case back with six deep-drawn grooves is the manufacture caliber DUW 6101.

Conclusion: if you want to buy an accurate wristwatch with an eye-catching case size and an innovatively designed automatic movement, you don't necessarily have to select a Swiss brand. The watch enthusiast who is guided by the aforementioned criteria will be just as satisfied with a NOMOS Club Sport, in which the new DUW 6101 caliber is installed, only without long delivery wait times, as well as at a price of only slightly more than $3,300. Added to this is the special quality of the steel case of this NOMOS, not only because it is waterproof to 984 feet (300 m), but the screw-down case back with fine thread and six deep-drawn grooves is particularly noteworthy.

Today this elaborate form of chipless processing is found almost exclusively in screw-down backs of watchcases from noble Swiss watch manufacturers, such as the Calatrava Pilot Travel Time from Patek Philippe or the Vacheron Constantin Automatic Ref. 87172.

According to NOMOS, none of the testers of this NOMOS Club Sport to date have noticed this special design of the screw-down case back.

14.4. Constructive Brainstorming, Two Examples

14.4.1. The Fine-Adjustment Mechanism of the New Rolex Caliber 3235

This new fine-adjustment mechanism is only a part of the improvements made by Rolex and may be regarded as an almost banal but clever idea. This simplification is one of the rules of technical mechanics and can be found in Dubbel's *Pocketbook*

Illustration 737
Comparison of a model from Rolex with one from another watchmaker

Feature	Rolex	NOMOS
Caliber	3235	DUW 6101
Diameter (mm)	28.5	35.2
Height (mm)	5.37	3.60
Power Reserve (hrs.)	70	42
Jewels	31	27
Balance Mounting	bridge	bridge
Price in USD	approx. 15,000 on the second-hand market	approx. 3,600

Illustration 738
NOMOS Caliber DUW 6101 with fine-adjustment mechanism removed. This in-house development is called the NOMOS Swing System.

of Mechanical Engineering (volume 1) under the heading "Strength Theory—Bending."

As shown in chapter 6 ("Rolex"), on page 217, the balance bridge of the caliber 3235 was greatly simplified compared to the caliber 3135, and many parts were saved. This simplification also significantly reduced assembly and adjustment times. The core of this innovation is the revised design of the balance bridge, which is now based on the principle of the "unilaterally clamped beam." As a result, the vertical play of the balance axle is adjusted only via the height adjustment on the right side. Since the balance staff is housed on both sides in a pivot bearing, the slight angular deviation at the upper bearing block of the shock absorber does not play a role.

The locking of the stud carrier has also been greatly simplified. In addition, the automatic assembly, including the rotor as well as its bearing, was redesigned. Since the rotor bearing previously used, with two synthetic rubies of different diameters, was too high, a ball bearing is now used. According to Rolex this reduces the movement height from 0.236 to 0.211 inches (6.00 to 5.37 mm). All the above measures should, one hopes, lead not only to functional improvements, but also to significant savings in parts production and assembly. However, the processes and design changes described here were never mentioned in the numerous reports in the trade press about the Rolex caliber 3235.

Illustration 739
Rolex caliber 3235. In the foreground is the newly designed fine-adjustment mechanism. The balance bridge is designed according to the "cantilever beam" system. The vertical play of the balance staff is no longer adjusted on both sides as before, but only on the right side.

14.4.2. The Use of Ceramics in Watch Movements

Another topic with which most readers are probably not familiar is the use of ceramic materials in watch movements.

Technical ceramics are used in a wide variety of industrial applications. Because of their great hardness, high wear resistance, enormous temperature resistance, insensitivity to acids, etc., ceramics are used to make cutting plates for lathe tools, heat shields on the nose of the space shuttle, artificial joints for humans, and many other things. Watch manufacturer RADO Watch Co. Ltd. in Lengnau, near Biel, in the canton of Bern, can be considered a pioneer in the use of ceramics. About thirty years ago, RADO began introducing watches whose case and band were made of ceramics. When the author first heard from the then IWC production director that they were planning to use all-ceramic pawls in the IWC automatic movements for the Pellaton drive mechanism, he started to look around to see who else in the watch industry was using ceramics or similar materials besides RADO.

The starting point for this analysis was the automatic group of Rolex calibers, which uses hard-anodized aluminum wheels. Hard anodizing is a ceramic aluminum oxide coating that is considered to be very wear resistant.

To determine possible wear processes, the gears in question from a new caliber 3135 movement and identical gears from a movement of the same type that had been in operation for twenty-five years were removed and examined under a Munich institutes' scanning electron microscope (SEM). Scanning electron microscopes are used for scientific investigations and material-relevant analyses.

Upon examining and comparing the new gears with those that had been in use for more than twenty-five years, I was astonished! Even the gears that had run for a quarter of a century showed no signs of wear and were indistinguishable from the new ones!

Illustration 740
Caliber C 52010 movement. The rotor bearing with eccentric is connected to the rotor by two through-bolts with round nuts.

The author informed the then production manager of IWC about these impressive findings and promised him the necessary support with the company's management. So far I have received no answer.

Instead, IWC continued its work with ceramic components, and after the pawls and the "automatic wheel," the rotor bearing, including the eccentric, was also made of zirconium ceramic. Whether the bearings are made of synthetic rubies or zirconium oxide ceramics is irrelevant for the steel axle running in them.

Producing such a multifaceted component—the eccentric, rotor holder, and bearing—in one piece from zirconium oxide ceramics already means a considerable technical effort. First of all, a mold is made in which this part in powder form must first be "thermally baked." The intended contour shrinks by almost a third of the original mass. This raw body must then be brought to the desired shape and the necessary dimensions in several steps by means of various diamond-grinding wheels. In terms of production technology, the whole process involves a great

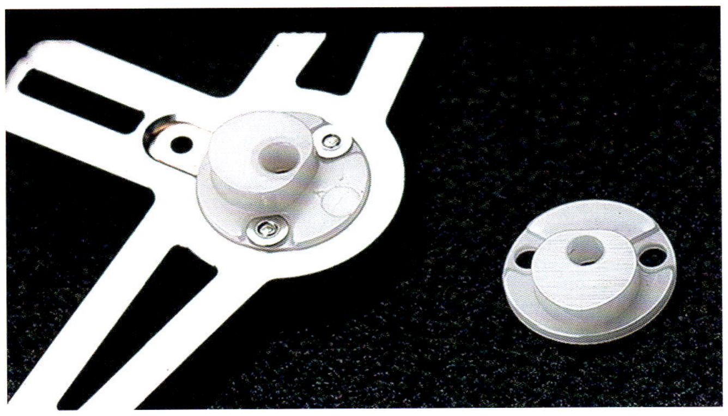

Illustration 741
Underside of the rotor with zirconium oxide bearing, which also contains the rotor mount and the eccentric in one piece. The rotor is held by two through-bolts with nuts, whereby these nuts are in the form of washers with internal threads (nonstandard parts). In front of it, a disassembled zirconium bearing can be seen. No one found an explanation for the small area attached to the narrowest part of the eccentric. It also makes no comprehensible sense with regard to the kinematics of the system.

Illustration 742
With the rotor removed, one can see details of the Pellaton winding mechanism, as well as the associated pawls and ceramic automatic wheel. The axle pin is firmly screwed in place and does not allow any deviating movements as the original Pellaton drive did (see also fig. 155, p. 91).

deal of effort that is often out of all proportion to the benefits, because the steel axle running in it is the weakest link in this wear chain. For this steel axle it is ultimately irrelevant whether it runs in bearings made of artificial rubies or zirconium oxide ceramics, as has already been explained.

Of course, a watch manufacturer as renowned as IWC never undertakes and takes responsibility for such solutions and decisions without good reason.

For this reason, the development-and-design department was contacted and asked about the reasons and facts behind the technical solution chosen here. They were happy to answer my questions.

One of the main reasons for this decision was the supplier's proposal to design the rotor bearing and the eccentric from a single part made of zirconium oxide. This company had also supplied the ceramic components used previously. The advantages cited by IWC in this regard included the following:

1. The tolerance addition
The tolerance addition that occurs when different components are assembled is eliminated.

2. Misalignment between eccentric and rotor
With the old design, there were often misalignments between the two ruby bearings when the eccentric was screwed to the rotor plate, so that the rotor was slightly skewed and did not run as easily. With the design chosen now, this error can no longer occur.

3. Marketing reasons
The marketing department wanted to increase the use of new types of materials; hence the slogan and label "Ceramic Technology."

4. Marketing strategy

Illustration 743
Images of new gears of the automatic device module made using a scanning electron microscope (SEM) at 100× and 300× magnification. The damage to the gear heads, some of which was slight but not relevant, probably resulted from transport or installation.

Illustration 744
Images of gears from an automatic device module that had been in use for around twenty-five years. Compared with the new gears, they show no signs of wear. Only the corners of some of the tooth tips show similar damage to the new gears. The tooth flanks themselves do not show the usual running-in marks.

Illustration 745
Pawl carrier with roller fork and the two ceramic pawls. *To the right*, the upper plate of the pawl carrier has been removed and the two pawls pressed back (rest position). An innovative, specially curved leaf spring allows two safe positions for the two pawls. In one they are pressed against the automatic wheel in a functionally correct manner; in the other they are held in the rest position. The half-curved sheet metal nose, which was added later, ensures that the longer pawl cannot slip off to the side—another small innovation.

The fact that the use of ceramics as a significant and modern material opens up convincing new marketing strategies can also be confirmed from a technical point of view.

Even if the manufacturing costs of these new components made of zirconium oxide ceramics are very high, they can still be accepted in the price range set by IWC for its "upper class" watches. After all, watches equipped with ceramic technology are in a similar price strata to timepieces from Rolex.

According to IWC, the great sales success also confirms the decision taken by the marketing department.

14.4.3. Alternatives

There are also alternatives for the solution chosen here, which will be shown both from a technical point of view and from the point of view of overall costs:

Illustration 746
Older version of the pawl carrier (caliber 85110)
Here the two pawls are still made of copper beryllium. The two teeth are laterally widened and engage the automatic wheel in an offset manner. This complex technique requires that the pawls be manufactured in the full material thickness of the teeth, after which the remaining length must be milled down.

1. Assembly of the rotor and eccentric
To guarantee exact alignment of the two ruby bearings during assembly of these two components, it is necessary only to insert a vertically guided plug gauge before screwing. This guarantees perfect running in the axle journal. The mounting is then carried out by means of two Allen screws, with two through-holes in the eccentric and the two threaded holes in the rotor plate.

2 Comparison of assembly times
Since it is not possible to use a fine thread on the IWC ceramic unit, this unit must be screwed to the rotor via two cylinder-head screws with washer-type nuts. Since this screw connection does not allow precise positioning, a suitable flange seat must still be provided on the ceramic unit or on the rotor. This also involves additional manufacturing effort.

Notes from IWC and Rolex

René Schwarz, former head of service at IWC and later also at Richemont, said that in his more than twenty-five years of practice, he had never once experienced a broken axle pin or damaged ruby bearing in the IWC caliber 8541. The same could be heard from Georg Ressel, former head of technology at Rolex Cologne. Also, on the Rolex caliber 3135, which uses a comparable rotor bearing, there was never a broken axle or damage to the rubies.

Unfortunately, such unspectacular statements are difficult to use on the marketing side. Advertising messages with new ceramic technology are simply better received.

Conclusion and Technology and Marketing Recommendations

From a technical point of view, the basic principle of the Pellaton drive could be retained, and the two pawls and the automatic wheel could be made of hard-anodized aluminum.

Since the interaction of the pawls and the automatic wheel does not depend on pitch accuracy, as is required of the gears of a watch movement, but only on functional reliability, there is nothing to prevent the use of hard-anodized aluminum components. These components are also much more resistant to breakage than those made of ceramic. Furthermore, it should be noted that the hard anodizing can be applied evenly in a layer thickness of approximately 10–20 μ, resulting in smooth functional surfaces.

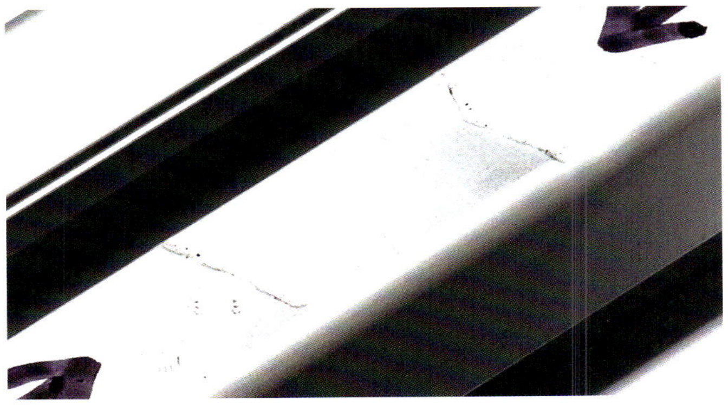

Illustration 747
A 1.77 × 1.77 in. (45 × 45 mm) aluminum profile from Bosch Rexroth AG after having been subjected to powerful hammer blows. Although the aluminum shows slight dents, no cracks have appeared in the anodized layer.

There is also the added benefit of being able to anodize these components in different colors, such as red, blue, or golden yellow. As with blued screwheads, additional focal points could be created for watches with sapphire display backs that would be visually well received by watch enthusiasts.

The entire technology could be realized at a fraction of the manufacturing costs of solid ceramic solutions. The higher breaking load and easier processing of aluminum would bring even more advantages.

IWC's other movements, which are still equipped with the copper-beryllium components used to date, could also be easily retrofitted with this new technology.

And should the marketing department continue to insist on using the slogan "ceramic technology," it can be loosely retained. Hard-anodized components are also included in the generic term "ceramic technology."

14.5. The Use of Outsourced Movements (*Ébauches*)

Watch manufacturers who either do not produce their own movement or do not want to equip all their watches with movements made in-house use raw movements from well-known suppliers, such as ETA SA Manufacture Horlogère Suisse in Grenchen, Canton Solothurn, Switzerland.

The caliber 2892A2 is often used for normal automatic watches. With an outer diameter of 1.007 inches (25.60 mm), a height of 0.141 inches (3.60 mm), a date display with quick correction, a stop-second function, bilateral winding, 28,800 A/h, twenty-one jewels, and a power reserve of about fifty hours, this movement has all the necessary features that are relevant for a wristwatch.

As a raw movement, this *ébauche* is offered in three quality grades and can also be supplied as a chronometer movement. This caliber is also used in the IWC Ingenieur with perpetual calendar Ref. 9240 (illus. 141, p. 85).

Another much-sought-after product from ETA is the chronograph movement 7750, which is used in numerous watches, is offered in two quality versions, and is also available as a chronometer movement.

Numerous watch manufacturers such as Breitling, IWC, TAG Heuer, Junghans, and others use this proven *ébauche* with an outer diameter of 1.18 inches (30.00 mm), height of 0.311 inches (7.90 mm), twenty-five jewels, 28,800 vibrations per hour, forty-eight-hour power reserve, date display, and three subdials.

This chronograph caliber may already be called a long-proven and robust product. Many users take this raw movement as a starting point to make their own modifications (i.e., to refine or specialize this movement with their own components). As an example we have chosen an original ETA 7750 and a chronograph movement of this series refined by IWC. The IWC model has a modified rotor, a completely redesigned fine-adjustment mechanism, and various circular-grained (i.e., surface-finished) components.

The fact that counterfeiters also use the chronograph caliber ETA Valjoux 7750 can be seen in the attempt to counterfeit a Rolex Daytona. On this watch the movement was rotated 90° counterclockwise to bring the Rolex arrangement of the three subdials into the correct position. On the dial the counterfeiter overlooked the fact that on the original Rolex, there is no "T" (for tritium) printed on both sides near the 6.

In summary, it can be said that there is nothing against the use of *ébauches* like the automatic movement 2892A2 or the

Illustration 748
ETA 2892A2 raw caliber (*ébauche*) ready for installation

Illustration 749
Simple version of the Valjoux 7750 chronograph caliber from ETA

Illustration 750
The IWC C.79350 chronograph caliber, a refined ETA 7750 version *ébauche*. Here ETA's own fine-rate-adjustment system has been omitted and replaced by a balance wheel with four weight screws. The surfaces of the balance cock and other components have a circular grain finish. The rotor weight support plate also now bears IWC lettering.

Illustration 751
Fine-adjustment mechanism of the ETA-Valjoux caliber 7750. All visible components are the same as the normal production standard.

chronograph movement Valjoux 7750 when it comes to the production of an accurate wristwatch with a reliable but inexpensive movement.

Theoretically, one could equip a Valjoux 7750 movement with a Breguet hairspring and a Gyromax balance and refer to this increase in value with a corresponding dial label.

The Porsche development department in Weissach also used to own an almost normal VW bus equipped with a 4-liter boxer engine from a 911 Carrera S, and, after the chassis and brakes had been upgraded, even received approval from the TÜV Stuttgart. The fact that this performance-enhanced VW bus with its approximately 400 hp scared off many a freeway user is not denied by the employees from Weissach. Why shouldn't a comparable story also be conceivable for wristwatches?

Illustration 752
A rather clumsy fake of a Rolex Daytona movement with dial. The company label is applied too thickly; the same applies to the tips of the crown. The "T" (indicating tritium) at six o'clock has never appeared on a Rolex Daytona.

Illustration 753
Fine-adjustment mechanism of the IWC version of the caliber 7750 with the IWC designation C. 79350. The surfaces of the balance cock and other components have a circular-grained finish. Fine adjustment of the rate is done via a Triovis regulator.

Illustration 754
Close-up of the Triovis precision adjustment system. By the cylinder head screw being turned, the rate can be advanced or retarded (+/–). Only in later IWC calibers was the Triovis system replaced by four weight screws in the balance ring (approximately 8× magnification).

Illustration 755
Patek Philippe Calatrava Ref. 5227G-001 with an 18K white-gold case, ivory dial, and 1.53-inch (39 mm) case diameter; the classic Calatrava par excellence. The automatic caliber 324 SC is encased in this watch. SC stands for *seconde centrale*, or central seconds.

15. Purchasing Old Wristwatches—Opportunities and Risks

15.1. Watch Purchase Checklist

Anyone who is prepared to invest a five- or six-figure sum for a wristwatch should plan their approach thoroughly. The following checklist has been created for this purpose. If a potential seller does not want to answer the suggested questions, they may not need to or may have something to hide. In any case, buyers should take their time and consistently work their way through the checklist, because a watch is quickly bought but often difficult to sell again. The reason is obvious. Also, the next potential buyer will possibly proceed according to this checklist. Here are the questions and inspection criteria:

Dial

- Is the surface lacquered, gold-plated, solid gold, enamel, etc.?
- Presentation: indices, Roman or Arabic numerals, Breguet numerals, etc.?
- Original condition, original replacement, or refurbished?
- If refurbished, by the manufacturer (with proof) or by a third party?
- If a replacement dial, when, and was it mounted during an overhaul by the manufacturer?

Movement (Manually Wound and Automatic)

- Are the caliber designation and movement number notation comparable with the watches listed in this book?
- Is the visual condition of the movement near mint?
- Does the movement show signs of wear? Are movement retaining or other screws missing, are the pinions and gears worn in, is the rotor bearing worn out, or are there rotor wear marks on the plates or on the case back?
- Is the movement's oil and grease retention inadequate?
- Is the rhodium plating (surface coating) already partially missing on plates and bridges, and is the brass already showing?
- Is there corrosion anywhere, or traces of water damage?
- What do the rate values look like (Witschi rate measuring instrument)?

Case (Two- or Three-Part, with Press-Fit or Screw-Down Case Back)

- Are reference and case numbers correctly present and positioned according to examples in this book?
- Do diameters and other dimensions still correspond to the original examples and dimensions given in this book?
- Are hallmarks and other indications inside the case back correctly included according to examples in this book?
- If repair marks are present in the case back, how many are there and can they be interpreted?
- Has the case back been sanded down or re-turned to remove repair marks?
- Is the material thickness still correct or has the bottom been turned down?
- Are all case edges still sharp or already rounded?

Function

- Can the crown be pulled into the various setting positions?
- Can the time and date be set accurately?
- Is the crown still sharp edged and thus little used, or already worn?

Model

- Is this model one of the classics listed here?
- Does the model have a design corresponding to today's taste?
- Was this model produced in large quantities or only in small numbers?
- For current watches: Does this model belong to the current limited editions, such as the "125 Years of Wempe" annual calendar?

Illustration 756
Close-up of the Swiss precious metal certificate for 18K gold with small Helvetia and material details (approximately 9× magnification). In contrast to the tampered-with base with engraved data, these indications are struck by means of a stamp. As a result, material compaction has occurred, which can also be detected by x-ray.

Illustration 757
Comparison of a genuine press-fit case back and one that was tampered with. The lower press-fit case back is from the rose-gold reference 96 on page 153 (illus. 250) and is original. All the information has been stamped, using different stamps. The yellow-gold case back on the upper left has been tampered with. The letters, numbers, and other markings are engraved (i.e., machined). Except for the incorrect "i" in "SWISS" and some other deviations, almost everything has been professionally copied. The difference is noticeable only to the expert. The layman can detect this manipulation with the help of a stereo microscope, or by means of x-rays. This manipulated (i.e., faked) case back was sold on eBay on January 26, 2010, for $531 (390.15 €) by the dealer, the Guildmark Group.

Papers and Other Documents

• Are purchase and warranty documents available and issued and stamped by the concessionaire?

• Is there a pedigree book extract from Patek Philippe? Was this issued after presentation of the watch, or only the reference, movement, and case numbers? Only in the first case is it possible to check if all information is correct and complete.

• Are there any other proofs of purchase from antique-watch dealers, repair or overhaul invoices from the manufacturer, or other documents that allow a reliable statement to be made about the condition and maintenance of the watch?

• Was the watch purchased at a public auction?

Strap and Buckle

• Is the strap original or a replacement from the secondary market?

• If the strap is original, has it been little used and does it still look appealing?

• If it is a gold watch with a leather strap, are the gold spring bars still mounted? (use a fine wooden pencil or a toothpick to look for the side with the "pin" on the spring bar)

• Is the original (gold) buckle still present on a leather strap?

• If there is a folding clasp of newer design, is it a high-quality and expensive one (retail price about $1,900 to $2,000)?

One or Multiple Previous Owners

• How many previous owners can be proven for this watch?
• Can statements about the origin and history of old watches be verified?

Rate Accuracy, an Important Statement

• Is there a current measurement report from a Witschi measuring instrument that provides information on the watch's rate, rate deviation, amplitude of the balance, and symmetry of the oscillating system (drop)?
• If this information is not available, can this measurement be made up for or carried out by a concessionaire with the appropriate equipment?
• Is evidence available of previous overhauls and associated rate tests?
• Are there any measurements of the case's watertightness (measurement report) available?

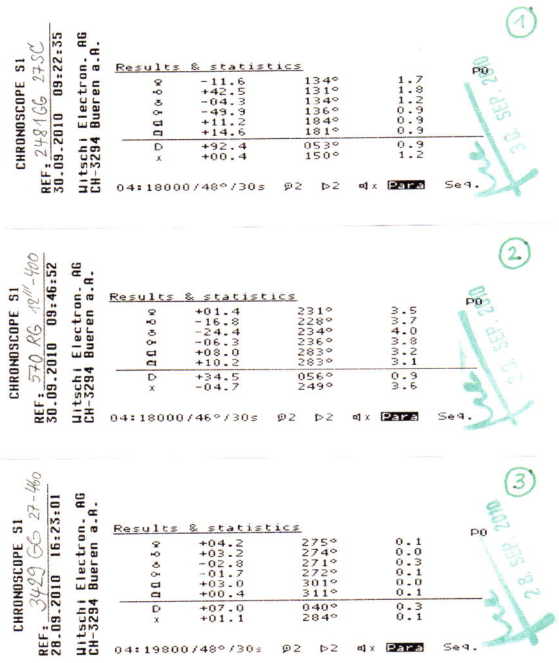

Illustration 758
A test on the Witschi electronic rate-measuring device is always the moment of truth for a watch. The watch in the upper measurement chart is in urgent need of an overhaul. The rate performance of the watch in the middle measurement chart is not bad but fails to achieve chronometer values. Here, too, the watch should be taken to a good watchmaker. The rate values in the lower measurement chart correspond to chronometer specifications and testify to the best craftsmanship.

The measurement charts shown reflect the rate accuracy of various Patek Philippe watches. First, the average rate in six positions over twenty-four hours is listed. In second place, the average amplitude is recorded. On the far right the decay (i.e., the symmetry of the oscillation system) is listed.

The first measurement record is for the reference 2481 in yellow gold shown on page 162 (caliber 27 SC, elevation angle 48°). The daily mean rate (x) is +0.4 sec. and thus almost exact, but the daily mean rate deviation (D) is 92.4 sec. and thus beyond good and evil. The amplitudes in the four vertical positions are only half the nominal value. This watch must have been waiting for an overhaul for many years.

The second measurement record is that for the reference 570 in red gold found on page 156 (caliber 12'''-400, elevation angle 46°). This watch lags 4.7 sec. daily. The average rate deviation of 34.5 sec. is not acceptable. The amplitudes in the vertical range are too low. The drop of 3.6 proves the lack of symmetry of the oscillating system. This watch would have to receive an overhaul immediately.

The third example shows the measurements for the reference 3425 (caliber 27-460, elevation angle 48°) shown on page 188. This watch was overhauled at Patek Philippe in Munich in 2003 and spent the following years in a safe. Despite this, this watch still reached chronometer values at the time of this measurement. It advances about one second per day, reaches a delta of 7.0 sec., and shows decent amplitudes. The symmetry values of the oscillating system are also still correct.

A Patek Philippe guarantees chronometer quality and should not exceed 8 sec./day for the average rate deviation in six positions. The amplitude should be about 280° or above in the vertical positions and about 260° in the horizontal positions. The drop should be 0.1 to 0.2 for an overhauled watch.

Legal Considerations

• For private purchases, is the watch free from third-party rights?
• Does the status of the watch comply with national customs and tax conditions?
• If the watch was imported from abroad (outside the EU area), have customs duties and import VAT been paid? If yes, are the corresponding official receipts available?

Purchase Contract

• Is an (informal) purchase contract offered by the seller?
• Does this purchase contract list all of the seller's relevant details, including address, item, tax status, etc.?
• Does this contract of sale include all technical data, selling price, and age of the watch?
• Does the seller provide a warranty and a right to return the watch within a certain period of time?
• Does the seller guarantee the watch's largely original condition on the purchase contract?

In Conclusion, General Considerations

• Does the buyer or watch enthusiast really have a lasting affinity and special preference for the model under consideration?
• Has the prospective buyer looked at the market and inquired about sales or prices of comparable models at other purchasing sources?
• Has the buyer informed themselves about the market position and the possible increase in value, as well as their level of awareness concerning the selected model before the transaction?

The potential buyer should not accept any compromises, because, after all, they must pay the required price without any ifs, ands, or buts!

PATEK PHILIPPE

MAITRES HORLOGERS A GENÈVE DEPUIS 1839

CERTIFICAT D'ORIGINE
ET DE GARANTIE

NOUS CERTIFIONS QUE LA MONTRE
Référence N° 3561
Mouvement N° 1116766
de calibre 27 460 A et 37 rubis,
réglée aux diverses températures et positions a été construite dans nos ateliers à Genève.
Le cadran est argenté, heures relief or
et le boîtier en or jaune 18 carats.

Nous garantissons cette montre contre tout défaut de fabrication.

Cette garantie ne comprend pas l'entretien périodique nécessaire à chaque montre. Cet entretien devra être fait par un horloger qualifié tous les ans pour la montre bracelet, tous les deux ans pour la montre de poche.

APOA

PATEK PHILIPPE

SWISS STORES LTD
JAMAICA
15-March-75

PRÉSIDENT

Illustration 759
Warranty certificate of the reference 3561 watch shown on page 203, illustration 311. This watch was sold by the company SWISS STORES LTD, Jamaica, on March 15, 1975.

Illustration 760
Patek Philippe automatic caliber 27-460 with a movement diameter of 1.06 inches (27.00 mm), a height of 0.18 inches (4.60 mm), and a rate of 19,800 vibrations/hour. This movement was the successor to the caliber 12'''-600 AT and was unveiled in 1960.

15.2. Purchase from an Antique-Watch Dealer

Buying a watch from an antique-watch dealer (vintage watch dealer) is often one of the most obvious ways to acquire an old wristwatch, such as a Patek Philippe for example. But here, too, rules must be observed, as we can list in the following.

Almost thirty years ago, a watch enthusiast was looking around for an old wristwatch and purchased a Patek Philippe reference 3514 in yellow gold from a vintage watch dealer in downtown Munich for exactly $6,170 (see p. 199). The following was handwritten on the invoice from this dealer: "1 Men's Watch Patek Philippe 18 ct. Automatic with leather strap & gold clasp 1960s." No reference was noted, nor were the numbers for the movement and case listed. About three months later, the same watch enthusiast bought another Patek Philippe automatic from the aforementioned dealer, this time opting for a reference 2552 in yellow gold. The invoice, once again for $6,170, was handwritten as before with the following text: "1 Men's Watch Patek Philippe 18 ct. Autom." Again, the most-important information that should not be missing when buying such an expensive timepiece was missing. In both cases the watch dealer, himself a watchmaker by profession, emphasized in the deepest tone of conviction that the watches were in excellent technical condition and that even the dials were still original. Freely following a quote attributed to Lenin, "Trust is good; control is better," the proud watch owner visited Patek Philippe's customer service in Geneva a few weeks after buying the second watch. There he presented the two watches and asked whether they needed to be serviced. Only after a lengthy discussion did the Patek Philippe service employee on duty agree to take the two automatic watches to the workshop and seek the advice of his colleagues.

After the owner of the two watches had been waiting for more than an hour, Christophe Harsch, the name of this service employee, returned with the two watches, which had now been opened, and recited the testimony of his colleagues. And this was devastating. In both cases the dials had been refurbished, but not according to type. On the reference 3514, made in 1965, the rotor touched the screw-down case back. Also, eight repair marks had been noted there—evidence of heavy use. On the reference 2552, year of manufacture 1957, the 12'''-600 AT automatic movement was in the worst condition imaginable. Including dial replacement, movement overhaul, and proper refurbishment of the case, costs of around $1,300 to $1,700 would have to be expected in both cases. For lack of an alternative, the watches were left at Patek Philippe in Geneva, and an overhaul was ordered. Five months later they were received. For the automatic reference 3514, the total cost was $1,410, with the overhaul of the movement costing $839 and the replacement of the dial in exchange $520. For the reference 2552, the overhaul costs were even higher, coming to $1,712, with $1,097 charged for the overhaul of the movement and $548 for the "best possible restoration of the dial." Because, after this experience, the owner was no longer able to fully enjoy these two watches, the reference 3514 was taken in payment by a

Illustration 761
Close-up of the previously shown caliber 27-460. In the foreground is the balance cock with Gyromax balance wheel and Breguet hairspring. The eight *masselotte* weights are integrated into the balance ring. On the upper left, one can see a part of the escapement wheel and the two synthetic ruby pallets of the escapement lever.

seller for $6,727 when buying a much more valuable Patek Philippe. The already antique reference 2552, which looked like new again after its stay in Geneva, was passed on to a Dutch deep-sea fisherman for $7,008. For the watch enthusiast from Munich, the damage was thus limited. And the new owners of these two watches could indeed be satisfied with their purchases.

By the way, the Munich watch enthusiast mentioned here was the author of this book. If he had the checklist mentioned here twenty years ago, he probably would not have bought these two watches.

All in all, buying from a dealer has clear advantages. You can hold the watch in your hands, examine its inner workings, and ask all the relevant questions. There are reasons why comparable watches are often more expensive at a retailer than when purchased online. If it is not goods on consignment, the dealer has had to buy and pay for the watches themselves. Operating costs, debt service, and profit margin then result in the additional price that the dealer needs for their business to succeed.

When you have found the watch you want at a dealer, you should not talk about its price, but first about its characteristics. Here it will certainly be advantageous to take the previously mentioned checklist in hand and go through the items one by one. After all, if a five- or six-figure dollar amount is up for discussion, the dealer should also take the necessary time for this potential business. Other customers are often happy to wait if they can follow an informative sales talk and thus learn something. And finally, a mutually satisfactory transaction is the best prerequisite for the next purchase. After all, once someone has been infected by the Patek Philippe virus, they will certainly not be satisfied with just one watch in the near future.

15.3. Buying at National and International Auctions

Well-known auction houses that also offer valuable men's wristwatches and have an international presence include Christie's, Sotheby's and Antiquorum, and Patrizzi in Geneva; Ineichen in Zurich; and Bonhams in London, Edinburgh, Los Angeles, New York, and many other cities. In addition, there are auctioneers Heritage in the USA as well as Dr. Crott in Mannheim and Henry's in Mutterstadt, both in Germany.

For the watch collector, the auction catalogs published prior to the auctions are real treats, particularly because these catalogs contain excellent photographs and occasionally in-depth information and can also be viewed via the internet. These insights allow a timely disposition of the intended purchases. Those who are on-site can personally examine the watches (lots) in question before the auction and ask questions of interest.

Illustration 762
Jaeger-LeCoultre Geophysic Chronometer Ref. E168 from the year 1958, with original guarantee certificate and box

Illustration 763
Jaeger-LeCoultre Geophysic Chronometer Ref. E168 in stainless steel. A total of just 1,390 examples of this model with screw-down case back were produced.

Those who can bid only by phone or internet have to rely on the descriptions in the catalogs.

If the respective descriptions of condition, which are often formulated in their own functional language, later prove to be not entirely accurate, one holds poor cards as a buyer. Especially because the auction company always auctions watches of third parties and thus acts only as an intermediary. Once a watch has been purchased and the purchase price paid, it may be difficult to complain about and receive acknowledgment of defects and associated faults that are discovered only later.

More than thirty years ago, when the author was still collecting high-quality wristwatches—more precisely only watches from Patek Philippe and Vacheron & Constantin with automatic winding—he visited numerous auctions at Christie's and Sotheby's, as well as Antiquorum in Geneva, London, New York, and Hong Kong. It is on such occasions that a professionally trained eye can detect the differences between the actual condition of a watch and its description in the catalog. After all, it is never possible to understand exactly what an old wristwatch has experienced or been through in forty or fifty years.

If you want to bid on a particular watch at an auction by telephone or via the internet, you should provide the auction house in good time with a list of relevant questions about the watch to be auctioned. If the auction house responds only with the routine "Condition Report," you should ask again. A telephone conversation with the responsible "watch expert" has also proven helpful. In such a dialog, the expert in charge has the watch in question in his or her own hands and can provide comprehensive answers to the questions posed. Here it can also do no harm to record the most-important statements in a telephone memo. Because if it is about a lot of money (e.g., a watch for $20,000 or $30,000), one should also be demanding of the auction house in question and never be satisfied with blanket statements.

Once the desired watch has been found, inspected, and purchased at auction, the only thing left to do is to pay for it and to arrange the exact handling of the transaction. If the auction was held in the European Union, the watch will be shipped to the bidder by insured transport after payment. In the case of international auctions, the auction houses often use specialized companies that take care of the transport and customs clearance. Charges for such additional services are often $150 to $250.

In Germany, customs duties are levied on the invoice value and on the transportation costs and are invoiced and collected immediately upon delivery or collection. These charges are currently 19% import sales tax, as well as customs duties between 0.30 and 0.80 euros per watch. If a watch lover has personally purchased a watch at an auction in Hong Kong or New York, for example, and its value is currently above $470, this watch must be declared and taxed at customs when entering the country. If you do not comply with these legal requirements, you are guilty of tax evasion.

A Special Case Should Be Mentioned Here in Brief

If a buyer wishes to receive their watch, which they have bought at auction at Sotheby's in New York, for example, in Germany only for examination and appraisal, and only six months later to forward this watch to Christie's in Geneva, they can also have the valuable piece imported temporarily. However, to do this they need an authorized forwarding agent who specializes in carrying out such transactions and is licensed by the customs authorities. The fee to be paid for this is not insignificant.

Illustration 764
Patek Philippe Ref. 3448 automatic with perpetual calendar in 18K yellow gold, the Geneva watchmaker's first automatic with perpetual calendar. On the right is the complete image with the noble wood box, watch, and pin pusher.

15.4. Buying at Watch Fairs

The largest trading place for buying, selling, and exchanging wristwatches is now the Munich Watch Fair. It takes place at regular intervals at the Hotel Infinity in Unterscheißheim, near Munich. Exact dates of this event and references to other watch exchanges can be found on the internet. In the opinion of the author, watch auctions and watch exchanges complement each other very well. At auctions, an abundance of previously inspected watches are available to a wide audience. Accordingly, the prices are often high. At watch fairs, rare pieces may appear that are discovered or recognized as such only by experts or specialized enthusiasts. Unfortunately, these opportunities are diminishing more and more. The extensive information available through the literature and the media, as well as the steadily growing interest in old watches, has ensured that such "lucky finds" are becoming rare. Nevertheless, watch fairs have many advantages for the collector of beautiful timepieces. One can look at the desired piece personally, compare price and condition with several other offers of the same reference, and finally decide on a watch. In some essential points, a purchase at a watch fair possesses clear advantages compared to an auction. Unlike at an auction house, at a watch fair the price can be negotiated and, if necessary, another watch can be given in payment. This is especially true if one's current budget is not sufficient to purchase the new dream watch and one can trade in another timepiece.

The risk of encountering an "embellished" or "fake" watch is likely to be several times greater at watch fairs than at auctions. Here, too, you should always take the provided checklist and work through it consistently before making a decision. If the seller in question does not understand, there are certainly reasons for this. Nobody has anything to give away nowadays, least of all dealers of antique watches.

15.5. Buying on the Internet

If one believes the statement by the head of department at Munich police headquarters that 40% of all fraud reports are cases from the internet, then particular caution is called for here. Despite this advance warning, it should not be overlooked that in no other purchase venue are opportunities and risks as closely linked as purchases on the internet. This applies to the items offered on eBay (www.eBay.com) just as it does to the offers that one can find via Google (www.google.com). The previously mentioned clock platform www.chrono24.com can also lead to a sought-after collector's item.

As a test, I searched for a Patek Philippe reference 2526 for a few hours at the previously mentioned internet addresses. No watch was listed on eBay anywhere in the world. The Google search, on the other hand, was more successful. After four opportunities were found, the search was canceled again. However, the most interesting example of the reference 2526 was found at the internet auction house LIFEAUCTIONEERS

Illustration 765
Patek Philippe Calatrava previously seen on page 422, with the original sales certificate. This document is always signed by the respective president of Patek Philippe.

at www.lifeauctioneers.com. This was a model for "Tiffany & Co" with movement no. 760,001 and case no. 674,917 that had been sold on June 30, 1953. This movement no. 760,001 tells us that this is the second movement of this famous caliber to be made. This watch was sold for $42,000.

At www.chrono24.com, six versions of the reference 2526 were offered at that time: five of them in yellow gold, one in platinum. While the prices for the yellow-gold versions ranged from $24,100 to $26,900, the Roman dealer had only the note "price upon request" listed for its platinum watch.

In principle the same rules apply to purchases on the internet as to other acquisition channels. Purchases on the internet generally require even-greater expertise. Those who do not want to experience a failure with watches from Patek Philippe or other brands should read this book carefully and make their decision only after a comparative evaluation.

The following recommendations should be strictly observed so that a purchase can be completed successfully.

Ask Further Questions

If the seller's quote does not include all the relevant information, then email the seller with the outstanding questions to be answered.

Checking Movement and Case Numbers

If it is a very expensive watch, have the movement and case numbers sent to you and check the data against the manufacturer's archives, such as those of Patek Philippe in Geneva.

Proof of the Watch's Actual Existence

To prove the existence of the watch in case of doubt, ask the seller for a picture showing the watch lying on the cover of a current daily newspaper.

Recording All of the Seller's Personal Data

In any case, before concluding the purchase communicate with the seller by phone, exchange addresses and phone numbers, and clarify the arrangement of packaging and shipping. You should also ask by phone whether the buyer can personally pick up the watch.

Caution for Countries with Risks

For countries with higher risks, such as Indonesia, it is better to refrain from buying altogether.

Immediate Purchase Confirmation

If you have placed your bid on eBay in the final seconds and have won the bid, you should contact the seller immediately and ask for confirmation of the agreed-upon conditions.

Payment via PayPal

When paying, one should choose to use only PayPal, an eBay company. With this method of payment, one has the possibility of complaint if the purchased watch does not correspond to the described characteristics or has not arrived for any reason. Within thirty days, all conflicts can be resolved and a possible reversal of the transaction can be carried out. The money should then not be lost even in the case of a wrong purchase. Prerequisite for this payment method is a PayPal membership, which is easy to acquire. It is important that both trading partners are certified members of PayPal. Payments via bank transfer or Western Union should be avoided at all costs.

Clarification of Shipping Method, Transportation Costs, and Importation Conditions

Before payment and shipment, the shipping method, costs, and customs declaration should also be settled. Although inherently much more expensive, the companies FedEx, UPS, and DPD (Deutscher Paketdienst) are alternatives to the post office. The three companies mentioned also take care of import and customs clearance for foreign shipments and, as commercial service companies, are usually much faster than the good old post office. There are also companies that specialize in transporting valuables. These can be found on the internet.

Importation Costs Such as Customs Duty, Import Sales Tax, etc.

Customs duties and import sales tax must be paid upon importation. If, for example, a watch is sent from the USA by USPS (United States Postal Service), it can often take weeks before you receive a message from the nearest postal customs office about the

Illustration 766
Patek Philippe Calatrava Ref. 5227G-001 from behind with opened dustcover. Behind the sapphire crystal case back, one can see the 324 SC automatic caliber with gold rotor reflected in the white-gold dustcover.

arrival of the expected shipment. You usually have to pick up the shipment there in person and pay import duties on the basis of the invoice issued by the buyer. To avoid cheating, customs now requires proof of payment (e.g., from PayPal) to calculate the import duties. Honesty is the best policy.

Receipt of Goods and Object Inspection

Once the shipment has arrived and the watch has been unpacked with witnesses, a thorough inspection should be carried out with the help of an expert in the case of expensive watches. If there are serious doubts about the authenticity or condition of the watch, you should immediately contact Patek Philippe and arrange an inspection.

Return Policy

When paying through PayPal, one has thirty days to resolve conflicts. In the worst case the watch goes back to the seller, and you should get the purchase price credited to your PayPal account.

Seller Rating

If the purchase of the newly acquired watch was successfully completed, you should let the seller know. With eBay, positive feedback is sufficient; with other sales, a friendly email.

For purchases in the higher price range (e.g., more than $11,000 or $22,000), you should always arrange a personal handover appointment (e.g., on the premises of a local bank or at the nearest Patek Philippe concessionaire). Here there is also the possibility of carrying out this purchase on the principle of "goods for money," thus eliminating unnecessary risks. A cheap air ticket to Hong Kong or New York may cost between $770 and $1,300, but it still comes out much cheaper than a big loss incurred on a supposedly good deal.

Illustration 767
The complete fine-adjustment mechanism of an IWC C. 8541 B automatic, with separate adjustment capability for rate and drop, Incabloc shock protection, balance with Breguet hairspring, and three *masselotte* weights. Such great technical effort no longer goes into most of today's movements (approximately 18× magnification).

16. The Expert Overhaul

Anyone who drives a passenger car and attaches importance to its reliability and good condition also treats their vehicle to the maintenance cycles recommended by the manufacturer. This principle applies both to the economical four-cylinder from Wolfsburg (Volkswagen) and the noble eight-cylinder from Untertürkheim (Mercedes-Benz).

If the engine of an automobile is operated at an average speed of 3,000 rpm, 180,000 revolutions are accumulated per hour. With three hours of operation per day and 365 days per year, this adds up to exactly 157,100,000 revolutions! If a watch operates at 28,800 vibrations per hour (A/h), 691,200 vibrations are generated per day—a watch does not take a break—and 252,288,000 vibrations per year. Various oils and greases ensure proper lubrication and keep wear within limits. In addition, the watch should be treated with deliberate care, avoiding unnecessary stress. Before you go swimming, it should be taken off and stored safely, even if the manufacturer promises water resistance to a depth of 393 feet (120 m). The farsighted golfer also takes off their watch when playing golf. Wrapped in a wool or silk scarf and safely stowed in a side pocket of the bag, even a valuable watch will survive eighteen holes undamaged. If you keep the watch on, even if it is a sports model, you risk putting the balance shaft and other sensitive components under excessive, shock-like stress with every swing of a golf club.

Finally, a word about temperature. Of course a good wristwatch can withstand several degrees below zero and also Sahara-like heat. However, a wristwatch (e.g., from Patek Philippe) was neither built nor is appropriate for these extreme conditions. The noble watches from Switzerland are intended to give their owners pleasure when worn and to convey the exact time in the most beautiful way. To ensure that these little marvels of micromechanics retain their special properties over decades, they should be given an overhaul every five to six years, even if they are used only occasionally. If the watch is worn constantly, service cycles of four to five years are advantageous. When the watch is open, a trained eye can recognize at a glance the oil level in the upper shock protection of the balance shaft. If there is no more lubricant there, it is also missing elsewhere. An overhaul is therefore unavoidable.

16.1. An Overhaul by Patek Philippe

The German branch of Patek Philippe resides in the heart of Munich. It has also been open to the public since 2005. However, its main focus is on servicing and supplying concessionaires in Germany. It thus bears the main responsibility for the German business. Anyone who would like to entrust their wristwatch to this branch but lives too far away can send it there by insured mail. They will immediately receive a certificate of receipt, and a few days later a cost estimate for the desired or necessary work. Today the majority of watches can be overhauled in the Munich atelier. Only some complications and older watches are sent on to Geneva for overhaul.

At the head office in Geneva (address: Salon Patek Philippe, Rue du Rhône 41, CH-1204 Genève, Tel. +41 22 809 50 50), which was restored to its former glory in 2006, watches can still

Illustration 768
Patek Philippe wristwatch with manual winding Ref. 565 in yellow gold, the second model with a screw-down case back, unveiled in 1938

Illustration 769
Patek Philippe Ref. 565. The open case provides a view of the manually wound 12'''-120 movement with small seconds. The movement number is 961,488, the case number 649,615. Above the case back is the dustcover.

be handed over personally for overhaul and repair. A watchmaker can also always be consulted there and asked about technical matters. Further service centers can be found on the internet (www.patek.com).

In Munich, anyone wishing to hand over their beloved wristwatch of our Geneva brand to a concessionaire can do so at Bucherer Jewelers, formerly Huber, at Residenzstrasse 11, opposite the State Opera House. An employee of Bucherer's associated workshop is trained to work on Patek Philippe and Rolex watches and can carry out up to a Level 2 overhaul of these valuable watches in-house.

But what are the advantages of a watch overhauled at Patek Philippe in Munich or Geneva? There are two arguments at the forefront here that are not so easy to refute. First, the watchmakers

Illustration 770
The standard service sticker was applied to the Patek Philippe Ref. 565 after the overhaul was completed.

Illustration 771
The circular-grained finish on the inside of the case back has the usual hallmarks and identification marking. The company name PATEK-PHILIPPE & C°, the small Helvetia, the reference number 565, and the case number 649615 all are clearly visible. On the left is the precious-metal hallmark for 750 gold. One should always keep a copy of the movement and case numbers in a safe place.

working at Patek Philippe are always and only concerned with the products of their house. Each action is repeated several times a day and can almost be carried out blindfolded. Second, there are no bureaucratic efforts or hurdles when exchanging movement parts. Here there is no need to extensively study caliber descriptions and laboriously dispose of all exchange parts. At Patek Philippe the watchmaker goes to the appropriate cabinet and takes out all the necessary spare parts. This step takes them only a few minutes! And many functional parts, which often still fully perform their tasks, are simply replaced as a preventive measure.

Finally, there is the word of honor of the former president Philippe Stern, which says that every watch, even those from previous production, must meet the highest standards of precision and accuracy. What more could the wearers of beautiful watches from Patek Philippe want?

After the issue of quality, the questions of price and delivery time are still relevant for the Patek Philippe customer. We can answer the first question with the numerous and precisely described repair examples, which allow interesting and detailed insights into such work. For the second question, the answer is not so easy for Patek Philippe. Depending on the type of watch and the extent of the overhaul, delivery times of five to eight months are common. And if another dial has to be restored because there is no longer a suitable spare part in stock, the wait times can be even longer.

Finally, there is the aspect of value retention. If a Patek Philippe watch has been restored in the company's own workshops and the corresponding invoice is available, this circumstance has a decisive influence on the resale value. In such a case, the equally valuable work of a concessionaire or its licensed workshop can take only second place.

16.2. An Overhaul by the Concessionaire

When overhauling wristwatches, Patek Philippe has set various levels. These include both the age of the watch and the complexity of its movement. These specifications even apply to Patek Philippe's own branches, where many, but by no means all, references may be serviced. This also applies to the Munich site.

Nevertheless, visiting a concessionaire and having an overhaul carried out there has some advantages. First, the distances are short at such a specialist company because the in-house workshop is usually in the same building as the store. One can therefore call the respective specialist and consult them. The watchmaker in question knows the specifics of the calibers and can competently give the desired information. For the repair they usually have the most-important spare parts on-site and only have to order special or rare components from Geneva. Depending on the size of the workshop, one or two employees are specially trained to work on Patek Philippe watches, but inevitably they also carry out overhauls on watches from other manufacturers.

Illustration 773
Tang buckle of a Vacheron & Constantin of 750 yellow gold. When exchanging the leather strap of a watch at the concessionaire, always ensure that the existing gold buckle is refitted to the new strap.

In this respect, a Patek Philippe watch has much in common with a fast, rear-engine automobile from Zuffenhausen (Porsche). With these high-quality requirements, only specially qualified experts should be allowed to lend a hand so that the required high brand level is maintained without compromise.

Illustration 772
Close-up of the dial of a Patek Philippe Ref. 2551 after an overhaul at the manufacturer's factory. The watch's condition may be characterized as almost new.

Illustration 774
Original old watchcases are sought-after collector's items, and one should always ensure to take the case with one when dropping off a watch for an overhaul. Our example is a valuable leather case from the 1950s in which Vacheron & Constantin used to deliver its valuable watches.

However, this challenge can mean a special opportunity for the concessionaire. If they have one or two talented watchmakers in their workshop and allow them to receive ongoing training at Patek Philippe in Geneva, they can build up the same level of expertise over time as is available at Patek Philippe itself. However, strategic thinking is required here on the part of the concessionaire.

Basically shorter delivery times should be feasible for comparable work at the concessionaire. They could also be more favorable in terms of overhaul costs if they do not charge according to predefined work values but according to the actual time required. Another advantage of the concessionaire is the marketing component. Through direct channels and personal contact with the customer, a level of relationship can be established that goes far beyond proven competence as well as exact rate values. And why shouldn't a customer who is satisfied with the overhaul of their watch also look around on-site and look for a new watch? Where a customer feels comfortable, the next business deal is sure to come about.

16.3. An Overhaul by a Trusted Watchmaker

The secret of success of the personally known watchmaker may often be understood like a special marketing strategy. Anyone who, in addition to professional competence and good work, also finds time for a personal conversation is very responsive to the special needs of a watch enthusiast. The prerequisite for a long customer relationship is always based on good and comprehensible watchmaking performance. A collector who also provided watches for this book had some watches overhauled at a Munich antique-watch dealer. When these were remeasured after execution and payment at Patek Philippe on a time scale or on a Witschi measuring device, the results were very disappointing. While the average rate was reasonably correct, the average daily rate deviation, the amplitudes, and the drop in six positions were in no way acceptable (see examples p. 425, fig. 748). Another watchmaker near Pforzheim has been successfully overhauling watches of outstanding brands such as Rolex, Patek Philippe, and Vacheron & Constantin for years. This watchmaker has also carried out excellent overhauls on some of the watches presented in this book, and has proven his special competence.

The watchmaker you trust has to live with a small handicap. If the movement of a watch contains no damaged parts and only a normal overhaul is to be carried out, this watchmaker can do a good job. If parts are needed that can be obtained from the spare-parts trade of the watch industry (e.g., screws, springs, etc.), the watchmaker you trust will also have no problems. Only if original spare parts are needed does it become tight for the watchmaker. In Germany, only Patek Philippe concessionaires

Illustration 775
Close-up of a watch movement with automatic winding, an IWC Caliber C. 8541 B with movement number 2,153,521. A trusted watchmaker could explain this movement in detail. The balance cock with separate settings for rate and drop, the balance with Breguet hairspring, *masselotte* weights, and much more can be seen.

Illustration 776
IWC automatic caliber C. 8541 B after removal from the case with the dial still in place. The hands and indices are 18K yellow gold. The SWISS label below six o'clock reveals that this dial was made prior to August 1973. The reason for this is described in chapter 12, "The Secret of the Dials," under the heading APRIOR on page 362.

receive spare parts, and these are still differentiated. This depends on which of the offered training levels the responsible employee at Patek Philippe in Geneva has already completed.

Basically, it depends on the knowledge as well as the aptitude of the respective watchmaker, because with our watchmakers the Gaussian distribution applies just as it does with doctors and engineers: only about 10% of all participants find a place on the far right.

But what do we learn from this? If an old wristwatch from Patek Philippe has not seen an expert for decades and work is needed on the dial, movement, and case, the author would entrust this beautiful piece only to Patek Philippe directly. This is because the same professions come together to make a watch as for a major overhaul. The watchmaker is responsible for the movement and the assembly of the watch. The case maker is responsible for the manufacture of the case and also carries out the repairs associated with the case. The *polisseur* takes care of the correct case surfaces, as well as the sharp edges. In Switzerland, each representative of these three professions completes his or her own training, which at Patek Philippe takes a full three to four years. There, attention is paid to the strict separation of these three areas of activity when carrying out overhauls, so that the company's same high standard of quality is maintained when overhauling a wristwatch as when producing new watches!

If a normal overhaul is due and you know the workshop manager of the nearest concessionaire personally, you can entrust your watch to them in good faith.

If there is work to be done on a watch where no spare parts are needed and watchmaking performance is paramount, the watchmaker you trust is the right address. And because most vintage-watch stores are run by watchmakers and there is often a wide range of older watches on offer there, you can often end up buying another watch. So each of these three mentioned ways has its opportunities and risks.

A word in conclusion.

With very battered cases, some watchmakers tend to simply polish out even deep scratches and other damage by using a buffing wheel. However, this might be the completely wrong way, because such a treatment of the case changes its original shape, and a lot of valuable precious metal is removed. In the example of the reference 3425 (see p. 188), the commissioned case maker has convincingly demonstrated how damage can be repaired and the watch restored to its original condition. This applies not only to the correction of the shape, but also to the restoration of the original sharp case edges. Unprofessional repolishing with a buffing wheel provides a superficial shine but is always associated with the disadvantage of rounded edges and high material loss (see p. 185). In this case, for all three of the service paths described, the working method explained on page 188 should be used by means of material laser deposition with subsequent correct shaping. The described technology with material laser cutting, turning, milling, and diamond finishing has been mastered by only a few companies. These can be found both in the Pforzheim area and in the watchmaking strongholds of French-speaking Switzerland.

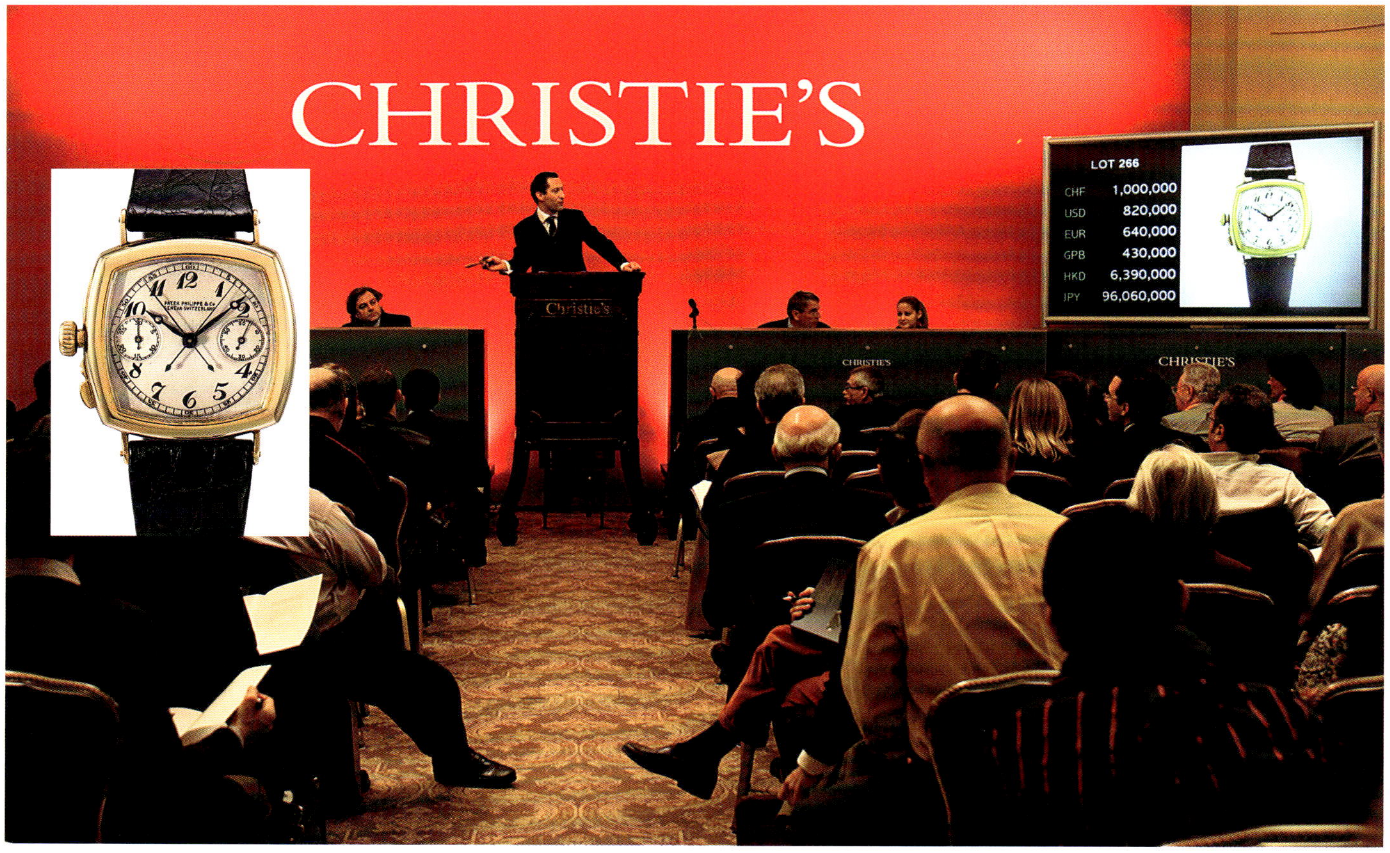

Illustration 777
An auction at Christie's in Geneva. A rare and unique specimen was called on November 13, 2006. The double chronograph with one pusher had been produced in 1925 for a left-handed person and sold on January 4, 1927. Then chief auctioneer Aurel Bacs looks at the display board showing the watch and the current high bid of one million CHF. In the end, the collector's item also shown on the left was knocked down for CHF 2,372,000.
Christie's photo

17. Value and Value Enhancement Potential

The example of a simple chronograph reference 1463 in stainless steel was offered in the preface. According to the German sales catalog from 1961, this simple-looking time recorder was available for exactly 1,350 DM. At a Christie's auction in Geneva on November 12, 2007, the impressive sum of $592,133 was paid for an example of this reference (lot 35, movement no. 868,840, case no. 660,184, produced in 1955, seller's signature FRECCERO, with steel bracelet, picture on p. 194). At the same auction house in Geneva, another example of this watch was offered on November 19, 2009, and sold for $514,689 (lot 85, movement no. 863.724, case no. 640.561, stainless steel, seller's signature WALSER WALD, produced in 1945).

If, in both cases, a savings account with a 5% annual interest rate or a deposit account with a guaranteed annual return of 10% were invested, the following amounts would have been obtained for these two cases (Kn = final capital, K0 = initial capital, n = years, p = interest rate):

Example 1 (after forty-six years)
Purchase from Freccero
(5% interest on a savings account or 10% return on a securities account)
$Kn = K0\ (1 + p/100)n = 1.350\ (1 + 5/100)46 = 12{,}736.25$ DM
$Kn = K0\ (1 + p/100)n = 1.350\ (1 + 10/100)46 = 108{,}242.34$ DM

Example 2 (after forty-eight years)
Purchase from Walser Wald
$Kn = K0\ (1 + p/100)n = 1.350\ (1 + 5/100)48 = 14{,}041.71$ DM
$Kn = K0\ (1 + p/100)n = 1.350\ (1 + 10/100)48 = 130{,}973.27$ DM

In example 1, the capital has increased by a factor of 9.43 and 80.18, respectively. In example 2, the values are a factor of 10.4 and 97.02. What would an investment advisor say to these figures?

And if we take reference 2585 in stainless steel as a further example, the banker would be amazed in this case as well (see p. 180). This watch was sold at Sotheby's in New York on October 28, 1996, for $11,500. On May 16, 2005, this watch appeared at Christie's in Geneva, was offered at 40,000 CHF, and finally sold for 104,400 CHF (approximately $85,195 USD). Subtracting the consignment fee and buyer's premium from this sum, the seller was still left with about 73,080 CHF ($59,636.50). Subtracting the previous investment of $11,500 in New York, the net profit comes to $48,136.50. This means a 4.2-fold increase in capital in just under nine years!

17.1. Watches from Auction Houses

The most-important auction houses were previously listed at the beginning of section 15.3. The price to be achieved for a watch depends on several criteria. First, a word about the trend of the time, which depends on the overall economic situation, on the one hand, and on the respective watch model, on the other.

It was not until the mid-1980s that the supply of valuable watches increased significantly. Prices of special wristwatches reached lonely peaks around 1990 and almost halved again with the collapse of the economy in 1993. Prices subsequently rose again, slowly but steadily.

Anyone wishing to buy wristwatches at auctions, in this case those of the Patek Philippe brand, must have a precise knowledge of the respective offer and be able to estimate the

Illustration 778
Lower portion of the dial of a Patek Philippe reference 3448 with automatic winding and perpetual calendar. The SWISS label below six o'clock proves that this watch was made prior to August 1973. Had this piece been made after 1973, APRIOR regulations would have required the Ó SWISS Ó (meaning Sigma SWISS Sigma) label. In the rarest case it could also be a replacement dial made as a one-off by Cadrans Flückiger SA after 2007. The watch has a sapphire crystal. The cost of such a dial was probably around $7,800.

demand for the watch that may be purchased. The participants in an auction are recruited from different interest groups. There are dealers who want to increase their supply and sell the watches won at auction as quickly as possible. There are also collectors who want to expand their collections. Furthermore, there are the rarity hunters seeking to track down and buy largely unknown but valuable watches. Finally, there is the group of bidders who bid on watches on behalf of third parties and who usually have to comply with the precise directives of their clients. This includes bidders sent by museums.

As a collector you should not expect any bargains at such auctions, but only solid purchases for your own collection that are in line with current prices and values. As a rule these are also not suitable for immediate sale. Because other participants in the auction would have won with a slightly higher bid if they had wanted to do so.

Final Advice

• If they are expensive watches, go to the auction a day early and take the opportunity to preview them.
• Take the checklist in this book with you and go through your most-important questions with the auction house expert. Only when you have come to the conclusion that you have found the right watch should you place a bid.
• Draw up a table of figures for the watches you wish to bid for, in which the last three to four of your own bids (including buyer's premium) are shown from left to right.
• Complete this bidding table with the three following bidding levels and mark them in red, because if you want to go one bid higher at the last moment, you already know their complete final price.
• But wait with your last bid until shortly before the hammer falls!
• If you don't want to take the auctioned piece with you immediately, you can also arrange to have your purchase insured and shipped. This will save you additional correspondence.
• An exchange is possible only in the case of a *mariage* or a forgery. Possible later repairs are at the expense of the buyer because the watch was bought as is.

Example: Lot 242 (+ 20 % surcharge)						
Without surcharge	2,000	2,200	2,400	2,600	2,800	3,000
With surcharge	2,400	2,640	2,880	3,120	3,360	3,600
Planned Highest Bid						

Thus, buying valuable old watches through auction houses can be a reasonably safe process, but not necessarily a very profitable one. Because if you want to have a watch you just bought at auction sold again two to three years later through an auction house, a consignment fee of roughly 20% will be deducted from the hammer price. And the new owner has to pay a premium of about 20%. In purely arithmetical terms, this adds up to about 40% of the hammer price, which is shared between your profit and the expenses incurred by the new owner. A good example of this is the reference 3454 shown on page 194.

Illustration 779
Patek Philippe reference 3970 P chronograph with perpetual calendar, with manually wound caliber CH 2770 Q movement and platinum case; a very sought-after wristwatch

17.2. Watches from Watch Fairs

Purchasing a watch at an international watch fair can often be a good opportunity to acquire a rare watch under favorable conditions, on the one hand, and with considerable potential for appreciation, on the other. A suitable example might be the reference 3433 (see p. 190). This watch was purchased by a collector at the Munich Watch Fair for $5,200. Critical to this Patek Philippe was the fact that an incompetent "manipulator" had mounted a dial without small seconds, and the watch could be identified only by the inscription on the inner case back. Afterward, Patek Philippe charged exactly $3,330.22 for the masterful restoration of this beautiful piece. With a total investment of $8,603.72, the collector had not done badly, because on the open market reference 3433 watches are not to be found for under $11,000.

For example, a dealer offered an unrefurbished reference 3433 for $13,830. After lengthy negotiation, he would have sold this watch for $12,085.

There were no other models of the reference 3433 on offer anywhere in the world at that time.

In spite of everything, when buying watches with great potential for appreciation, the checklist already mentioned several times still applies. What is the use of a favorable purchase if, due to hidden technical defects, the watch cannot be resold at a profit, but only at a loss.

Illustration 780
Close-up of the chronograph caliber CH 2770 Q with balance cock, Gyromax balance wheel, and Breguet hairspring. The slotted hole with cylinder head screw, which is aligned radially to the balance axle, is used to adjust the hairspring stud carrier (in French: *Le piton mobile*).

Illustration 781 (right)
Detail enlargement from the photo below, showing the hand of the wearer of the Rolex Sea Dweller Ref. 126.600.

Illustration 782
The proud owner of a new Rolex Sea Dweller and successful businessman Josef W. has reason to smile. His global companies produce state-of-the-art manufacturing equipment.

17.3. Watches from the Internet

One thing must be said in advance: where the highest returns are tempting, the greatest risk also awaits! A watch enthusiast from Munich purchased a Patek Philippe reference 2438 in yellow gold in Australia and paid $36,000 for it. When the watch was paid for via PayPal with four partial transfers and arrived a week later, the first doubts came to the supposed bargain hunter. Seeking help, he turned to the author for advice. At first glance it was obvious that it was a mediocre *mariage*. Only the caliber 27 SC movement was authentic. The perpetual-calendar module

Illustration 783
For watches that are, for example, forty years old or older, the authenticity or originality of the dial should be verified. If the dial of the watch was declared original at the time of sale, but it later turns out that this statement is not true, the term "hidden defects" applies. This means that the purchase can be changed (i.e., reversed) within one year, or even longer if necessary. With private purchases, caution is required, because you cannot always assume that the seller can still perform this reversal.

was taken from a watch from another manufacturer. The case and the dial were cheap imitations. Despite these findings, the watch had to be sent to Patek Philippe in Geneva, and they had to be asked for a statement. Only with neutral evidence can a transaction be reversed at PayPal. After returning the watch and waiting about twenty days, the Munich collector had his money credited back to his PayPal account. Behind this story was a small tragedy, as was later learned. A Korean civil engineer, who worked in Australia, had purchased this watch from a dealer during a stay in Los Angeles. In complete ignorance, he had allowed himself to be persuaded by the dealer to purchase the collector's item, which was described as a rarity and a bargain, for $25,000. However, because he had become insecure in the meantime and also needed money for his upcoming wedding, he wanted to get rid of this *mariage* again. Unfortunately without success, as we learned here.

We have already learned in detail how we should proceed with internet purchases in section 15.5, "Buying on the Internet." Now two more examples of how to earn a modest but interesting percentage return.

A collector purchased a reference 2508 in yellow gold in the USA. With an eBay price of about $6,676 and $1,269 in import fees, this watch had cost about $7,945 (see p. 166). At www.chrono24.com, two more examples of this reference were offered in Europe shortly before the editorial deadline. These offers were at $19,775 and $20,874. If the collector now wants to resell the watch at a premium of about 50% after the speculation

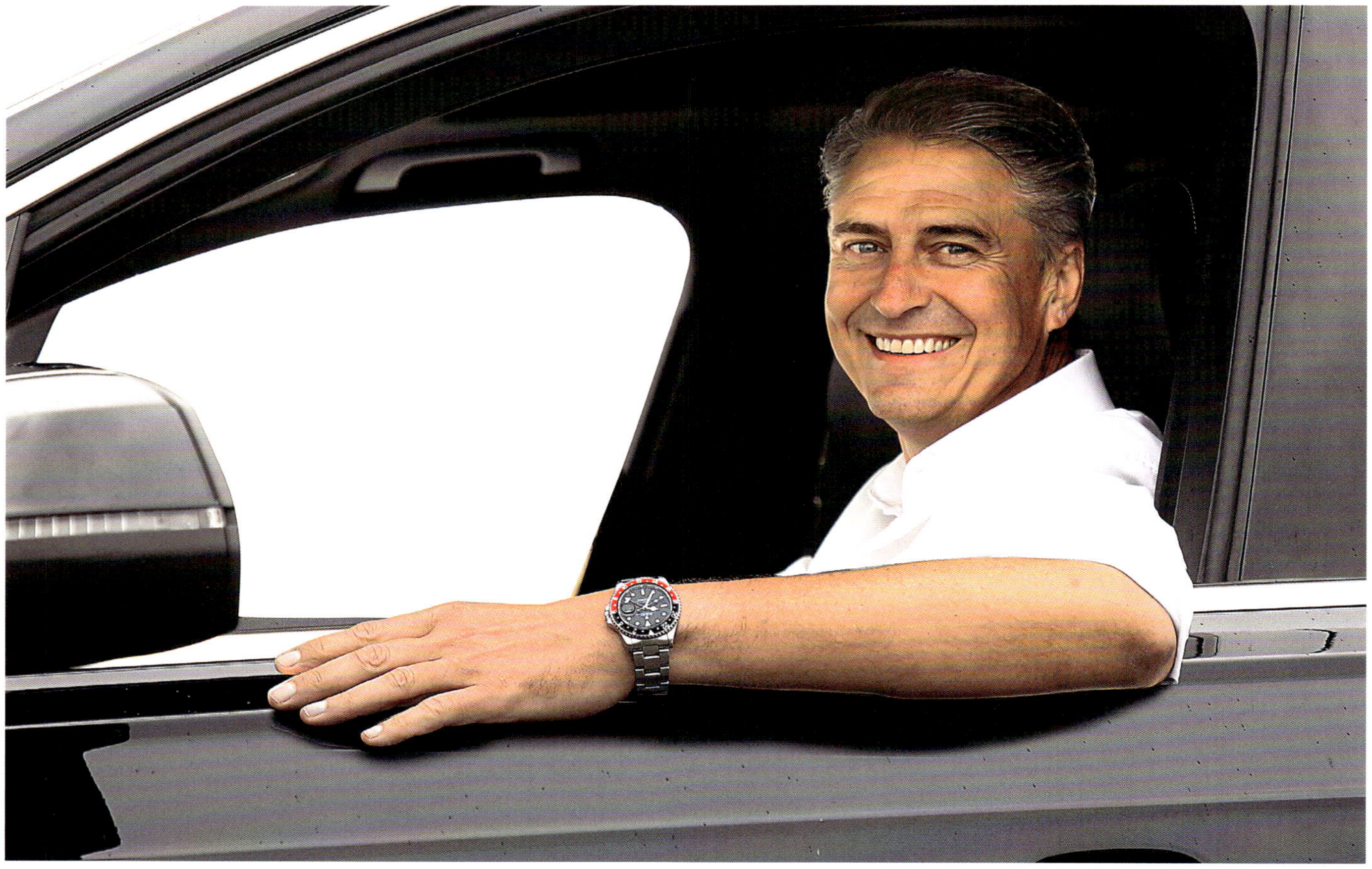

Illustration 784
The author's dentist, Dr. med dent. Dietmar H. from Munich, is also considered a great watch enthusiast. In addition to his work as a sought-after dentist, he is also highly regarded as an expert appraiser.

period of twelve months has expired, he will have to generate only $12,085 in a direct sale. This should not be difficult, because in addition to those offered by www.chrono24.com, three reference 2508 watches were also listed on eBay in the USA: two with yellow-gold cases and one in white gold. Their immediate sale prices were $19,800, $18,950, and $40,000, of course still without overhaul costs and import duties.

One of the current and highly sought-after cult watches is the manually wound reference 570. A collector from Heidelberg purchased a rose-gold reference 570 with small seconds on eBay in the USA for $6,766 (approx. 5,100 €; see p. 156). Including transport and import fees, he had to pay a total of around $6,741 for this piece. At www.chrono24.com in summer 2010, only one rose-gold example of the reference 570 was listed. The seller Interwatches from Miami, Florida, wanted $18,500 for it, of course without transport costs and customs duties. The next offer of a reference 570 in rose gold, but with "price on request," was already above the last price mention of $25,818 and corresponded exactly to the model shown on page 156. This offer came from a seller in Lugano.

At the same time, only a reference 570 in rose gold and small seconds could be found on eBay. However, this listing was also by the dealer Interwatches as before at www.chrono24.com. The buy-it-now price here was also $18,500.

If our watch collector adds another $1,757 for an overhaul at Patek Philippe, his total investment in this watch comes to around $8,499. If he wants to pass on this overhauled piece after the speculation period of twelve months with a 50% surcharge, he would have to ask only $12,744 for this beautiful piece in a direct sale. Some time ago, the dealers Bachmann & Scher offered a white-gold reference 570 with caliber 27 SC movement for $25,818, while the company PrimetimeWatches offered a yellow-gold version at an immediate sale price of $14,062. In neither case had an overhaul been performed at Patek Philippe.

17.4. Private Sales

A purchase from a private individual cannot be unproblematic. However, if a valuable Patek Philippe turns up in an inheritance liquidation and the liquidator's asking price matches your budget, strike immediately! In the case of several watches, the liquidator will first go to a dealer and have them appraise the watch collection. As a rule the liquidator will not agree to an immediate sale because at least two offers are required for higher sums.

Different rules apply to transactions with friends and acquaintances. Accept an offer only if the offer price is clear and comprehensible and you do not get the impression that your counterpart wants to do a lucrative business deal with you. You do not do business among friends and acquaintances, but pass on a favorable price one to one. Trust strengthens friendship.

If an acquaintance has to part with some valuable watches for financial reasons, do not take advantage of this situation. Ask some dealers for the current purchase prices of these watches and offer this price to your acquaintance. If they agree to your asking price, you don't have to feel bad. The acquaintance has immediately received a fair price for their timepieces and can get out of their temporary financial hiatus. And the buyer has received watches at prices that they can realize again in a later sale or trade-in. But even with such a purchase, one should consistently follow the checklist.

Illustration 785
Classic car and watch enthusiast Hanns-Peter K. can look to the future with satisfaction. Both his beautiful 250 SE convertible and the remarkable Patek Philippe Ref. 3445 can be considered good investments. In the event of a resale, the two cult objects dating to 1966 could fetch more than double their cost price.

Keyword Index

Bibliography

Brückner, Michael. *Uhren als Kapitalanlage*. Munich: FinanzBuch-Verlag, 2007.

Ehrhardt, Roy, and Joe DeMesy. *Patek Philippe*. Book 1, *Retail and Vintage Prices*. Umatilla, FL: Heart of America, 1995.

Goldberger, John. *Patek Philippe Steel Watches*. Bologna, Italy: Damiani, 2010.

Hampel, Heinz. *Automatic-Armbanduhren*. Munich: Callwey, 1992.

Huber, Martin, and Alan Banberry. *Patek Philippe Genève Armbanduhren*. Geneva, Switzerland: Patek Philippe SA, 1998.

Kahlert, Helmut, Richard Mühe, and Gisbert L. Brunner. *Armbanduhren*. Munich: Callwey, 1996.

Mehltretter, J. Michael. *Die Noblen aus der Schweiz*. Königswinter, Germany: Heel, 2006.

Mehltretter, J. Michael. *Patek Philippe Armbanduhren, Kultobjekt & Wertanlage*. Königswinter, Germany: Heel, 2011.

Meis, Reinhard. *A. Lange & Söhne: Eine Uhrmacher-Dynastie aus Dresden*. Munich: Callwey, 2001.

Negretti, Giampiero, and Paolo De Vecchi. *Patek Philippe, Komplizierte Armbanduhren*. Cologne: Könemann, 1997.

Osterhausen, Fritz von. *Wie kaufe ich eine alte Armbanduhr*. 2nd ed. Munich: Callwey, 1997.

Patrizzi, Madeleine, and Osvaldo Patrizzi *Collezionare Orologi Patek Philippe*. Geneva, Switzerland: G. Mondani, 1995.

Patrizzi, Osvaldo. *Collecting Patek Philippe Wristwatches*. Vols. 1 and 2. Geneva, Switzerland: G. Mondani, 2004.

Stern, Michael. *Handbuch für das Uhrmacherhandwerk*. Königswinter, Germany: Heel, 2010.

Theorie der Uhrmacherei. Neuchâtel, Switzerland: Swiss Federation of Technical Colleges, Watchmakers of Switzerland, Training and Educational Program, 2001.

Wehrle, Martin, K. *Royal Oak*. Le Brassus, Switzerland: Audemars Piguet, 2012.